The 1988 C-14 Dating Of The Shroud of Turin: A Stunning Exposé

JOSEPH G. MARINO

Foreword by Dr. Kenneth E. Stevenson,
original spokesman and editor for the
Shroud of Turin Research Project

The 1988 C-14 Dating Of The Shroud of Turin: A Stunning Exposé

Copyright © 2020
Cover Art copyright © 2020 by Miblart, Ternopil, UKRAINE
Cover Design by Joseph G. Marino
Front cover photo, Telegraph Group Limited, London, 1988

ISBN: 978-1-7348130-3-6

ALL RIGHTS RESERVED

No part of this publication may be reproduced in whole or in part, or stored in a retrieval system, or transmitted in any form or by any means, electronic, mechanical, photocopying, recording, or otherwise, without written permission from Joseph G. Marino

Printed in the U.S.A.

Every effort has been made to seek permission for use of and properly credit all copyrighted sources included in this book. However, in some cases we were unable to locate the copyright owners or received no response when we did contact them. Should you find your copyrighted materials in this book without proper attribution, please contact the author immediately and the proper corrections will be made in the next printing.

> "There will be hell to pay when the truth comes out."
>
> Raymond N. Rogers, Shroud of Turin Research Project chemist, commenting on the 1988 dating of the Shroud of Turin

PRAISE FOR
The 1988 C-14 Dating Of The Shroud of Turin: A Stunning Exposé

"By far the most thorough assessment of all available data pertaining to the 1988 carbon dating of the Shroud of Turin."

"This book should serve as a stern warning to all of us that a single datum should never be allowed to fly over the preponderance of data pointing in a different direction."

August D Accetta, MD
Founder of the Shroud Center of Southern California

"The author illuminates the uncertainty, confusion, irony and tragedy surrounding the carbon dating of the Shroud."

Mark Antonacci, author of "Resurrection of the Shroud" and "Testing the Shroud at the Atomic and Molecular Levels"

"I can think of absolutely no one besides Joe Marino who would take the years of incredibly painstaking research necessary in order to assemble, compile, and record all of these data and minutia, plus many different interpretations from multiple angles. What a favor for those interested in Shroud of Turin enquiry!"

Gary Habermas, Ph.D, co-author, "Verdict on the Shroud" & "The Shroud and the Controversy"

"Shroud of Turin authority Joe Marino and his wife Sue Benford are credited with advancing the most important reason why many (including me) regard the 1988 carbon dating of the Shroud as fatally flawed---namely, the fabric sample selected from a damaged/contaminated Shroud edge that had almost certainly been repaired by a Middle Ages textile artisan using the skill of "invisible repair" favored by kings to mend their favorite capes and blankets. The inexplicable twists and turns of the numerous meetings and arguments held between carbon dating specialists, Vatican officials, and Shroud scientists over a period of years somehow resulted in flawed decisions that left the Shroud dating open to controversy are challenging to follow, but Joe has exhaustively walked the trail for us through virtually all the meetings---both public and even many closed door---by quoting other participants so readers can see for themselves how the dating controversy evolved and remains, until more thorough tests are performed using samples uncontestedly from the original cloth using modern technologies."

Ed Prior, retired NASA scientist

"Without doubt, what Joe Marino has documented and assembled here will quickly be acknowledged as the definitive reference on what actually occurred during the 1988 radiocarbon dating of the Shroud. A critical volume for every Shroud scholar."

Barrie M. Schwortz, official Documenting Photographer for the 1978 Shroud of Turin Research Project

"Joe Marino has concisely articulated the many cautionary tales surrounding the Carbon-14 dating of the Shroud of Turin. This work, resounding with clarity, is a must-read for every serious inquirer into the Shroud's deep mysteries."

Cheryl H. White, Ph.D., Professor of History, Shroud researcher

ACKNOWLEDGEMENTS

Thanks to Stephanee Killen of Integrative Ink for her expert formatting, attention to detail, and kindnesses throughout the process.

In appreciation for all of the tireless individuals and groups that have put so much time and effort into studying the Shroud.

I would like to thank Annette Cloutiér for her work on editing and the Index.

I want to express gratitude to Giorgio Bracaglia, overseer for the Holy Shroud Guild archives, for supplementary information he provided to me by phone and email.

I'm indebted to Russell Williams for his assistance in getting the cover produced and for suggestions on the book description on Amazon and on the book's press release.

I would like to thank Michael Lewis Kowalski for his input on the book's press release.

Thanks to Paula Ann Mitchell of Mitchell Communications for all her help on the press release and social media video for the book.

I'm thankful to Massimo Paris, a native Italian, for clarifications regarding some translation of Italian documents.

A tip of the hat to Ed Prior, with whom I co-authored a Shroud article, for the nudge to get me to tackle writing this book.

I'm appreciative of a neighbor, Dr. Carl Sharp, for his valuable support.

I would like to thank Cindy and Steve Sheltmire for a key suggestion on the cover and for their enduring friendship and encouragement.

This book is dedicated to all of the Shroud researchers whose sources I have cited in this work, as well as to my late wife, M. Sue Benford.

TABLE OF CONTENTS

Foreword xiii

Special Contribution by Dr. Eric J. Jumper, Ph.D. xv

Preface xix

PART I 1
   PRE-APRIL 21, 1988, DAY OF SAMPLE EXTRACTION

PART II 343
   DAY OF THE SAMPLE EXTRACTION, APRIL 21, 1988

APPENDIX TO PART II 381

PART III 391
   POST-APRIL 21, 1988

Conclusion 711

Key Figures 713

References 723

Index 737

FOREWORD

By Dr. Kenneth E. Stevenson

In the course of my forty-three years of research into the Shroud of Turin, it has been my privilege to work with awesome men and women whose contributions to Sindonology have been extremely valuable in bringing the Shroud out of the realm of "relics" and into its proper place as perhaps the most important archaeological artifact of all time. In fact, the members of the Shroud of Turin Research Project (STURP), starting in 1978, determined to follow the facts, produced a final report in 1981, which to this day stands as an incredible example of the effectiveness of a coordinated scientific approach. To this day, the single "scientific report" that would seem on the surface to contradict the findings of STURP, was the 1988 C-14 testing under the auspices of the British Museum.

Enter Joe Marino and his wife, the late Sue Benford. Unphased by the cacophony of voices claiming that the Shroud was a forgery, they opened an entirely new area of Shroud research: "invisible reweaving". While those who readily accepted the C-14 results, ignored the report from STURP, Sue and Joe examined what could possibly have produced a result that was diametrically opposed to three and a half years of peer-reviewed science. After all, STURP using space age technology, had subjected the Shroud to five twenty-four-hour days of the most advanced non-destructive testing known to man. The conclusions of STURP as well as their suggested protocols for C-14, were completely ignored by a team who loudly trumpeted that the Shroud was a "fake".

Those of us who knew the facts were hard pressed to prove that the dating results were suspect. But Joe and Sue took their case to the wider world for review. While their suggestions were initially met with a healthy dose of skepticism, it wasn't long before their theory was confirmed by STURP chemist, the late Ray Rogers, and others.

"The 1988 C-14 Dating Of The Shroud of Turin: A Stunning Expose" will surely take its rightful place as an incredibly important and thorough compendium of all the evidence demonstrating how the C14 testing was hopelessly compromised. Beyond the fact that peer-reviewed science has confirmed that the dating results were clearly due to cotton fibers, Marino's extremely well-documented research takes the reader behind the scenes, revealing the unbelievable machinations and bias that accompanied the now-failed attempt to debunk the Shroud. Kudos for the most complete and impressive refutation of the C-14 dating. The case for authenticity takes a giant step forward with the publication of this, indeed, "Stunning Expose."

SPECIAL CONTRIBUTION BY DR. ERIC J. JUMPER, CO-FOUNDER OF THE SHROUD OF TURIN RESEARCH PROJECT

Eric J. Jumper, Ph.D.
Roth-Gibson Professor of Aerospace and Mechanical Engineering
University of Notre Dame
Co-Director of the Shroud of Turin Research Project at the time of the 1978 Testing of the Shroud in Turin

I co-directed the on-site examination of the Shroud of Turin in 1978. As director during my shifts, mostly at night, I had copious opportunities to examine the Shroud up close, at a distance, under a microscope and via X-Rays as they were being developed in our work spaces. In particular, I studied the image fibrils of the threads of the cloth containing image. First, the cloth was decidedly linen; the only cotton fibers that I observed on the Shroud were obviously not part of the Shroud, but incidental due to contamination. Second, the images were made up of multiple types from blood images to contamination from water damage, presumably due to putting out the burning container holding the Shroud at the time of the 1532 fire. Then there were the body-only images that were present only on the upper-most fibrils making up the threads of the cloth. I was the principle author of a summary paper describing the various stains and images we found on the Shroud. In the concluding remarks of that paper, I suggested that not finding anything on the Shroud that suggested an obvious artistic rendition and, in fact, finding evidence consistent with the cloth having been used to enfold a dead, crucified body, the Shroud should be subjected to Carbon-Dating tests. I added the comment that if the date of the cloth was shown to be inconsistent with a first century date, it could not be authentic, but that if it were found to be first century, it still would not prove the authenticity of the Shroud being the burial cloth of Jesus.

Once it appeared likely that permission to perform Carbon Dating would be given, as many of the original team members as well as some added expertise gathered to discuss the best way to remove the samples for dating. Dr. Robert (Bob) Dinegar was selected to head up the canvassing of various small-sample dating laboratories around the world and to develop the protocol for collecting the samples from the Shroud. In fact Dr. Dinegar headed up the first inter-laboratory calibration, comparing how consistent dates were from one laboratory to another; but, more importantly, Bob was put in charge for precisely where and how the samples were to be collected. This later responsibility was critical because we knew that no matter how the date came out there would be dissenters. In a day-long meeting we brainstormed all of the scientific objections we could think of for objecting to the outcome of the dating. A protocol was developed and submitted to Dr. Gonella, who represented the Archbishop of Turin, under whose authority the permission for dating would be given. Much of the controversy and political intrigue surrounding what happened next is contained in this book, but let me get to the main outcome. Dr. Dinegar and, in fact, all of the American team members of the Shroud of Turin Research Project, were systematically eliminated from participation in collecting and distributing the dating samples; further, the protocol for collecting the samples, after having been accepted, was completely ignored! Why was this so critical?: because the first directive was that the samples should not have been taken in the corner of the Shroud where the previous, so-called "Raes sample" was taken many years earlier. There were many reasons for that, not the least of which was that area of the Shroud had a completely wrong color signature from the rest of the Shroud. In my mind that could have been due to contamination, perhaps by people touching the Shroud there, over its centuries of history. Nonetheless, I had no reason to believe that the corner was not part of the original Shroud; my only concern was contamination.

When the date, or should I say dates, of the Shroud were revealed, the combined date was 1260-1390. I was somewhat surprised, because I knew about historical investigations that suggested that it could be presumptively shown to go back to the 600's; but, in all honesty, I was relieved that the Shroud efforts, which had consumed so much of my off-work time was finally over with a definitive test. I had done a calculation to determine just how much "modern," i.e., 16th Century carbon would have to have been present due to contamination; it was on the order of half the mass of the sample. So I accepted the findings. The one concern I had was in the range of date difference in the various laboratories; they seemed large compared to my memory of the inter-calibration Bob had performed. Since that time, some 32 years ago, I have staunchly declared that I was personally certain that the Shroud could not be of 1st Century origin and therefore was not "authentic." I was aware of the reweave theory, but like so many theories suggesting the dating was in error, I attributed it to the lunatic fringe. My first hint of belief was when I was made aware of Ray Rogers' paper that appeared in print shortly before Ray's death. I spent several evenings talking to the person that assisted Ray in the testing of a sample of the Shroud used to date the Shroud. I became convinced that the suspect corner of the Shroud that our protocol had demanded not be used was actually modified, probably in the way suggested by so-called invisible reweaving. Most convincing was the fact that Ray actually examined a thread that was half linen at one end and half cotton at the other. As I have stated, none of the members of the team that directly examined the Shroud at the time of testing or any of the samples taken at the time contained any cotton. Further, since the identified line where the reweaving was suggested was at a diagonal to the sample from which the samples were cut and distributed, the seriation of the amount of "modern" cotton associated with the specific lab's date aligned exactly with the dating from oldest to youngest dates. This help to suggest why the range of dates was larger than expected from the inter-calibration. I now state that I am no longer certain that the Shroud could not be first century. The material in this book discusses not only what I mention here, but many other facts that lend credibility to the hypothesis that the corner of Shroud where the samples for dating were taken are presumptively from a "modern" reweaving.

PREFACE

This book was originally a three-part article that completed a trio of writings by the author concerning circumstances surrounding the controversial 1988 radiocarbon (C-14) test on the artifact known as the "Shroud of Turin," or "Turin Shroud," or just "The Shroud," believed by many to be the actual burial cloth of the historical Jesus. The test, performed by three different laboratories, was said to have shown that the Shroud was produced in AD 1260-1390, many centuries too late to be connected with Jesus. I believe those results have been blindly accepted by most people, who have little idea how complex, chaotic, and incompetent the whole enterprise was. However, as shown in my article (co-authored with Ed Prior) "Chronological History of the Evidence for the Anomalous Nature of the C-14 Sample Area of the Shroud of Turin" (https://www.shroud.com/pdfs/chronology.pdf), C-14 dating can be problematical, as there are unknown factors, in some cases, that impinge on the reliability of the test. That same article also gives ample evidence that the area from which the C-14 sample was taken was anomalous *vis-à-vis* the rest of the cloth, very possibly from repairs made to that corner. In this book, I've actually reproduced verbatim some of the material from that article that fits perfectly with the context and subject matter here. The second article, "Bibliography of Major Sources Pertaining to the Events of the 1988 C-14 Dating of the Shroud of Turin" (https://www.shroud.com/pdfs/marinobib.pdf), is self-explanatory. The original three-part article, which I have revised and expanded into this book, is accessible at http://newvistas.homestead.com/C-14PoliticsPt1.html.

After the dating results were announced on October 13, 1988, many, including some who had believed that the Shroud was authentic, unquestionably accepted the results. The late Cardinal Anastasio Ballestrero of Turin said there was no reason to doubt the results, but later would say he believed that there had been a Masonic plot to discredit the Shroud. The Vatican also later came out publicly to say that the C-14 results were strange in light of all the other data that had been collected by the Shroud of Turin Research Project (STURP), which performed direct testing on the Shroud for five days, analyzed all the data for three years, and published their findings in highly-respected peer-reviewed scientific journals. The members of STURP came from some of the most prestigious and respected institutions in the country. Most of them worked in the U.S.A.'s nuclear or space programs. Despite this, their scientific credentials were maligned, ostensibly because they had left open the possibility that the Shroud could be the authentic burial cloth of Jesus, a stance many other scientists believed improbable, despite their own lack of serious investigation into the cloth.

Ironically, and despite all the problems with the C-14 dating, the labs may have actually been reasonably accurate in the date they came up with, insofar as a theory, such as the area from which the sample was taken from had been anomalous due to a reweave or other plausible explanations, might reasonably explain why a first-century cloth produced a medieval date; but even if that were the case, the various negative actions by Church officials and the C-14 labs' representatives need to be exposed and documented for history. As will be shown by the many statements cited throughout the book, mistakes, questionable decisions and discrepancies abounded before, during, and after the sample was taken on April 21, 1988, by both Church officials and representatives of the three labs, and I believe the evidence presented here will emphatically show that. Politics played a huge role in the process from beginning to end. Take note of how many times in this book the words "politics" or "political" appear. If the Shroud is ever tested again, hopefully such mistakes can be minimized or totally avoided. Actually, one should not be surprised at all that politics were involved, given that politics played a role in the Jesus' crucifixion. Politics being involved in his putative burial cloth would actually be an example of history repeating itself.

In an interview by the *Atlanta Journal* of the late Rev. Albert (Kim) Dreisbach, a proponent of the Shroud's authenticity, he said regarding the C-14 dating of the Shroud, "What we have not seen -- but will see -- is the rest of the story." The reader will find in this book most of the rest of the story, as we are now past the thirtieth anniversary of this celebrated test. What many people don't realize is that events decades before the actual dating were crucial factors in what has become an ongoing saga.

At the time of this writing, I have been studying the Shroud for over forty-three years, and have one of the best personal English-language collections of materials in the world, and especially have extensively read material regarding the 1988 dating. During those years, I have been privileged to obtain many rare documents, which enable me to more easily connect many dots and give a comprehensive treatment of the chaotic C-14 dating process. I hope that it will be useful not only for current sindonologists (those who study the Shroud of Turin) but for future sindonologists for whom it will invariably be harder to obtain the more rare printed materials that I have. I have also made use of some of the correspondence I've had with some of the major participants. The situation is so complex that it's a necessity to examine all the sources cited here to get the fullest picture possible of the events. Part I will look at the years leading up to the dating. Part II will analyze the actual events of the day when the sample was extracted. Part III will recount the aftermath, which continues to this day.

Some notes on format: rather than using some combination of the traditional footnotes, endnotes and references, I have decided on a modified format to keep the reader from having to do a lot of flipping back and forth between the text and the end of the book. There will be a chronological informational entry, the source used (reference), and with some entries, I've added my comments, which may cross-reference other sources and fill in some additional information. Regarding the references, I have also added an alphabetical list of them at the end, for those who want to see a compilation of them. I will leave out a few that are difficult to list as traditional reference entries (but they will be specified in the source section of the entry). In the text, for the sake of brevity, I don't always use the title of the person or use their complete last name, e.g., "Cardinal Ballestrero" is sometimes just "Ballestrero" and "Riggi di Numana" is sometimes just "Riggi." No disrespect is intended by the omission of a title. There are also two forms of month/day/year throughout. In the United States, e.g., October 1, 1986 is written as

10/1/86, whereas in Europe it's 1/10/86. Quotes will utilize both forms. In fact, in one translated document, an Italian used both forms! At times, I insert bracketed italicized comments in a quotation for clarification. Where there is an error in a passage I've quoted or a scientific symbol that is unfamiliar to almost one-hundred percent of readers, it's designated with a *"[sic]*," which specifies I have reproduced the word or symbol exactly as it was in the passage. Also, keep in mind that there are differences between American and non- American spellings of certain words. I have not changed British spellings of certain words, such as "mediaeval" as opposed to the U.S. spelling of "medieval." In addition, be aware that radiocarbon dating can be referred to as "radiodating," "Carbon 14," "C-14," "C14," "14C," or "^{14}C." Milligram is abbreviated as either "mg" or "mg.", depending on an author's preference. Similarly, centimeter is either "cm" or "cm."

Just as prosecutors in a criminal case interview all the major witnesses in a case to compare their testimonies, I have looked at multiple accounts of the major players in the proceedings of the Shroud dating via various documented sources. Of course, certain individuals are aware of some facts that others are not. Many of the key events recounted will have different perspectives from the principal figures, which will be listed in multiple entries with the same month/year designation. The entries are in chronological order, but because that some sources gave specific dates, some just gave approximations and some entries cover multiple months or years, it was impossible to attain a strict or exact listing of events by specific dates. It wasn't practical to do separate chronological entries by individual sources, since the amount of material varies among the various sources. There simply was no easy way to lay out the massive amount of (confusing) information. So, while there will be some duplication of accounts of some events, the duplication will actually help serve as confirmation of the events and enable the reader to compare and contrast the recollections. The reader is also encouraged to make use of the "Key Figures" section at the end of the book, as well as the Index, to facilitate understanding.

It will be useful to identify the various groups that were involved in the process. There were many tensions among some of the groups, which was not a good recipe for the cooperation needed to perform a rigorous test.

1. The Vatican. The Shroud is owned by the living Pope

2. The Pontifical Academy of Sciences, the Pope's advisors on scientific matters.

3. The Archdiocese of Turin. Turin is the Italian city in northwest Italy where the Shroud has been since 1578 (except for a few years during WWII when it was moved for safety reasons). The archbishop of Turin is normally appointed by the Pope as "Papal Custodian of the Shroud." Some of the archbishops have had non-clerical (lay) scientific advisors, as was the case during the 1988 dating process.

4. Centro Internazionale di Sindonologia in Turin. Sometimes referred to as "Turin Centro" or just "Centro." The group has a museum dedicated to the Shroud, and is involved in the many day-to- day activities related to it. It recently changed its name to "Centro Internazionale di Studie Sulla Sindone." The secretary at the time of the 1978 testing was the late Fr. Piero Coero-Borga, who was not pleased that STURP was allowed to do testing.

5. The Holy Shroud Guild. The oldest Shroud organization in the United States. One of its priests, the late Fr. Peter Rinaldi, of the Salesian order, was born in Turin, and acted as a mediator between the Turin authorities and STURP. The other main priest was the late Fr. Adam Otterbein of the Redemptorist order. The archives of the Guild contain many important letters of individuals involved in the dating process.

6. The Shroud of Turin Research Project (STURP). The American group that had studied the Shroud for five days around the clock in October 1978. In 1984, they sent to the Church authorities a proposal for twenty-six multidisciplinary tests, including C-14.

7. Association of Scientists and Scholars International for the Shroud of Turin, Ltd, (ASSIST). This group was formed in 1982. It had no real impact on the dating process, although it wasn't due to a lack of effort.

8. Various C-14 laboratories. Originally, it was thought that as many as seven different labs would take part in the testing. Eventually, it was reduced to only three.

9. The Italian government, specifically the "Ministry of Cultural Heritage and Activities and Tourism," which under national regulations, all restoration work is subject to prior approval. The Shroud chapel at St. John the Baptist Cathedral was included under their jurisdiction.

I also want to delineate the main sources used in this book. The full bibliographic citation can be found with each of the actual entries.

1. A book by the late Dr. Luigi Gonella and the late Giovanni Riggi di Numana. Gonella was the scientific advisor to the Cardinal of Turin, Anastasio Ballestrero. Riggi di Numana (also sometimes referred to as just "Riggi") was the scientist who actually cut the sample that was used for the C-14 dating of the Shroud. For the description of activities from the Gonella/Riggi book (and all other non-English sources), I used the Google translation program, which, while not perfect, is quite good overall. In a few instances, I consulted with a native Italian speaker for clarifications.

2. The computer archive of documents and correspondence of Gonella, who kept meticulous records from the late 1970s through the early 2000s. The archive includes copies of letters among many of the 1988 participants. This archive is very rare -- there are only a few researchers in the world who have one, but the author managed to get access to a copy.

3. A book by the late Dr. Harry Gove. Gove was the co-inventor of the Accelerator Mass Spectrometry (AMS) method used to date the Shroud. Gove was heavily involved in the planning for the C-14 test and was determined to get STURP eliminated from taking part and was successful. Ironically, however, his own lab was not one of the ones chosen to take part in the actual dating.

4. A book by the archaeologist William Meacham. Meacham was a participant in the 1986 Turin Planning Meeting that was called to set the protocols for the dating of the Shroud. Unfortunately, all the important recommendations from the meeting were ultimately ignored.

5. An extensive article by Meacham given at a by-invitation-only symposium held in Turin in 2000.

6. A book by Turinese medical doctor Dr. Pier Luigi Baima Bollone, a.k.a. Baima Bollone, who is a member of the Turin Centro and has had close ties to all of the main Italian participants in the dating.

7. A book by Giorgio Bracaglia, overseer of the archives of the Holy Shroud Guild. I also possess, on CD, some of the archives. He also provided additional information via documents not found on my own CD, as well as some information by email and phone.

8. A book by the late Rev. David Sox, who had associations with both STURP, which hoped to be involved in the C-14 dating, and Dr. Harry Gove.

9. A book by German authors Kersten and Gruber, who interviewed many of the principal characters involved in the dating process.

10. Two books by Italian researchers Petrosillo and Marinelli, who interviewed many of the principal characters involved in the dating process.

11. An extensive paper by Marinelli that used multiple Italian and French sources and contained more than two-hundred and fifty footnotes.

12. A doctoral dissertation published in 1989 by H. Laverdiere. The author interviewed many of the principal characters involved in the dating process.

13. An Italian Shroud documentary available on DVD. The director, Francesca Saracino, had access to the aforementioned archive of the scientific advisor the Cardinal of Turin, Luigi Gonella.

14. Articles by high-profile, international writers from the prominent Italian Shroud journal, *Collegamento Pro Sindone*.

15. A series of writings by a French group called "The Catholic Counter-Reformation in the XXth Century," which conducted exhaustive research into the Shroud dating for almost twenty years, including in-depth interviews with many of the principal characters and produced their findings in both French and English articles and documents. *[Note: despite the word order of "Catholic Counter-Reformation," the group abbreviates their initials, both in cross-references and in their web site URL as "CRC" and not "CCR."]*

16. Multiple newspaper articles, whose authors interviewed many of the principal characters involved in the dating process.

17. Various other books and articles.

A large part of the history of any event is the relevant documents. With many events, previously unknown or undisclosed documents are uncovered -- and then the history of the event must be rewritten. In the case of the Shroud, more and more documents and eyewitness reports have come to light since the original event in 1988. This book has been compiled from the most comprehensive set of documents ever assembled regarding the Shroud C-14 testing.

I believe that history will not be kind to those Church officials and individuals associated with the three labs. As the late STURP scientist Ray Rogers once said, "There will be hell to pay when the truth comes out." It is my hope that the Vatican will allow new testing in the near future. Science and technology have made enormous advances since the 1978 STURP study and the 1988 C-14 testing. New testing would allow the Church to both make up for the significant errors due to their own actions made in the 1988 dating, and also to take advantage of the scientific and technological advances to learn more information about this enigmatic cloth, which continues to draw the interest of many scientists, researchers and Christians and non-Christians alike. Negative comments in this book about individuals or institutions are made only in the context of attempting to find the truth about the Shroud.

If you don't think such dishonesty in science is possible, please note that if you Google "dishonesty in scientific research," you will get over two and one-half million hits. It exists, and given that the truth about the Shroud has ramifications for peoples' worldviews, the potential for dishonesty is perhaps elevated. In a 2019 article about the Shroud (https://www.academia.edu/40824797/Archaeology_and_the_Shroud?email_work_card=view-paper), the article's author said in a footnote (pg. 3), "There are numerous examples within Shroud research in which scientists and researchers have developed data that they profess to be the end of a particular debate concerning the Shroud, only to have that data be discovered as flawed or otherwise overstated. This paper will not attempt to discuss those cases, as there have already been volumes written on these discrepancies, some of which were fueled by well-meaning mistakes, others by pure ego and malice." In this book, one might not find many well-meaning mistakes, but one will find examples of various researchers from whom there is an abundance of data that is "flawed or otherwise overstated," plus numerous "discrepancies," including ones fueled by "pure ego and malice." And there are definitely "volumes" to recount. If one dismisses the accumulated evidence presented in this book, I dare say that intellectual dishonesty is being exhibited.

It is not known if there will be a second edition, so I have set up a page on my web site in order to be able to disseminate corrections, additions, etc. The URL is: http://newvistas.homestead.com/The-1988-C-14-Dating-of-the-Shroud-of-Turin--a-Stunning-Expos-.html.

There is a second online aspect as well. Although this book is being sold only in print, those who buy a copy can contact the author to get a free PDF of the book. The main reason for that is to make it easy for the reader to access some of the URLs listed in the book. However, because the PDF will match the print formatting, in which the links are not hot, the reader will need only to copy and paste the URL into a browser to access the site. It is one extra step beyond a hot link, but it's better than the alternative of having to manually type the URL as shown in the book, into a browser. The author can be contacted at **JMarino240@aol.com** to get the no-charge PDF.

Part I:
Pre-April 21, 1988, Day of Sample Extraction

Prefatory entry: According to a colleague of Harry Gove, the Shroud C-14 dating included, "deception, outright lies, low cunning, misrepresentation, and a pathological hunger for publicity as well as solid science and technology, faith that passeth all understanding, and, on Gove's part, tenacity and determination rarely encountered."

Source: Foreword by D. Allan Bromley in Gove, Harry. Relic, Icon or Hoax: Carbon Dating the Turin Shroud (Bristol and Philadelphia: Institute of Physics Publishing), 1996, pg. x.

Comments: Notice that Bromley set off the negative aspects completely from Gove. Bromley apparently saw things through the lenses of Gove. Bromley seemingly even was sarcastic with his King James "passeth all understanding" remark, as if to say, "How can any thinking person believe the Shroud could be authentic after the results of the Shroud dating?" After reading the material found in Part I, the readers can decide for themselves if Gove is blameless in the aforementioned negative aspects that Bromley did not ascribe to Gove. Bromley also echoed Gove's assertion that STURP's findings were not significant, and even claimed that the "group attempted to block the carbon dating of the shroud and, when that failed, attempted to move in and take credit for as much of the subsequent activity as possible." This was nonsense. STURP wanted to do a C-14 test, but in conjunction with numerous other multi-disciplinary analyses. And given the sad state of events that ensued, why would STURP want to take any credit for them?

1947. The method of radiocarbon dating (C-14) was invented by chemist Willard F. Libby, who received the Nobel Prize for Chemistry in 1960 for his work. The method measures the activity of the radioisotope carbon-14 still present in the sample. By comparing this with a modern standard, an estimate of the calendar age of the artifact can be made. It quickly became a staple in archaeology.

Source: [General knowledge.] See for example, https://en.wikipedia.org/wiki/Radiocarbon_dating.

1955. Various scholars brought up the idea of doing a radiocarbon dating test on the Shroud. A Turinese Salesian priest, the late Luigi Fossati, commented, "We do not concur with this idea. The Shroud, as history records, has suffered too many vicissitudes, including exposure to fire and water. In these conditions, it might not offer the necessary elements for research by the scientists."

Source: Petrosillo, Orazio and Emanuela Marinelli. The Enigma of the Shroud: A Challenge to Science, (San Gwann, Malta: Publishers Enterprises Group), 1996, pg. 19.

Comments: I have read in multiple Italian books and Italian articles that Libby himself said that the Shroud because of its particular history would not be a good candidate for C-14 dating, but none of them give a specific citation for the quote.

1961 *(December)*. A British Shroud researcher inquired to P.J. Anderson of the Harwell C-14 laboratory about the possibility of having the Shroud dated by the C-14 method. Anderson said, "The history of the Shroud does not encourage one to put a great deal of reliance upon the validity of my carbon 14 dating. The whole principle of the method depends upon the specimen not undergoing any exchange of carbon between its own molecules and atmospheric dioxide, etc. The cellulose of the linen itself would be good from this point of view, but the effect of the fires and subsequent drenching with water [...] and the possibility of contamination during early times, would, I think, make the results doubtful. Any microbiological action upon the Shroud (fungi, moulds, etc., which might arise from damp conditions) might have important effects upon the carbon 14 content. This possibility could not be ruled out."

Source: Published in a communication by British Shroud researcher Vera Barclay in *Sindon* (Journal of the Centro Internazionale di Sindonologia, Turin), December 1961, pg. 36, as cited in Wilson, Ian. The Shroud: The 2000-Year-Old Mystery Solved, (Sydney: Bantam Press), 2010, pg. 92.

Comments: In the Proceedings of the Twelfth Nobel Symposium at Uppsala University, titled "Radiocarbon Variations and Absolute Chronology" (1970), authors Save-Soderbergh and Olsson quote a Prof. Brew saying, "If a C14 date supports our theories, we put it in the main text. If it does not entirely contradict them, we put it in a footnote. And if it is completely 'out of date', we just drop it." Although it has to be acknowledged that this statement was made about twenty years before the 1988 Shroud C-14 dating and that progress in the technique was made in that period, that's somewhat negated by the fact that the AMS technique used in 1988 on the Shroud was only about eleven years old at the time.

1973 *(November)*. On November 22nd, the Shroud was exhibited before a limited gathering of journalists and specialists. The next day, the Shroud was shown on television all over Europe. On November 24th, a secret commission of experts examined the Shroud, and a small sample from one corner was extracted for study by Belgian textile expert, the late Prof. Gilbert Raes.

Source: Wilson, Ian. The Shroud: the 2000 Year Old Mystery Solved, (London: Bantam Press), 2010, pg. 308.

1976. Italian textile expert Riccardo Gervasio published a meticulous study on the restorations and repairs undergone by the Shroud. He found in the upper corners inlays of fabric with reinforced seams in overlock, from medieval times, to support and partially patch the original, which had literally frayed at the edge during the centuries.

Source: Siliato, Maria-Grazia. Contre Enquete Sur Le Saint Suaire, (Paris: France Loisirs), 1998, pg. 40.

Comments: It is clear based on this, that corners of the Shroud should have been avoided when considering places from which to take samples for a C-14 test.
See also entry under "1979 *Fall*."

1977 *(May)*. The Accelerator Mass Spectrometer (AMS) method of C-14 dating was invented at the University of Rochester by scientists from General Ionex Corporation, Lawrence Berkeley National Laboratory (Berkeley, California), and University of Toronto. The leading figure in the group was the late Dr. Harry Gove of the University of Rochester. The first use of this method for C-14 dating took place at this facility.

Source: Gove, Harry. Relic, Icon or Hoax?: Carbon Dating the Turin Shroud, (Bristol and Philadelphia: Institute of Physics Publishing), 1996, pg. 320.

Comments: Sometimes the word "Tandem" is added before "Accelerator" and then the acronym "TAMS" can also be used.

1977 *(June)*. The University of Rochester issued a press release regarding this new dating procedure and articles were published in the *New York Times* and also *Time* magazine. After seeing the article in *Time*, Episcopal priest David Sox sent a letter on June 24th to the University to see if this method could be used to date the Shroud of Turin. In his letter, Sox informed Gove that the late Walter McCrone, a microscopist from Chicago, wanted to use the already-removed Raes sample from 1973 to carbon-date the Shroud.

Source: Gove, Harry. Relic, Icon or Hoax?: Carbon Dating the Turin Shroud, (Bristol and Philadelphia: Institute of Physics Publishing), 1996, pg. 320.

Comments: Sox would eventually be significantly involved in the Shroud C-14 dating process and published a book that was already printed when the official results were announced on October 13, 1988. That fact indicates there were leaks about the results.

1977 (July). On July 6th, Gove replied to Sox that it could be used on the Shroud, but not immediately. Gove wrote, "By coincidence, on the same day I replied to Sox I got a call from Walter McCrone. We had never met and obviously he had no knowledge of the Sox letter. He told me, in a rather secretive manner (the shroud seems to bring out conspiratorial traits in people) that he had an important piece of cloth whose age he was eager to know. In particular he said he would like to know whether it was two-thousand years old. Presumably he thought I couldn't guess to what cloth he was alluding. Had he called before we received Sox's letter I would, indeed, not have known. I had begun to suspect that Sox and McCrone were cooking up some scheme to obtain Raes' shroud samples. I told McCrone he should write to me and he did."

Source: Gove, Harry. Relic, Icon or Hoax?: Carbon Dating the Turin Shroud, (Bristol and Philadelphia: Institute of Physics Publishing), 1996, pg. 18.

Comments: As we shall see, Gove himself was not immune from having conspiratorial traits. It will be also clear that he was pretty good himself at cooking up schemes to be involved with dating the Shroud. Schemes by various individuals were to be prevalent through the whole Shroud C-14 dating process. What did Gove think of McCrone? He wrote (pg. 19), "I sometimes think that McCrone dreamed of becoming history's greatest iconoclast. Having, in his view, demolished the authenticity of the Vinland Map he saw the chance to do the same to the Turin Shroud!"

1978 (May). Gove wrote, "On 11 May 1978 I phoned the Reverend H. David Sox at his home in London to find out what the latest interest was in carbon dating the shroud. He told me that the samples removed from the shroud in 1973 for examination by Professor Gilbert Raes had been in the hands of the Turin ecclesiastical authorities ever since Raes had mailed them back. He and McCrone had been working together to arrange for these samples to be made available for carbon dating."

At the end of May, McCrone, the microscopist who claimed at first he believed the Shroud could be authentic, but later believed the Shroud had been produced by an artist, discussed with Harry Gove, the inventor of the AMS method that was eventually used on the Shroud, carbon dating the Shroud. According to Gove, McCrone "mentioned that the two pieces removed in 1973 came from the hem of the shroud and thus might be of more recent vintage."

Source: Gove, Harry. Relic, Icon or Hoax?: Carbon Dating the Turin Shroud, (Bristol and Philadelphia: Institute of Physics Publishing), 1996, pp. 20 and 22.

Comments: Although McCrone later accepted outright the results of the 1988 testing because it supposedly proved his theory about an artist having produced the Shroud, it's interesting that he, and at a relatively early date, believed that the area from which the sample was actually taken might have contained repairs. McCrone was chastised by some for trying to deal directly with King Umberto of Italy, who owned the Shroud before bequeathing it to the Vatican upon his death in 1983, instead of going through the appropriate channels [for example, see Kenneth Stevenson and Gary Habermas' _The Shroud and the Controversy_, (Nashville: Thomas Nelson Publishers), 1990, pp. 46-47]. Gove discussed throughout his book many of the political and personal issues pertaining to McCrone's involvement in the Shroud.

Gove himself was heavily involved in Shroud politics; he was lobbying, by his own admission in his book, to eliminate STURP, the team that tested the Shroud in October 1978, from being involved in any new testing. One STURP member was quoted in a late 1970s Shroud article as saying about Gove, "His ego arrived about a half-hour before he did."

1978 *(June).* On June 29th, Gove met David Sox for the first time. Sox, at the time, was the General Secretary of the British Society for the Turin Shroud (BSTS) and was very keen to see a C-14 test performed on the Shroud. Gove wrote, "He can occasionally be mildly effeminate with a slightly shrill laugh. He has a tendency to be a name dropper in conversations. On the other hand he does know many important people and he knew a lot about the Turin Shroud and the people associated with it." Sox suggested to Gove that he should attend the Turin scientific congress planned for October 7-8, 1978. It would give Gove a chance to meet with Fr. Peter Rinaldi, a Turin-born Salesian priest stationed in the United States. Among those that Sox believed might have influence in who would participate in the C-14 dating, "Rinaldi was *'numero uno'* and, in all modesty, he said that he was number two." Gove added, "As we were both to later learn, Luigi Gonella in his capacity as science advisor to the archbishop of Turin on matters of the shroud was 'numero uno' and all the other numbers as well!" In addition, Gove noted, "Walter McCrone would be there and wanted to play a major role in the dating operation."

Source: Gove, Harry. Relic, Icon or Hoax?: Carbon Dating the Turin Shroud, (Bristol and Philadelphia: Institute of Physics Publishing), 1996, pp. 22-23.

Comments: It's hard not to come away with the impression that Gove was saying some very critical (and unnecessary) things about Sox, but because of his contacts would be a very useful person to him. Gove, who would constantly be at odds with Ballestrero's advisor Gonella during the C-14 process, apparently had a strong dislike for the latter going back to the Turin congress. It also seems that Gove was concerned by the possible competition from McCrone. On page 21 of his book, Gove recounted a call he had received from McCrone, who "gratuitously donned the mantle of leadership in getting the shroud dated despite lacking expertise in the field of carbon dating." Already at this early stage, individuals were jockeying for positions in a potential C-14 test.

1978 *(July).* After an article on the Shroud is published in the popular journal *Science* and many negative letters-to-the-editor were written (for giving attention to a supposedly religious object), Gove recalled his own thoughts: "As a scientist, can one justify spending any of one's professional time on a religious artifact? My main justification was that dating the shroud captured the public imagination and it would be a tremendous boost for this new and publicly unknown technique. Other than this, it would be difficult to argue that dating the shroud served any scientific purpose."

Source: Gove, Harry. Relic, Icon or Hoax?: Carbon Dating the Turin Shroud, (Bristol and Philadelphia: Institute of Physics Publishing), 1996, pg. 25.

Comments: Gove's logic about the Shroud dating not serving any scientific purpose apart from promoting the AMS method seems strange. Whether or not the Shroud is a "religious" object, it was an archaeological object that should have been on equal footing and valued as any other object that Gove's lab might have been asked to date. And Gove's emphasis on the promotion of AMS shows that Gove was not immune to Bromley's observation of "pathological hunger for publicity."

1978 *(October)*. The Second International Congress on the Shroud was held in Turin. Both the University of Rochester and Brookhaven National Laboratory (Upton, New York) submitted papers describing how the Shroud could be carbon dated using minute samples with this new method. The Shroud was on public display in the Cathedral of St. John the Baptist from August 27 through October 8 and was followed by the congress on October 9-10, and then by five days of non-destructive tests by members of STURP as well as a few other scientists.

Source: Gove, Harry. Relic, Icon or Hoax?: Carbon Dating the Turin Shroud, (Bristol and Philadelphia: Institute of Physics Publishing), 1996, pg. 320.

Comments: When discussing the STURP 1978 testing in his book (pp. 6-7), Gove related, "I believed STURP's members to be so convinced it was Christ's shroud that I was determined to prevent their involvement in its carbon dating, if that were ever to come about. I feared the most important measurement that could be made on the shroud would be rendered less credible by their participation. Fortunately in this I was successful." Gove here clearly indicated that he believed that C-14 was more important than the sum total of all other testing, which approaches the heights of arrogance. The STURP members came from some of the most prestigious U.S. institutions, including Los Alamos National Laboratory, Sandia Laboratory, Jet Propulsion Laboratory, Air Force Academy and others, most of which were and are involved in the U.S.' nuclear and space programs. It should also be pointed out that the STURP team was mainly composed of nominal Christians, Jews and agnostics.

On page 29 of his book, Gove stated, "The truth of the matter was that I did not care whether the Turin Shroud was carbon dated or not [...]. My further concern was that if it ever were done it should be done by the people who knew what they were doing and were dispassionate and not under the control of true believers." I don't believe for a second that Gove didn't care whether it was carbon dated or not. There is plenty of evidence in this book that he did. What was Gove's status *vis-à-vis* religion? On page 36 of his book, Gove supplies a writer's quote about himself: "He doesn't believe in miracles [...]. But Harry Gove's not 'born again'. He's a once-a-year Episcopalian. He doesn't believe the shroud came from the corpse of Jesus Christ [...]." If he doesn't believe in miracles, why is he even a once-a-year Episcopalian?" And if he didn't believe (before C-14 dating) the Shroud wrapped Jesus, how is that any less dispassionate than a scientist who believed the Shroud could have been Jesus' burial cloth?

On page 35, Gove wrote, [...] "the STURP team would work their 'non-destructive' space age scientific wonders on it for what turned out to be five days and nights." Their spokesman, press liaison man and chief of security, was Ken Stevenson. Stevenson was a rather abrasive black man, a graduate of the Air Force Academy, and, at that time, with International Business Machines [...]. Like Jackson and Jumper, he had an exaggerated view of the importance of this mission, the overwhelming need for security and secrecy, and of his own importance in STURP." Gove's use of quotation marks for "non-destructive" obviously indicated he believed some tests were harmful. While Gove was critical of McCrone wanting to play a major role in the C-14 process despite lacking expertise, he apparently had no problem making scientific judgments about the expertise of scientists that could send men into space and make nuclear bombs. His phrase "space age scientific wonders" is clearly sarcastic. Regarding Stevenson, what does it add describing him as "black?" I know Ken and have seen him interact with numerous people. He is anything but abrasive. As far as Stevenson's "exaggerated view," Gove uses no qualifiers such as "my perception was," and just states it as fact. It's just one of many instances documented throughout this book where Gove passionately disliked anyone or anything associated with STURP.

On page 38, Gove mentioned that the late Bishop John Robinson, in his paper at the congress, made a plea that the Shroud should be radiocarbon dated using the new (AMS) method. There was a line of people waiting to make comments on Robinson's talk. Well-known Shroud author Ian Wilson offered to make some remarks and to introduce Gove, who described it as "an offer I was pleased to accept -- after all, Wilson was one of the stars of the Congress." When Wilson introduced him, Gove noticed considerable media activity. He gushed, "When I reached the podium it was as if I were some sort of Hollywood celebrity [...]. When I later saw the CBC *Man Alive* programme I noted that my brief moment of glory was faithfully recorded for the pleasure of the viewing public -- and I must confess, mine." Again recall Bromley's phrase in the foreword, "a pathological hunger for publicity," which was aimed at other scientists but not Gove. And there's more.

On page 39, Gove wrote, "During the break that followed, I was besieged with questions by the press. Their attention outrivalled *[sic]* what they had bestowed on Eric Jumper after his show and tell involving the cardboard cutout of the shroud image. I thought to myself that these spit and polish young Air Force officers with their vaunted space age technology were being outshone -- and by a professor of physics old enough to be their father, who just happened to have superior technology at his command." Notice that Gove used the demeaning phrase "show and tell" to describe Jumper's presentation. Gove did not clarify why his technology, only recently created, was superior to the technology that took vehicles into space.

On page 41, Gove related that when he returned home to Rochester, he "was met by reporters from the three local TV stations and the two local newspapers." I was beginning to feel like a real celebrity --something that happens very infrequently to professors of physics."

On page 42, Gove stated that after another trip out of town, when he returned the mayor of Rochester was on the same plane. A local TV station started filming and interviewing Gove as the Mayor walked by. Gove noted that "he must have wondered who I was to be taking the spotlight from him." The reporter asked Gove when he would get Shroud samples to date. Gove remembered, "It was all very heady stuff. I began to think that maybe the best thing that could happen to me and my laboratory would be to continue getting all this free publicity without ever having to actually date the shroud."

1978 *(October)*. In a paper by Gove presented to the conference, he stated, "It would be preferable to obtain threads from several places throughout the material."

Source: Marinelli, Emanuela. "The Setting for the Radiocarbon Dating of the Shroud." Presented at 1st International Congress on the Holy Shroud in Spain -- Valencia -- Centro Español de Sindonologia (CES), April 28-30, 2012, pg. 2. www.shroud.com/pdfs/marinelliv.pdf.

1979 *(February)*. The Rochester and Brookhaven groups sent a letter to archbishop of Turin offering to date the Shroud using only milligrams of cloth.

Source: Gove, Harry. Relic, Icon or Hoax?: Carbon Dating the Turin Shroud, (Bristol and Philadelphia: Institute of Physics Publishing), 1996, pg. 320.

Comments: According to Gove (pg. 149), this offer was "sidetracked" and was resubmitted in August 1979.

1979 *(February)*. The very month that Rochester and Brookhaven sent the letter to Turin to offer to date the Shroud, Gove gave a talk titled

"Dating the Shroud of Turin" at a church in New York State. According to the priest at St. John's Church, the late Fr. Vincent Donovan, "Gove's tone was very condescending throughout much of his lecture." Gove assumed that the letter Sox had sent him a few years before was an invitation directly from Turin. He said, "I did not go to them, they came to me," implying that Sox had been representing Cardinal Ballestrero. Gove also ridiculed some of the Catholic Church's Shroud scholars. Bracaglia wrote, "In the end, he closed his presentation with self-adulation and reminded the audience, that he, with his stature, was not among the 40 scientists who had been admitted to the testing room during the 1978 examination."

Source: Bracaglia, Giorgio. Uncovering the Paradox within the Archives of the Holy Shroud Guild, (Honeoye, N.Y.: Holy Shroud Guild), 2019, pg. 193.

Comments: Phrases such as "condescending tone," "they came to me," "self-adulation," "his stature" -- all were part of a recurring pattern of Gove puffing himself up -- while constantly criticizing practically every other scientist and Church official involved.

1979 (March). STURP held a meeting in Santa Barbara, California to discuss their preliminary findings obtained from the October 1978 testing. Dr. Walter McCrone claimed the Shroud was a painting based on his studies. Gove gave a presentation and said he believed that the C-14 test was the only test worth doing at the present time. He asserted that it would settle once and for all whether the Shroud could have wrapped Jesus or was forged recently and therefore was a relic, icon or intentional hoax. He mused, "During this whole meeting, I had been taking notes as if what this motley mixture of scientists, priests, ministers, and peacetime warriors were reporting provided significant information regarding the real question of the authenticity of the shroud. They seemed to me to be a group of kids playing with expensive toys, hoping they would reveal some ultimate truth -- a truth of which most of them were already convinced. I summarize it to show the kind of information STURP was collecting and how little it mainly signified."

Source: Gove, Harry. Relic, Icon or Hoax?: Carbon Dating the Turin Shroud, (Bristol and Philadelphia: Institute of Physics Publishing), 1996, pp. 51, 53-54.

Comments: Considering that C-14 dating is not always accurate and has been known to even date objects into the future, Gove's belief that a non-first-century date would settle the question once and for all was not warranted, and the fact that after more than thirty years later the debate is as intense as ever, bears that out. Gove obviously believed that he was the only adult in the room.

1979 *(April)*. Gove phoned the late Robert Dinegar of STURP and expressed concern that McCrone would be involved in the C-14 dating. Dinegar assured Gove that he would not. Dinegar was also told that the C-14 scientists wanted to be independent from STURP and that the Shroud sample should not be blind *[i.e., identified as the Shroud when tested].*

Source: Gove, Harry. Relic, Icon or Hoax?: Carbon Dating the Turin Shroud (Bristol and Philadelphia: Institute of Physics Publishing), 1996, pg. 56.

Comments: It is not clear why Gove did not want the Shroud sample to be blind. The result would be more believable if it were.

1979 *(May)*. Toward the end of the month, Sox called Gove and suggested that he should assure Dinegar that the C-14 scientists realized that STURP had spent an enormous amount of time and money on the Shroud and something could be arranged to let them collaborate on the C-14 testing. Gove then stated, "I had very ambiguous feelings about STURP. On the one hand they were too convinced in their hearts that the shroud was Christ's burial cloth. On the other hand they had good connections in Turin and could be useful in obtaining a shroud sample for dating -- if only they could be prevented from playing any other role."

Source: Gove, Harry. Relic, Icon or Hoax?: Carbon Dating the Turin Shroud, (Bristol and Philadelphia: Institute of Physics Publishing), 1996, pg. 57.

Comments: Just as Gove was with Sox, Gove was only amenable to STURP mainly based on their usefulness to him in helping to arrange a C-14 test. Although Gove constantly expressed his opinion that STURP as a whole believed that the Shroud wrapped Jesus, Sox said in his book The Shroud Unmasked, (Basingstoke, Hampshire: The Lamp Press, 1988, pg. 69), "Heller *[STURP member]* says that in his interviews only three STURP team members, John Jackson, Robert Bucklin, Barry *[sic]* Schwortz (one of the photographers), beleive *[sic]* that it is probably the authentic burial cloth of Jesus of Nazareth; 'The rest of us have to say that we do not know'." Gove was so fanatical in his loathing of STURP that while discussing one of the their members, James Druzik, Gove stated in his book (pg. 291), "[A]s an art conservator, he would make an excellent member of a team of art experts that should be assembled to examine the shroud -- only if he were willing to shed his STURP connections, of course."

1979 *(August)*. The Rochester and Brookhaven proposal from February was resubmitted to the Vatican. Gove stated that STURP later backed this.

Source: Gove, Harry. Relic, Icon or Hoax?: Carbon Dating the Turin Shroud, (Bristol and Philadelphia: Institute of Physics Publishing), 1996, pg. 149.

Comments: Gove wrote (pg. 149) that at a planning meeting in Turin in fall 1986, Gonella related that this proposal suggested using the 1973 Raes sample (given to the late Belgian textile expert Gilbert Raes for analysis) but Turin found out that the chain of evidence indicating the samples were genuine was broken. Raes later mailed the samples back to Turin. Gove further related that when they had offered in the late 1970s to date the Shroud, "we had no idea that such incredibly sloppy control had been exercised over the Raes samples by the archbishop of

Turin -- almost as if they had been cut from a dishrag." Ironically, Gove would later characterize the 1988 dating as "a shoddy enterprise."

1979 *(August)*. According to Bracaglia, when the Rochester/Brookhaven proposal was resubmitted, Dinegar suggested doing a C-14 test as soon as possible using the Raes samples already available. Dinegar was confident that the test would be able to date it accurately plus or minus one-hundred and fifty years.

However, Gonella wrote to Fr. Rinaldi, "The Bishop tends to soothe crying people and appease those who protest. He was under the fire of protests from the Centro, so he did several things to appease Coero *[Coero-Borga]*, namely, downplay the Commission (i.e., 'me'), postpone C-14 without any answer to Gove (an ambiguous situation which may well explode someday -- a negative answer with some sort of justification is perhaps better than no answer, though I hope in no negative answer because then it would be difficult to reverse it)." Based on Gonella's communication to Rinaldi, it appears that the Cardinal did receive the Gove proposal but did not respond in order to appease the Turin Centro. Presumably, the Turin Centro, which was composed of both clergy and laymen and had a fair measure of control over the Shroud, was not a big fan of Gove. As we shall see, they did not view any American in a favorable light.

Source: Bracaglia, Giorgio. Uncovering the Paradox within the Archives of the Holy Shroud Guild, (Honeoye, N.Y.: Holy Shroud Guild), 2019, pp. 181-182.

Comments: Gove asserted in his book (pg. 312) that Coero-Borga never delivered the proposal to Ballestrero. Gonella took blame for it in a letter to Gove. Discrepancies and subterfuge were already surfacing nine years before the actual C-14 test.

1979 *(Fall)*. Three STURP members, R.W. Mottern, R.J. London and R.A. Morris published a paper titled "Radiographic Examination of the Shroud of Turin -- a Preliminary Report." According to the late Swiss archaeologist Maria-Grazia Siliato, they looked at the corners that Italian expert Gervasio alluded to in his 1976 report. According to Siliato, the three STURP members "had carried out radiographic examinations of this part: they had seen that the homogeneity of the cloth presented considerable disparities, parts of 'low density' that is to say, frayed and deteriorated, neighboring with others, of a 'high density' that is to say heavily restored: the discovery was published in [a] scientific journal."

Source: Siliato, Maria-Grazia. Contre Enquete Sur Le Saint Suaire, (Paris: France Loisirs), 1998, pg. 40.

Comments: Given that Gervasio's report was in 1976 and the Radiographic report was reported in 1979, one might think that the data should have steered the authorities away from taking the samples from a corner in 1988, but there would be many other instances when significant data could have been examined, but weren't.

In a preview of things to come, Siliato would go on to observe (pg. 41), "The visible memory of numerous repairs, the very obvious discrepancies in weight [of each of the samples], rightly suggest that the fragments used for carbon-14 dating of the Shroud by the three laboratories are all very loaded, in very variable proportions, with textile materials foreign and indeterminate [...]."

1980 (March). Fr. Rinaldi approached the late Monsignor Jose Cottino, spokesman for Cardinal Ballestrero, seeking permission to speak to Cardinal Ballestrero regarding carbon-dating the Shroud. Cottino told Rinaldi that the Cardinal didn't want to do any other tests until the 1978 findings were completed. Meanwhile, the Turin Centro was not at all pleased that Ballestrero had put Gonella in charge of the project. Subsequently, Ballestrero acceded to the Centro's wishes and focused on just conservation of the Shroud.

Source: Bracaglia, Giorgio. Uncovering the Paradox within the Archives of the Holy Shroud Guild, (Honeoye, N.Y.: Holy Shroud Guild), 2019, pp. 182-183.

Comments: It would be several more years before all of STURP's papers were published.

1981 *(May)*. Fr. Otterbein was getting very concerned that the conflict between the Centro and STURP was drastically deteriorating. Otterbein wrote a letter to STURP members Tom D'Muhala and Dr. John Heller requesting that the latter or someone from the group to go to Turin to try and work things out. Otterbein felt that without this intervention, STURP would suffer consequences, since Ballestrero had made it clear that any future testing must be a "collegial decision." Surprisingly, Rinaldi did not agree with Otterbein and instead blamed Dr. Baima Bollone and the Turin Centro for the problem. Rinaldi insisted that it was the Italians who were acting childishly and suggested that STURP should continue on their path. That of course, drew support from STURP, but it did not mesh with Ballestrero's wishes.

Source: Bracaglia, Giorgio. Uncovering the Paradox within the Archives of the Holy Shroud Guild, (Honeoye, N.Y.: Holy Shroud Guild), 2019, pp. 182-183.

Comments: Even though Fr. Rinaldi was born in Italy and would presumably have had a better feel than Fr. Otterbein about the Turin dynamics, it appears that Fr. Otterbein was the one who was right. It's not clear that if STURP had taken Fr. Otterbein's advice rather than Fr. Rinaldi's, they would have had more input in the dating process, but it certainly didn't help their cause to have sided with Rinaldi in this instance.

1982. Talks about conservation for the Shroud began. Heller informed Otterbein that discussions were underway about improving the conservation methods for the Shroud. Per the Cardinal's request, Heller and Gonella were organizing the team of conservators. There were about twelve people desired for the proposed committee, but they were acting as their own agent. At that point in time, the future of STURP was unclear; there was a possibility it could actually dissolve. Otterbein believed that this proposed committee would need the Holy

Shroud Guild to be a liaison among all the various parties. The Guild would pay for the transportation and lodging of the American scientists in return for the committee keeping them informed about the progress on general-policy decisions so that they could fend off any possible feud between the Centro and themselves.

In his original meeting with Heller, Otterbein did not discern that part of the strategy of the new conservation committee was to be independent of Church control, either perceived or actual. Heller clarified in a letter to Otterbein that he had no interest in trying to get STURP, the Centro and the ecclesiastical authorities on the same page. Meanwhile Gonella wrote a letter on July 15th to Otterbein that Ballestrero had not decided if taking a sample for C-14 dating would be part of the conservation procedures and wanted to get advice from the Pontifical Academy of Sciences, the Pope's scientific advisors.

Otterbein knew that the conservation committee to be headed by Gonella and Heller was not a good idea. In a memo from Otterbein, he acknowledged understanding their position as scientists, but knowing about the inner workings of ecclesiastical circles, he knew it to be "impractical and unrealistic." Otterbein knew that the committee would never get off the ground without any ecclesiastical involvement, despite the fact that Ballestrero had requested they form it.

Another Shroud research group arose that went by the name of "Association of Scientists and Scholars International for the Shroud of Turin, Ltd," (ASSIST). The group invited Dr. Baima Bollone of Turin to join; other Europeans joined as well. Otterbein hoped that the group would be a unifying force by promoting a synergy within the existing groups.

Since the Centro was focused on conservation, the Guild realized that to have a direct link to Ballestrero, it would be necessary for them to align themselves with the Centro's interests. They were hoping that Dr. Heller would be a good-will envoy representing the Guild. Otterbein discussed this in a letter he sent to Dinegar and reminded him that the Cardinal had said that future tests would be collegial.

Sources: Bracaglia, Giorgio. <u>Uncovering the Paradox within the Archives of the Holy Shroud Guild</u>, (Honeoye, N.Y.: Holy Shroud Guild), 2019, pp. 183-184, 186-188.

Comments: Otterbein's memos were written for documentation purposes. It's certainly ironic that after Ballestrero had insisted that future tests be collegial, one C-14 scientist ultimately dictated the testing in 1988, which was only one test; as we will see, that test was supposed to have been done in multi-disciplinary context.

1982. At that point, Dinegar was not aware of Gove's attempt to control the dynamics of the C-14 test. Dinegar pressed on with his assignment from STURP to enlist five different C-14 labs to do the Shroud dating. With the help of his friend Dr. Garman Harbottle, Dinegar had produced a proposal that was more solid than the original 1979 proposal. Harbottle issued a preliminary report; he advocated for a blind test and also acknowledged the Church's concern for using as little material as possible. Harbottle claimed that if a minimum of three labs were used, the results would be adequate to deal with an outlier (a result outside the normal data pattern).

Dinegar was aware that the Guild had some concerns about STURP's status. Otterbein wrote a letter to Dinegar that for credibility, the test should be performed by an international group that was independent from STURP. Individual members of STURP could take part, but it would have to be as part of the international group.

Source: Bracaglia, Giorgio. Uncovering the Paradox within the Archives of the Holy Shroud Guild, (Honeoye, N.Y.: Holy Shroud Guild), 2019, pg. 193.

Comments: The labs supposedly came up with results with no outlier in the 1988 test. Various researchers questioned the data over the years, and information discovered in 2017, via a Freedom of Information Act request, showed that the raw data was massaged in a way as to look better than it was. More details will be supplied in a 2017 time-frame entry.

1982. A C-14 intercomparison test sponsored by the International Study Group, affiliated with *Nature* was conducted among twenty labs. The test was carried out because they realized "the normal procedure for assessing errors […] has long been neglected in the radiocarbon dating field." The study showed various outliers, defined as "an observation that lies outside the overall pattern of a distribution." The group concluded that a variability existed "not entirely explained by the quoted errors." A more thorough report in 1983 (Baxter, M.S., *PACT* **8**, 123) confirmed "systematic bias and unexplained variability." In the same study, Reider Nydal asserted, "the main reason for using more than one sample […] is generally to get some idea of the magnitude of other sources of error rather than purely statistical ones."

Source: International Study Group, "An Inter-Laboratory Comparison of Radiocarbon Measurements in Tree Rings." *Nature* (London), **298**, 619 (1982).

Comments: This should not be confused with the similar study mentioned below in the next entry and also a 1989 study mentioned later. This report points out that errors in C-14 dating are common, which is a good reason to use more than one sample, which the Shroud dating process did not do. See also the C-14 scientist, the late Garman Harbottle's comments under "1988 *(April)*."

1982. At a conference held in Bradford, England, Dr. Robert Otlet of the C-14 lab in Harwell, England suggested that the British Museum provide textile samples to labs who were interested in dating the Shroud in order to gauge their accuracy. This became known as the "laboratory intercomparison test."

Source: Gove, Harry. Relic, Icon or Hoax?: Carbon Dating the Turin Shroud (Bristol and Philadelphia: Institute of Physics Publishing), 1996, pg. 321.

Comments: This test was carried out in 1983 as outlined immediately below.

1983. Dr. Michael Tite of the research laboratory of the British Museum informed five other C-14 labs at University of Arizona, Bern/Zurich (Switzerland), University of Oxford (England) and University of

Rochester (all of which used the new AMS method) and Brookhaven and Harwell (which used the older proportional-counter method) that he was willing to coordinate the intercomparison tests that Otlet had suggested.

Source: Gove, Harry. Relic, Icon or Hoax?: Carbon Dating the Turin Shroud, (Bristol and Philadelphia: Institute of Physics Publishing), 1996, pg. 321.

1983. For the intercomparison tests as described in the entry above, each of the six labs received two 100 milligram samples. The labs were informed of the origin of the samples but not of their ages.

One sample was Egyptian linen from B.C. 3,000 and the other was Peruvian cotton from A.D. 1,200. The Zurich lab, which used a new pretreatment, was off by one-thousand years on the Egyptian sample. In addition, all six labs dated the Peruvian cloth between A.D. 1400-1668. That sample was then replaced, without explanation (in February 1984), with another Peruvian cloth from A.D. 1,000-1,400. (The summary section for the pre-replacement sample stated, "The variation between samples is higher than expected on the basis of quoted measurement error. The 95% confidence interval is too late by 300 years, centering on ca *[circa]* AD 1500.) Tests for outliers were done but as with aforementioned 1982 study, non-statistical factors are at work." They also wrote, "The numbers of dates are small so that only high deviations can be rejected as definitely discordant."

Source: Burleigh, R., M. Leese and M. Tite. "An Intercomparison of Some AMS and Small Gas Counter Laboratories."*Radiocarbon* **28**, No. 2A, 571-577 (1986).

Comments: There were also a small number of dates with the Shroud dating. The results of the study were not actually released until the Trondheim C-14 conference in 1985. When Gove discussed this test in his book (pp. 77-78) he noted that the British Museum, which would, in fact, oversee the 1988 Shroud dating, had selected a sample in this 1983 study that was much more modern than they had believed. Gove stated, "I was surprised they had not verified the age of their first sample but agreed to measure the new one. Even venerable institutions like the British Museum can err." Indeed. In fact, the British Museum exhibited for twenty years the "Piltdown Man," which was presented as the fossilized remains of a previously-unknown early-human, despite doubts about it since 1915. In 1953, it was proved to be a hoax, and the British Museum finally had to admit its error.

1983. A French C-14 scientist, Dr. Jacques Evin, who would later be involved in supplying a controversial extra control sample for the 1988 Shroud dating, wrote in an article "It is evident that the samples to be dated should contain carbon. However, the mere presence of this chemical is not sufficient to produce a valid result. After many years of C-14 dating it has been shown that numerous results are in error or are erroneously interpreted, often because of lack of knowledge about the value of the material selected for dating."

Source: Evin, J. "Materials of Terrestrial Origin Used for Radiocarbon Dating." *PACT* **8**: 235-276 (1983), on pg. 235.

Comments: Evin's comment makes the failure of the labs not having done a chemical characterization of the three Shroud samples in 1988 inexcusable. A "control" sample is one of a known date that is also dated at the same time as the main object, in order to gauge the accuracy of the labs' results.

1983. Evin also told Bro. Bruno of the "Catholic Counter-Reformation in the XXth Century" in 1983, "Consult the scientific publications, independent of all journalism, propaganda and speculation, and you will note that there is still not a single list of results obtained from the accelerator method. Now, until that method has produced an impressive number of results on ordinary samples, taken under easy conditions and already dated by other methods, I consider, we consider, we professionals that it is not ready. Therefore, it must not be used to experiment on the Holy Shroud. That is elementary. When I am having trouble with my machines that is not the moment to carry out a crucial, difficult, spectacular experiment awaited by the world at large. I must first of all make my adjustments."

Source: De Nantes, Georges (Abbe) and Brother Bruno Bonnet-Eymard. "The Scientific Rehabilitation of the Holy Shroud of Turin: Part Four -- the Carbon Affair," *Catholic Counter-Reformation in the XX*th *Century*, May 1989, no. 219, pg. 11. (See http://crc-internet.org/ for the group's website**)**. For some pertinent pages relevant to the material cited from them in this book, see**:** http://crc-internet.org/our-doctrine/catholic-counter-reformation/holy-shroud-turin/ii-conclusion-new-trial/**,** http://crc-internet.org/our-doctrine/catholic-counter-reformation/holy-shroud-turin/appendices/, and http://web.archive.org/web/20001002025902/http://www.crc-internet.org/may00.htm.

Comments: Keep in mind that the accelerator (AMS) method had only been invented in 1977. Notice that Evin considered the accelerator method only as a <u>confirmatory</u> test, i.e., support for objects whose date was <u>already known</u>. C-14 dates, no matter what method is used, are never accepted in isolation but always in the context of other data. In the case of the Shroud, however, all the data collected by STURP and others was discarded and the C-14 results from a method only about eleven years old were pontificated as the sole arbiter of the cloth's date. It should also be said that by 1988, Evin did believe that the test would be reliable (http://www.shroud.com/pdfs/ssi27part3.pdf). It is frustrating for people all around the world interested in the Shroud that the Vatican apparently rushed into the C-14 dating, and after numerous problems with procedures were revealed, have not attempted after more than thirty years to try to rectify the situation.

1983. Heller wanted Gonella to come to Connecticut for a casual meeting to continue the discussions from the year before regarding the conservation committee. Gonella was unable to make it, but sent a letter containing some of his current thoughts. He stressed that conservation was to be the focus of the next investigation of the Shroud. Otterbein wrote a memo and stated that Ballestrero suggested that the committee should include both art conservationists and textile experts.

With the recent formation of ASSIST, there was a possibility that STURP could enlist them for help. But STURP had been a self-contained group and was hesitant to go outside of themselves, which was perceived by many as an unwillingness to collaborate with others. But Jackson had outlined in a previous letter how he would engender cooperation among groups. He had actually proposed the creation of a group of representatives from all of the major Shroud groups that existed at the time of the 1978 examination. Each group would maintain some autonomy, but would actively try to resolve any conflicts among groups, much like the United Nations.

Heller asked fellow STURP member D'Muhala, who had raised much of the funding for the 1978 examination, to raise $100,000 for the conservation project. Earlier in the year, Heller had also been looking for funds to finance his book Report on the Shroud of Turin, released May 1983. According to Bracaglia, Heller's mistake was that he assumed the friction between STURP and the Turin authorities was due to the Church's unwavering belief in the sanctity of the Shroud. He clearly didn't realize that friction was caused by such cultural factors as heritage, nationalities and religious faiths. He showed a lack of appreciation for the roles of the Guild and the Centro in getting STURP access to the Shroud in 1978.

Ballestrero had caused problems with various actions. By limiting the role of the Centro in the 1978 testing, he neglected their proud history of having had an ecclesiastical role and responsibility toward the Shroud and minimized their past contributions. He alienated them by assigning Gonella, who was not associated with the Centro, as manager of the project. But the Centro succeeded in getting the Cardinal to accede to at least some of their demands.

Fr. Otterbein wrote a candid letter to Jackson laying out why he was concerned about STURP overseeing the C-14 test. Others were concerned as well. The BSTS agreed that cooperation among groups was needed and that multiple C-14 experts would be needed. Otterbein reminded Jackson that Ballestrero had stated that international interaction was necessary. Dinegar, meanwhile, accepted the advice of the Guild; he informed Otterbein by letter that STURP, even though in transition, would have Harbottle of Brookhaven as their main advisor. STURP completed their reorganization in 1983 and expressed their desire to work under an international umbrella.

In April, the Symposium on Archaeometry was held in Naples, Italy; Harbottle chaired the carbon dating segment. While in Italy, Harbottle wanted to discuss with Gonella the possibility of getting permission of Ballestrero to initiate an intercomparison test among six labs that could be involved in the Shroud dating. However, King Umberto of Italy, who actually owned the Shroud at the time, had recently died. Gonella knew that the timing wasn't right, so nothing materialized from the discussion.

Source: Bracaglia, Giorgio. Uncovering the Paradox within the Archives of the Holy Shroud Guild, (Honeoye, N.Y.: Holy Shroud Guild), 2019, pp. 188-192, 194-195, 201. Also, some details were learned via a phone call with Bracaglia on December 7, 2019.

1984 *(January)*. On January 4th the late Edward Hall, head of the Oxford lab, wrote a multi-page letter to Prof. Porter, an academic of the Pontifical Academy of Sciences to encourage the Vatican to allow C-14 testing. A copy of the letter was sent to the Academy president, the late Prof. Carlos Chagas, and also to the Vatican. But the Vatican copy only contained the first page, which criticized STURP, but without saying why. Chagas wrote Cardinal Ballestrero that everything was going fine, but then wrote the Vatican saying there were problems with Ballestrero.

Source: *The Night of the Shroud (La Notte de la Sindone)*, documentary directed by Francesca Saracino, 2011. In 2016, it was revised and retitled "Cold Case: The Shroud of Turin," which is available at amazon.com. I have a review copy of the original version, which has an English voiceover. The revised version has English subtitles. (The material cited here can be found between approximately the fifteen- minute and thirty-five-minute range on the original version review-DVD.)

Comments: Chagas' strange action here was the first of many to surface through the time of the sample-taking in 1988.

1984. STURP sent a proposal of twenty-six tests, which included C-14 dating, to the Vatican, who forwarded it to the Pontifical Academy of Sciences, the body that advises the Pope on scientific matters, and to the Congregation for the Doctrine of the Faith. Gonella wrote, "For reasons that Cardinal Ballestrero and I were never able to understand, a deployment formed aiming at excluding any research that was not the radiocarbon dating."

Source: Marinelli, Emanuela. "The Setting for the Radiocarbon Dating of the Shroud." Presented at 1st International Congress on the Holy Shroud in Spain -- Valencia -- Centro Español de Sindonologia (CES), April 28-30, 2012, pg. 5. www.shroud.com/pdfs/marinelliv.pdf.

Comments: Gove related in his book (pg. 84) that Chagas' assistant, Vittorio Canuto of the Pontifical Academy of Sciences informed him in 1985 of STURP's twenty-six test proposal. Gove opined, "Carbon dating is by far the most important test and certainly the test that should be made first." Perhaps Gove's biased opinion, along with the fact that at some point Chagas would advise the late Pope John Paul II with an unfounded belief that STURP could harm the cloth in their proposed second round of tests, were enough to have the Pope ultimately decree that only the C-14 test would be done -- a tragic mistake.

The STURP proposal can be viewed at http://www.mondosindone.com/Site/documenti/DSS001_02%20-%20Sturp%201984.pdf.

1984. According to two German authors, Kersten and Gruber, "A total of twenty-six researches was *[sic]* proposed, to obtain definite answers to eighty-five crucial lines of inquiry. Besides the C-14 dating a diverse range of physical, chemical, optical and other valuable experiments was offered. The race to the best start positions for a new analysis of the linen began. Everyone wanted to take part. Many different interests were involved. There were some who wanted nothing to do with a radiocarbon dating test: hardened sceptics pointed out that some 20 percent of the entire material consisted of biologically foreign matter, which could influence the sensitive dating result considerably. The STURP researchers intended the dating to be just one part of their comprehensive, interdisciplinary analysis. Yet other researchers wanted just the C-14 test, and this at any price."

"The tug-of-war began. Sox *[David Sox, an American Anglican priest, the author of several Shroud books who had been involved in helping persuade the Vatican to agree to radiocarbon dating the Shroud]* calumnised the STURP researchers as 'militarily organized religious fanatics'. In fact they were respectable scientists of the most diverse convictions, even numbering some agnostics. But it was no longer a case of fair treatment or balanced argument. Here motives were at play which were not always transparent, and all too often revealed their less noble origins. At the Trondheim congress Gove spoke the lines fed to him by Sox, proposing that the STURP group be entirely excluded. The six laboratories agreed. Perhaps the STURP people seemed suspect to them. After all, they had already laid out a whole range of results that spoke for the authenticity of the cloth. Could preconceived judgements have been at play in this phase of the decision-making, which found a cloth in which Jesus lay unacceptable for science? It is difficult to understand. The STURP researchers carried out their investigations in a perfectly neutral manner and based their judgements on the facts. Science itself does not doubt the historicity of Jesus. So there should be no need for mental dislocation before being able to accept an object that was connected with him. After all, it was not a question of some paranormal object, things occult or even a miracle, with which science as we know currently has its problems. On closer inspection the decision of the radiocarbon lobby against the collaboration of STURP seems to be based on professional vanity."

Source: Kersten, Holger and Elmar R. Gruber. Jesus Conspiracy: The Turin Shroud & The Truth About The Resurrection, (Rockport, MA: Element), 1994, pp. 39-40.

Comments: This account suggests that Gove's anti-STURP crusade was fueled by Sox, but the authors do not elaborate. Was Sox perhaps trying to influence matters by his own initiative, or was he possibly being manipulated by other individuals or groups?

1984. By this time, Cardinal Ballestrero had already heard a C-14 proposal by Gove; Chagas and Gove were lobbying for that test only. He had also heard from the STURP team, which wanted the C-14 test to be one of twenty-six different tests. Gonella wrote that the Cardinal was engaged in a dark struggle in which he had trouble discerning who were friends and who were enemies. He felt that every step he took was in danger of being misrepresented and felt that some were taking various actions without his knowledge.

Gove sent the results of the 1983 "intercomparison laboratory test" at the British Museum to both Cardinal Ballestrero and to Prof. Chagas of the Pontifical Academy. STURP had sent their twenty-six-tests proposal to the Vatican, which sent it via the Secretariat of State, Cardinal Casaroli, to Prof. Chagas and also to the Congregation for the Doctrine of the Faith. For the STURP proposal, Chagas did not consult the whole body of the Pontifical Academy of which he was the head, or even consult with Cardinal Ballestrero, but instead only sent it to Canuto. (Had he sent it to the whole group in the Academy he could have had the advice of twenty-four Nobel Prize winners in Physics and Chemistry!) Canuto sent it to Gove.

Source: Gonella, Luigi and Riggi di Numana, Giovanni. Sindone: il mistero continua, (Milan: Fondazione 3M), 2005, pp. 32-33.

Comments: It was already clear at this point that politics would play a crucial -- and out-of-proportion -- role in the eventual dating process. It certainly is striking that the Cardinal had to categorize some of the key players as "friends" and "enemies."

We will see later that Gove gives conflicting information regarding the timelines regarding his first interactions with both Canuto and Gove -- and the possible reason for it.

1984. Garman Harbottle from the Brookhaven lab described the decisions made at the 1982 laboratory intercomparison test in Bradford England: "At that time there was considerable pressure to proceed with a dating proposal through STURP involving Rochester, Brookhaven, and Harwell -- there were **political** *[my bolding]* consideration*[s]* both within and outside the Church that suggested it might 'now or never.' The Tucson machine was, I believe, not yet routinely operating. The Oxford machine was bogged down in a seemingly -- endless series of technical failures. (A year later, at the Naples Conference in the spring of 1983, it was still not operating.) Despite all this, I proposed that in order to include as many laboratories as possible in order to do the scientifically best, and, to laymen most convincing job of dating the Shroud, the actual dating exercise should be postponed until Tucson (Dr. Long) and Oxford (Dr. Hall) got their machine on line (Harbottle, personal communication)."

Source: Laverdiere, H. "The Socio-Politic of a Relic: Carbon Dating of the Turin Shroud," (Doctoral thesis) 1989, pg. 65. Accessible via free download at http://ethos.bl.uk/OrderDetails.do?did=1&uin=uk.bl.ethos.531916.

Comments: So at the time of the C-14 dating in 1988, both Oxford and Arizona had only six years earlier had either not been routinely running or were having numerous problems. One C-14 lab head also didn't think Zurich should be picked because they had a significant dating error during the 1982 pre-test (see appendix to part 2). Harbottle also said (pg. 70) "We felt at the time of the Brookhaven/Rochester/STURP proposal [...] that a result obtained by one laboratory or one methodology would be sure to be challenged" *[emphasis in original]*.

1984. When STURP submitted their proposed twenty-six tests, the abstract for it was published in the Shroud journal, *Shroud Spectrum International*. Ballestrero then asked all groups to submit any proposals to that journal; the BSTS and ASSIST soon did so. Archaeologist William Meacham, however, sent his proposal directly to Ballestrero. It addressed Meacham's many concerns about testing procedures and past proposals. Dinegar was fearful that Meacham's views would delay the C-14 test. Many other factors would actually also contribute to a delay.

Since King Umberto bequeathed the Shroud to the living Pope, protocol stipulated that the Shroud would be in the ultimate care of the Pontifical Academy of Sciences, under Chagas, whose point man was Gove.

Source: Bracaglia, Giorgio. Uncovering the Paradox within the Archives of the Holy Shroud Guild, (Honeoye, N.Y.: Holy Shroud Guild), 2019, pp. 201-202.

Comments: Here is an excerpt from the STURP proposal re C-14: "We request permission to take samples from several different non-image locations on the Shroud. Two hundred milligrams (200mg) from each of the burned areas under and around the patches (excluding the dorsal shoulder foldmark intersection area will be trimmed away. This material can be removed without affecting the visual appearance of the relic or causing damage to the fabric structure. Conservatively, this quantity will provide three samples for each participating laboratory. We also request 200-mg samples from the presumed side strip, one patch, and the lower right-hand corner of the Shroud. Each such sample is the equivalent of about two square inches (10 cm^2) of material. We also request a 200-mg-sample from the Holland-backing cloth. This latter forms an internal standard of known date to go with our unknown. An external standard will come from two samples from other sources, both of known origin and date." (http://www.mondosindone.com/Site/documenti/DSS001_02%20-%20Sturp%201984.pdf, pg. 48.) Note that the proposal would not have visually affected the appearance of the cloth; this consideration would become important in terms of the sample that ultimately was selected.

1984. STURP, in another of their work packages, expressed concern about repairs that had been done on the Shroud. They wrote, "Old mends. X-ray examination of the old repairs indicates that, while the repair as a whole seems to be serving its purpose, there are many areas in which patches appear to be pulling away and might be causing damage at the present time. Some of the repairs are crude and may not be needed. Since the twist direction and number of twists per unit of length of the mending threads are not known, it is not possible to gage *[sic]* if the forces of expansion and contraction from relative humidity are pulling adversely between patches, mending threads, and the Shroud linen itself. Twist will also play a role along with weave when considering purely mechanical movement. Transverse compression, longitudinal extension and compression, torsion, bending, and frictional restraint are a few types of fiber deformation. With x-ray alone it is not possible to determine these relationships, nor can the areas of extreme pyrolysis be directly observed. Therefore, any attempt to evaluate the virtues or dangers of old repairs can be made only from direct observation. Again, a stereomicroscope is needed to undertake this examination [...]."

Source: "Formal Proposal For Performing Scientific Research On The Shroud of Turin." Submitted By The Shroud of Turin Research Project, Inc. http://www.mondosindone.com/Site/documenti/DSS001_02%20-%20Sturp%201984.pdf, pp. 32-33.

1985. Archaeologist William Meacham, who would take part in the Turin planning meeting in Turin in 1986, wrote, "Gove through his side-kick Vittorio *[a.k.a. "Victor"]* Canuto had begun courting the president of the Pontifical Academy of Sciences, a Brazilian biochemist named Carlos Chagas. (I would later meet Chagas at the Turin C-14 conference; he was a genteel and distinguished man in his seventies.) The purpose of this courtship was to bring the Academy into the issue, as a counterfoil to the complete control theretofore wielded by Gonella on behalf of Cardinal Ballestrero, so that ultimately the C-14 dating could be done free and clear of STURP. From Gove's slanted and self-idolizing book, it is also clear that he developed other more sinister plots and objectives, which he pursued with the tenacity of a bulldog in a china shop. Whilst remaining part of STURP's C-14 team he worked to undermine and replace it with his own plan. Later, he began to press for all other STURP testing to be blocked while the dating went ahead."

"The fact that Gove's side-kick had a close relationship with Chagas proved most useful in advancing these goals, as did Gove's knack for ringing advantage out of every situation. For example, at the conclusion of the Trondheim session, when 'it was suggested' that Gove draft the protocol, it was also agreed that the protocol, after being reviewed by the six laboratories to be involved, 'would be presented to the authorities in Turin or the Vatican.' When Gove sent round his draft for comment by the six laboratories, he indicated that he would send it to the Pontifical

Academy of Sciences. Later, Gove relates that Chagas 'seemed to be particularly pleased that the Trondheim Group had emphasized [sic!] the importance of the involvement of the Pontifical Academy.' *[Author note: the "[sic!]" is Meacham's.]* This was a far stretch from one actually happened at Trondheim, but Chagas fell for it."

"It is not easy to understand why Gove became so obsessed with dating the Shroud. From the account in his book it is clear that he made hundreds of phone calls, many to Brazil or Rome, sent hundreds of letters, faxes and cables, and spent thousands of hours plotting strategy with his sidekick. He made frantic efforts to contact ambassadors, senators, highly placed Cardinals, and anyone else who he thought might be of use. The simplest explanation, that there was funding to be had, does not seem to fit, and Gove specifically rejects that as a motivation -- not, of course, that his rejection of it would mean very much! He would say that, wouldn't he? But it does seem rather that it was all about glory, no matter which way the dating went. It was literally a win-win game for the C-14 folk: if the Shroud date came out 1st century, they were the ones who revealed that this to the world; if it turned out to be medieval, they were the ones who proved it could not be Christ's."

"For Gove, there may have also been two other factors: the AMS was his baby (or so he ardently believed anyway); and after he began to lock horns with Gonella it became personal. Gove's published comments on Gonella, and some directed at me also, are defamatory; he quotes his side-kick as remarking: 'How is the Pope to know that Gonella is a second-rate scientist and that the rest of us are super scientists?' According to Gove, Gonella was 'unprofessional,' 'an obscure polytechnic lecturer nobody had heard of,' 'a troublemaker of the first order,' 'malicious,' 'small minded' etc., *ad nauseam*. All through the book Gove spares no effort to portray Gonella in the worst possible light. Ironically, in almost all of the issues described with such scorn by Gove, Gonella was right and Gove was dead wrong. But on the one that mattered the most, the one that would lead to a flawed dating of the Shroud, Gove essentially ignored it and Gonella made a huge error."

"The intense media interest in the subject must also have attracted Gove and some of the other C-14 daters. Teddy Hall of Oxford supposedly offered the BBC exclusive coverage of his dating of the Shroud sample, for a hefty fee. After the results were announced Hall obtained a very handsome endowment for his lab from a group of businessmen. Garman Harbottle of Brookhaven may have also coveted the attention and glory. In one of his letters to Dinegar, Harbottle pointed out (as cited by Gove) that 'STURP had done all the experiments, written all the scientific articles, been on all the talk shows and starred in all the films on the Turin shroud.' Sounds as though he wanted in on the action! Gove professed a disdain for the press, but then seemed to relish each interview or reporter's query. He stated that all the media attention was 'heady stuff.' Overall, one gets the impression that Gove considered the Shroud dating enterprise to be his, and anyone who got in the way should be crushed. Unlike the others, for him it probably was not about the money, and maybe in the final analysis it was not even about the public attention or the glory, but rather his absolute fixation on and monolithic drive towards his objective, positively reeking of arrogance and dogmatism, bulldozing everything and everyone aside. In a military man or politician, this might be an admirable quality. In a scientist, it was pathetic. And he has the audacity to denigrate, over and over again, those he suspected of being 'true believers.' He said that he had dealt with STURP for 10 years and 'regard them as a pack of religious zealots.' This of course is totally absurd, and merely serves to indicate how blinkered Gove's perception of reality was [...]."

Source: Meacham, William. The Rape of the Turin Shroud: How Christianity's most precious relic was wrongly condemned and violated, (Lulu.com), 2005, pp. 64-66.

Comments: Recall that earlier Gove had said, "The truth of the matter was that I did not care whether the Turin Shroud was carbon dated or not [...]. Meacham claimed that Gove was obsessed with dating it. If it isn't clear already, the reader should soon be able to gauge which assertion is more accurate.

1985. Italian researcher Franco Faia, who was able to get access to Gonella's correspondence files after his death, said in an interview "Gove was supported by Canuto and had Chagas in his pocket. He had him in his pocket. At one point Gonella said, 'He seems to be dominated by Gove.' It was hard to understand. And he was the president of the Pontifical Academy of Sciences with all those Nobel Prize winners."

Source: *The Night of the Shroud (La Notte de la Sindone)*, documentary directed by Francesca Saracino, 2011. In 2016, it was revised and retitled "Cold Case: The Shroud of Turin," which is available at amazon.com. I have a review copy of the original version, which has an English voiceover. The revised version has English subtitles. (The material cited here can be found between approximately the fifteen- minute and thirty-five-minute range on the original version review-DVD.)

Comments: One has to wonder if Gove's apparent influence over Chagas was a factor in his aforementioned schizophrenic attitude toward Cardinal Ballestrero. The phrase "in his pocket" is secondarily defined as, "very close to and closely involved with someone," but actually has the primary meaning of, "dependent on someone financially and therefore under their influence."

1985 (May). Meacham sent a document dated May 10, 1985 directly to Turin's archbishop. Among other things, he said, "Radiocarbon (C-14) dating is without doubt the most important scientific test to which the Shroud of Turin will ever be subjected. The eventual C-14 date will have an enormous impact on the general public view of the Shroud's authenticity. In many respects it will represent the culmination of the scientific investigation into Christianity's holiest relic. The Shroud is, on the other hand, less than an ideal material for C-14, with many possibilities of contamination over the centuries. For these reasons, then, a very careful and specially devised set of procedures should be adopted in the collection and pre-treatment of samples, in the selection of C-14 technology, to be employed, and in the interpretation of laboratory results."

"The proposal here submitted is not a technical plan for C-14 dating. Rather, it is an organizational scheme designed to achieve the highest possible reliability of C-14 measurement of the Shroud. It is imperative that this crucial test be conducted with all the rigorous controls and stringent procedures that science can bring to bear on the problem, in order to achieve in the end a broad academic consensus on the validity of the results. The inclusion of all relevant expertise in the planning and conducting of C-14 testing is highly important and should not be left to the fortuitous circumstances of personal contacts or technical preferences -- it should be mandated by the authorities responsible for Shroud."

"It is my firm belief that C-14 dating of the Shroud would best be carried out under the supervision of a body of internationally prominent scholars, which I shall call the Shroud Dating Panel, appointed by the ecclesiastical authorities with a view to insure that expertise from all relevant fields of science are included. It would be the duty of this Panel to study all aspects of the procedures, to make consensus decisions, to collect and submit samples for the C-14 tests, to examine all aspects of possible contamination or other distortion of the laboratory measurement, and to interpret the results."

"Of greatest importance in planning the C-14 dating of the Shroud is the inclusion of experts from all relevant fields. Clearly, the Panel should be composed of eminent radiocarbon scientists, archaeologists and art historians; it could include representatives of the bona-fide research groups which have submitted C-14 proposals for the Shroud, as well as scholars not affiliated with those groups. It would be chosen without regard to personality, nationality or other rivalries which have plagued self-appointed Shroud research groups until now. It could I believe be one of the most prestigious groups ever organized for the purpose of a single scientific test, and it would certainly give a scientific imprimatur to the eventual results and the significance attached to them."

"Once appointed, the Panel would plan and implement a C-14 dating program, subject to the obvious limitation of disfigurement of the relic. The terms of reference for the Panel would include determining the minimum amounts of material required, and reaching agreement with the ecclesiastical authorities concerning the site or sites on the relic to be sampled. Certainly such a program would require at least 1-2 years and several meetings of the Panel, before results could be made known to the public. But for reasons which I hope to make clear below, such a course of action seems much the more preferable to simply selecting one or more of the C-14 proposals from various groups which have submitted applications. At the very least, the Panel could serve to review and modify these proposals as necessary. The principal reasons for the establishment of a supervisory Panel are, therefore 1. the enormous and universal importance that will be attached to the results of C-14 dating; 2. the complexity of technical issues surrounding the dating of Shroud samples; 3. the need for all relevant expertise to be consulted on these issues, both in designing the testing procedure and in interpreting the results to the general public."

"The problems involved in C-14 dating the Shroud are considerable, in spite of some claims to the contrary. According to Dinegar for example, STURP made a detailed study in 1978-82 of the possible dating of the Shroud, and concluded that 'sample preparation procedures can insure no error in date due to foreign contamination accreted over the centuries.' This is absolutely not true; even the most sophisticated pre-treatment methods cannot prove that lignins or hydrocarbons are not trapped in the structure of the cellulose, that occluded carbonates are not interfering with the level of C-14 activity, that ion exchange with older or younger materials has not taken place, etc. It is a very serious error to proceed with C-14 dating on the assumption that it is an infallible method [...]."

"In conclusion, I would urge the Church to take all steps to guarantee that this most important of tests is conducted according to the most rigorous scholarly procedure that can be devised. The appointment of a Panel as suggested in this paper would represent an effort to exclude, as far as is humanly possible, the effects of personality conflicts or cliques, national rivalries, and simple human error from the pursuit of the truth of the Holy Shroud."

Source: Meacham, William. "A Proposal to Cardinal Ballestrero, Archbishop of Turin, for the Planning and Implementation of Radiocarbon Dating of the Holy Shroud."

Comments: Sadly, Meacham's advice was ignored and the labs carried forth with the idea that the C-14 method (at least for the Shroud!) was infallible, and the very things he detailed in his conclusion that would prevent the truth occurred.

1985 *(June)*. In Trondheim, Norway, the 12th International Radiocarbon Conference was held. The British Museum presented the laboratory intercomparison test results. Gove organized a meeting of representatives of the British Museum, the six labs and one representative from STURP. It was suggested that the Pontifical Academy of Sciences be contacted, and that Gove produce a C-14 dating protocol for the Shroud.

Source: Gove, Harry. Relic, Icon or Hoax?: Carbon Dating the Turin Shroud, (Bristol and Philadelphia: Institute of Physics Publishing), 1996, pg. 321.

Comments: On page 9 of his book, Gove indicated that he thought it would be good to establish contact with the Pontifical Academy of Sciences. He said that he was able to contact Chagas for the first time several weeks after the Trondheim conference via Canuto. But there is a discrepancy regarding the timing of Gove's first contact with Chagas. According to Gonella, as cited in one of the entries above, Gove had sent Chagas in 1984 the results of the 1983 "intercomparison laboratory test" at the British Museum to both Cardinal Ballestrero and to Prof. Chagas, well before mid-1985. There will more about this discrepancy later in the book.

In his book, Bracaglia (pg. 202), said, "Gove used the Trondheim meeting to try to cement the role of the British Museum in coordinating the carbon dating, under the supervision of Dr. Tite."

1985 *(June)*. According to Gove, Dinegar stated that he was representing STURP, which he said had been responsible for the intercomparison test and which would arrange the Shroud C-14 test and the announcement of the results. Gove asserted that STURP played no role in the intercomparison test and "had no credentials for playing any role at all in the carbon dating -- if it ever took place." The next day, according to Gove, Dinegar made "an impassioned plea that STURP had to be intimately involved with this carbon dating enterprise. As chairman of the meeting, I found his discourse intensely annoying, especially in light of the offensive remarks he had made the previous day, but I didn't know how to silence him."

Source: Gove, Harry. Relic, Icon or Hoax?: Carbon Dating the Turin Shroud, (Bristol and Philadelphia: Institute of Physics Publishing), 1996, pp. 81-83.

Comments: Once again, Gove exhibited his distaste for STURP. Dinegar's take on the situation can be found immediately below.

1985 *(June)*. Dr. Robert Dinegar, who was STURP's key person regarding a C-14 proposal, took notes at the meeting (copy of which sent to author) and made the following observations. *[There are individual entries for June 23rd through June 26th, which includes both my summaries of Dinegar's notes and direct quotes from him.]*

1985 *(June 23rd)*. Dinegar told Rainer Berger of UCLA that the real problem with dating the Shroud was not in doing the experiments but in the "**politics**" *[my bolding]* of obtaining samples. He added that STURP could handle getting the samples and that the labs would do an excellent job in determining the date. There was a discussion about an incorrect date that one of the labs had gotten in the 1983 intercomparison test held by the British museum in 1983. Hans Oeschger of the University of Bern came forth and said he didn't mind identifying his lab as the one who had gotten this particular date at the time. He revealed they didn't know why it happened as they had gotten good dates before. It apparently caused the lab a lot of embarrassment.

1985 *(June 24th)*. "Various topics were brought up. Robert Otlet of Harwell (England) said that the participants should act like a team, not individuals seeking the spotlight. All others agreed but wanted to be able to do it 'their way.' Otlet was worried about the premature leaking of the results. Others were bothered by the perceived distrust from the Church authorities. Otlet emphasize that how much Harwell profited by the association with the Shroud project. The matter was then quickly dropped."

1985 *(June 25th)*. Gove then asserted that there was no further place for STURP; the labs were capable of doing the job independently. Gove stated that the British Museum had the international reputation to parcel out the samples and collect the data; but STURP had no standing as a reputable scientific body. Dinegar agreed with Gove's points about the reputation of the British Museum and the labs but took offense at his characterization of STURP.

An argument ensued about a conflict of interest for the British Museum, since Edward Hall of the Oxford lab was a trustee. The Zurich lab expressed significant concern about this. Dinegar noted that Zurich had showed a general lack of knowledge about the Shroud. Gove did not hide his annoyance that Dinegar been given the opportunity to speak up, and he even started to walk out before Dinegar convinced him to return to his seat. Gove turned the discussion to the procurement of the samples, which body would pass on the recommendation and what Church body would actually give the approval. Gove stated that he was tired of dealing with clerics, Gonella and STURP. He said he would the deal directly with friends on the board of the Pontifical Academy of Sciences. He then went into a tirade about how he had been ignored by the Church and Gonella in 1978. He claimed the letters that he wrote were never delivered to the appropriate persons and that he had refused to work with STURP in 1979. Dinegar told Gove that the latter had been treated badly in 1978, but that not joining STURP was Gove's own decision. At a STURP meeting in Santa Barbara in 1979 Gove said he was going to act without any STURP involvement and get material for the C-14 labs.

Comments: After the meeting, Otlet gave Dinegar a ride and said that he didn't want to appear to be taking sides and so would not say anything more about the situation at the meeting. Otlet invited Dinegar to come to England to discuss things further, but Dinegar told him he saw no point in going all that way to discuss Gove's obvious disdain for STURP, or any of the other volatile issues that had been raised. Dinegar wrote that it was obvious that 1) Gove had come to Trondheim determined to get other C-14 labs to follow his lead; 2) Gove's stance was that STURP was not qualified to be involved in the C-14 testing; 3) although Dinegar didn't experience any direct hostility from the other participants, no one was pro-STURP enough to show support. Dinegar then resolved that Gove would never be involved in any STURP effort.

1985 *(June 26th)*. At breakfast in the hotel, Dinegar crossed paths with Gove. Gove invited Dinegar to sit with him. Gove said to Dinegar, "I guess you're peeved with me." Dinegar replied that "peeved" was not the word. Gove chuckled and said, "Worse than that, eh?" Dinegar made no comment. Gove then told Dinegar that he had misunderstood him, who now told Dinegar that there was a place for STURP in the testing. Gove next complimented STURP on the work they had done. Dinegar practically fell out of his chair. When Gove left the table for some coffee, Dinegar remarked to a companion of Gove that based on Gove's words of the 25th, any reasonable person would have concluded that STURP was being excluded. "The companion nodded, smiling weakly."

When Gove returned, he brought up the point about the British Museum coordinating the samples extraction and the collating of the data. Dinegar replied that the British Museum was qualified, but not exclusively so. Gove then asserted that Dinegar had recommended the British Museum for the Shroud dating. Dinegar pulled out a copy of the proposal sent to Gove on February 14, 1984 and told Gove to read it again. Dinegar pointed out that the British Museum would do it for the intercomparison tests, not the Shroud tests. Gove looked at it, admitted that the document in hand was the one he had received and confirmed what Dinegar had just said. Gove told Dinegar that his remarks on the 25th about STURP had been an effort to "take a stand" within the C-14 coalition, which Dinegar interpreted to mean that Gove wanted to appear hardnosed and in charge. Gove also said that many in the Shroud community were worried about some religious statements made on the Shroud by some members of STURP. Gove then reminded Dinegar that it would take considerable funding to be involved in the Shroud dating (mostly travel) and inquired as to the state of STURP's finances. Dinegar responded that STURP would be cooperative with the C-14 labs, that STURP had tried its best to separate religion from science in their analyses, and that they believed they could raise the funds needed to participate. Dinegar ended his diary notes with "he who has the threads calls the shots."

Source for June 23rd-26th entries: Dr. Robert Dinegar's handwritten notes, a copy of which was sent to the author. There is also a reproduction of the notes in the Holy Shroud Guild archives per Giorgio Bracaglia, who oversees the Guild's web site (https://www.holyshroudguild.org/index.html).

Comments: Dinegar's mini-diary shows how prevalent politics were in the planning for the Shroud C-14 test. The diary brings out many different significant points. The remarks by Oeschger indicate that C-14 datings can sometimes be wrong even when the reason is not fully clear. Gove claimed that STURP had no international reputation. That was absurd. The claim belies the fact that the STURP members came from some of the most prestigious U.S. institutions, including Los Alamos National Laboratory, Sandia Laboratory, Jet Propulsion Laboratory, Air Force Academy and others, most of which were and are involved in the U.S.' nuclear and space programs. It should also be remembered that the STURP team included nominal Christians, Jews and agnostics. There had been some concern about the British Museum being the overseer of the testing, since the head of the Oxford lab, Edward Hall, was a trustee. That concern was borne out when after the test Hall retired, and the Oxford lab was given a million-pound donation for a new chair, which was filled by none other than Michael Tite of the British Museum, who had been the overseer of the Shroud dating. Ironically, the donation was made on the Good Friday [!] following the dating ("Turin Shroud Professor raises 1 million *[pounds]* for Oxford Post" in *Daily Telegraph*, March 25, 1989).

Gove clearly had not forgotten what he considered to have been slights from the Church in the 1978 time period. It's amazing, though, that Gove could openly make anti-STURP statements in a group meeting one day and then shortly thereafter tell Dinegar that he (Gove) was misunderstood and that there was a place for STURP in the testing. I cannot fathom how Gove could have said his original statements were simply misunderstood when the first statement clearly contradicted the second statement. Gove seemed to be the consummate politician: saying whatever was the most opportune for him, regardless of the fact of blatant and conflicting statements.

Fr. Otterbein wrote Dinegar a letter on July 15, 1985 (**Source**: Luigi Gonella archives, of which the author has a copy), saying, "I am sure that it was a difficult session. As I told you, I never trusted Gove but thought that working with five other labs, he would have to toe the line. However, he showed his true colors and obviously is still a big problem. I had thought that as a result of all your work, the C-14 labs were well organized and willing to cooperate with each other, but apparently we still have *prima donnas* to contend with and professional rivalry. It is too bad."

"Although I have been hoping for a simple solution to problems, e.g., the need for cooperation among the various interested groups, such a solution looks more and more improbable and unrealistic. For example, when the Cardinal asked for bridge building between the scientific groups, I realized that the Cardinal was not qualified to evaluate the various plans, nor did he know who was best qualified to do each test. It would be great, if the scientists could get together and agree on the best program and the most qualified personnel. Now it looks more and more as though a Panel of Vatican Academy scientists will have to do such a job, and that WILL TAKE TIME! (...)."

1985 *(June)*. The German authors Holsten and Gruber wrote "While at Trondheim the representatives of the six laboratories drafted a protocol, which was accepted by all:"

1. The British Museum would assume the coordinating role in the investigation, and act as "guarantor" for a correct performance.

2. STURP members could cut the specimens from the cloth, so that they would not feel completely excluded.

3. The British Museum would provide two control specimens from cloths of known age. All three specimens, including those of the Shroud, were to be unraveled, so that they could no longer be identified.

4. The British Museum would receive a written assurance from the laboratories that they would inform no one of the results except those authorized by the Museum.

5. The laboratories could use the methods they considered best for preparing the samples, but were to keep a precise record of all the details.

6. The results were to be communicated to the Holy See before publication.

"Dinegar of STURP, who was present in Trondheim, insisted that the radiocarbon dating should only be done after the analysis of the fibre components proposed by the STURP researchers Heller and Adler. The C-14 specialists threw out his suggestion, claiming that any further examination would be absurd as long as the age of the cloth was not settled. Obviously the STURP people were to be portrayed as fantasizers and undermined. Sox was even angry that they were allowed to be present at the sampling, while Gove declared in a polemical speech that he would abandon everything if STURP had any role other than the one allotted. At least now it was clear where the lines were drawn. Nothing remained of the team spirit for which scientists are often praised, nor of their dedication to the pursuit of knowledge, or the search for a variety of techniques to do justice to a complex problem. At Trondheim the declaration of war was on the table."

"The next stage was the involvement of the Pontifical Academy of Sciences. It is not clear who approached the scientific experts of the Vatican, but their involvement is only logical when one considers that after Umberto's death in 1983 the relic passed into the hands of the Vatican. What is strange is the unconvincing role that their president, Prof. Carlos Chagas, was to play. The Vatican agreed to a conference between the Academy and the scientists who were interested in the dating test. Chagas was to organize it. Strangely enough he delegated this task not to a member of the Academy, but to Vittorio Canuto, an astrophysicist at the NASA Institute for Space Studies in New York, who was his private adviser. Together they decided against the STURP proposals and in favour of the Trondheim protocol."

"At this point one has to ask what possible motives could have lain behind this. STURP members had already presented a lot of excellent research, their dedication and their thorough command of the subject matter were known. Their own proposals included a C-14 dating test as part of a more comprehensive project, and this would certainly have increased the significance of the individual experiment. As far as the Turin cloth was concerned, Gove and the radiocarbon specialists were considered to be complete amateurs. Most of them did not even know the general facts about its history and the research already done. The 'test run' under the watchful eye of the British Museum had turned out to be a disastrous farce, leading one to expect any number of problems, and the behaviour of the researchers themselves had so far been marked by pride and conceit."

Source: Kersten, Holger and Elmar R. Gruber. Jesus Conspiracy: The Turin Shroud And The Truth About The Resurrection, (Rockport, MA: Element), 1994, pp. 40-42.

Comments: According to the Gonella/Riggi book (pg. 33), although Gove was willing to let STURP cut the sample, he did not want any appearance that STURP initiated the process, because they were "prejudiced." The question is: why was STURP more prejudiced than Gove? From all indications, it's because STURP did not pronounce after their 1978 study that the Shroud was a fake. As noted above, Gove and Canuto did not want STURP to be able to do their multi-test protocol. Unbelievably, Gove called STURP's proposals "unscientific" (Gonella/Riggi book, pg. 34).

1985 *(June)*. According to Bracaglia, the Trondheim meeting was disastrous for STURP. Surrounded by people favorable to Gove, he set the agenda in a way that was designed to minimize STURP's role. Dinegar asserted at the meeting that STURP would be able to procure the Shroud samples for the dating. According to prominent C-14 scientist Henry Polach, who attended the Trondheim meeting, STURP was not regarded favorably there. He wrote, "What I can observe as well of what colleagues who wish to be involved tell me, the whole issue is a scientific mess. My view is that those involved seek publicity, are emotional about the dating of the shroud and let their emotions color their scientific judgement. Further, not a single competent radiocarbon dater has been invited to comment on the dating procedures, methodology, the chemistry of the sample, and the precision that was required and achievable."

Gove and his colleagues cast STURP as religious zealots, who would only interfere in the "science" of radiocarbon dating. Gove appealed directly to Chagas to eliminate STURP. However, Chagas continued to get input from various individuals so that the best possible advice could be given to the authorities. Chagas asked for advice from two Americans on the Pontifical Academy of Sciences, Canuto and Victor Weisskopf. Both men recommended that Gove be allowed to lead the dating project. Weisskopf wrote to a correspondent, "I cannot take the scientific efforts seriously as long as no studies are made as to the radioactive carbon content of the Shroud. Only such a study can give an unambiguous answer as to the age of the Shroud."

Another American, the late archaeologist Eugenia Nitowski, who was a Carmelite Nun known as Sister Damian, also had concerns about STURP's involvement, not with C-14 test, but with some of the other tests they were proposing. She agreed with STURP that the image was caused by dehydration of the fibrils. As such, she did not think that STURP should be allowed to use ultraviolet light on the Shroud, which would cause yellow discoloration and extensive depolymerization.

Source: Bracaglia, Giorgio. Uncovering the Paradox within the Archives of the Holy Shroud Guild, (Honeoye, N.Y.: Holy Shroud Guild), 2019, pp. 223-225.

Comments: Later we will see Gove ridicule STURP's claim that the Shroud samples needed to be characterized -- after Gove quotes the statement, he added "(whatever that means)." It's what Polach referred to as "the chemistry of the sample." One wonders how someone like Gove could have been ignorant on that point. Again and again, Gove openly showed his bias against STURP. Weisskopf was another scientist who put his unwavering confidence in the C-14 test alone as the final arbiter of the Shroud's dating, despite the well-documented complexity of the Shroud.

1985 *(June)*. After the meeting in Trondheim, "Gove told Canuto that if STURP had any role beyond arranging for the sample removal he would pull his lab out of the whole enterprise."

Source: Sox, David. The Shroud Unmasked, (Basingstoke, Hampshire: The Lamp Press), 1988, pg. 98.

Comments: According to Bracaglia's book (pg. 208) Ballestrero had been informed by the Vatican that he was not authorized to make independent decisions about the Shroud dating. Since that was the case, the perception that Gonella had a lot of influence was not accurate. Gove, possibly feeling emboldened by that, had openly criticized Gonella and other Church officials. That would come back to haunt him.

1985 *(July)*. On July 8th, Chagas had Canuto review STURP's twenty-six-tests proposal. Chagas reported as his own Canuto's report, which criticized STURP's proposal to a C-14 in conjunction with twenty-five other tests, which included analysis for conservation methods and also for identifying pollutants on the cloth.

Source: *The Night of the Shroud (La Notte de la Sindone)*, documentary directed by Francesca Saracino, 2011. In 2016, it was revised and retitled "Cold Case: The Shroud of Turin," which is available at amazon.com. I have a review copy of the original version, which has an English voiceover. The revised version has English subtitles. (The material cited here can be found between approximately the fifteen-minute and thirty-five minute range on the original version review-DVD.)

Comments: Kersten and Gruber stated in their book (pp. 40-42) that Chagas and Canuto had already decided in June at the Trondheim conference. It's not clear whether Chagas' directive to Canuto to review STURP's proposals was a change of mind or just for show. In any event, STURP's proposals were (temporarily) approved , per next entry.

1985 *(July)*. On July 13th, Cardinal Joseph Ratzinger, later Pope Benedict XVI, approved STURP's proposal for the twenty-six tests. On July 22nd, just eight days after Cardinal Ratzinger approved STURP's plan for twenty-six tests, Gove met in New York with Canuto and Monsignor Celli, Vatican ambassador to the United Nations. Canuto then informed Church officials that the tests STURP proposes were "dangerous."

Source: *The Night of the Shroud (La Notte de la Sindone)*, documentary directed by Francesca Saracino, 2011. In 2016, it was revised and retitled "Cold Case: The Shroud of Turin," which is available at amazon.com. I have a review copy of the original version, which has an English voiceover. The revised version has English subtitles. (The material cited here can be found between approximately the fifteen-minute and thirty-five minute-range on the original version review-DVD.)

Comments: It is known from the Gonella archive that Cardinal Ratzinger had approved STURP to have two full weeks to do the twenty-six tests. The question: what individual or group was powerful enough to eventually override Cardinal Ratzinger's approval?? If STURP had been given two full weeks with the cloth, there's no telling how many significant revelations would have resulted. Had STURP been allowed to perform their twenty-six tests, we undoubtedly would not be having the controversy we currently have.

1985 *(July)*. Gove wrote to Dinegar on the 15th, "Dear Robert, Enclosed please find a memorandum on the informal meeting of the people interested in dating the Shroud of Turin, held in Trondheim on June 25, 1985, and a suggested protocol for further action toward actually carrying out such a dating measurement. At the Trondheim meeting, it was suggested that I prepare a first draft of such a protocol and provide copies to all the appropriate people. The people to whom the enclosed material was sent are listed below. Please share your copy with anyone else in your group you wish."

"I would be glad to get any comments you may have on the documents, including suggestions for changes, additions, etc., and would appreciate receiving them as soon as possible. As soon as a revised version of the protocol can be prepared, I will send it to the President of the Pontifical Academy of Sciences, Professor Carlos Chagas, Federal University of Rio de Janeiro, Brazil, for comments and suggestions by the Academy. I would also like to send copies to three other members of the Academy whom I know personally, namely Professor A. Bohr, Dr. G. Herzberg and Professor V. Weisskopf."

"I look forward to hearing from you."

Yours sincerely,
[signed]
H.E. Gove

HEG/sb

Sheridan Bowman	E.T. Hall
Janice Boyd	Garman Harbottle
Shirley Brignall	Robert Hedges
Richard Burleigh	A.J.T. Jull
Nicholas Conard	Hans Oeschger
Robert Dinegar	Robert Otlet
Douglas Donahue	Meyer Rubio
David Elmore	M.S. Tite
Harry Gove	Willi Wolfli
J.A.J. Gowlett	

July 10, 1985

MEMORANDUM

"SUBJECT: Trondheim Meeting RE: Shroud of Turin"

"On June 25, 1985, representatives of the six laboratories which participated in the British Museum intercomparison tests, along with a representative of the British Museum, met informally in the Division of Physics building at the Norwegian Institute of Technology (NTH) in Trondheim, Norway. The results of the tests had been presented the previous day at the 12th International Radiocarbon Conference. The meeting was organized by Harry Gove (University of Rochester) and a meeting room was kindly arranged by Reidar Nydal, Chairman of the Organizing Committee for the Conference. The purpose of the meeting was to discuss what, if any, the next steps should be which might lead to radiocarbon dating of the Shroud of Turin. It was the expressed interest of the six laboratories in dating the shroud that led to the inter-comparison tests conducted by the British Museum. The following people attended the meeting:"

Sheridan Bowman	The British Museum
Janice Boyd	SYNCOM/NORDA
Shirley Brignall	University of Rochester
Nicholas Conard	University of Rochester
Robert Dinegar	SYNCOM/STURP
Douglas Donahue	University of Arizona
Harry Gove	University of Rochester
J.A.J. Gowlett	Oxford University
Robert Hedges	Oxford University
A.J.T. Jull	University of Arizona
Robert Otlet	Harwell
Willi Wolfli	ETH – Zurich

"There was another representative from Zurich whose name was not recorded. Robert Dinegar primarily represented the Shroud of Turin Research Project, Inc. (STURP), but is stated by Garman Harbottle of Brookhaven National Laboratory, to be his co-principal investigator for C-14 dating the Shroud of Turin. Harbottle, who did not attend the Trondheim meeting, heads the group at Brookhaven which participated in the British Museum intercomparison tests. He has recently become a member of STURP."

"The original offer to date the Shroud of Turin using very small samples and using the tandem accelerator mass spectrometry and the small proportional counter technique was initiated by a group from the University of Rochester and made jointly by them and the Brookhaven National Laboratory. The offer was made in 1979 directly to Archbishop Anastasio Cardinal Ballestrero of Turin. It is not known for certain that it was ever seen by the Archbishop, since it was addressed to a recommended intermediary."

"In the fall of 1978, a team of U.S. scientists, all members of the non-profit organization, the Shroud of Turin Research Project, Inc. (STURP), participated along with European scientists in a week long series of non-destructive measurements on the shroud. These tests were permitted by the Turin ecclesiastical authorities after a prolonged period of complex negotiations between these authorities and the STURP leadership. No member of the STURP team had any expertise in radiocarbon dating, and, although the Rochester offer to date the shroud using TAMS *[normally known by "AMS"]* was announced at a two day scientific conference preceding the STURP tests, it was made clear by Church authorities in Turin no shroud material would be made available for carbon testing. STURP, however, has always indicated its strong support for radiocarbon measurements on shroud cloth."

"STURP formed a carbon-14 dating committee with Robert Dinegar as chairman. He approached various carbon dating laboratories suggested by Gove concerning their interest in participating in carbon-14 measurements on the shroud. Positive responses were received from six laboratories, namely the University of Arizona, Brookhaven National Laboratory, Harwell, Oxford University, the University of Rochester and the Zurich/Bern group."

"The involvement of the British Museum had its genesis at the 1982 Archaeometry Conference in Bradford, England. At that meeting, people representing some of its six laboratories listed above suggested that the British Museum be asked to provide them with samples of cloth of dates known to the Museum, but of ages unknown to the six groups for a laboratory intercomparison test. The Museum agreed and the results were presented at Trondheim. The results suggest that all six laboratories are well-qualified to date the shroud."

"The meeting in Trondheim of the people listed at the beginning of this document to consider what the next steps might be was chaired by Harry Gove. The following points emerged."

1. Some concern was expressed, particularly by the chairman, over STURP's future role in the enterprise, since none of its members

had any radiocarbon expertise except for the recent membership of Harbottle in STURP.

2. If samples of the shroud should be made available, they should be provided to all six laboratories, but in a non-identifiable form. Other blind samples of cloth of known age should also be provided to the six laboratories.

3. These samples of known age should be provided by whatever institution established their ages. The same institution should be given the shroud samples for distribution to the six laboratories. There was general agreement that the British Museum would be an appropriate and acceptable institution. It is not at all certain, however, that the Museum would be willing to play this role. The Museum's Board of Trustees would have to concur. Two of the six laboratories are located in Britain (Harwell and Oxford) as, of course, is the Museum, and the director of one of the laboratories is a British Museum trustee. Some of the people from Britain at this meeting expressed some concern at this apparent over-emphasis on British participation. This concern, however, did not appear to be shared by anyone else at the meeting.

4. It was agreed that the chairman of the meeting (Harry Gove) compose a document outlining a suggested procedure or protocol for dating the shroud and send it in draft form to all the people present at the meeting, and also to some of the other people from the six laboratories and the Museum not present at the meeting.

"A draft of the suggested procedure mentioned above is attached to this memorandum. In preparing it, some of the following points were considered by its author (in the following I will use the first person singular)."

A. I had originally thought that, having demonstrated their ability to date cloth several millennia old, the six laboratories should be provided with known shroud samples. I was convinced otherwise at the meeting. The attached protocol recommends blind test of the shroud material and of other cloth samples to be provided at the same time. The age of the shroud almost certainly lies in the range 600 (its known historical age) to about 2000 years (or somewhat older if the cloth were a substantial period of time before zero AD). At least one sample of cloth of known age should be provided. If the age of the sample lay between 600 and 2000 years, the shroud age would remain unknown to the six laboratories, unless, by chance, its age coincided with that of the sample. There is some argument that to avoid this possibility of coincidence, two samples of known age plus the shroud sample be supplied to the laboratories. However, the fewer samples to be dated the better, since the measurements are very time consuming. Two non-shroud samples should be the maximum number supplied.

B. The age of the samples supplied should be known unambiguously to the supplier. If this requires carbon dating of a large sample of cloth by the conventional method, it should be done. If the British Museum were involved, this could be carried out in their radiocarbon laboratory. All the cloth samples, including the shroud sample, should be supplied to the laboratories unraveled and cut into small lengths (~ 1 or 2 mm). At least 2 cm^2 of cloth from each sample should be made available to each laboratory.

C. The good offices of STURP should be used to obtain samples of cloth from the main body (on which the image is imprinted) of the Shroud of Turin. People supervising the removal of these samples from the shroud must include at least one representative from three groups: a) the six laboratories, b) STURP and c) the British Museum or whatever other institution agrees to play the role described in this document. The shroud samples should be handed over to the representative of this latter institution. The crucial role of arranging for provision of shroud samples should conclude STURP's participation in the enterprise.

D. The British Museum (or alternative institution) should prepare the shroud and two other blind samples as described in B above and send them to the six laboratories. Prior to supplying the laboratories with the samples, the Museum should send an agreement to be signed by the head of each laboratory group that the results will not be revealed to anyone other than the designated British Museum official. The agreement should contain the statement that no information that the measurements are in progress should be supplied to the press. I must confess I am considerably dismayed by the publicity that has already been given to the intercomparison test paper presented at Trondheim. My fears of wide press coverage of the British Museum paper at Trondheim fortunately were not realized, but it is clear that the Shroud of Turin is a very newsworthy item.

E. Each of the six laboratories should use whatever method they choose for preparing the cloth samples for measurement. Each laboratory should provide a detailed explanation of the preparation procedure and of how they arrived at both the mean values and the uncertainty in the mean values they quote for the three samples. The results should be sent to the British Museum (or its alternate) for analysis. The six laboratories should be directly involved in a major way in this analysis. The Museum should then send the final results with the samples identified to the six laboratories. The participants should continue to refrain from any contacts with the press.

F. The results of the measurements and the paper on the laboratory intercomparison tests should be sent by the most expeditious channels simultaneously to the President of the Pontifical Academy of Sciences, whom I would consider the most appropriate representative of the Pope in this particular matter, and to the Archbishop of Turin, over the signatures of representatives of the six laboratories and the British Museum. A press release should be issued by the same group of people at the same time. Every effort should be made to inform the Vatican, Turin and the world press simultaneously.

G. The attached protocol, along with the paper on the laboratory intercomparison tests, should be sent to the President of the Pontifical Academy of Sciences, soliciting comments on the proposed procedures and any suggested changes in the procedures. This academy comprises scientists of high repute in diverse fields and seems to me the logical group to consult on scientific matters of interest to the Vatican. I should note that the Shroud of Turin is now legally the property of the Vatican, although, it continues to reside in Turin, under the stewardship of the Bishop / Archbishop / Cardinal of Turin.

"A PROPOSED PROTOCOL FOR DATING THE SHROUD OF TURIN"

"Six laboratories, three in Europe and three in the USA, indicated a willingness to determine the age of the image-imprinted fabric of the Shroud of Turin, using very small samples of cloth. Two of these laboratories use small gas counters and measure the radioactive decay of Carbon-14 in the sample and four laboratories use tandem electrostatic accelerators as ultrasensitive mass spectrometers to measure by direct atom counting the amount of Carbon-14 present. The latter technique was first developed in 1977 while the former has been in use for several decades. The six laboratories participated in a laboratory comparison using textile samples supplied by the British Museum whose ages were known to the Museum, but not to the laboratories until after their results were reported to the Museum. The results of this intercomparison were presented at the 12th International Radiocarbon Conference held in Trondheim, Norway, in June 1985, and will be published in the journal Radiocarbon. A copy of the paper is included in Appendix I. *[This appendix is not reproduced in this book.]* The results of the measurement show that all six laboratories can measure the age of textiles several millennia old to an accuracy of ± 150 years or better are therefore qualified to date the shroud cloth with sufficient accuracy, should samples be made available. The laboratories involved are AERE Harwell, the University of Arizona, Brookhaven National Laboratory, Oxford University, the University of Rochester, and Zurich/Bern."

"The following protocol for dating the Shroud of Turin is suggested."

1. A letter signed by senior representatives of the six participating laboratories should be sent to the British Museum requesting their participation in the Turin Shroud measurements as described below. Such participation would require approval of the Museum's board of trustees.

2. The good offices of the Shroud of Turin Research Project Inc., (STURP) would be used to arrange for appropriate samples of cloth from the main body of the shroud to be removed under the supervision of at least one representative from three groups: a) the six laboratories, b) STURP and c) the British Museum, along with any other representatives the Vatican and Turin authorities may designate. The samples would be handed over to a representative of the British Museum.

3. The British Museum would provide two additional samples of textiles, (both of whose ages are known accurately ± 100 years or better) and unambiguously to the Museum. All three samples including the shroud material would be unraveled and the threads cut into short lengths (1 or 2 mm long) at the Museum to render them as undistinguishable as possible. At least 2 cm^2 of cloth from each sample so prepared should be sent to each of the six laboratories.

4. Prior to supplying the samples, the British Museum should obtain from the six laboratories a signed agreement not to reveal their results to anyone other than the designated British Museum official and to provide no information whatsoever to the press.

5. Each of the six laboratories should use whatever method they choose for preparing the cloth samples for measurement. Each laboratory should provide a detailed description of the preparation procedure and how they calculated their mean values and their uncertainties. The results should be sent to the British Museum for analysis. The six laboratories should be directly involved in this

analysis. The British Museum should send the final results with the samples identified to the six laboratories. The participants would continue to refrain from any contacts with the press or with anyone not directly involved in the measurements and their analysis.

6. The results of the measurements should be sent by the most expeditious channels available to appropriate representatives of the Vatican and of the Catholic diocese in Turin. An appropriate Vatican representative might be the president of the Pontifical Academy of Sciences, whose membership is listed in Appendix II. *[This appendix is not reproduced in this book.]* The submission should be signed by representatives of the six laboratories and the British Museum. A joint press release would be issued by the same group of people, delayed appropriately to ensure that delivery of the results to Turin and the Vatican has taken place.

Source: Cover letter dated July 15, 1985 and memo dated July 10, 1985 from Harry Gove to Robert Dinegar. Luigi Gonella archives, of which the author has a copy.

Comments: Gove stated that Harbottle had become a member of STURP but as we will later see, there was a difference of opinion about that.

Regarding the 1982 intercomparison test, Gove failed to mention in the memorandum and the protocol that Zurich had been off by one-thousand years on one of the samples.

Virtually none of this protocol ended up being used.

1985 *(July)*. Canuto sent a report dated July 24, 1985, titled "REFEREE'S REPORT TO PROF. C CHAGAS" regarding STURP's twenty-six-tests proposal. It read, "This Proposal is interesting but somewhat difficult to evalucte *[sic]* for it is not written in the standard format. This reviewer is therefore somewhat at a loss and forced to hunt for the good, valuable parts and aspects of this Proposal, which, regrettably enough, are submerged in details, not all of which are truly needed or even relevant."

"First of all, even a cursory reading leaves one with the distinct impression that something is missing since a time-honored procedure calls for a special presentation of the state of the art in the field, followed by and equally-detailed and convincing discussion and justification of why and how all the new study is to be carried out. Regrettably, the proposal fails in this respect. One cannot believe that all the 26 items of study that are being proposed are new. This is factually not the case, as the available literature shows. One is therefore left with the first unanswered question of why 26 Items are being proposed. Not only does this seem unrealistic but it may not even be desirable since, clearly, different items must have different priorities. Together with the above lack of overall justification of why the problem has been the proposed, a lack of a clearly-stated relative importance along with 26 items is a grave omission that does not help the understanding of this newly-proposed enterprise."

"Dealing with an artefact of immense value, even if only considered from a strictly historical point of view, leaving aside any religious connotations, it is certailnly [sic] not the use of 26 items that will convince any reviewer of all the desirability of undertaking such a large and complex research program. The authors of the Proposal have failed on this score not because of lack of knowledge, but rather because of a desire to over-convince the reader of their possibilities and ability to do the job. The net result is rather amorphous, all-embracing Proposal that victimizes, by drawing it into almost anonymity, the only item of objective interest, Itm [sic] 6, the dating with the C-14. These very important points are not made in the Proposal. The first truly scientific meeting of the Shroud is bound to capture the world attention and yet it is presented as just one more item, rather than being emphasized as it deserves."

"At a recent meeting (1985) on C-14 working a meeting held in Norway (Trondheim), the representatives of the only six groups in the world capable of carrying out the C-14 dating with the required, recognized authority and competence, did informally discuss the dating of the Shroud. Dr. Gove, one of the leaders of one of the 6 teams, has since then distributed a Proposal for the dating of the Shroud to all the other members. It is expected that in a short time, they will send back annotations, corrections and suggestions. It is the opinion of this Reviewer that such a Proposal ought to be read with great attention."

"It is important to note that these six groups have recently calibrated their techniques by dating two samples provided to them by the British Museum, BM. The results have been very satisfactory, thus providing credibility and strength to their different methodologies (the names and affiliations are given below). In the light of the above, it is the opinion of this Reviewer that in order to ensure maximum scientific credibility to any dating process, one *[has]* to make the result public. It is at least surprising not to find this crucial steps *[sic]* clearly discussed in the Proposal. They are very simply glossed over as technicalities to the worked out whereas we believe they are of crucial importance. It is the firm belief of this Reviewer that only following the above Protocol, or a very similar one, can one assure the world at large that the final date has the necessary scientific credibility, a *sine-qua-non* condition for such an enormously important undertaking."

"As discussed in the Proposal, all previous handling of samples of the Shroud have been conducted in an irresponsible manner, to say the least. This is to be avoided at all costs in the future. Only an internationally-recognized group of laboratories, called upon by the proper Church authorities, can lend credibility to the project. It is to be avoided at all costs that this Project viewed as a biased enterprise, as it will, will rightly or wrongly, if STURP were to take charge or even look as if it is initiated the whole project. STURP has done very meritorious work regarding the Shroud and it should be complimented for its enthusiasm, initiative and perseverance. However it does not have the required scientific ability which as stated above rests on six laboratories in the world. At this particular junction, when one contemplates what many may consider the 'critical experiment,' it is imperative that all, even minimal shades of bias caused by passionate involvement, will be avoided so as to be able to stand by the outcome as a truly a scientific results."

"The conclusions and recommendations of this Reviewer when the summarized as follows:"

1. If the process of dating the Shroud is deemed necessary by the proper Church authority (ies), the Protocol described above must be adhered to.

2. The Pontifical Academy of Sciences must oversee the overall Project, an idea discussed informally and generally agreed upon at the Norway Meeting.

3. STURP should play the initial important role of obtaining, under proper supervision and instruction of the British Museum, the necessary amount of cloth material. STURP has experience in this process.

4. The British Museum must be brought into the overall operation for the undisputed international standing in this field and operations.

"The scientific process of dating must rest on the shoulders of the six laboratories listed below."

V. M. Canuto
New York, N.Y., July 24, 1988

"The six laboratories for C-14 dating"

1	Brookhaven Nat. Lab.	Dr. Harbottle
2	Bern-Zurich Coll.	Dr. Oeschger
3	Harwell Nuc. Physics Labs	Dr. Otlet
4	Oxford	Dr. Hall
5	Arizona (Univ. of)	Dr. Donahue
6	Rochester, N.Y.	Dr. Gove
Consultant	Univ. Washington, Seattle	Dr. Stuiver
Consultant	British Museum	Dr. Tite

Source: Luigi Gonella archives, of which the author has a copy.

Comments: While Canuto does praise STURP for their previous work, he also made several strange statements. He said that for C-14, STURP "does not have the required scientific ability which as stated above rests on six laboratories in the world, despite the fact that Harbottle of Brookhaven, one of the six labs on Canuto's list, was assisting STURP as early as 1982. He also stated that STURP could obtain the material for the dating, but "under proper supervision and instruction of the British Museum." STURP knew more about the Shroud than any other group in the world. Why would they need supervision and instruction from the British Museum?

1985 *(August)*. After several drafts, the final version of the protocol was submitted to the six C-14 labs, the British Museum and to the President of the Pontifical Academy of Sciences.

Source: Gove, Harry. Relic, Icon or Hoax?: Carbon Dating the Turin Shroud (Bristol and Philadelphia: Institute of Physics Publishing), 1996, pg. 321.

Comments: Note that although in June, Gove invited a representative from STURP to a meeting, STURP was not mentioned as having been sent the final version of the protocol. Given Gove's disdain for STURP, it was not likely just an oversight. Gove did finally send Dinegar a copy in September as noted below. Numerous other examples of Gove's negative attitudes and actions towards STURP will be documented throughout this book.

1985 *(September)*. Gove wrote to Dinegar and the latter's daughter (Dr. Janice D. Boyd), who was assisting her father, "I enclose a revised version of the Trondheim protocol. *[I was unable to find a copy of the revised version in the Gonella archive but see final summary by Gove in "Nuclear Instruments and Methods in Physics Research" as reproduced under "1986 (September and October)."]* I have received replies from representatives of all six laboratories, the British Museum and STURP (the latter from the two of you) and have tried to incorporate most of the changes suggested. Let me comment on those changes stimulated by the two of you."

1. I have substantially broadened STURP's participation (see 1, 2, 5 and 6).

2. I have not included STURP as a distributor of samples to the three US laboratories, because it seems unnecessary and inexplicable.

3. I see no reason to involve Stuiver or Muennich -- the fewer outsiders the better. In any case, they could not substitute for the British Museum.

"I have given up my original idea of sending this final (I hope) version of the protocol to the president of the Pontifical Academy of Sciences for approval by that organization. That could involve a large number of people and probably be very time consuming. I will however, send Professor Chagas, the president of the Pontifical Academy a copy of the protocol just to keep the academy informed."

"There is a feeling, and it is one that I share, that STURP, Inc. accept an auxiliary role in the process of determining a radiocarbon date for the Shroud of Turin. In particular, no member of STURP or spokesman for STURP should make statements to the media regarding the C14 operation or its outcome except to echo what has been agreed at the final press release, and only following that press release. Some of the statements made by members of STURP following the 1978 tests were not acceptable to most scientists outside STURP, and suggested an emotional involvement in the Shroud of Turin that could have been interpreted as a lack of impartiality."

"STURP must acknowledge the primacy of the C14 test. If the date of the shroud cloth is medieval, interest in making other scientific measurements will decline precipitously. If the shroud is of Roman date, all the high technology tests that STURP can muster should proceed with alacrity. However, a linen cloth bearing the incredibly detailed image of a crucified man conforming so exactly to the scriptural details of Christ's crucifixion and 2000 years of age may not be so readily available for the kind of scientific testing carried out in 1978. C14 measurements should have been item #1 on STURP's recent list of further tests it would like to pursue. I would still like to receive the full STURP proposal for some 20 or so tests which has recently been considered by Professor Chagas."

"In any case, I hope that STURP now feels it is being fully involved in the shroud dating as proposed in the protocol."

Yours sincerely,
[signed]
H.E. Gove

Source: Letter of September 17, 1985 from Harry Gove to Robert Dinegar. Luigi Gonella archives, of which the author has a copy.

Comments: Gove clearly is trying to flaunt every speck of power he believed he had: "STURP must acknowledge ..." and "C14 measurements should have been item #1 (...)." And his statement "However, a linen cloth bearing the incredibly detailed image of a crucified man conforming so exactly to the scriptural details of Christ's crucifixion and 2000 years of age may not be so readily available for the kind of scientific testing carried out in 1978" seems to imply that a normal group of scientists would not have found the cloth so mysterious. See Dinegar's responses in entry immediately below.

Gove was so concerned about STURP leaking information to the media if they were involved, but yet it ended up that the C-14 labs chosen would leak information about the results.

1985 *(September)*. Dinegar wrote the following reply to Gove's letter above: "Dear Harry, Thank you for your letter of and copy of the protocol. I agree that you have increased STURP's participation. I believe that the Board of Directors of STURP would accept 1, 2, 5, and 6. Number 5, however, has a rather vague sentence, 'the six laboratories and STURP should be directly involved in this analysis.' I will not ask for a full explanation now, but this point will have to be clarified. We, of course, agree with the Labs and STURP receiving the calibrated dates at the same time. Since number 6 is in our proposal, we agree."

"Concerning your cover letter: The paragraph beginning 'There is feeling ...' is in my opinion, clearly unnecessary. The reference to something that happened following the 1978 tests need not be brought up in 1985!! STURP emphasizes freedom of speech -- as I am sure all the Labs do -- and anyone is free to rebutt [sic] any statement made by someone else. As the official public relations spokesman for STURP, I shall see to it that no emotional announcements are made. I cannot keep an individual from speaking out, be he a member of STURP OR THE LABS!! No member of STURP (except Gar[man] Harbottle) will pose as a C-14 expert; I shall, however, feel free to answer any general question on the subject, calling on expert assistance if available and necessary. The sentence beginning 'In particular ...' is unclear and unacceptable as well as unnecessary!!!"

"I now turn to the paragraph 'STURP must acknowledge ...' Again, this declamation adds nothing to the spirit of cooperation. Besides, we do agree on the unique importance of the C-14 date!!! The reason C-14 testing was listed number 6 in our proposal was because WE LISTED THEM IN ALPHABETICAL ORDER! Our listing had nothing to do with importance! We have recently ranked the experiments, but had not done so when we submitted our request. The reviewer -- Dr. Canuto -- and I discussed this very thing at length -- even so far as to suggest separating C-14 and the rest of the tests. With all our correspondence on the subject I cannot understand why you ever thought otherwise."

"You did not receive a copy of the full proposal -- only the C-14 experiment -- because the Board voted to limit the number to itself and the Church authorities until we had a decision. I am asking the president of STURP (Tom D'Muhala) to send you one. Thank you for the revised protocol -- it appears workable."

Sincerely,

Robert H. Dinegar

Source: Letter of September 23, 1985 from Robert Dinegar to Harry Gove. Luigi Gonella archives, of which the author has a copy.

Comments: Dinegar took Gove to task for assuming STURP didn't hold C-14 in high regard and forgetting about previous correspondence that emphasized that. Gove's mistrust of STURP no doubt blinded him. Gove sent another letter to Dinegar in late November, Dinegar responded back in early December and Gove sent a final response several days after. See further below for texts of those letters.

1985 *(October)*. On the 7th, Gove wrote to the British Museum and lobbied to be appointed as the overseer of the Shroud dating.

Source: Gonella, Luigi and Riggi di Numana, Giovanni. Sindone: il mistero continua, (Milan: Fondazione 3M), 2005, pg. 34.

Comments: Note that Gove approached the British Museum and not any Church official. Clearly, too many people were involved in the process, and too many of them were attempting power grabs.

1985 *(October)*. Gove and Shirley Brignall of the University of Rochester met in New York City with Dr. Carlos Chagas, president of the Pontifical Academy of Sciences, and Dr. Vittori (Victor) Canuto, a NASA astrophysicist and scientific aide to Chagas, to discuss the possibility of holding for all interested groups a workshop on dating the Shroud.

Source: Gove, Harry. Relic, Icon or Hoax?: Carbon Dating the Turin Shroud, (Bristol and Philadelphia: Institute of Physics Publishing), 1996, pg. 321.

1985 *(November)*. On November 12th, Canuto informed Gove that Chagas had received permission from the Vatican and Turin archdiocese to hold the workshop, which was scheduled for June 9-11, 1986. Gove then wrote Dinegar that STURP's "good offices" were no longer needed, as the Pontifical Academy of Sciences would take over.

Source: Sox, David. The Shroud Unmasked (Basingstoke, Hampshire: The Lamp Press), 1988, pg. 98.

1985 *(November)*. Cardinal Casaroli wrote to Chagas on November 18th asking him to assist Cardinal Ballestrero in organizing a planning meeting for procedures for dating the Shroud. On the 26th, Gove wrote to STURP that Chagas had looked at STURP's proposal and Gove's proposal and that Chagas preferred Gove's. According to Gove, Chagas sent the material to Secretary of State, Cardinal Casaroli, and recommended that Gove's proposal be adopted. Chagas also supposedly obtained permission to hold a planning meeting in Turin on June 9-11, 1986.

Source: Gonella, Luigi and Riggi di Numana, Giovanni. Sindone: il mistero continua, (Milan: Fondazione 3M), 2005, pp. 34-35.

1985 *(November)*. Gove wrote to Dinegar, "Dear Bob: I would like to report to you some significant developments that have occurred in the last four months regarding radiocarbon measurements on the cloth of the Turin shroud. Before summarizing the events, let me give you, as the presently fashionable phrase is, the bottom line. The President of the Pontifical Academy of Sciences, Professor Carlos Chagas, has obtained agreement from the appropriate Vatican authorities and Cardinal Ballestrero of Turin to hold a workshop in Turin starting June 9, 1986, involving representatives of the six laboratories, the British Museum (if it will serve as the coordinating institution) and appropriate Rome and Turin representatives. This workshop, as I understand it, will be under the auspices of the Pontifical Academy and will have as its purpose a discussion of procedures to be followed to obtain a value of the age of the shroud cloth. The question now is how best to date the shroud and not whether permission to do so will be granted. It is clear the Trondheim protocol will be the cornerstone of the discussions. I have been invited to prepare a proposal to the National Science Foundation's Cooperative Science Program with Western Europe, to fund travel and other expenses to cover the costs for the U.S. participants in this workshop. A representative from each of the U.S. laboratories will be invited. The NSF cannot supply funding for the European participants, but I am sure a source of funding for such expenses can be found."

"This remarkable development had its genesis in a phone call I received on July 22, 1985, less than two weeks after I had mailed you the first version of the Trondheim protocol, from a Dr. Vittorio Canuto. He is someone I had never heard of, and before I briefly outline this conversation and subsequent events, I will give you some information about him."

[At this point there was a handwritten note at bottom of page: "See last page." Content will be detailed at end of entry.]

"Canuto is a theoretical astrophysicist working at NASA's Institute for Space Studies in New York. He has been involved in various ways in the past with the Pontifical Academy and knows Professor Chagas quite well. I believe be serves in some fashion as a scientific advisor to the Vatican's Holy See Mission to the United Nations. Coincidentally, he comes from Turin (some 20 years ago) and he knows some of the people there associated with the shroud."

"In the July 22 phone call, Canuto said he had recently received a letter from Chagas asking his opinion to a proposal made by STURP to make some 26 tests on the shroud to follow up measurements they had made on the shroud in 1978. Item six on the list was radiocarbon dating. I have never seen this somewhat arcane document except for item six on which I was consulted by STURP, a fact known to Canuto. He wanted whatever information I could give him. I told him about the British Museum paper presented at Trondheim, our informal meeting there and the subsequent composition of the first draft of the Trondheim protocol. He urged me to send him copies of the relevant documents so he could inform Chagas. I decided that since there seemed to be a consensus at our Trondheim meeting that the Pontifical Academy of Sciences was our best Vatican contact, I would send Canuto the letter, memorandum and suggested dating protocol of July 10, 1985 which I sent you. My letter to Canuto containing this and other information was dated July 25, 1985. Canuto sent the documents to Chagas."

"Professor Chagas requested Canuto and another person in the U.S. (a member of the Pontifical Academy) *[it was Weisskopf per Bracaglia book, pg. 224]* to review both our and the STURP proposal. Both reviewers favoured ours. Chagas passed all the information on to the Vatican Secretary of State with his own recommendation favouring our proposal."

"On October 14, 1985, Professor Chagas was in New York in connection with the 40th anniversary of the founding of the United Nations. Shirley

Brignall and I met with him and Dr. Canuto for about two and a half hours on that date at the Holy See Mission to the United Nations in New York. Professor Chagas seemed particularly pleased that the Trondheim group had emphasized the importance and need for involvement of the Pontifical Academy. He said he was leaving for Rome in about three days and among other things would try to arrange a meeting of representatives of the six laboratories and the British Museum (he stressed his hope that the British Museum would be involved) in Rome or Turin to discuss procedures for radiocarbon measurements on the shroud. He was singly successful in arranging such a meeting as I outlined at the beginning of this letter."

"In the next little while I will apply to the NSF for funds for the Turin Workshop to pay travel and living expenses of the U.S. participants. The meeting must also be attended by a representative of each of the three European laboratories and of the coordinating institution and funds for support of these delegates should be sought. I suppose, but do not know that funds might be available from other sources that the Pontifical Academy of Sciences can tap. I would be willing to raise the question with Professor Chagas if you wish."

"It is now of rather considerable importance that the British Museum agree to be the coordinating institution. I will renew our invitation of October 7, 1985 to them to do so as soon as possible."

"I will keep you informed of further developments and welcome any comments you may have. I have sent this letter to a representative of each of the laboratories, the British Museum and STURP with copies as listed below."

Yours sincerely,
[signed]
H. E. Gove

HEG/sb

pc: Dr. Richard Burleigh, British Museum
      Dr. Vittorio Canuto, NASA
      Prof. Carlos Chagas, Brazil
      Prof. E.T. Hall, Oxford

"P.S. Handwritten note, also dated November 26, 1985 from Gove to Dinegar: "Bob, I might have some problems getting NSF funding for your travel (or whoever would be the STURP representative) to Turin. Could you find another source of funding?" [Dinegar then handwrote for STURP members he would send copies to, "Can we fund this?"]

Source: Letter of November 26, 1985 from Harry Gove to Robert Dinegar. Luigi Gonella archives, of which the author has a copy.

Comments: Gove told Dinegar that he only learned of Canuto via his letter of July 22, 1985, but Gonella related in his book (pg. 33) that Canuto had sent to Gove in 1984 a copy of the results of "intercomparison laboratory test."

See Gove's shameless admission in following entry that he mislead Dinegar about trying to get NSF funding for STURP's representative.

1985 *(November and December).* Chagas informed Gove that he should contact the United States participants and also discuss arrangements for possible financing from the National Science Foundation (NSF) for one representative of each lab for travel expenses. On November 26th, Gove communicated by letter to all the labs regarding the recent events. On the copy of the letter to Dinegar, Gove noted that he might have trouble getting funding for Dinegar and asked if Dinegar would be able to find another source. Gove then stated, "This was distinctly tongue in cheek because I had no intention of asking the NSF to include him in the grant […]. By this time, I was finding Dinegar's actions on behalf of STURP intensely annoying. I felt, naively as it turned out, that I now had a golden opportunity to eliminate STURP once and for all. On 6 December 1985 I wrote to Dinegar saying that I had been trying to rationalize a role that STURP might now play in the carbon dating the shroud. I noted that my initial patent reluctance to have STURP further involved in any way had been somewhat modified. I realized that STURP had an obvious interest in the shroud because of the years of effort put into its investigation. Their contacts in Turin might be important in obtaining samples. Now, however, the need for STURP's good offices no longer existed."

Source: Gove, Harry. <u>Relic, Icon or Hoax?: Carbon Dating the Turin Shroud,</u> (Bristol and Philadelphia: Institute of Physics Publishing), 1996, pp. 86-88.

Comments: Gove's recounting of the issue of NSF funding for Dinegar and the Dinegar response to the Gove letter immediately above and Dinegar's letter in the entry immediately below all indicate that Dinegar believed Gove's offer was serious. Also see below in the second "1986 *(May)*" entry Gove's interaction with Dinegar regarding the postponement of the Turin workshop. Gove's desire to slight/eliminate STURP knew no bounds.

1985 *(December).* Dinegar wrote to Gove, "Dear Harry, Thank you for your letter outlining the background and latest developments in carbon-dating the Shroud. I think a meeting on the best methods to us in obtaining a date is a fine idea. STURP plans to be there and I, at least, will be present. I cannot believe the date is cast in concrete at this early time so I am asking that it be moved a week or two later. The reason for this is that I will be delivering a paper in Beijing, China during the preceding week. This makes getting to Turin by 9 June a little close, but it can be done."

"Your comment about problems with getting NSF funding is rather strange -- we are equal members of the project as are the six labs and the British Museum. Your proposal clearly indicates this. I'm sure a proper request will eliminate any unequal treatment of the organizations involved. Incidentally, I could not find STURP specifically mentioned as a participant in this workshop. If I overlooked this, please excuse me; if my comment is correct, I'm sure you will correct this oversight in the future."

"I and the others in STURP await further developments along this line."

Sincerely,

[signed]
Robert H. Dinegar

Source: Letter of December 3, 1985 from Robert Dinegar to Harry Gove. Luigi Gonella archives, of which the author has a copy.

Comments: Absolutely no one should be surprised that Gove did not mention STURP as a participant in the workshop.

1985 *(December)*. Gove wrote to Dinegar, "Dear Bob, Since the events described in my letter of November 26, 1985 transpired, I have been trying to rationalize a role STURP might play in subsequent events. My initial obvious reluctance to have STURP further involved in any way with carbon dating the Turin Shroud was somewhat modified by knowing that STURP had an obvious interest in the shroud because of the years of effort they had put into its investigation and mostly by the assumption that their contacts in Turin would probably be important in obtaining samples of shroud cloth for dating. It seems to me this latter need of STURP's good offices no longer exists. Professor Chagas, President of the Pontifical Academy, has obtained the agreement of Cardinal Ballestrero to hold a workshop in Turin to discuss details of procedures to be followed in dating the shroud and this obviously will include obtaining samples. The people directly involved in the measurements will be there along with appropriate Vatican and Turin officials. STURP's presence at this workshop and all subsequent activities can only be as an interested bystander. I am sure many other groups would like to play this role as well, e.g., the many shroud societies, McCrone Associates, ASSIST, etc. I am afraid that STURP has been accidentally and unwittingly pre-empted. I would be hard pressed to explain why STURP is involved any further in the carbon dating process particularly since some people already have reservations about STURP's impartiality concerning the genuine nature of the shroud."

"I think it would be best for STURP to wait until the shroud's age has been established and then to renew their request to perform the other 25 tests, whatever they may be. I assume they would want to proceed with the tests whatever its age turns out to me. Certainly if it is appreciably younger than 2000 years, STURP should have no difficulty in getting permission for any tests they can invent."

Source: Letter of December 6, 1985 from Harry Gove to Robert Dinegar. Luigi Gonella archives, of which the author has a copy.

Comments: I wonder if Gove really thought that anyone would actually believe his statement, "I am afraid that STURP has been accidentally and unwittingly pre-empted." Notice his last sentence when he brings up STURP's proposals using a pejorative term "for any tests they can *invent* [my italics]."

1985 *(December)*. Gove wrote STURP again on December 6th that they were not needed because Chagas was arranging for a meeting, which would include the delivery of the samples to be used in the testing. Gove again trumpeted the assertion that STURP had to be excluded because it was not impartial but was open to STURP doing tests only after the C-14 dating had been done. According to Gonella, Cardinal Ballestrero was unaware of these developments.

Cardinal Casaroli wrote to Cardinal Ballestrero on December 23rd that Chagas wanted to hold the meeting on June 9-11, 1986. STURP, meanwhile, wrote Ballestrero about the letters from Gove that was taking the direction of doing only the C-14 test only instead of the battery of tests STURP wanted, which Gonella said sound scientific practice warranted and explicitly desired by the Congregation of the Faith. Cardinal Ballestrero then wrote to Cardinal Casaroli to pin a date down for the meeting and also instructed Gonella to meet with Gove and Chagas.

Source: Gonella, Luigi and Riggi di Numana, Giovanni. Sindone: il mistero continua (Milan: Fondazione 3M), 2005, pg. 35.

Comments: As documented above, much happened between 1984 through the end of 1985 regarding the power plays going on to determine the planning of the C-14 test and possibly the STURP twenty-six-tests proposal. Let's review some of the events to see if any of the dots can be connected.

* 1984: Gove sent Chagas the results of the 1983 "intercomparison laboratory test."
* 1984: STURP sent a copy of their twenty-six-test proposal, including C-14, to Chagas, who sent it to Canuto, who sent it to Gove.
* At the Trondheim conference in June 1985, the six labs determined that the British Museum would assume the coordinating role in the C-14 dating.
* At this point, Gove's lab at Rochester was still under consideration as one of the labs that would take part in the testing.
* STURP was reluctantly given the task of cutting the sample for the C-14 "so that they would not feel completely excluded."
* On July 8th, Chagas had Canuto review STURP's twenty-six-tests proposal.
* On July 13th, Cardinal Ratzinger approved STURP's proposal for twenty-six tests.
* On July 15th, Gove sent Dinegar a draft of the Trondheim protocol and told him he would send to Chagas.
* Gove informed Dinegar that he heard from Canuto for the first time in a phone call on July 22nd.
* But according to the documentary *La Notte de la Sindone*, Gove actually met with Canuto in person on July 22nd in New York along with Monsignor Celli.
* Another discrepancy about the Gove/Canuto connection: Gonella said in his book with Riggi that Canuto had sent to Gove in 1984 a copy of the results of "intercomparison laboratory test."
* Canuto after the July 22, 1985 meeting then informed Church officials that the tests STURP proposed were "dangerous."
* Gove informed Canuto that the former would pull out of the C-14 dating if STURP had any role beyond excising the sample.
* It was supposedly via Canuto that Gove made contact with Chagas for the first time. Gove, on pg. 133 of his book, said the first contact was when he sent Chagas a letter. But Gove had already sent to Chagas in 1984 a report on the 1983 intercomparison laboratory test. It was around this same in 1984 that STURP had sent a copy of their twenty-six-tests proposal to Chagas. (Gove also planned

to send Chagas a copy of the Trondheim draft protocol before he supposedly had made his first contact him on July 22, 1985.)
* On July 24th, Canuto sent a report to Chagas criticizing the STURP twenty-six-tests proposal.
* On July 25th, Gove sent Canuto a copy of the Trondheim protocol.
* In August, Gove sends the final draft of the Trondheim protocol to the six labs, the British Museum and Chagas, but STURP is not sent one. In September Gove finally sent one to Dinegar.
* On October 7th, Gove wrote to the British Museum and lobbied to be appointed as overseer of the dating.
* On October 14th, Gove met with Chagas and Canuto in New York to discuss a planning meeting for a C-14 test.
* On November 12th, Canuto informed Gove that Chagas had received permission from the Vatican and Turin to hold the C-14 workshop, planned for June 9-11.
* On November 18th, Casaroli wrote to Chagas asking him to assist Ballestrero in organizing the Workshop.
* On November 26th, Gove informed STURP that Chagas had looked at STURP's proposal (it was actually Canuto and Weisskopf who reviewed it) and that Chagas preferred Gove's. However, Gove did not tell Dinegar that he had actually seen a copy of the STURP document. Chagas then made that recommendation of Gove's protocol to Casaroli.
* On December 6th, Gove wrote to Dinegar that the former was trying to rationalize a role that STURP might play in the C-14 dating. He concluded the "need of STURP's good offices no longer exists."

Several questions can be asked: How was Cardinal Ratzinger's approval of STURP's tests eliminated? Why did Gove assert that his first contact with Canuto was by phone on July 22nd, when Gonella stated that on that very day Gove actually met with Canuto in New York and had also said that Canuto had sent Gove in 1984 the results of the "intercomparison laboratory test"? Did Gove specifically try to hide the fact that he had met with Canuto or that they had actually communicated in 1984? Why did Gove assert that his first contact with Chagas was after his supposed phone call with Canuto on July 22, 1985 when Gove had, in fact, been in contact with him in 1984?

Based on Gove's account, he told Dinegar in November, recounting the events post-July 25th, that Chagas had asked Canuto and Weisskopf to review the Trondheim protocol and the STURP proposal that included C-14, and that they both preferred Gove's. Then Chagas made that recommendation to Casaroli. But Gove didn't reveal to Dinegar that he had actually seen a copy of STURP's proposal. Given Gove's animosity toward STURP, and what had become a coalition between Chagas, Gove and Canuto (which will be expanded on later), there is no doubt that Gove would have favored the rejection of the STURP proposal and the acceptance of his Trondheim protocol and thus have had an impact on Canuto's and Weisskopf's reviews. I believe this would explain why Gove would have wanted to put forth the impression that he was in contact with both Chagas and Canuto later than he claimed – so that it would appear that he didn't influence their decision about the Trondheim protocol and the STURP proposal. An argument on the other side is that Gove normally didn't try to conceal his aims. No doubt there are unknown pieces of information that would illuminate all of this.

1985 and 1986. Archaeologist William Meacham wrote, "During 1985 and early 1986 I had a lengthy correspondence with Gonella over the possible pitfalls of C-14 dating, amplifying on his concerns in my earlier paper on the subject [...]."

"[...] In this correspondence with Gonella, I continued to hammer the issue of sampling strategy and possible contamination. I pointed out that 'the edges are anomalous in my view because they were much more subject to handling, in more intimate contact with wooden boards when the Shroud was mounted, and may have been treated to prevent or repair unraveling.' But I thought a sample from the edge could still be used, as long as it was not the only one: 'The sampling strategy that I would hope for would aim at four distinct areas -- the charred cloth, the adjacent uncharred cloth, an edge sample (the size of Raes' *[textile expert who was given threads in 1973 for examination]*, for CO_2 proportional counting), and a thread sample (the size of Baima's, for AMS*)' [Baima Bollone of the Centro Internazionale Di Sindonologia in Turin]*. In comparing to these other samples previously removed, I hoped to soften his opposition to 'punching holes all over the cloth' as he put it once. But I sensed that the correspondence was not actually sinking in. In one of his replies he asked me what part of the Shroud I thought would not be affected by the forms of contamination I had enumerated; in another he stated that such concerns needed to be quantified, otherwise they would be merely speculative. He failed to understand that these concerns are normally dealt with by archaeologists and geologists **after** *[bolding in original]* rogue dates have occurred. What was needed for the Shroud was an approach that minimized the likelihood of getting an aberrant date. He wondered 'why should the Shroud be considered any more likely to be contaminated than any other sample routinely dated by C-14?' I responded that the principal and very important difference is that the Shroud is unprovenanced, has been in so many different handling situations, in contact with so many diverse substances, subject to such extremes in temperature and humidity, unlike an object that has been buried in a stable matrix for several thousand years. And of course, another huge difference was that, unlike ordinary samples from an excavation, with the Shroud it would be very difficult to go back and collect more samples to study the problem once it had arisen. Ultimately, these same questions were discussed at the Turin meeting."

Source: Meacham, William. The Rape of the Turin Shroud: How Christianity's most precious relic was wrongly condemned and violated, (Lulu.com), 2005, pp. 68-69.

Comments: In retrospect, the decision to carbon date samples from only one region of the Shroud --- a region so subject to handling and stress that it was an obvious candidate for earlier undocumented reweave/repairs --- was a serious mistake. For a thorough treatment of the possibility of a reweave, see my paper

- * "The Invisible Reweave and Other Challenges to the Turin Shroud's C-14 Medieval Dating: A Review," accessible: https://www.academia.edu/40272184/The_Invisible_Reweave_and_Other_Challenges_to_the_Turin_Shrouds_C-14_Medieval_Dating_A_Review.

* The corresponding PowerPoint is at: https://www.academia.edu/40271611/_Powerpoint_for_The_Invisible_Reweave_and_Other_Challenges_to_the_Turin_Shrouds_C-14_Medieval_Dating_a_Review. Each slide in the PowerPoint is numbered at the lower right and can be matched with the slide number shown in the text.

The "invisible reweave" is also known as "French reweave." My paper has been on the academia.edu site since September 2019. For what it's worth, I know that at least seven skeptics of the theory are aware of the paper, and as of the time of this writing (mid August 2020), not one has tried to refute the evidence contained in it.

1986. Meacham wrote, "The point that I hoped to drive home is that there are many things that can go wrong with C-14 dating; for some the cause is unknown, while the others are grouped under the terms 'rogue dates.' It is important for anyone wishing to understand the normal archaeological use of C-14 to know that a single date or even a series of dates on a single object or feature is seldom if ever cited to answer important questions about the age of a culture or a site. To put a single radiocarbon date in the position of being the ultimate arbiter of the age of the Turin Shroud is a blatant departure from the way C-14 is normally used. There are simply too many pitfalls. This was not a position that went down well with the hotshots from the radiocarbon labs."

Source: Meacham, William. The Rape of the Turin Shroud: How Christianity's most precious relic was wrongly condemned and violated, (Lulu.com), 2005, pg. 59.

1986. An article in the journal *Radiocarbon* noted, "[T]he dating of the Shroud of Turin would now be possible in principle although it is generally agreed that any such measurement ought not to be undertaken by a single laboratory, or even by use of one technique alone."

Source: Laverdiere, H. "The Socio-Politic of a Relic: Carbon Dating of the Turin Shroud," (Doctoral Thesis) 1989, pg. 69. Accessible via free download at http://ethos.bl.uk/OrderDetails.do?did=1&uin=uk.bl.ethos.531916.

Comments: Although the Turin authorities did choose more than one lab, they did not take the advice to restrict the test to one technique. Tite was actually a co-author on the article from which this quote was taken.

1986 *(January)*. Gove submitted to the National Science Foundation's Cooperative Science Program with Western Europe a proposal to cover the expenses of three United States labs -- Arizona, Brookhaven, and Rochester -- to attend the workshop.

Source: Gove, Harry. Relic, Icon or Hoax: Carbon Dating the Turin Shroud (Bristol and Philadelphia: Institute of Physics Publishing), 1996, pp. 321-322.

1986 *(January)*. On the 17th, Gove contacted Tite to inform him about Chagas' various preparations for the Turin workshop. Tite told Gove that when the samples were taken, a textile expert should be present to insure that a sample from the image area was taken and was not from a rewoven section. Tite would try to suggest an appropriate expert.

Source: Gove, Harry. Relic, Icon or Hoax?: Carbon Dating the Turin Shroud, (Bristol and Philadelphia: Institute of Physics Publishing), 1996, pp. 90-91.

Comments: The sample was not ultimately taken from the image area and despite the presence of two textile experts, there is strong evidence that the area from which the sample was taken may, in fact, been rewoven. The late STURP scientist, Al Adler, discussing his view after the dating that a sample taken from the side of the cloth wouldn't necessarily give the same date as one taken from the middle of the cloth remarked "So you can talk all you want about how reproducible the date is, but you can't talk about how accurate it is. You have no way of knowing if the area you took the C-14 sample from represents the whole cloth. That's an area which has obviously been repaired. There's cloth missing there. It's been rewoven on the edge. They even cut part of it off, because it was obviously rewoven on the edge. The simplest explanation why the date may be off is that it's rewoven cloth there. And that's not been tested." [From: Case, T.W. *The Shroud of Turin And The C-14 Dating Fiasco: A Scientific Detective Story,* (Cincinnati: White Horse Press, 1996, pg. 73.]

There are two peer-reviewed papers that indicate that the C-14 corner was rewoven:

* Rogers, Raymond N. "Studies on the radiocarbon sample from the shroud of turin." *Thermochimica Acta*, Vol. **425**, No. 1/2, 20 January 2005. (http://www.sindone.info/ROGERS-3.PDF);

* Benford, M.S. and Joseph G. Marino. "Discrepancies in the radiocarbon dating area of the turin shroud." *Chemistry Today*, vol. **26**, no.4 (July/August 2008). (https://www.shroud.com/pdfs/benfordmarino2008.pdf).

There are also three peer-reviewed papers that indicate the raw data from the labs (not released until 2017 and not publicized until 2019) show the labs' medieval dates results were suspect, which is consistent with the theory of a rewoven sample:

* Riani, Marco *et al.* "Carbon Dating of the Shroud of Turin: Partially Labelled Regressors and the Design of Experiments." *Stat Comput* DOI 10.1007/s11222-012-9329-5, 2012. Accessible at http://www.riani.it/pub/RAFC2012.pdf.

* Casabianca, Tristan *et. al.* "Radiocarbon Dating of the Turin Shroud: New Evidence from Raw Data." *Archaeometry*, 2019, doi: 10.1111/arcm.12467). (https://www.academia.edu/38607635/Radiocardon_Dating_of_the_Turin_Shroud_New_Evidence_

From_Raw_Data) *[Note: "Radiocarbon is misspelled as "Radiocardon" in the URL].*

* Walsh, Bryan and Larry Schwalbe. "An instructive inter-laboratory comparison: The 1988 radiocarbon dating of the Shroud of Turin." *Journal of Archaeological Science: Reports,* v.**29** (2020) 102015. https://www.sciencedirect.com/science/article/pii/ S2352409X19301865).

In addition, the day I was to submit the manuscript, a new paper by Paolo Di Lazzaro *et al.,* titled "Revisione propositiva dei risultati di radio-datazione della Sindone di Torino" ("Proposal revision of the radio-dating results of the Turin Shroud"), was published as a technical report by Di Lazzaro's ENEA, the Italian Agency for New Technologies, Energy and Sustainable Economic Development. It notes "an unexpected and anomalous dependence between the radiocarbon age and the position of the dated subsamples." See: https://www.academia.edu/43936359/ Revisione_propositiva_dei_risultati_di_radio_datazione_della_ Sindone_di_Torino.

1986 *(January)*. Late in the month, Gonella went to the Pontifical Academy of Sciences to talk to Chagas. Gonella described STURP's 1978 examination, the campaign being waged against STURP to exclude them, expressed the Cardinal's concerns about having six labs participate (especially regarding the amount of material that would be needed). Chagas told Gonella that Turin considered STURP's involvement **politically** *[my bolding]* necessary.

Source: Gonella, Luigi and Riggi di Numana, Giovanni. Sindone: il mistero continua, (Milan: Fondazione 3M), 2005, pp. 35-36.

Comments: This is the one instance where the "politics" was appropriate. Simply put, STURP had the most experience with the cloth and should have been involved in any testing. Alas, they were not. According to the Gonella/Riggi book (pp. 45-46), Chagas requested Ballestrero on January

25th for some formal changes to the planning meeting, and Ballestrero approved them.

1986 *(February)*. Gove and Brignall met with Gonella in New York City to discuss the upcoming workshop in Turin regarding the C-14 dating of the Shroud. Gonella wanted the workshop to be held at his own institution, the Turin Polytechnic. Gove pointed out that the workshop was being organized by the Pontifical Academy of Sciences and suggested it should be held in their Vatican headquarters. Gonella expressed his opposition to having six labs involved.

Source: Gove, Harry. Relic, Icon or Hoax?: Carbon Dating the Turin Shroud (Bristol and Philadelphia: Institute of Physics Publishing), 1996, pg. 322.

1986 *(February)*. A draft dated February 16, 1986 titled "NON-DISCLOSURE AGREEMENT -- AGE OF SHROUD OF TURIN," put together by Gonella, was produced. It read, "The undersigned, as the representative of the listed institutions, recognizes the need to prevent premature disclosure of the results of radiocarbon dating of the Turin Shroud. The considerable public interest in this artifact and particularly in its age will provide strong pressure for such untimely disclosure and the concomitant challenge to resist it. The undersigned agrees that the appropriate and only time to first present the results to the public is as soon as possible after complete analysis of the results of the measurements has been carried out by the six laboratories at AERE Harwell, University of Arizona, Brookhaven National Laboratory, ETH Zurich, Oxford University, and University of Rochester and the British Museum. At that time representatives of these seven institutions will compose a mutually agreeable public information document. Before release to the public, the appropriate authorities in Turin and the Vatican will be informed expeditiously of its contents. The undersigned solemnly promises not to reveal any information regarding the age of the Shroud of Turin until this public release. The promise applies equally to all members of his/her institution who may have knowledge of the shroud's age."

"Because the radiocarbon measurements and their analysis should be carried out in as serene an atmosphere as possible, public information even that measurements on the shroud are in progress should be strictly circumscribed. The undersigned therefore also agrees not to reveal that measurements and analysis of data relating to the age of the Turin Shroud are being undertaken."

"Deliberate premature disclosure by an institution involved could only be motivated by a desire for unseemly publicity for that institution. The repute of the seven institutions taking part in the measurements is such that the possibility it would happen is vanishingly small."

Source: Luigi Gonella archives, of which the author has a copy.

Comments: Despite the draft alluding to the negative impact of a premature disclosure and a later signed non-disclosure agreement, a premature disclosure was eventually made, with both Arizona and Zurich violating the agreement not to let insiders in.

1986 *(February)*. The same day that the above draft of the non-disclosure agreement was produced (presumably by Gonella as well), another document titled "POSSIBLE ITEMS FOR DISCUSSION AT THE TURIN WORKSHOP JUNE 9, 10 AND 11, 1986" was produced. It read:

1. Agreement on Non-Disclosure
    a) Is there a need for such an agreement?
    b) What time period should it cover?
    c) Wording of the non-disclosure agreement
    d) Who should sign it?
    e) How binding can it be?

2. Sample Removal and Delivery
    a) Sample size required.
    b) Sample location on shroud.
    c) Removal procedure -- who will do it?
    d) Witnesses of removal
    e) Packaging after removal

       f) Delivery to British Museum in London, how and who witnesses. (Not from 6 labs.)
       g) Possible role of a textile expert.

3. Control Samples Provided by British Museum
    a) How many per lab?
    b) What material -- linen, cotton?
    c) Source of controls -- Peru, Egypt?
    d) What ages? How determined?
    e) Witnesses required? (Not from 6 labs.)

4. Sample Preparation and Distribution by British Museum
    a) How to render controls and shroud samples indistinguishable?
    b) Cleaning of samples by BM?
    c) How to ship samples to laboratories.
    d) Witnesses required? (Not from 6 labs.)
    e) Should all six labs get a shroud sample? Should BM decide this*?

*If one of the six labs got 3 control samples only, and no shroud sample, prior disclosure problems would be obviated.

1. Sample Preparation and Measurement Procedures by the Six Laboratories
    a) Shroud sample -- mold, oil, handling, smoke, water damage.
    b) Control samples.
    c) Each lab should describe its cleaning procedures.
    d) Each lab should describe measurement procedures and time required.

2. Data from Six laboratories Delivered to British Museum
    a) How should labs quote ages
    b) How should labs quote uncertainties?
    c) How is data delivered to BM?
    d) Analysis of data -- who involved -- BM, six labs, others?

e) Comparison of average control sample ages with their known ages.
f) How to deal with outliers -- if any.
g) Should labs be identified during analysis?
h) Witnesses of analysis.

3. Public Disclosure
   a) Jointly composed public information document.
   b) Information of shroud age supplied to Turin, Vatican, Pontifical Academy.
   c) Joint press release.

"A key question is whether an impartial witness with impeccable credentials (if such there be) should be present during several of the above procedures to maximize public acceptance of the credibility of the whole complex operation."

Source: Luigi Gonella archives, of which the author has a copy.

Comments: The answer to 1a is "Yes, but it wouldn't do any good." The answer to 1e is "It wasn't binding at all."

It's a shame that so much planning was put into this document and the workshop itself (rescheduled for late September/early October) and virtually all of it ended up being ignored, including the part of finding an impartial witness. Instead, everyone was basically just expected to accept the labs' version of events.

1986 *(February)*. Gonella met with Harbottle in the U.S. on the 15th to discuss some technical matters, including the possibility of using carbonized material from around the burns so as not to have to cut additional material from the Shroud. Harbottle warned Gonella that Gove was extremely interested in the publicity he would get from dating the Shroud, and knew that it could enable him to more easily get allocated funds for his lab. Harbottle also said that Gove was a person who was used to managing things on his own, without ever "giving the ball to others."

Gonella met with Gove in Heidelberg on February 17th. Gonella wrote that Gove was evasive on technical issues, always bringing the subject back for the need to exclude STURP because they were not objective. Gove brought up the fact Dr. John Jackson of STURP wore a cross around his neck during the 1978 examinations. Gonella and Gove also discussed the merits of the new AMS dating versus the old "small-counter" technique that required a larger sample size. Gonella also discussed the need for labs not to know the results of the other labs, which seemed to concern Gove.

The next day, Gonella contacted Chagas in Brazil, expressed his concern about the attitudes of some of the potential participants, communicated Harbottle's warnings about Gove, and asked Chagas if the planning meeting could be held on Gonella's home turf at the Polytechnic of Turin, so as to have a more scientific/technical focus and better avoid the interest of journalists. This was a strictly personal letter, but Canuto informed Gove about it. Gove protested Gonella's suggestion and wanted Chagas to pick a more appropriate venue. Cardinal Ballestrero felt that the evolving proposal was getting too far away from what was originally envisioned and did not feel comfortable leaving the decision to Chagas. Ballestrero decided that the meeting should be held at his Turin residence and informed the Vatican as such.

Source: Gonella, Luigi and Riggi di Numana, Giovanni. Sindone: il mistero continua, (Milan: Fondazione 3M), 2005, pp. 35-36.

Comments: Gonella didn't elaborate on why Gove seemed concerned about the proposal that labs not know the results of the other labs, but it's very strange that Gove would not have been on board with that. While Jackson was known to be a devout Catholic, Gove unfairly extrapolated Jackson's faith to the whole STURP group.

1986 *(February)* Gonella wrote to Chagas: *[Note: see 1st paragraph of entry comments for explanation of punctuation that was added due to problems interpreting the handwritten letter]*: "Dear Prof. Chagas, Following your suggestion in our phone conversation of this afternoon, I'm writing to you about my talk with Gove yesterday and the other Shroud problems. I hope to be able to be clear both in words and this handwriting *[written in English]*, but I'm not so sure as I'm hit by a big, persistent toothache which is impairing my faculties."

"As I told you by phone, my conversation with Gove was somewhat puzzling. We left as good friends (which was the primary purpose of the visit, in the directions I had from the Cardinal) -- as we were a few years ago anyhow when I had first met him -- and I hope I've been able to convey [?][?] reference frame for the Shroud tests, but I had the impression that he is a perturbed man with strong [?] feelings about the project and his role in it. He kept oscillating between saying that he is just one out of six and cannot speak for the others, and that he has nothing against anybody -- and almost in the same breath -- saying that he cannot accept this one or the other to be present at the workshop, or that he 'cannot justify before the others certain presences.' He somehow seems to miss the point that it is not up to him to decide which work is to be done in the Shroud meeting [?]. In one breath he was saying that he deals only with C-14 and doesn't want any interference by anybody who is not a C-14 expert and in the next he starts heavy criticism on any other measurement proposed on the Shroud, calling these useless or absurd without waiting to hear of their context or aims."

"As it was quite clear he was against any involvement of STURP at any stage because he deemed them biased, or 'not impartial.' I asked him to please state *[tell]* me explicitly what did he find biased in STURP's work -- then he told me that the public cannot accept them as unbiased because everybody now did see photographs of the '78 tests showing Jackson wearing a big cross at his work -- 'and a really big one.' At this I sort of exploded -- asked him if he found my [?] disquieting with my coat of arms -- no he says -- but Jackson's cross shows the extent of his *[Jackson's]* feelings. Asked him if he found anything to object in Jackson's papers on the Shroud -- no he said, because it was quite evident in those papers that Jackson wrote with absolute objectivity because he was leaning over backward to correct for his feelings. I was almost speechless. Another point was statements by Stevenson/Habermas in their book 'Verdict on the Shroud' that STURP did officially repudiate, [expelling ?] Stevenson from membership because he had accepted ascribing to team interpretation his own. (Stevenson, by the way, was no scientist, but was on the team to help with the handling of the press*.) This is the genesis of the 'reserves' *[reservations]* against STURP voiced out by certain sections of the academic world. Gove, after my remonstrances, said that he agreed that Jackson had the right to wear any cross he wanted, but that the public could not thereafter accept him as unbiased. I told him that it was just postures like his own (Gove's) that [?] views on the public, because science is to be evaluated by method and if fellow scientists judge the credibility of their fellow scientists by the color of their feelings, instead of the intrinsic worth of their scientific work, it is the end of science."

"As I was telling you in Rome last month, I'm very disturbed by these rumors against STURP, not because of their existence; there are always rumors, and they could expected from certain quarters against anybody who did not proclaim the Shroud to be a fake), but because they being picked up by fellow scientists without checking them, without [balancing ?] them against the published work of the accused. I told Gove that it was quite evident he too had deep feelings, maybe of different color and his argument against STURP could easily be turned against himself [?] we [?] stick to the narrow path of scientific rigor -- that he will be subjected to the same kind of rumors and *a priori* judgments [?] or not [?] if he were to date the Shroud at the 1st century A.D. He seemed impressed and I hope he will reflect."

"Apart from the feelings, I explained *[to]* Gove that for us in Turin the Shroud problem was multi-faced and had to be tackled within a multidisciplinary way. While he was interested only in C-14 -- and the 9-11 June workshop was *[to be]* dedicated only to C-14 -- the duty of the Cardinal as custodian of the Shroud (hence my duties as his advisor) were mainly concerned on the conservation issue and the image-forming mechanism which is directly tied to [?] problems that keep existing whatever be the Shroud's age. The general protocol of the Church of Turin on *[the]* Shroud test is that we [?] appoint any scientist to any expertise or task, and don't tell them what [?] we take into examination proposals by scientists, to be caused [?] their sole and complete responsibility, and accept or reject them on the strength of the scientific stand of the proponents in their own field and of the safety of the Shroud itself. As a matter of fact we have to examine STURP's proposal of 26 'work packages' & a British proposal of 5 packages & 1 half-baked French proposal. C-14 is package #6 in the STURP proposal and is being considered because of STURP's initiative. Of course this proposal is different from the other measurements as it calls for non-STURP operators, and therefore is worth of special handling, and also, even more so, because for it the Holy See asked the advice and patronage of the Pontifical Academy, that we in Turin most gratefully accept. It will therefore be handle by itself in a special way -- [?] always in the frame of the general protocol stated above."

("Correct me, Professor Chagas, if I'm wrong or overstepping.) The original STURP protocol considered already the parallel work of the six labs with blind dating procedure (by the way, it would have been quite easy for STURP to have proposed to carry out the C-14 dating only by their own member, Harbottle of Brookhaven, and e.g., Arizona and Zurich handling [?] thing by themselves, but Dinegar, acting on the general cooperation [?] of STURP, had asked Gove a few years ago whether he was still interested, and together they picked the 6 labs who were both [?][?] interested in). The 'Trondheim protocol' is an improvement on the [?] 'package 6' for the introduction of the British Museum (that would not have been included in the format of package 6 for historical reasons and anyhow is gladly accepted by STURP). I asked Gove if he saw any differences between the 'Trondheim protocol' and 'package 6' (that he helped to draft) beside the introduction of the British Museum, but (three times) evaded the answer by changing *[the]* subject. *[The next line is completely unreadable.]* The interaction of STURP at several stages of the proceeding, apart from the obvious presence throughout of STURP member Harbottle, is quite useful for fitting the C-14 test with the other ones and for expediting for the C-14 sampling *[--]* the expertise on the Shroud that STURP gained in 5-10 years of scientific studies on the subject."

"As I see it, C-14 dating numbers 3 phases: 1) cloth-related excision phase: to cut the sample (where, how, how much); 2) sample and data handling '**political**' *[my bolding]* phase: how to distribute samples for a blind test, analyze data and 'publish them;' 3) measurement phase.

Phase 3 is up to 'the six labs'. Phase 2 is to be carried out by BM *[British Museum]*, by general agreement. Phase 1 is the one which I'm more involved in, as it calls for duties as advisor in re: safety of the Shroud. It is in this phrase that interaction with other tests may occurr *[sic]*, and therefore STURP involvement is necessary. This I tried to explain *[to]* Gove -- and he seemed to understand and accept. Except of course when he fell under a fit of anti-STURP feelings."

"A point on which we are very much on the air is the actual amount of material needed. STURP's 'package 6' asked for an outrageous amount that Turin cannot accept, as it is utterly unjustified (as I already had told them -- useless to go into the reasons Dinegar arrived at those figures to please the non-STURP labs). Trondheim's protocol skips over it [?] speaking of '1cm^2 per lab' without further expounding the matter. I asked Gove why Trondheim's protocol did not address the issue -- he said it did with the request of [?] [?] asked only 1cm^2. It turned out that plus [?] came out of Gove's estimate of a [?] of the Shroud cloth of [?mg]/cm^2 and a carbon content of 25% -- the density being estimated on his [?] measurement on a photograph of Raes samples. I told him he was very off from the four different evaluations made by others point to about 20mg/cm^2 [?] he said that I this case he had to ask for more material. I asked him why they were asking for the same amount for all labs, given that it is well known that the accelerator technique requires less material than the small counter technique -- he said in order to treat them all on the same [?][?]."

"I objected to this (and told him so)! It is not matter of distributing candies to children -- I do not see why we should give anyone more than another only because someone else needs more for technical reasons. I therefore asked him to make a clear proposal about the quantity of material he really needs and ask the others to do the same thing (he said he could only speak for himself); let us start from the [?] and work out the cm^2 on the basis of reliable data. I do not want anybody to work with a straightjacket, but I'll fight for every single milligram of Shroud if there is no sound technical reasons for employing it -- this is my bound duty as advisor for the Shroud safety. He seemed to accept the positions of mine, and he told me he would proceed *[with a]* new estimate."

"I did not tell Gove, but tell you here, that I found the 'Trondheim protocol' rather sloppy. Maybe if Gove had read STURP's papers and [talked ?] to them he would have started on sounder figures. We'll have to thrash it out at the Turin workshop. It will be better if all people will arrive there with reliable data and not half-baked assumptions. How reliable are the data is debatable, but if we collect them objectively we'll be able to appraise the situation with known uncertainty. I phoned STURP's textile expert, Prof. Drusik at Los Angeles, to please collect all data relevant to the issue. Harbottle, whom I saw on Saturday, tells me his (and Harwell's) small counter technique needs 10 mg of <u>carbon</u>, and the accelerator technique claims to be able to work easily with 2 mg. Gove told me they could work certainly with 5 mg., and even 2, though he would not like it, and might even sketch it down to 1 but definitely would not like it."

"As I remember, in Rome I was telling you about the possibility that the actual measurements were carried out by only 3 labs out of 6 -- and you told me that in your opinion all 6 should have a sample if they were asking no more than 1 cm^2 each, but if they were asking for more we ought to consider limiting their number. Gove gave me a draft of possible agenda for the Turin workshop, that I suppose he is mailing [?] too, where it is taken into consideration the idea of providing one of the labs, of course unknown except for the BM, with a blank instead of a Shroud sample, in order to eliminate the possibility of premature disclosure of results. I'm leaving it to others/you and the labs to judge on the **political** *[my bolding]* advisability of such [?]. [?][?] I find it a [?] from the point of view of concerning the Shroud [?]."

"Another point is the possible utilization of the charred material at the holes edges beneath the patches. An archaeologist expressed horror at the idea, because -- he says -- there is ion exchange. Harbottle, who had his doctorate in ion exchange says he thinks it would be perfectly OK, as most C-14 dating in archaeology is made on charred material. He is collecting data and also will make an experiment. Gove said he was not against using charred material, in principle, and will think over the possibility. The point is that with charred material there would not be any [?] question of quantity:

there is enough and to remove it (of course under close supervision of experts) will only enhance the safety of the Shroud."

"When we met in Rome, it was clear that you would have liked very much to be present at the tests on the Shroud (or at least part of them). Moreover the Cardinal was favoring the idea of having a workshop and the work on the Shroud as close as possible to each other [?][?] press-security reasons. I had marked down that your schedule was to be in Italy all the month of June, but not available in the period 17-21 June. On the strength of this I asked STURP whether it possible for them to begin the work on the Shroud (of course [?], [?][?] the limits, it was approved) within the month of June. They told they could, with a big but possible effort, begin testing on June 23rd, and retargeted their organizational effort on this date. Of course it would be easier for them to have time up to July 15th or so, but it can be done and they are quite willing to do it. These dates of course are top-secret because we do not want the press breathing down our necks. Keep in mind that STURP is to work round the clock for two weeks. We did not yet decide how much to accept of the proposed work -- but I do not think much will be denied, as most packages are very interesting. The excision of the samples for C-14 can be made at anytime during the week, except perhaps the very first days if preliminary measurements are required to ensure optimization of the sites."

"Please tell us as soon as possible your wishes on the matter. For the Turin workshop we need to decide whom to invite, who will be conveyor (who might be a different person than the Chairman), the venue and the agenda. For the latter, most of the points listed by Gove in his draft are good and some may be added -- provided it is clear that it is not Gove to fix the agenda nor to act as conveyor. For the venue, would the Dept. of Physics of the Polytechnic of Turin be acceptable? For press-security a group of physicists coming to it would *[next phrase unintelligible]* if the meeting were to be at any religion-connected building. Of course a visit to the Archbishop would be made at the proper moment -- and after all it is matter of a scientific meeting. It goes without saying that the chairmanship is obviously yours. As the meeting is in Turin, would it be presumptuous or instead, appropriate for me to act as conveyor, as scientific advisor to the Cardinal? Please tell me what you think and suggest."

"During my talk with Harbottle, most devoted to technical matters on C-14, he saw fit to tell me something about the labs' setup that it is better for me to report to you. He is certainly not very happy at the stance that Gove is assuming on the issue, practically presenting himself as spokesman for all -- he warned me [?] to be wary of the well-known tendency of Gove to 'pick up the football and carry it all by himself', and that there are unfavourable comments at Trondheim on the 'Trondheim protocol', etc. Gove is considered sort of an outsider by the C-14 people, as he is a nuclear physicist while the others devoted all their life to C-14 dating. Harbottle, being a Brookhaven man, is out of the Universities struggle for funding: he told me that Rochester and Arizona are engaged in a struggle for survival, as the NSF is sharply curtailing the funds for nuclear physics labs with every expensive machine that yield little additional results, as the field seems now almost played out. Beside the Arizona machine keeps breaking down (this prickled my ears -- would it be appropriate at the Turin workshop to ask for a reliability status report of the seven labs? It would not look good if the results were stopped for several months owing to technical bugs of one or two labs.) In such a situation, the dating of the Shroud (especially if it came out to be accepted by the newspapers as the 'debunking of a legend' will loom high in the eyes of congressmen and ensure funding, and maybe plain survival."

"Let it be clear that I don't hold these things against Gove, because I feel it quite proper for a head of department to look for relevant jobs to enhance the prestige of his lab. What demands watching is his possible use of the issue for an internal fight among labs by means of presenting himself in a position of initiator and policeman of a joint enterprise -- suggested by his effort in eliminating from the picture the fellow scientists of STURP who (in close collaboration with Arizona) were *de facto* the real originators of the project [...]."

Source: Handwritten letter of February 18, 1986 from Prof. Luigi Gonella to Prof. Carlos Chagas. Luigi Gonella archives, of which the author has a copy.

Comments: The left margins on all the pages of the reproduced letter were somewhat cut off. I believe it was only one or two words in most cases. I have added a question mark (one for each word) in brackets to signify words I can't identify. In some cases, I've made an educated guess, so in cases like, that there will be a word in brackets along with a question mark.

Gove certainly seems to have come off as a control freak and also seemed somewhat schizophrenic in his responses to Gonella's comments and questions. One can easily see the limited approach Gove wanted to take versus the multi-disciplinary approach that STURP wanted to take.

Several of things that Gonella feared: premature release of the labs and delays due to technical bugs in one of the labs, did, in fact, occur with the 1988 dating. At that point, it was planned that STURP would have two full weeks to perform multi-disciplinary tests on the Shroud

*Gonella told Chagas that Ken Stevenson was not a scientist. Stevenson wrote in a 1991 paper presented at a Shroud conference in St. Louis (see: https://www.youtube.com/watch?v=w07Mkn9_-7M&feature=youtu.be&list=UUSv7BD9sKjIcA24ct1Hz_Aw), "[...] I will tell you clearly that I do have an engineering degree, and that as a part of that engineering degree there were two full semesters of aeronautical engineering, two full semesters of electrical engineering, two semesters of mechanical engineering, two full semesters of physics, four full semesters of chemistry, eight full semesters of higher math. And as 22-1/2 credit hours per semester for four years, I submit to you I have a scientific engineering degree."

1986 *(February)*. Gonella informed Gove that STURP planned to perform their twenty-six tests two weeks after the June workshop. Chagas later informed Gove that Gonella felt the University of Rochester was trying to take charge, and that Gove's agnosticism was at odds with his interest in the Shroud.

Source: Sox, David. The Shroud Unmasked (Basingstoke, Hampshire: The Lamp Press), 1988, pp. 100-101.

Comments: Most scientists involved with the study of the Shroud at the time were interested in it mainly from a scientific point of view, so it seems odd that Gonella would make a statement like that. However, it is clear that Gonella and Gove did not like each other [despite Gonella's contention (to Chagas) that he and Gove were friends], and it's likely that both would latch onto anything to criticize the other. An article in the *Rolling Stone* in 1978 described Gove as a "once-a-year Episcopalian."

1986 (March). Sox wrote that two clear camps had developed: the Pontifical Academy and Gove in one and STURP and Gonella in the other. There were five main sticking points: 1) After the meeting with Gove, Gonella met with Monsignor Celli, Vatican Ambassador to the United Nations to push the STURP camp. 2) On the way back to Turin, Gonella planned to visit Tite at the British Museum. 3) Despite the fact that the Pontifical Academy of Sciences normally held workshops in Rome, Gonella insisted that it be at his own institution, Turin Polytechnic. 4) Gonella and others in Turin insisted on STURP's involvement in the workshop. 5) STURP-appointed textile experts were the best choice.

Source: Sox, David. The Shroud Unmasked (Basingstoke, Hampshire: The Lamp Press), 1988, pg. 101.

Comments: Recall that Gonella, in his handwritten letter to Chagas on February 18th, asked Chagas if the Polytechnic would be a suitable venue for the workshop. Here, Sox claimed that Gonella "insisted" that it be at the Polytechnic. Clearly, Sox had a bias against Gonella.

1986 (March). At a Shroud symposium held in Hong Kong, Meacham presented a paper in which he tried to warn everyone of putting too much stock in the C-14 date. He wrote, "Reviewing recent Shroud literature of all persuasions, I find little awareness of the limitations of the C-14 method, an urge to 'date first and ask questions later,' and a general disregard for the close collaboration between field and laboratory personnel which is the ideal in archaeometric projects. Regarding the Shroud, consultations should take place among archeologists *[sic]*, historians, conservationists, cellulose chemists and of course radiocarbon scientists in order to formulate a specific C-14 sampling and dating procedure."

"Even among social and physical scientists, there are numerous misconceptions about the radiocarbon method of dating; among journalists and the general public there are of course many more. But among specialists who frequently make use of the test, it is *not [italics in original]* considered as a method which produces an 'absolute date' for every sample that can be measured [...]."

"[...] To measure Shroud samples, one must therefore consider every possible type of contamination and attempt to identify and counter them all, before the measurement is made and a 'radiocarbon age' assigned. Clearly, this result can only be considered as a possibility, at best a good probability, but hedged by many uncertainties. It would not be an absolute calendar date, and it would not 'prove' the Shroud to be authentic or a forger. Rather, it would be one further piece of evidence to be evaluated in the light the total complex of data about the Shroud."

He further stated, "I doubt that anyone with significant experience in the dating of excavated samples would dismiss for one moment the potential danger of contamination and other sources of error. No responsible field archaeologist would trust a single date, or a series of dates on a single feature, to settle a major historical issue, establish a site or cultural chronology, etc. No responsible radiocarbon scientist would claim that it was *proven [italics in original]* that all contaminants had been removed and that the dating range produced for a sample was without doubt the calendar age. The public and many non-specialist academics do seem to share the misconception that C-14 dates are absolute."

Meacham also cited a quote from a paper presented in 1985 at a C-14 conference in Trondheim, Norway, in which the R.A. Johnson *et al.* wrote, "The existence of significant indeterminant errors can *never [italics in original]* be excluded from any age determination. No method is immune from giving *grossly [italics in original]* incorrect datings when there are non-apparent problems with the samples originating in the field. The results illustrated [in this paper] show that this situation occurs *frequently [italics in original]*."

Source: Meacham, William, editor. "Radiocarbon Measurement and the Age of the Turin Shroud: Possibilities and Uncertainties" in *Turin Shroud: Image of Christ? Symposium and Exhibition of*

Photographs, March 3-9, 1986, Proceedings. (Hong Kong: Turin Shroud Photographic Exhibition Organizing Committee), pp. 41, 42, 43 and 53. www.shroud.com/meacham.htm.

Comments: Recall the statement by Fr. Fossati in 1955 in which he mentioned the many vicissitudes the Shroud had undergone in its history, which certainly could have introduced "significant indeterminant errors" and "non-apparent problems" into the equation. The C-14 scientists who dated the Shroud were not field archaeologists and had no business ascribing a 95% confidence rate to the C-14 dating results. They arrogantly felt no need for the archaeologists, historians, conservationists and cellulose chemists of which Meacham spoke. In the scientific literature, there are many other similar statements to the one made by Johnson *et al.* and by Meacham, yet it seems that most who believe the Shroud is a forgery, have no problems accepting that 95% percent rate. One exception was the Jesuit priest, Fr. Robert Wild, who in an article in *Biblical Archaeology Review [vol. 10, no.2 (1984), pg. 38]*, said, […] "test results showing a late date would be attributed to contamination, a not unreasonable suggestion in light of everything the Shroud has been through."

Meacham, in a paper delivered to a conference in Turin, supplied some concrete numbers regarding the questionable reliability of C-14 dates: "Rogue dates are common in archaeology and geology […]. Such has been my experience as an archaeologist who has excavated, submitted and interpreted more than one hundred carbon 14 samples from Neolithic, Bronze Age and Early Historical sites. Of these dates obtained, 78 were considered credible, 26 were rejected as unreliable and 11 were problematic […]." **[Source**: "Thoughts on the Shroud ^{14}C Debate" in Scannerini, Silvano and Piero Savarino (eds), The Turin Shroud, Past Present and Future: International Scientific Symposium, Torino, 2-5 March 2000, *(Torino, Effata Editrice, 2000), pp. 443-444.]*

To give an example of a possible unique source of contamination for the Shroud, Meacham wrote (pg. 50 of his Hong Kong paper), "Whereas all radiocarbon laboratories advise against placing a paper label in contact with the sample for the few weeks in transit from field to lab, the Shroud has had a backing cloth for 450 years!"

It's worth repeating that Meacham observed that a C-14 dating for the Shroud later than the first century would not prove that the Shroud was not from that time, which is reasonable given the Shroud's chaotic history and all of the lack of rigor in the testing that has been documented in one of my own articles cited in the preface. Meacham also commented in the *Shroud Spectrum International* version of his paper (pg. 17), "It is a very serious error indeed to proceed with C14 dating on the assumption that it is an infallible method." Meacham, who had been invited to the 1986 Turin workshop, stated this in a submission sent to Cardinal Ballestrero. It is clear that the "date first and ask questions later" approach was a total disaster.

1986 (March). Tensions began to mount among STURP, ASSIST, and the BSTS. STURP, based on Meacham's report, Gove's statements at Trondheim and Gonella's willingness to proceed with the C-14 test in just a few months possibly pressured STURP to rush. Dinegar suggested to STURP they try to procure the needed samples for the C-14 testing before C-14 labs did. Shortly after, Gove sent a letter to Dinegar saying that Ballestrero and Chagas had approved a workshop in Turin, which would formulate the process for excising samples for the C-14 dating. Thus, Gove wrote to Dinegar that "STURP's good offices are no longer needed." Chagas had submitted to Canuto and Victor Weisskopf all of the C-14 testing proposals and recommended Gove's Trondheim report from 1985 as the basis for the workshop that was (originally) scheduled for June 1986.

It was understandable why STURP was concerned about the close relationship between Chagas and Gove. Rinaldi wrote a letter to Jackson with a plea to delay the operation, hopefully until the summer or fall of 1987. Rinaldi and Otterbein felt that the delay would buy enough time to review all the submitted proposals and coordinate the collaboration that all the parties had agreed to. ASSIST was not prepared as a group to prepare to meet Jackson's schedule. Wilson also objected on behalf of the BSTS to the new dates. Wilson was persuaded by Meacham's arguments about possible contamination that had been raised at the Hong Kong symposium.

Rinaldi wrote a sindonologist that he and Gonella had received a communication from Ballestrero that the scientists who had submitting dating proposals could proceed after approval by a review board headed by Gonella, who felt the testing could begin in October. Rinaldi believed that based on Ballestrero's communication, that the Vatican had trusted the dating enterprise to Ballestrero. Otterbein was not in favor of Turin being in charge of the dating, since Gonella was closely associated with STURP, which undoubtedly would rekindle criticism about a lack of international cooperation. Otterbein wrote to Rinaldi, "Hence, I do not feel that I can support or get involved in the testing program this year. I fear that the non-acceptance of the results will do more harm than good, and I consider the participants of STURP in the program this year involves the serious danger of physical harm to the participants." Meanwhile, Rinaldi wrote to Ballestrero to warn him that starting the C-14 dating process in the next few months would be dangerous. Rinaldi asked Ballestrero to allow more time to facilitate getting international cooperation.

Source: Bracaglia, Giorgio. <u>Uncovering the Paradox within the Archives of the Holy Shroud Guild</u>, (Honeoye, N.Y.: Holy Shroud Guild), 2019, pp. 206-207, 222-224.

Comments: Bracaglia had written in his book (pg. 207) that Ballestrero had been told he was not to make any independent decisions. In this entry, it now appears that Turin had, in theory, regained the power. But as the reader knows by now, very few decisions were unchanged in the process from beginning to end.

Dinegar had previously talked of fear of terrorists trying to disrupt the planning meeting. Otterbein here talked of the fear of physical harm to participants. Later, we will see Rinaldi speak of "guerilla tactics." The dating of the Shroud was not your typical ho-hum scientific experiment!

I knew both Otterbein and Rinaldi. I had assumed that as President and Vice-President of the Guild respectively, and as fellow priests, their views would be similar. But as we have seen, they did not agree on everything. In fact, I was told only recently by someone in the know that the two didn't even get along that well, which was rather shocking to me. Why do I mention this? Because it's only fair to acknowledge that even people I know that were working in the best interests of the Shroud had conflicts, not just the Turin authorities or the C-14 labs' representatives.

1986 (March). On the 21st, Fr. Enrico di Rovasenda, who had the title of "director" at the Pontifical Academy of Sciences, wrote to Cardinal Ballestrero that he had sent to Chagas the final agreement for the meeting. There were subsequent communications and Ballestrero realized that Chagas was more aligned with Gove than anyone in Turin. On the 25th, Fr. di Rovasenda again wrote Ballestrero that a journalist in New York had telephoned di Rovasenda to ask for details of the meeting organized by the Pontifical Academy of Sciences.

Source: Gonella, Luigi and Riggi di Numana, Giovanni. Sindone: il mistero continua, (Milan: Fondazione 3M), 2005, pp. 36-37.

Comments: Why was Chagas more aligned with Gove than anyone in Turin? Was it simply a philosophical and/or scientific agreement -- or was there something more going on?

1986 (March and April). Petrosillo and Marinelli wrote extensively about the planned workshop meeting. "There was a definite conflict between two major groups: 1) Gove and the Pontifical Academy of Sciences (ostensibly Chagas) and 2) Gonella and STURP. Gove believe there had been absurd complications and subsequently wrote a twelve-page letter to Chagas, asserting that the six labs didn't have 'an irresistible desire to date the Shroud,' as he perceived Ballestrero to believe. But Gove also believed that if labs were chosen to date the Shroud, they should have the right to dictate how the test was done. Gove trusted that Chagas could smooth the way for that to happen."

"Around this time, Gonella wrote a confidential letter to Chagas. In the letter, Gonella accused Gove of trying to get funds from the National Science Foundation and thus presenting himself as leader of the six labs, with the specific intent to increase the prestige of his Rochester lab. Gonella believed that the amount of material being requested was too large. He also felt that the Trondheim protocol insufficient and that the

C-14 test needed to be done in conjunction with the STURP-proposed multidisciplinary test package. Gove was upset and to make things worse, Dinegar wrote him a letter saying that the Vatican Secretary of State and the Pontifical Academy had accepted the STURP proposal for dating the Shroud and they would be responsible for excising the samples."

"In April, invitations were sent to the various participants with the scheduled workshop set for June 9th."

Source: Petrosillo, Orazio and Emanuela Marinelli. La <u>Sindone: Storia Di Un Enigma</u>, (Milan: Rizzoli, 3rd edition, 1998), pg. 136.

Comments: To follow up on this thread, go to the Petrosillo/Marinelli entry for "1986 *(May)*."

1986 *(April)*. Chagas sent letters of invitation to various groups to attend the planning workshop to be held on June 9-11 in Cardinal Ballestrero's Palace. Chagas had previously met with a British reporter, Peter Jennings, and informed him of the meeting. Jennings later published an article that publicized the meeting.

Source: Gove, Harry. <u>Relic, Icon or Hoax?: Carbon Dating the Turin Shroud</u> (Bristol and Philadelphia: Institute of Physics Publishing), 1996, pg. 322.

1986 *(April)*. According to Kersten and Gruber, "In spring 1986 *[on April 1st per another source]* Canuto committed the indiscretion of passing Gove a confidential letter from Prof. Luigi Gonella, the scientific adviser to Turin's Cardinal Ballestrero. Gove passed it to his friend Sox, who actually published it. In this letter Gonella accused Gove of trying to secure research funds for his institute in Rochester from the National Science Foundation, the largest science funding trust in the USA, by posing as the director of the six laboratories. In the midst of these intrigues Dinegar announced to Gove that the Vatican Secretary and the Pontifical Academy of Sciences had accepted the STURP project! To complete the confusion, on 13 April an article by the journalist Peter

Jennings appeared in Britain claiming that Chagas had personally spoken out against a radiocarbon test. After the diplomatic tug-of-war between the parties, probably themselves not knowing by now who was on who's [sic] side, the conference of carbon experts -- the conclave of the carbonists -- took place in late September 1986. Representatives from the six radiocarbon laboratories and other scientists, including some from STURP, took part. But no one was invited from the Turin Centro Internazionale di Sindonologia, the largest association of sindonologists."

"While Gonella continued to insist that the taking of specimens should be integrated with the whole STURP programme, Gove was adamant. The carbon test could not be postponed any longer, and any further undertaking would just delay things. The representatives of the institutions pointed out that the test should be a so-called blind test. In a blind test the researchers do not know which piece is the actual test object and which samples are control specimens. In this way conscious or unconscious manipulation, or prejudicial treatment of the different specimens, was to be prevented. Gonella made it clear that it was not Turin but the laboratories which were insisting on a blind test, 'so they would feel protected from the press'."

"After three days the participants came to the agreement that seven institutions should be involved in the dating, with five using the AMS technique and two using the other methods. The British Museum was to supply a control sample. Original cloth and control samples were to be handed over to the individual laboratories after being unthreaded so as to be unrecognizable. Three institutions were to assume the supervision: the Pontifical Academy of Sciences, the British Museum and the Archiepiscopal Ordinariat of Turin. At a certain date after the examination all the laboratories were to pass on the data from their experimental results to three institutions for statistical analysis: the Pontifical Academy of Sciences, the British Museum and the Istituto di Metrologia G. Colonetti (IMGC) in Turin. In this way stringent scientific criteria were to be ensured, guaranteeing the credibility of the enterprise."

"These proposals appear to have been well thought out, although the additional STURP projects were still not out of the way. But the C-14 lobby had won priority for themselves."

Source: Kersten, Holger and Elmar R. Gruber. Jesus Conspiracy: The Turin Shroud And The Truth About The Resurrection (Rockport, MA: Element), 1994, pp. 42-43.

Comments: Perhaps the best summary of Kersten and Gruber's description of these Shroud activities is their phrase that "the behaviour of the researchers themselves had so far been marked by pride and conceit." All too often, it appears that Shroud research has been set back by the inability of some of the participants to put aside petty concerns and to focus on the scientific method.

1986 *(April)*. On April 1st, Canuto had just seen a letter that Gonella sent to Chagas, although Chagas would not actually let Canuto read it. *[This was the letter of February 18th reproduced above.]* Chagas said that the letter was very unprofessional and extremely critical of Gove. Gonella felt that Gove was attempting to get funding from the National Science Foundation and since there was such fierce competition among the "NASF" *[sic]* labs, he was trying to get involved in something that would make him stand out. According to Gonella, nuclear physics was a dying field, and C-14 dating was a significant factor in the eyes of U.S. Congressmen, who would be voting for appropriations. In Gonella's eyes, Gove was acting as leader of the six labs to bring prestige to the University of Rochester. Gonella had more to say: the other labs were annoyed by Gove. The sizes for the samples being asked for were too large, and no distinction had been made between the AMS labs and the proportional-counter labs. The Trondheim protocol was not well thought out. Gonella also made the point that the C-14 tests should be done in the context of the other proposed STURP tests. Gove was incensed to the point that he almost withdrew but on April 3rd, he received from Dinegar a STURP letter addressed "Dear Colleague." Dinegar reported that STURP would be obtaining samples. Gonella had notified STURP that their proposal for C-14 had been accepted by the Vatican and the Pontifical Academy of Sciences. STURP would be allowed to extract the samples. Dinegar added a note to Gove's letter: "A 'slip' now could have disastrous consequences." On April 13th an article in *Our Sunday Visitor* by Peter Jennings reported "Vatican May Allow Carbon Test on

Shroud." Jennings mentioned in the article that he called Chagas, who said "I would as soon the test did not occur."

Source: Sox, David. The Shroud Unmasked, (Basingstoke, Hampshire: The Lamp Press), 1988, pp. 102-104.

Comments: According to the Gonella/Riggi book (pg. 45), Chagas told Gonella he wasn't against the testing, but only that he did not need dating for his faith, which didn't match with his quote to Jennings, "I would as soon the test did not occur."

1986 (April). On April 2nd, Gove called Canuto to say he was disturbed that Gonella would write a letter to Chagas accusing him of unprofessional behavior. Gove wrote, "My impulse was to say to hell with the whole thing except that some pretty high-class people in high-class institutions were now involved. A fevered imagination might conclude that the people connected with some or all of the institutions were motivated by lust, venality, cupidity, self-aggrandizement or other unprofessional concerns, but few rational people would take such a charge very seriously."

Source: Gove, Harry. Relic, Icon or Hoax?: Carbon Dating the Turin Shroud, (Bristol and Philadelphia: Institute of Physics Publishing), 1996, pg. 106.

Comments: I believe there were, and currently are, more rational people than Gove thought, that took, and still take, that charge very seriously.

1986 (April). Harbottle informed Gonella on April 2nd that Gove had air-expressed a letter to Chagas that threatened Rochester's withdrawal if Gonella had any influence pertaining to the planning meeting. On April 5th Chagas phoned Gonella and said that there were problems holding the meeting at the Turin Polytechnic. Gonella replied that the meeting would be held in the Cardinal's residence. Chagas then told Gonella that he would be nominated as vice-president of the meeting and wanted

Gonella's recommendations for those who should attend the meeting, what should be put on the agenda, and mentioned that the samples might be supplied at the meeting. A few hours later, STURP's Tom D'Muhala phoned Gonella to say that Gove wanted the samples to be taken at the conclusion of the meeting. When Ballestrero was informed of all this, he dictated a telex to Gonella that confirmed the date and location of the meeting, specified that STURP would be involved, gave the list of persons that were to be invited, listed the draft agenda (which included the number of samples to be taken), and stated that any other decisions would be made after the meeting. On the 9th, Chagas called Gonella to ask if the samples would be taken during the meeting, and Gonella replied they would not. Jenning's article in *Our Sunday Visitor* appeared on the 13th, based on interviews with Chagas and Dr. Burleigh from the British Museum, revealed the British Museum's specific involvement, supplied a list of the attendees, and mentioned some of the proposed procedures.

Source: Gonella, Luigi and Riggi di Numana, Giovanni. Sindone: il mistero continua, (Milan: Fondazione 3M), 2005, pg. 38.

1986 (*April*). On the 15th, Chagas sent invitations for the planning meeting to be held in Turin on June 9-11. The invitation was extended on behalf of Chagas alone but specified that the meeting would be held under the "high patronage" of Cardinal Ballestrero. Chagas said that he would make decisions on the procedures to be submitted to the Cardinal, but did not list the specific meeting agenda. Chagas did not inform Ballestrero of the invitation. Gonella only learned of it in May from a phone call from one of the guests who sent a photocopy of the text. STURP was also not invited. Gonella wrote that Ballestrero couldn't accept the proposal as such, and expressed his complaints by phone to Secretary of State Casaroli.

Source: Gonella, Luigi and Riggi di Numana, Giovanni. Sindone: il mistero continua (Milan: Fondazione 3M), 2005, pg. 39.

1986 *(May)*. According to Petrosillo and Marinelli, "The workshop meeting set by Chagas for June 9th was temporarily canceled. Gonella disclosed that Chagas hadn't even invited the Cardinal to attend. *[!]* Ballestrero aired his grievances to the Vatican and the Secretary of State proceeded to postpone the meeting. The question then became who would be directing the project. Ballestrero requested Gonella to go to the United States to meet with Gove. Chagas then proposed the workshop be held in Rome. Gove was then informed that the meeting had been postponed because of conflicts between the labs and the fact that STURP had not been invited. Gove also learned that Fr. Borga and others in Turin appealed to the Pope via the deputy Secretary of State to block preparations for the C-14 test. On the 27th, Gove received a telex that the Holy See and Ballestrero would not be blocking the test."

"Gonella met with Chagas and Canuto in Rome, suggesting that they announce in the upcoming new invitation to the workshop meeting that only two of the six labs would be chosen to do the Shroud C-14 test. Canuto convinced Chagas that the invitation couldn't be formulated in a way that the labs could find offensive. Chagas then gets Gove to prepare the draft agenda. Chagas uses the term 'dating labs' instead of specifying 'six.' In the letter being prepared to be sent out to the participants, Gove is designated as 'secretary' for the meeting. Gonella didn't like the proposed agenda: Turin hadn't even been consulted, he himself was not even mentioned and disputed the statement that the Pontifical Academy 'thank Professor Gove for organizing the meeting.' Sox, meanwhile, commented in his book that the labs wanted autonomy, STURP saw the C-14 test as a continuation of their 1978 testing and the Pontifical Academy believed its prestige would override all other considerations. Chagas had wanted the meeting in Rome, which wasn't going to happen. But he didn't want it to be held at the Turin Polytechnic where Gonella taught."

Source: Petrosillo, Orazio and Emanuela Marinelli. La Sindone: Storia Di Un Enigma. (Milan: Rizzoli, 3rd edition, 1998), pp. 136-137.

Comments: I can't help but wonder if Gonella gave Canuto an earful for having revealed to Gove the contents of his (Gonella's) confidential letter to Chagas about Gove back in February.

It's amazing that Ballestrero, the Pontifical Custodian, had to air grievances to the Vatican. With all the infighting between groups and individuals, what could possibly go wrong?

To continue with this thread, go to the Petrosillo/Marinelli book "La Sindone ..." entry for "1986 *(September/October)*."

1986 *(May)*. Cardinal Casaroli informed Cardinal Ballestrero on May 13th that the meeting was postponed due a decision from up the ladder. Three days later, Chagas telegraphed all the invitees that the meeting had been postponed due to organizational problems. On the 19th, Gove sent to Casaroli and Chagas a telegram of protest with the signatures (in addition to his) of Prof. Donahue (Arizona), Dr. Harbottle (Brookhaven), Prof. Hall (Oxford) and Sir David Wilson (Director of the British Museum). The telegram attributed the postponement to the Vatican/Archdiocese of Turin and asserted that the Pontifical Academy of Sciences was damaged, and it demanded the revocation of the postponement and warned of "extremely deleterious consequences" if it weren't revoked. Chagas then telegraphed to the guests on May 27th that the postponement was only temporary.

The day before Chagas had informed the guests on the 19th that the meeting had been postponed, Ballestrero had sent a memo to Casaroli and suggested that the former, along with the Pontifical Academy, should reorganize it. The memo touched on the guests who would be invited, the British Museum's involvement, and also the number of labs that would be involved in the dating. Gonella noted that Sir David Wilson was not listed as one of the guests. It turns out that his name had been added to the Gove telegram without his knowledge to give more weight to it!

Source: Gonella, Luigi and Riggi di Numana, Giovanni. Sindone: il mistero continua, (Milan: Fondazione 3M), 2005, pg. 39.

Comments: The fact that Gove added Sir David Wilson's name to the telegram without his knowledge was simply dishonest and deplorable.

1986 (May). On May 16th, Gove received a call that the workshop planned for June 9-11 was postponed per the Pope's orders. The reasons given were the Chagas comment in the Jennings' article, conflicts of interest, and STURP's criticism that Chagas had not invited all the individuals that they had recommended, which included the late Steven Lukasik of STURP, Meacham and Giovanni Riggi di Numana (normally just referred to by most as "Riggi"), STURP's only Italian member. Gove later learned that the late Fr. Piero Coero-Borga of the International Center of Sindonology in Turin and another unnamed individual had put heavy pressure upon Cardinal Martinus Somolo, Undersecretary of State, to stop the preparations for the C-14 test. On May 27th, Gove was sent a telex that read "Holy See and Cardinal Ballestrero have no intention not to date Shroud. Adjournment of meeting should only be considered temporarily postponed. Contact you in future."

Source: Sox, David. The Shroud Unmasked, (Basingstoke, Hampshire: The Lamp Press), 1988, pp. 104-105.

1986 (May). Regarding the postponement of the workshop in June, Gove wrote, "I decided that, rather than supinely letting this pass, I would try everything I possibly could to discover why the meeting had been postponed and to get the order for postponement rescinded. Despite all my efforts, I failed and the meeting was indeed postponed."

Source: Gove, Harry. Relic, Icon or Hoax?: Carbon Dating the Turin Shroud, (Bristol and Philadelphia: Institute of Physics Publishing), 1996, pg. 119.

Comments: Gove exhibited here the tenacity that his colleague Bromley alluded to in his foreword to Gove's book, but one can question whether that situation was an appropriate place for it. Did he really think he was in a position to get the postponement order rescinded??

1986 (May). On May 17th, Gove called Gonella in Turin to find out more information regarding why the workshop was postponed.

Gonella sounded flustered. He said the initial action had been taken by the Vatican. However, he admitted that Turin also had input, but did get specific. Gonella believed the first reason was the Peter Jennings article in *Our Sunday Visitor*, which listed the time, place and purpose of the workshop based on information from Chagas. Gonella said this had upset Cardinal Ballestrero. The second reason was that Chagas had not invited all the people that the Cardinal had suggested. (Gove added that Gonella had actually made the suggestions.) Thirdly, it was the Cardinal's meeting, not a meeting of the Pontifical Academy, but Chagas apparently had refused to concede this. Gonella said that "Chagas was trying to play too dominant a role."

Gove told Gonella that the postponement was a serious matter. There were a number of prestigious scientific institutions involved that would probably perceive that the Catholic Church was afraid of getting the Shroud dated. Gove urged Gonella to have the Cardinal reverse the decision. Gonella said he would pass on Gove's comments to the Cardinal. Gonella continued to place all the blame on Chagas. Gonella insisted that Gove should get information on all this from Chagas, but Gove informed him that he had talked to Chagas shortly before and that he didn't know any details. Gonella told Gove the postponement would not prevent the Shroud from being dated.

On May 18th, Gove discussed the matter with Canuto and decided that as many people as possible that were to have been involved in the workshop should sign a cable that would be sent to Cardinal Ballestrero, Cardinal Casaroli at the Vatican, and Chagas. Canuto advised that Gove should first consult Chagas. When Gove contacted Chagas, he agreed that this cable should be sent.

On May 19th, Gove sent the cable to Ballestrero, Casaroli and Chagas. It was signed by Gove, Donahue from Arizona, Harbottle from Brookhaven, Sir David Wilson from the British Museum and Edward Hall from Oxford. Gove called Canuto, who had read the contents of the letter to Ambassador Celli, who said the wording was too strong for him to pass along to Cardinal Casaroli. But Celli believed the cable was the best way to proceed.

On May 20th, Gove heard from Canuto that Fr. Rovasenda, who in the past had been the director of the Pontifical Academy of Sciences, said that it was actually Pope John Paul II who had given the order for the postponement. Gove then got the idea of trying to get the assistance of various ambassadors to the Vatican.

On May 22nd, Gove learned from Canuto that he had called Cardinal Casaroli, who said he didn't even know about the postponement. *[!]* He said that another branch of the Vatican State Department was handling matters. Gove believed it was Cardinal Somolo, the Undersecretary of State. The next day, Canuto said that Fr. di Rovasenda had reported to Cardinal Ballestrero that a group from Turin involved with the Shroud didn't want the C-14 dating to take place and were putting some pressure on the Pope. Ballestrero said he was not opposed to the dating and plans were made for May 26th for a meeting between Chagas and Cardinal Somolo. Canuto believed that individuals from the International Centre of Sindonology in Turin had Cardinal Somolo inform the Pope of their displeasure about the test. Canuto asked Gove if he had informed Dinegar of the postponement, and Gove replied that he had not. Canuto said he would call Dinegar to see what he knew about it. Canuto called Gove later in the day and said he was unable to contact Dinegar. Canuto expressed the thought that only he and Gove seemed concerned about the postponement. Gove wrote, "Chagas, on the other hand, was probably eating good food in Rome apparently unconcerned."

On May 28th, Gove phoned Chagas in the Vatican. Chagas said if the workshop was held in the Vatican, holding the workshop as planned wouldn't be a problem. Gove then asked why it wasn't held in the Vatican; Chagas told him that Cardinal Ballestrero insisted that it be held in Turin. Gove asked if the International Sindonology Center in Turin was connected with the archdiocese. Chagas replied that the Center was fighting the archdiocese. Gove asked why Cardinal Ballestrero just didn't ask them to stop. Chagas said "that it was very difficult to understand some aspects of Italian **politics** *[my bolding].*"

Source: Gove, Harry. Relic, Icon or Hoax?: Carbon Dating the Turin Shroud (Bristol and Philadelphia: Institute of Physics Publishing), 1996, pp. 120-127.

Comments: This is further evidence of politics, especially the Italian aspects, rearing its ugly head. Keep in mind how much material is available just from what is publicly known. There is undoubtedly much more that has never seen the light of day.

Note Gove's dig at Chagas, whom he perceived to be "probably eating good food in Rome apparently unconcerned" about the Workshop meeting postponement.

Piecing together information from some previous entries and also some information mentioned by Dinegar in a conference presentation (see: https://www.youtube.com/watch?v=NGRrt3d9Rao&feature=youtu.be&list=UUSv7BD9sKjIcA24ct1Hz_Aw between 7:40 and 11:45), some rather shocking behavior of Gove can be revealed, summarized below in bullet form:

* Late 1985: Gove told Dinegar that he was trying to get NSF funding for STURP's representative to attend the Turin workshop meeting.
* Gove admitted in his 1996 book that he had no intention of getting funding for STURP and openly said he was trying to get rid of STURP "once and for all."
* Late 1985: Dinegar informed Gove that he had to give a paper in China shortly before the scheduled meeting in Turin in June but should be able to make it. Dinegar believed that Gove was serious about the offer regarding funding (and was not "tongue-in-cheek"). Dinegar told Gove what he had said about the funding was "strange." Dinegar also complained to Gove that STURP was not specifically mentioned in the proposal.
* Late May 1986: Canuto asked Gove if he informed Dinegar about the postponement and Gove told him "no." Canuto tried to contact Dinegar but was not successful.
* Late May 1986: Dinegar was leaving for China, from where he would fly direct to Turin for the workshop. By chance, before he left, he talked to another participant from the Turin workshop who informed him that the meeting was postponed.
* So, Gove was prepared to let Dinegar fly from China to Turin, where Dinegar would have discovered the meeting was not taking place at that time. Given that Gove had already mislead Dinegar about the funding and was constantly admitting he was trying to

eliminate STURP, I don't think it's unreasonable to conclude that Gove's failure to inform Dinegar was likely a willful act.

1986 *(June)*. Chagas wrote a letter to Dinegar regarding the rescheduling of the workshop for September 29-October 1, 1986. Chagas indicated that Ballestrero wanted to finalize the number of samples to be taken, the sizes, and how they would be distributed to the labs. There was to be a full discussions on methodology, on possible contamination, whether charred sections of the Shroud could be used as dating samples, and how the publication of the results were to be made. Chagas asked Dinegar to keep the information confidential "as this is required for the complete success of this difficult enterprise." Twenty-two experts were to be invited.

Source: Bracaglia, Giorgio. <u>Uncovering the Paradox within the Archives of the Holy Shroud Guild</u>, (Honeoye, N.Y.: Holy Shroud Guild), 2019, pp. 215-216.

Comments: No representative from the Turin Centro was invited, which was a significant oversight. I remember talking to Dinegar on the phone before the planned workshop. When I inquired about it, he would not divulge any information to me; he was literally concerned about terrorists attempting to interfere! Fr. Otterbein had also expressed fear of some participants being physically harmed. The Shroud is definitely a volatile object.

1986 *(June)*. Ballestrero instructed Gonella on June 24th to meet with Chagas to try to straighten things out. When Gonella met with Chagas at the latter's office, Canuto was there and launched into a harsh criticism of Ballestrero, saying about the latter, "who does he think he is?" Chagas refused to reduce the number of guests. Casaroli and Ballestrero did not like the inclusion of some people to the meeting they felt were unnecessary, such as Canuto and Gove's secretary, and also did not like the fact that each lab would have two representatives there instead of just one. Chagas refused to exclude all of them, saying that they would take it personally. However, because Ballestrero was reluctant to continue to delay the meeting out of fear that it could appear that the Church feared the results of the test, he forged ahead.

Source: Gonella, Luigi and Riggi di Numana, Giovanni. Sindone: il mistero continua, (Milan: Fondazione 3M), 2005, pp. 44-45.

Comments: In answer to Canuto's question about Ballestrero (after which Gonella inserted a "!"), it's fair to point out that the Pope did appoint the Cardinal as Custodian of the Shroud.

I know a scientist, Ed Prior (co-author of a joint article with me), who, like Canuto, worked for NASA. He emailed me on July 16, 2020, "I did communicate with Canuto directly; since we were both published scientists, I hoped he would share with me the background story of the Shroud carbon dating episode. I particularly did not understand how, after so many meetings with scientists and Shroud authorities; the shocking decision was made to select only one sampled region instead of several from various areas of the Shroud. And the one sampled region was arguably the worst that could have been selected: the edge of the cloth, where it is well known that church officials held the Shroud with dirty hands (unlike today, there were few bathrooms, if any, down the church hall in 1400 AD) at exhibitions over the centuries. Also, materials engineers can tell you that the Shroud edge is where maximum stresses pull on the fabric as it is held up---causing the greatest risk of tears and damage. There is evidence uncovered [...] showing that is exactly what happened in the past, and that the Middle Ages' art of 'Invisible Reweaving' (performed by skilled artisans to repair a King or Queen's favorite bed cover or cape, etc.) was used to repair the Shroud's edge. This renders an edge sample from the Shroud useless for carbon dating to determine if the original cloth actually is from the 1st Century."

"Unfortunately, I did not learn anything useful from Canuto. I believe if the STURP people had been allowed to perform the Shroud sample selections as originally planned, they would have chosen less contaminated and unrepaired, uncontested original Shroud fabric (based on their original STURP analysis of the cloth) -- and today there would be no controversy about the Shroud's age. I personally believe mistakes were made by the majority of those involved in the planning of the carbon dating process. Although the 3 universities may not have participated in the final Shroud sample selection, they should have known it was unwise to select only one Shroud region -- and particularly the edge! Before I semi-retired from 'ShroudWorld,' I emailed the universities that a dating redo was called for; only the then-head of Oxford's C-14 lab replied and agreed a final carbon dating with careful sample selections would settle the controversy. A tenth of a gram from an uncontested region would do it, but nothing happened. I suspect the problem may not be with academics (although notoriously reluctant to admit any mistakes) as much as it is with the Vatican, which so far has refused to authorize another C-14 test with far better technology and accuracy than the 30-year old '1260-1390!' Or, did they?"

"Today there are not only more accurate technologies that require only a tiny fraction of a gram of cloth for a carbon dating; there are a number of companies (not just universities) that perform the dating service for a price professionally, quietly, and ask no questions. The Vatican has already used such companies to carbon date ancient bones in a European church that some believe may be John the Baptist's. The bones dated to the 1st Century. It would not be difficult for the Vatican to quietly send the same company several selected Shroud milligram sized fragments (far from the edge!), along with a group of other tiny samples from Egyptian or Jewish tombs, some modern cloths, etc. -- all too small for the company to determine whether any have the Shroud's distinctive weave. If I were Pope, I'd order that immediately, but confidentially. Maybe he already has....."

1986 *(June)*. Discussions began on the wording of the letter that would be sent to reschedule the postponed workshop. Gonella met with Chagas and Canuto in Rome and suggested saying that only two of the six labs would be selected to do the dating, using burnt samples. Lukasik of STURP would attend, as well as another friend of Gonella, Jacques Evin of the University of Lyons. Another Frenchman, Jean-Claude Duplessy of the lab at Gif-sur-Yvette, was also to be invited.

After Gonella left the meeting, Canuto convinced Chagas not to utilize Gonella's wording regarding the six labs, believing they would find it offensive. Canuto then asked Gove to ready the draft agenda, and Chagas insisted on using the words "the dating laboratories" instead of mentioning the six. Gove was designated as workshop secretary. Gonella reacted with various points: 1) Turin had not been consulted over the agenda. 2) Gonella was not mentioned as the moderator. 3) The words 'the Pontifical Academy thanked Professor Gove for organizing the meeting' should be deleted.

Confrontations were brewing. The C-14 labs insisted on decision-making autonomy; STURP felt that the C-14 test was a continuation of their investigation of 1978 and the Pontifical Academy of Sciences thought its prestige would trump all other factors.

Source: Sox, David. The Shroud Unmasked, (Basingstoke, Hampshire: The Lamp Press), 1988, pp. 104-105.

Comments: According to Laverdiere, in her dissertation (pg. 70), Duplessy had been invited because of "his experience in dealing with cloth." This is another source of expertise that Turin ignored.

1986 *(June).* On the 25th, Chagas wrote to Ballestrero, "Most Reverend Eminence, Thank you first of all for the amiable telephone conversation this morning, which has given me the possibility regarding the dating of the Holy Shroud. While I confirm that my complete adherence to His intention -- and also of the Holy See -- to date the Sacred Sheet samples, I am pleased to give Your Eminence assurance of my personal commitment to the preliminary meeting, in which all norms to be observed for the Sacred Cloth, of which you are Venerated Custodian, will be specified. I certainly cannot commit six laboratories, which I have already invited, without guaranteeing to all six that the dating operations will be carried out in the ways agreed to you. Hereby I would like to introduce to you Prof. Vittorio Canuto of the NASA Institute of New York, who in addition to being Scientific Consultant to the Representative of the Holy See to the UN, and also appreciated as a councilor of mine and of the Pontifical Academy of Sciences. Prof. Canuto scrupulously and with great competence collaborated in the definition of the choice of laboratories and in the development of the procedures to be followed. He carries with me in my name the letter (in English and Italian translation) that I submit to your attention for the invitation which he sent promptly to the interested laboratories, as well as to three other scientists, whose clarification Prof. Gonella left me, who yesterday -- in your name -- gave me your revered message. We had a conversation with him which I hope has clarified all the points that were under discussion."

"I also have the obligation, Most Reverend Eminence, to underline the important circumstance that the six qualified laboratories, already invited, represent the only six institutes that give complete guarantee for having already dated a single ancient fabric, using microquantities of samples, with results calibrated and judged exhaustive under the supervision of the British Museum. I therefore look forward to your kind and attentive nod of consent, to be addressed by the Pontifical Academy of Sciences, to authorize the transmission of the invitations for the meeting to be held in Turin from 29 September to 1 October 1986."

[signed]
Carlos Chagas

Source: Luigi Gonella archives, of which the author has a copy.

1986 *(June and July)*. Gonella came to Rome to meet with Chagas and Canuto to discuss a revised workshop. Gonella had suggestions for a letter that would be sent to all the individuals and institutions that were going to be involved previously. The letter would include that only two of the six laboratories would be selected to perform the testing, that scorched cloth would be used and that for data analysis, the Turin Egyptian Museum would be used. He also said there should be additional invitations to C-14 scientist Jacques Evin, a person from the J. P. Getty Museum and STURP chemist Alan Adler. Canuto indicated to Gove that Chagas seemed to approve this but Canuto "convinced him that such a letter was ridiculous." Canuto and Chagas then composed a draft of a new letter, which was then sent to Cardinal Ballestrero and accepted.

On July 2nd, Gove and colleague Brignall received invitations from Chagas to the rescheduled workshop, which was now to be held on September 29-October 1. Chagas asked them to keep it confidential for the time being.

On July 14th, Gove phoned Canuto. Gove recounted, "Canuto began by saying (modestly) that he could only credit himself with the saving the situation. He said a memo had been sent to the Vatican from Turin accusing Chagas of violation of confidentiality because of the Jennings interview and making other baseless charges."

On July 22nd, Gove replied to Chagas' workshop invitation. Gove said that since the main reason for the postponement of the June meeting was Gonella's objection to the relatively insignificant Jennings article, he thought he would mention that Edward Hall had showed him (Gove) some correspondence he had received from the British author Ian Wilson, who at the time was the chairman of the BSTS. From the material that Hall had, it was clear to Gove that Gonella was informing Wilson about the workshop details. Wilson then shared the information with various people.

Wilson had sent a letter to Hall on June 26th and included three confidential memos from Wilson to individuals mostly in Britain but also including Gonella and Tom D'Muhala of STURP. The first memo stated that Gonella was giving STURP the approval for carrying out in July their proposed tests. The second stated that Gonella had phoned from Turin and explained that the planned testing by STURP was then postponed due to complications related to the C-14 test as well as some logistical problems STURP was having. In the third memo, Wilson said no reason had been given why the Turin workshop in June had been postponed. In his letter to Hall, Wilson put the blame on Chagas for the postponement and indicated that he found it difficult to understand why or how long the stalling would be.

Gove said that this was "a serious breach of confidence and was remarkable evidence of double standards on Gonella's part." Gove felt that it was inexcusable for Gonella to be critical of Chagas for his interview with Jennings and then to provide sensitive information to Wilson. (However, Gove acknowledged that Wilson did not make public any information that Gonella gave him.)

Gove told Chagas that a significant amount of effort needed to be devoted to planning the workshop and that an agenda needed to be agreed on by seven institutions; he offered to prepare one. Gove ended the letter by accepting Chagas' invitation to participate in the rescheduled workshop.

On July 25th, Gove and Canuto talked on the phone. Canuto suggested that Gove and Brignall come to New York for a meeting that would include Ambassador Celli. Canuto informed Gove that he thought Gonella would be trying to produce an agenda for the workshop even though he wasn't asked to do so by Chagas. In June at the meeting in the Vatican with Chagas, Canuto and Gonella in attendance, Gonella had been asking technical questions. Chagas had responded that they would be discussed at the workshop. It was believed that this may have caused Gonella to return to Turin to prepare an agenda.

Source: Gove, Harry. Relic, Icon or Hoax?: Carbon Dating the Turin Shroud, (Bristol and Philadelphia: Institute of Physics Publishing), 1996, pp. 131-134.

Comments: Gove did not elaborate regarding his phone call on the 14th on the "baseless charges" that Canuto asserts that Turin made to the Vatican about Chagas.

1986 *(July)*. Following Ballestrero's lead to proceed with the rescheduling of the planning meeting, invitations were sent from Rome on July 2nd to hold the meeting from September 29-October 1. Gonella commented that two years had passed since Cardinal Ballestrero had proposed a meeting to discuss a multidisciplinary research proposal, and now they were only discussing radiocarbon dating, and in a very tense climate, with unclear alternative proposals.

Source: Gonella, Luigi and Riggi di Numana, Giovanni. Sindone: il mistero continua, (Milan: Fondazione 3M), 2005, pg. 45.

1986 *(August)*. On August 4th, Gove received a letter from Chagas, who thanked him for his letter and call of July 22nd. Chagas said he expected some difficulties with the agenda but had no doubts they would succeed. Gove was informed by Chagas that he knew he had to be careful around some people associated with Cardinal Ballestrero, but he felt confident he would prevail because he believed he had clear support from the Pope himself.

Gove and Brignall flew to New York on August 6th and met with Canuto and Ambassador Celli to discuss the workshop. Gove had intended to submit the agenda in a letter to Chagas but Celli suggested that Gove recommend to Chagas also showing both the Vatican Secretary of State and the Pope, so Gove did so when the letter was sent on August 8th. Gove informed Chagas in the letter that it had been discussed with Celli and Canuto. If Chagas approved of the agenda, Gove said he wanted to send copies to the heads of the six laboratories, the British Museum and the textile expert Mechthild Flury-Lemberg. Comments would be solicited before a final document was submitted. Gove also requested that copies be sent to Cardinal Casaroli and Pope John Paul II and anyone else that Chagas deemed appropriate.

On August 26th, Gove called Chagas. Chagas indicated he approved of the agenda but proposed a minor change. Instead of referring to six labs, Chagas suggested just using the phrase "carbon dating laboratories" without specifying the number. Chagas knew that Gonella was opposed to a large number of labs being involved, so wanted to avoid controversy. Gove continued to press to leave "six" in. Chagas related that he was going to send the agenda that very day. He said that instead of sending a copy to Cardinal Casaroli, he would send it rather to Undersecretary of State Cardinal Somolo, who actually knew more about the workshop and had frequent contact with the Pope. Gove asked Chagas if he was going to ask for comments from those it was sent to, and Chagas said he would not be doing so.

The next day, Gove called Chagas again and told him he was sending a letter regarding the changes that had been discussed in their conversation of the previous day. Chagas said he would not send out the agenda until he had received the letter. Chagas was then told by Gove that he had been caught off guard regarding the conversation about "six" labs and had agreed too quickly. Gove again expressed the desire to have "six" in the agenda. He also said that if the Turin authorities wanted to limit the sample size to such a small amount that it would not be enough for six labs and/or if only several of the six were deemed better suited to perform the test, that it should be discussed at the workshop. Gove felt that six labs had participated in the British Museum test, so six would be appropriate for the Shroud dating. Finally, Gove suggested that Chagas "might somewhere indicate on the agenda that I had been involved in its preparation. I confess that I wanted some explicit recognition from Chagas that I had put considerable effort into organizing the workshop."

Source: Gove, Harry. Relic, Icon or Hoax?: Carbon Dating the Turin Shroud, (Bristol and Philadelphia: Institute of Physics Publishing), 1996, pp. 135-137.

Comments: Once again, Gove, by acknowledging he wanted explicit credit for organizing the workshop, blatantly admitted his desire for recognition.

1986 *(September)*. Gove phoned Chagas on September 2nd and asked if he had received his letter of August 27th. Chagas replied in the affirmative and then asked if he could be frank. Chagas said that he didn't know Gove very well and vice versa, but he had the impression that Gove did not trust him, possibly because he was a Latin American or in a different field of science. Gove recounted in his book that Chagas' statement surprised him. Gove was concerned whether Chagas had any influence in the Vatican and also whether he could hold his own with Gonella. Gove felt the trust question could be attributed to not agreeing to the deletion of the word "six" in the proposed agenda. Chagas went on to say that he still wanted to keep it out. He did think that six should participate, but that the question should be settled in Turin, so Gove finally gave in. Chagas indicated that he wanted to be the chairman for the last two sessions, which dealt with the disclosure to the public regarding the results and conclusions. Gove had actually planned to let Gonella chair one of those sessions. Granting Gove's request for explicit recognition, Chagas "said he would add some words like 'the Pontifical Academy of Sciences has asked Professor Gove to organize the workshop and is grateful for his efforts'."

After the phone call, Gove drafted a letter to Chagas to express some of his feelings about the Shroud dating process and Gonella's involvement. Gove said that he "was not used to getting this heavily involved in an activity that fails because of extraneous and irrelevant machinations of individuals with highly questionable motives." Gove said that Chagas' perception that Gove lacked trust in Chagas' leadership because he was a Latin-American or in a different field of science was just simply wrong. Gove went on to say that perhaps the real problem was that Chagas lacked trust in him and the role he was playing in the whole process. If that were the case, Gove felt that perhaps it would be best for the whole enterprise if he would withdraw completely.

Before sending the letter to Chagas, Gove ran it by Canuto on September 3rd. Canuto said that he didn't understand what Chagas was saying but advised Gove not to send the letter offering to withdraw, which would be playing into Gonella's hands. Gove didn't send the letter to Chagas but sent a draft to Canuto, adding that he (Gove) was distressed by Chagas' comments and felt that Chagas' lack of trust in Gove set a very bad tone for the workshop. If that were so, Gove repeated that perhaps it would be best if he withdrew.

On September 5th, Chagas called Gove. Chagas didn't say he had talked to Canuto but it was clear to Gove that he had. Chagas said that he hoped his frank, private comments wouldn't be taken too seriously. Chagas told Gove that he had the highest regard for him a scientist and that he (Gove) "was more key to the success of the shroud enterprise than he was." Chagas confirmed that the agenda would mention the Academy requesting Gove to organize the agenda and that Chagas would chair the two aforementioned sessions. Gove wrote, "I was quite pleased with this phone call." A few days later, Gove received a copy of the agenda.

Canuto called Gove on September 22nd. Canuto related that Gonella had called Monsignor Dardozzi, the vice-Chancellor of the Academy on September 17th, and the discussion lasted one and one-half hours. Gonella, according to Canuto, "had been blown out of the water" by several points in the agenda: 1) that the Academy had thanked Gove for organizing the workshop; 2) that Turin was not consulted on the agenda (Canuto said that Monsignor Celli's involvement negated that, but Gove said that since Celli was from the Vatican, he would be suspect in Turin.); 3) that he (Gonella) was not listed as being a moderator at any of the sessions.

Source: Gove, Harry. Relic, Icon or Hoax?: Carbon Dating the Turin Shroud (Bristol and Philadelphia: Institute of Physics Publishing), 1996, pp. 137-140.

Comments: As we shall see later, the agreed-upon protocol was ignored at the time of the sample-taking on April 21, 1988. Gove's exchange with Chagas about the apparent lack of trust between the two is telling. Chagas brought up the issue, acknowledging that the feeling was probably mutual. Yet, Gove ends up telling Chagas that the real problem may have been only that Chagas didn't trust him. When Gove complained about "extraneous and irrelevant machinations of individuals with highly questionable motives," he was oblivious to the fact that others attributed that very description to him. He clearly saw the world only through Gove-colored glasses. Regarding the organization of the workshop among the Vatican, the Pontifical Academy of Sciences, Turin, and Gove, it seems as if the left hand did not know what the right hand was doing, not a great recipe for an important scientific test.

1986 *(September/October)*. According to Gonella, on the 9th of September, Chagas sent all the guests a circular with an agenda of the meeting prepared by Gove, and his thanks to the latter for having "organized the meeting on behalf of the Pontifical Academy of Sciences"; Chagas also appointed moderators for the various sessions (without having questioned them) and declared that he was responsible for the meeting with Cardinal Ballestrero and the Holy See; the agenda did not mention the fundamental question of the number of samples to be taken. Turin only found out about it on the 18th, and Ballestrero immediately made his complaints to the Secretary of State for this "umpteenth violation" of established procedures and of its own. Chagas informed Ballestrero on the 20th that the agenda in question was only a proposal. Although the British Museum had been mentioned as the likely entity to analyze the data after the testing, Gonella suggested the G. Colonetti Institute in Turin be involved as well. When Gonella and Chagas discussed matters in their June 24th meeting, Chagas opined that a Turin institution would be seen as suspect by many. It was revealed that in May, a French C-14 lab, Gif-sur-Yvette had submitted an independent proposal to date the Shroud, and so senior scientist Jean-Claude Duplessy was invited to attend.

At this point in time, the guest list included Duplessy, Tite (British Museum), di Rovasenda, Dardozzi (the latter two from the Pontifical Academy), Canuto and Flury-Lemberg (both per Chagas), Dinegar, Adler and Lukasik (STURP), Damon and Donahue (Arizona), Hall and Hedges (Oxford), Wolfli (Zurich), Harbottle (Brookhaven), Otlet (Harwell), and Gove and Brignall (Rochester). In addition, Jacques Evin from the University of Lyon was invited as an advisor, as was archaeologist William Meacham from Hong Kong.

On the 24th, Casaroli sent a letter to Ballestrero requesting that he clarify the aims of the research and to remember that the proposals would be submitted to the superior authority, and recommended that the works be serene. Gonella met with Tite on the 28th. Gonella asked Tite about having the G. Colonetti Institute assist the British Museum with data analysis and Tite was happy to do so.

The question of religious affiliation was brought up and the following designations were publicized: Tite, Hall and Hedges (agnostic); Donahue, Evin and Lukasik (Catholic); Damon (Quaker); Dinegar (Anglican); Meacham (Methodist). The meeting, in English, was audio recorded and produced seven-hundred papers of information, which were sent to the 3M Foundation's "Technical-scientific Documentation Center on the Shroud."

Gonella noted that the labs, not Ballestrero, requested that the test be "blind." Gonella also noted that having seven labs perform a dating was unprecedented. The amount of material to be taken was paramount on the agenda. The idea of seven labs, but less than seven samples, was discussed. Gove stated he was in favor of seven labs and seven samples. Hall preferred five labs. Harbottle only specified more than two. Meacham said that he was not aware of many cases that were limited to just two or three labs.

Chagas then gave Gove the floor to discuss the size of the samples. Everyone was aware that Gove had Chagas' support. In short order, the wishes of Turin were ignored. Gonella wrote that he was torn between the duty to minimize the damage to the Shroud (and maintain the multidisciplinary nature of the research) and the directive not to appear biased against certain requests made by the labs. The discussion of the quantity of material was very confusing, because some spoke of grams of carbon before treatment, others after treatment and others of grams of tissue, and it was not clear how many measurements each laboratory intended to make; at a certain point a total of 12.5 cm^2 came out, as a provisional estimate. Although that was twice the amount that Chagas had proposed in a January meeting, Gonella was fine with it. The idea of using carbonized material was discarded *a priori*, despite the favorable indications by some, on the grounds that the mere fact that its validity had been discussed would have made the result suspicious to the public.

Chagas then announced that a consensus had been reached to use seven labs. Gonella protested that no such consensus had been agreed upon. Chagas, visibly annoyed, said that Gonella could lead a discussion the following morning. Gonella stated then that from the metrological *[physics]* point of view, he saw no advantage for the accuracy or reliability of the measurement in the use of more than two or three samples; he asked laboratories to explain what scientific advantages they found there, otherwise they could not justify such a withdrawal professionally. Next, Gonella emphasized that those responsible for the Shroud would react negatively to a request of over 250 mg, if not scientifically justified. Lukasik, on behalf of STURP, declared himself in agreement with Gonella, who presented in writing a proposal to use three samples of fabric and others of charred material, letting STURP, with the permission of the Turin authorities, performing the necessary steps.

The representatives of Brookhaven, Zurich and Oxford said that Gonella was right on the metrological level -- more samples did not improve the measure, but in an enterprise like the dating of the Shroud, it was necessary to take into account the reactions of the public, for which more laboratories would have provided more credibility; the representatives of Harwell and Gif-sur-Yvette supported this last point; Gove did not reply but objected to the idea that STURP would excise the samples. Hall asked what role the Turin archdiocese would have in the certification of the samples. Gove said that the institutions that have the absolute respectability and enjoy the confidence of the public were the British Museum and the Pontifical Academy of Sciences and saw no reason that the archdiocese should be involved. Gonella protested that the responsibility for the chain of evidence and the safety of the Shroud was clearly the responsibility of the archdiocese, and no one could have excluded Turin from the control of the collection and distribution of the samples. Chagas and Damon both were of the opinion that all three entities should be involved. Gonella emphasized that although the archdiocese was responsible for certifying the samples, they would not be involved in analyzing the labs' data. Gonella then suggested also bringing in the Colonetti Metrological Institute into the process, with which Tite agreed. (According to Chagas, the Pontifical Academy had no statistical competence.)

Chagas then asked Gonella about the excision of the samples and potential problems. The latter replied that the withdrawal had to be appropriately placed in the context of the other exams proposed for the better identification of the site and the availability of experts and equipment; Turin found it prudent to make use of STURP's experience on the Shroud and maintain a multidisciplinary research framework; the other exams proposed by both STURP and other groups were held pending the outcome of that meeting. Tite asked if STURP had textile experts and if Flury-Lemberg could participate. Chagas proposed to STURP to request the latter's collaboration, and STURP accepted. Chagas asked when STURP could undertake the work; he was informed they had gotten permission *[Cardinal Ratzinger had approved in July 1985 STURP getting two weeks to perform tests on the Shroud.]* Chagas suggested May 1987. Hall expressed the opinion that matters shouldn't become too complicated.

So, at the end of the second day of work there was nothing left to summarize the discussion to define the remaining points, including the size of the samples, but Gove intervened to ask that any other testing besides C-14 be postponed until a result was obtained, as they were "all agreed" that the C-14 was the important test and if the Shroud were only six-hundred-years-old, why even do the other tests? Chagas, who only minutes before had advocated for multi-disciplinary tests, did an abrupt about-face and backed Gove.

At this point, Harbottle intervened in support of Gove, saying that he had seen with amazement that morning that in 1978 when the Shroud had been subjected to X-rays and illuminations that could be dangerous. Gonella thought the intervention odd, since precise details of those tests had been in the literature and no concerns had been previously raised, plus the fact that Harbottle had been attempting for years to officially join STURP. Despite Gonella's protest that the discussion had been diverted from the objectives of the meeting and previously-defined positions were being questioned, Chagas gave the go-ahead to Gove, and the following morning *[October 1st]* he was engaged in a confused and inconclusive diatribe on points already discussed. Matters concerning control samples and the blind-testing were discussed; Gove was not in favor of blind-testing but the other labs were. Gove also brought up the possibility of getting the testing samples right then, using "just a pair of scissors."

Chagas touted Riggi as an expert on the handling of the Shroud fabric *[he had assisted STURP in their 1978 examination and was present because he assisted with logistics of the Turin planning meeting]*, asking him for advice regarding the upcoming testing. Chagas then said that it was now his opinion that the responsibility for the collection should go to the radiocarbon laboratories, and its execution entrusted to Flury-Lemberg. The question of what area of the cloth should be utilized from which to take the sample was raised. Flury-Lemberg said only one was needed as the cloth was the same everywhere. Gonella asked her if as a textile expert she thought the edges could be more contaminated, as some critics had suggested, and she replied "no," because contamination should be visible to the naked eye and would have minimal effect on dating. Di Rovasenda asked if the fire that had charred portions of the cloth could have affected the non-carbonized parts for dating purposes; Harbottle and Otlet both said it wouldn't. With regard to the blindness of the examination, it was observed that it was necessary to make the material unrecognizable, and to talk about fraying the sample, but everyone agreed against a shredding, because this would have made the treatment problematic.

The Turin meeting had left several questions unresolved, starting with the actual amount of the samples to be taken and showed how much Chagas depended on Gove. The progress of the meeting could have been completely different if the prestige of the Pontifical Academy of Sciences deployed in support of the lines proposed by Ballestrero, which Chagas had been commissioned to help, rather than oppose. However, a good relationship was established with the laboratories (except for the open enmity of Rochester), and the defined points still allowed a healthy multidisciplinary approach to the operation. The public was informed of the Turin meeting by a press release from the Custodian of the Shroud on October 4th, as requested by the guests. Gove and his friends were not satisfied with the meeting in Turin because they were not happy with the multi-disciplinary approach, and they immediately started to work to distort the conclusions. On October 6th, as soon as he returned to the USA, Canuto phoned Gove that Chagas shared the concerns expressed by Harbottle about the danger of other examinations on the Shroud and solicited letters from the laboratories against the STURP exams; he added that the press release of Cardinal Ballestrero was in error when speaking of pontifical approval because the Pope had not yet seen the Turin protocol. The day after Gove wrote to Chagas asking that the STURP exams be postponed until the dating was completed, and then phoned the other six laboratories asking them to write in the same vein. (Gonella found out about this via being copied on a letter sent to the six labs on April 15, 1987; he informed Turin at that time.)

As the meeting drew to a close, Chagas asked Canuto to draw up a summary report with assistance from Gonella and Donahue. Canuto produced a report of the meeting with the title "Conclusions and Procedural Steps", which was often cited as the "Turin protocol" and was sent by Chagas to the Secretary of State on October 8th. Gonella noted it actually had no documentary value because it was not signed by anyone and was not even seen by the participants. *[!]* The phraseology used tended to pass over critical points and actually contradicted positions that had been taken. After the initial draft, Canuto had Donahue adjust the grammar and promised Gonella to show him the revised copy in the evening at the hotel. The final text contained significant variations of which Gonella asked for the necessary corrections that were promised, but not made.

Source: Gonella, Luigi and Riggi di Numana, Giovanni. Sindone: il mistero continua, (Milan: Fondazione 3M), 2005, pp. 45-56.

Comments: Gonella inserted a "!" after the sentence about Chagas having thanked Gove for organizing the meeting. Gonella also note that if seven labs participated, the cited 12.5 cm^2 figure would have to be doubled again. STURP's opportunity to do tests in May 1987 never materialized. Chagas' backtracking ("Chagas, who only minutes before had advocated for multi-disciplinary tests, did an abrupt about-face and backed Gove") was striking, but not surprising -- remember Italian researcher Franco Faia asserted that Chagas was in Gove's pocket -- and this incident seems to prove that. Gove's nonchalance about getting the samples then and there was also striking. Flury-Lemberg's assertion that the cloth was the same everywhere was ludicrous, as it completely ignored the idea that repairs could have been made that were not visible to the naked eye.

1986 *(September and October).* Cardinal Ballestrero addressed the workshop participants: "While warmly welcoming all of you in the city of the Shroud, I wish to begin by thanking you for your participation in this meeting. A very particular thanks I owe to Prof. Carlos Chagas, President of the Pontifical Academy of Sciences, for the care and competence he devoted to prepare the meeting. I do not think it useless to point out a few data that are at the basis of this meeting."

"Already at the time of the last Exhibition of the Shroud several investigating groups and scientists from all parts of the world had expressed the wish that a Carbon 14 test be made on the Shroud to date the Cloth. Authorization was then asked of, and obtained from, Umberto di Savoia, who died a short time later, leaving the precious Shroud Cloth the property of the Holy Father John Paul II. Upon my repeated requests the Holy Father eventually approved that this Carbon 14 test be made, and also upon my personal request, after having appointed me Pontifical Custodian for the conservation and the veneration of the Shroud, the Pope involved the Pontifical Academy of Sciences as scientific consultant."

"Prof. Luigi Gonella, of the Turin Polytechnic, scientific and technical consultant of Turin Bishopric since the time of the last Exhibition of the Shroud, is here present not only for his specific scientific competence but also as my personal assistant for what concerns the responsibility bestowed upon me as Pontifical Custodian of the Shroud."

"It seems to me worthwhile to state that this research, desired by the Church to be of a purely scientific character aimed at dating the Shroud cloth, does not mean, nor could it, address any issue of faith related to the death and resurrection of Jesus Christ."

"It is my hope that the works of this meeting will proceed not only with the competence that may be expected by a gathering of such eminent specialists, but also with a capacity and willingness for comparison and integration as befitting to the method of Science in our times."

"As my specific competence is not in the field appropriate to an active participation in the forthcoming discussions, I shall not take part in your talks, leaving to my assistant, Prof. Gonella, the task of liaison with me."

"I hope that these days of study will bring out such conclusions to allow presenting a valid and acceptable project for at last carrying out the radiocarbon dating of the Shroud cloth, a test that, owing to the uniqueness and singular character of the object, certainly could not be easily repeated."

"The project coming out of this meeting, including concrete operative proposals, will be submitted to the Higher Authority of the Holy See, as it is explicitly requested in the letter sent to me on Sept. 24 by the Secretary of State: '[...] the proposals and observations that will emerge in the discussion will have to be submitted to the Higher Authority'."

"And as a final thought, at this point, it is only left to me to wish you all a serene and fruitful work."

"LIST OF PARTICIPANTS"

1. Prof. Carlos CHAGAS, President, PONTIFICAL ACADEMY OF SCIENCES, VATICAN CITY
2. Prof. Alan D. ADLER, CHEMISTRY DEPARTMENT, Western Connecticut State University
3. Mrs. Shirley L. BRIGNALL, DEPARTMENT OF PHYSICS AND ASTRONOMY, The University of Rochester
4. Prof. Vittorio CANUTO, N.A.S.A.
5. Prof. Paul E. DAMON, DEPARTMENT OF GEOSCIENCES, The University of Arizona
6. Ing. Don Renato DARDOZZI, Co-Director, PONTIFICAL ACADEMY OF SCIENCES
7. Dr. Robert H. DINEGAR, LOS ALAMOS NATIONAL LABORATORY
8. Prof. D.J. DONAHUE, DEPARTMENT OF PHYSICS, The University of Arizona
9. Prof. Jean-Claude DUPLESSY, Directeur, CENTRE DES FAIBLES RADIOACTIVITES Laboratoire mixte CNRS_CEA

10. Dr. Jacques EVIN, LABORATOIRE DE RADICARBONE, Universite Claude Bernard Lyon
11. Dr. Mechthild FLURY-LEMBERG, Head, TEXTILE WORKSHOP, ABEGG-STIFTUNG (Switzerland)
12. Prof. Luigi GONELLA, DIPARTIMENTO DI FISICA, Politecnico di Torino
13. Dr. Harry E. GOVE, DEPARTMENT OF PHYSICS AND ASTRONOMY, The University of Rochester
14. Prof. E. Teddy HALL, RESEARCH LABORATORY FOR ARCHAEOLOGY AND THE HISTORY OF ART, Oxford University
15. Prof. Garman HARBOTTLE, DEPARTMENT OF CHEMISTRY, Brookhaven National Laboratory
16. Dr. Robert E.M. HEDGES, Director, RADIOCARBON ACCELERATOR UNIT, Oxford University
17. Dr. Steve LUKASIK, J.P. GETTY CONSERVATION INST.
18. Prof. William MEACHAM, CENTRE OF ASIAN STUDIES, HONG KONG
19. Prof. Robert L. OTLET, ISOTOPE MEASUREMENTS LABORATORY, Harwell Laboratory (United Kingdom)
20. Rev. Enrico di ROVASENDA, Director, PONTIFICAL ACADEMY OF SCIENCES, VATICAN CITY
21. Prof. M.S. TITE, RESEARCH LABORATORY, The British Museum
22. Prof. Dr. Willy WOLFLI, INSTITUT FUR MITTELENERGIEPHYSIK, (Zurich, Switzerland)

Source: Opening address by Anastasio Cardinal Ballestrero, Archbishop of Turin, to workshop attendees. Copy in possession of author.

Comments: The attendee list here counts twenty-two participants. Gonella's account immediately below mentions twenty.

1986 *(September/October)*. Chagas, and not Gonella, acted as moderator. Gonella noted that the media was "unduly interested in carbon dating" and stated that X-ray fluorescence measurements in 1978 showed that the Shroud was not a painting. STURP wanted that measurement repeated, and Gonella supported STURP doing all other twenty-five tests, in addition to the C-14. Chagas countered that X-ray fluorescence might damage the Shroud, and suggested that the "obtrusiveness" of C-14 dating would be "mild by comparison."

Gove was the only one to bring up the point of STURP's role in the C-14 dating and said that connecting it to the desired multi-disciplinary tests was only delaying performing the C-14. Gonella had insisted that the sample taking for the C-14 test would be done as part of STURP's twenty-six-test package. STURP's endorsement of Flury-Lemberg was not firm -- Dinegar said they would need a few hours to decide. Shortly after this, Gove ran into an agitated Flury-Lemberg, who said she was invited to dinner by STURP, but said she was under pressure to formally join the group and sign their secrecy agreement. Gove was angry about this and told her there was no need for this.

If seven labs were going to be involved, there was the question of how much sample material should be taken. Dinegar asserted that there were four-hundred cm. of charred cloth underneath the patches put on after the 1532 fire by the Poor Clare nuns. But nearly everyone was leery of using such material. The late Willy Wolfli *[also spelled as Wöelfli or Wölfli by some]* from Zurich said there was no firm evidence that charring affects samples, but Hall from Oxford, while agreeing that charring might have little effect, stated that uncharred material should be used to be on the safe side.

The scientists suggested five to ten mg (in carbon weight) for the AMS labs and ten to fifteen mg for the proportional-counter labs (if seven would be involved), but Gonella was not in favor of seven labs. Chagas replied "The more measurements, the better. This is the measurement of the century."

Another disputed point was the question of blind testing. Gonella reminded the labs that the Turin authorities didn't think it was necessary. Gonella said "We trust the scientists. It is you who have this idea which you feel will protect you from the press." He felt the matter should be discussed further, but the labs decided to keep blind testing as part of their procedures.

Gonella also brought up the matter of how many people would be physically present for the taking of the samples. He suggested having representatives from the labs observe on closed-circuit television. Chagas, with Archbishop Ballestrero present, summarized what he considered to be the agreement coming out of the workshop. That agreement was never released by Turin, but it was published in a scientific journal via Harry Gove. Chagas concluded "If the Shroud must be dated, it must be done right. It must be done now. This may be the last chance."

Source: Sox, David. <u>The Shroud Unmasked</u> (Basingstoke, Hampshire: The Lamp Press), 1988, pp. 106-108.

Comments: Sadly, many, including Gove himself, would conclude the test was not done right. "Blind testing" refers to the labs supposedly not knowing which sample was the Shroud when they tested it (along with control samples). But because of the Shroud's unique weave, the labs did know.

After providing the information in this entry to Bill Meacham, he emailed me on July 16, 2016 and said regarding Dinegar's comment that everyone was leery about using samples from underneath the patches, "Exactly the opposite was true. Everyone including STURP was happy with the idea of using that material, but I argued against it, and finally after discussion got support from Otlet. See pp. 73-74 of my book."

According to an article in the *Catholic Counter-Reformation In the XXth Century* (No. 295, April 1997, pg. 19), the protocol report would end up being <u>eight-hundred</u> (800) typewritten pages! I bet the person who typed it up wasn't too pleased when they ultimately virtually did not follow most of it. Meacham, who attended the meeting, confirmed to me in another email that the report was, in fact, that long.

1986 *(September/October)*. There was a heated discussion in the Turin Workshop regarding the size and number of the samples, how the samples would be certified, and the use of control samples. It was agreed by everyone that Mechthild Flury-Lemberg, a Swiss textile expert, would be responsible for extracting the sample.

While Gonella wanted the C-14 tests to be performed in the context of the other tests that STURP proposed, Gove insisted that no other tests should be performed on the Shroud until a result was known from a C-14 dating. Archaeologist William Meacham from Hong Kong, who regularly would use multiple samples from a site when he employed

C-14, proposed that samples should be taken from various places on the cloth, but Flury-Lemberg vehemently objected, claiming that the borders couldn't be more contaminated than the rest of the cloth. STURP advised taking samples from at least three different areas of the cloth.

Meacham, as an experienced archaeologist, took seriously the issue of contamination and proposed taking a thread from the middle of the cloth, between the front and back images, a small piece from the edge adjoining where the 1973 sample was taken, a piece of charred material, part of the side strip, and a piece of the backing cloth that had been sewn on in 1534. Microchemical tests, mass spectrometry, micro-Raman would then be done, as well as the appropriate pretreatment for impurities.

Source: Marinelli, Emanuela. "The Setting for the Radiocarbon Dating of the Shroud." Presented at 1st International Congress on the Holy Shroud in Spain -- Valencia -- Centro Español de Sindonologia (CES), April 28-30, 2012, pp. 5-6. www.shroud.com/pdfs/marinelliv.pdf.

Comments: It's amazing that Flury-Lemberg asserted that the borders couldn't be more contaminated than the rest of the cloth. That's something that couldn't be known without chemical analysis. Gonella requested that corrections be made to the report but that wasn't done, according to the "Night of the Shroud" documentary (original version). Petrosillo/Marinelli)] noted, "Meanwhile Chagas was playing a sort of double game. He kept Gove informed about the difficulties that had arisen with STURP. He even told him that Gonella did not think much of him: his interest in the dating was considered suspicious for he seems to have been an agnostic" ["Enigma …" book (pg. 29]. As mentioned previously, another writer had described Gove as a "once-a-year Episcopalian."

1986 *(September/October)*. Petrosillo and Marinelli wrote, "Representatives of the six labs as well as Jean-Claude Duplessy of the French Gif-sur-Yvette lab participate in the workshop. Shockingly, no one from the Turin Centro was invited. Ballestrero announces that the Pontifical Academy will be his consultants and Gonella is designated as his representative, but the Pontifical Academy will organize and direct the project. One of the major topics is the location from which the sample will be taken. Gonella proposes samples from underneath the patches sewn in after the 1532 fire. Hall proposed using non-carbonized material to be on the safe side and everyone agrees. Another major topic is the amount of sample to be excised. The consensus was five to ten mg. for AMS labs and ten to fifteen for the proportional-counter labs. Gonella disagreed and asserts that two labs are sufficient, which would require smaller amounts. Chagas countered by saying that the Pontifical Academy had invited seven labs and the more the better. The labs suggested a blind analysis -- double-blind, triple-blind? Gonella said that Turin didn't consider blind testing necessary but the labs want it for public perception reasons."

"Chagas asked his private consultant to draw up a summary of proposals, a text not signed by anyone but which was presented to the public by Gove and others as an 'agreement' with ecclesiastical authority behind it. Gove would in 1987 publish in a scientific journal this so-called agreement, but according to Gonella, no agreement was reached but only a proposal that had to be subsequently modified or approved or rejected."

"After three days, it was agreed that a minimal amount of Shroud material would be taken, which would be sufficient to guarantee a scientifically-rigorous result and to maximize the credibility for the public. For these reasons, it was decided that the testing would be done by seven laboratories, using both methods of dating: five AMS and two proportional counters. The sample was to be taken from an area of the Shroud that's not useful for providing other useful information and should not include any burnt material. The samples will be such that they are sufficiently large to allow an appropriate pre-treatment process. 'The suggestion of preventive analyses to ascertain the presence or absence of contamination was rejected because the cleaning methods were sufficient to eliminate them,' according to Gonella. In addition to the Shroud sample, each lab would receive two control samples from the British Museum. For logistical reasons, the Shroud samples would be taken immediately before other interdisciplinary tests performed by other groups. Flury-Lemberg was to be responsible for choosing the material to be sampled and for the excision. Seven samples containing 50 mg. of carbon would be removed. In addition, a single 'dummy' sample would be prepared by the British Museum. The Shroud samples and the dummy sample would be distributed to the seven labs, whose individual samples would not be identified. Three certifying institutions would be responsible for distributing the samples: the Pontifical Academy of Sciences, the British Museum and the Archbishopric of Turin. Sampling would be conducted such that representatives of the seven labs would be completely aware of all aspects. The samples would be delivered directly to the representatives of the seven labs, which at that point would be responsible for them. Once the seven labs obtained results, they would submit them to: the Pontifical Academy of Sciences, the British Museum and the G. Colonetti Institute of Metrology in Turin."

"The results would be kept in sealed envelopes until a specified date, when they would be opened for statistical analysis. When this stage was completed, representatives from the three institutions and of the seven laboratories would meet in Turin with the aim to make a public announcement. It was hoped that the samples could be taken in May 1987 and that the results would be ready at Easter 1988. It was planned that results would be published in a scientific journal as a collaborative work. So, the Cardinal was given an operational proposal but Gonella deemed it unacceptable. If the labs requested twenty samples, was the Cardinal to give them twenty samples? Gonella said he was professionally disgusted with the scientific deficiencies of the motivations."

"Ballestrero and Chagas declared that if the Shroud was to be dated, it had to be done well. Gove noted that the procedures were designed to avoid suspicion of other samples being switched with the Shroud. At that time, there were still plans for other tests to be done besides the C-14. Gove asserted that for logistical reasons (but didn't specify what they were), the C-14 test should be done before the others. Bonnet-Eymard claimed the labs planned to eliminate chemical analysis of the threads, which are actually burned during the process. This would have fatal consequences for the results interpretations. Laboratory representatives would be allowed to be present when the samples were taken; each lab would work independently. Both C-14 methods were to be used: the proportional --counter, even though it normally used a bigger sample size, and the AMS, which hadn't been used extensively yet on fabrics. If all seven labs came up with a similar date, there presumably would be little grounds for dispute. And even if only one made a mistake, it could be disregarded."

"It was planned that the results would be analyzed jointly by the British Museum and the G. Colonetti Institute. Bonnet-Eymard asserted the Pontifical Academy was to be the guarantor of everything, along with the British Museum and Colonetti and that Riggi made a serious omission in his book <u>Rapporto Sindone</u> by not mentioning this. A summary document was submitted to Ballestrero, who ensures that he will deliver to the appropriate authorities. The following day Turin issued a press release, but was silent about some key details such as the numbers of laboratories and samples as well as the size of the samples. Within a week, Gonella said that some of the labs started writing to the Vatican requesting that no other tests besides the C-14 be performed. Gonella felt pressure. He realized that any objections could be perceived as the Church trying to stall scientific inquiry. He added that the labs had not been satisfied with the Turin workshop because the other tests were still being considered."

Source: Petrosillo, Orazio and Emanuela Marinelli. <u>La Sindone: Storia Di Un Enigma</u>, (Milan: Rizzoli, 3rd edition, 1998), pp. 138-142.

Comments: There was clearly a lack of leadership and coordination in the whole project.

The "dummy" sample was apparently meant to resemble the Shroud sample and was envisioned as part of the proposed "blind testing," which eventually never materialized.

Note that even if both methods of C-14 were used, there were possible deficiencies for both: the old proportional-counter method always used sample sizes much larger than what they would have for the Shroud and the AMS, which was only about eleven-years old at the time, had not tested fabrics extensively. This is another reason that a stand-alone C-14 test without other multidisciplinary tests was a no-win situation for those that believed that the Shroud could be first-century.

To continue with this thread, go to the Petrosillo/Marinelli entry for "1986 *(November)*."

1986 *(September/October)*. Meacham wrote, "Day Three saw more discussion on the issue of possible dummy samples, and Gove threatened to withdraw from the project if he could not be assured that his lab would be receiving a real Shroud sample. Chagas gently chided him, reminding that he came from a democratic country and surely he would abide by the consensus reached. Gove somewhat sheepishly withdrew his threat. For a moment Gonella had a look of barely suppressed joy, but it was not to last. Gove made the snide comment that 'STURP seems to be on some sort of crusade to prove the Shroud authentic.' A few minutes later, Gonella opened with a comment 'Apparently Gove does not like the STURP people.' Chagas weighed in at this point and said that we should keep the discussion to the issues at hand, and not let it denigrate into personal or emotional attacks. I was tempted to point out that the time for such an intervention by the chairman would have been after Gove's offensive remark, but held my tongue. This incident would come up later in correspondence with Chagas."

"The discussion turned to the question of statistical analysis again, and I began to realize that what the labs were most concerned about was obtaining matching results from each lab, or if not, then at least the one outlier could be quite easily identified and rejected before calculating the radiocarbon date. This concern and seemed to me misplaced; what was much more important was ensuring that a rogue sample was not chosen. It would be a tragedy if every lab got the same result but it was wrong because there was something inherently wrong with that particular sample chosen. Adler agreed that this was the biggest worry. We've raised the subject again of sampling sites, and now a consensus was forming that only one sample needed to be taken, cut into seven pieces, and distributed to the seven labs. I pleaded for a minimum of two sample sites, even if one of them was the charred material under the patches. At this there was much chortling from Gove and Canuto, and Chagas dryly remarked that I had spent so much energy arguing against the charred that he was mystified to see me 'reverse' my position. Adler tried to argue the point further, but Gonella would not support us. At this late stage the chairman's control of the meeting was the deciding factor, and he stated that a single sample was the consensus and move on. I felt a *frisson* of the anxiety, that the future reputation of the Shroud could be in jeopardy."

Source: Meacham, William. The Rape of the Turin Shroud: How Christianity's most precious relic was wrongly condemned and violated (Lulu.com), 2005, pg. 75.

Comments: Meacham's statement, "It would be a tragedy if every lab got the same result but it was wrong because there was something inherently wrong with that particular sample chosen" was prophetic.

It's a shame that Gonella did not support Meacham's and Adler's recommendations to use at least two sample sites.

1986 *(September and October)*. While addressing the gathering, Gonella said that the first formal proposal to date the Shroud was by STURP in October 1984. Gove objected and said that he had made the first proposal in 1978 and sent a formal letter to Cardinal Ballestrero in 1979. Gonella countered by saying the 1979 proposal was for the use of the Raes samples. Gonella then added that STURP had proposed taking new samples and also doing other tests; that started the chain of events that led to the Turin workshop. Gonella indicated that he only learned from STURP about the Trondheim protocol of July 1985 and subsequent revision in September 1985. He said that proposal was never formally submitted to Turin and so did not count. According to Gove, "Gonella seemed hell-bent on historical revision."

Gonella then explained how a seventh lab was added as a participant. In April 1986, J.C. Duplessy of the Gif-sur-Yvette AMS lab submitted an independent proposal to the date the Shroud, which led to his invitation. STURP had proposed many other tests. Gonella further said that Turin was possibly more interested in these other tests over C-14 because they addressed conservation, which he claimed was the major priority. Gove commented in his book that Turin had totally ignored conservation for at least 300 years. Gove asserted that decisions on conservation are based on a knowledge of an object's age.

STURP suggested that they, in close conjunction with Turin, should extract and distribute the samples. The Trondheim proposal had suggested using the British Museum as the coordinating institution, but Gonella felt that other institutions would be just as good. Regarding blind testing, Turin didn't request it -- it was the C-14 scientists who wanted it.

Gonella expressed concern regarding the amount of material that would be needed if seven labs were involved. He said that they must distinguish between scientific and **political** *[my bolding]* motivation. Gove asked Gonella to define "political;" the latter said it was anything not scientific. Gove wrote that Gonella considered the former's motives to be **political** *[my bolding]* because Gonella had written in a letter to Chagas that Gove was using the opportunity to date the Shroud "as a lever for extracting more support funds from the NSF." Gove then added "I wondered about his personal motivations as much as he did about mine."

Chagas then addressed the participants and explained how the Academy had become involved. Chagas had been given in 1984 STURP's proposal for the Shroud C-14 dating but felt that the Trondheim protocol was better. If the Shroud dating was carried out, it must be done right and it must be done now since this might be the last chance. The Turin workshop was meant to only address the C-14 dating and not, as Gonella seemed to be suggesting, additional tests as well.

Edward Hall from Oxford said he thought blind testing was needed only to convince people outside. He felt that the British Museum was needed to coordinate matters "to make it look convincing." Michael Tite from the British Museum indicated they would be willing to assist and would not be upset if not asked.

Further comments were made on blind testing, the number of labs to be involved and whether every lab should get a Shroud sample (i.e., some labs would only have received non-Shroud samples that looked like the Shroud and/or control samples that would gauge lab dating accuracy). Gove remarked that he thought people at his lab would lose motivation if they didn't actually have a Shroud sample. Several other scientists thought it was a good idea that not every lab would get a Shroud sample. Al Adler from STURP advised that samples should be taken from various parts on the Shroud. The archaeologist Meacham said that it was unusual to have so many labs involved. After a discussion about the cloth weight per square centimeter, Gove realized that all labs would be able to tell which sample was the Shroud, which made moot the whole question of blind testing.

Flury-Lemberg asserted that there was no need to take samples from various places on the cloth because the Shroud was the same from one end to the other. Chagas then asked if all seven labs should get a Shroud sample or whether several would just get control samples. Harbottle advised that seven be used since as the British Museum interlab comparison test had shown, "a hundred things could go wrong." Harbottle also suggested that the new AMS method as well as the older proportional-counter method be used. Chagas again asked if all seven labs should get a Shroud sample. Harbottle first commented that he thought six labs should get a Shroud sample and one should get a "dummy" Shroud sample. He thought that two control samples should also be provided to each lab. Harbottle also said that four labs would not be enough. After advice from Duplessy, he later suggested one Shroud sample and one control sample for each of the seven labs. Flury-Lemberg advised that it would be impossible to get a dummy sample that could pass for the Shroud. During this part of the discussion, Gove wrote that Gonella was periodically making "incoherent teeth grinding sounds." Gonella was also clearly unhappy that seven labs were being considered and that Harbottle said that four labs were too few.

The following day, Gonella presented a slide show that highlighted the tests that STURP had carried out in 1978. It showed the Shroud on a frame designed by STURP under bright lights for various tests. Gove wrote that Chagas perceived that the bright lights were intrusive, and several felt that the slide show had a negative impact. Gonella seemed to be favoring only two labs being involved. During a break, Gove struck up a conversation with STURP member Steven Lukasik, who had recently joined the group after having become fascinated with the Shroud after having read an article in *Readers' Digest*. Lukasik worked for Northrup, which built fighter planes for the U.S. Government. Gove described him as "one tough, steely eyed individual -- an unreconstructed cold warrior in his position of vice president for research at Northrup and clearly another 'true believer' as far as the shroud was concerned."

Woelfli from Zurich discussed the outlier in the British Museum intercomparison tests and indicated that it was due to human error, not in his lab, but he took responsibility for it and said such things could happen again. He said one problem with small samples is that the standard cleaning procedure is not always sufficient.

Hall made the argument that they were dealing with people from outside the scientific circle. Hall thought seven labs would be appropriate because they could perhaps say that six out of seven labs agreed rather than two out of three agree. Thus the more labs the more convincing the results would be. Gove wrote that he found this comment interesting because when it was finally decided to use only three labs and Hall's was one of them, Hall then made the comment that he agreed because the more labs involved the greater the chance that one or more could make a mistake with a resulting wrong answer.

Gonella said that Hall and Harbottle had given good reasons, **political** [*my bolding*] but good, for using seven samples and labs. He added that the C-14 dating had slowed down a plan for other tests, not the other way around as Gove claimed. Gonella felt that seven samples and labs were not for scientific reasons, but for aiming for greater acceptance by the public.

Chagas said that STURP was interested in other tests besides C-14 but this workshop was only devoted to the latter. If STURP wanted to be associated with the C-14, it was not his decision. He had discussed it with several people who maintained that any future tests on the Shroud depended on its age. The Academy had chosen seven labs, so seven labs should participate. C-14 dating of the Shroud was the measurement of this century; STURP might want to make further measurements next century. With the exception of Gonella, the consensus was that seven labs should get one Shroud sample and two control samples.

The question of blind testing came up again. Most participants seem to favor it. It was then agreed that only six labs would get a Shroud sample (with one kept in reserve) and the seventh would get the dummy Shroud sample, in spite of the fact that Flury-Lemberg asserted that it would be impossible to find a sample that closely resembled the Shroud. Gove was opposed to this but knew that he would be able to know if his lab had a real sample or dummy sample just by weighing the threads. Chagas then led a discussion on the procedure for the cutting of the samples and how they would be transported to the British Museum and then the labs. Additional questions were how would the results be statistically analyzed and how would they be made public? It was decided that the cleaning of the samples would be left to the labs themselves. Regarding which lab would get a dummy sample, Gove stated he was not clear who would make that decision. He wrote that he had come to the conclusion that if it was up to Gonella, Rochester would not get a Shroud sample. Gove wanted the British Museum and the Pontifical Academy of Sciences to decide on the distribution of the samples. Who would oversee the taking of the sample? Would it be someone chosen by STURP? Chagas said that, with all due respect to STURP, Flury-Lemberg was the best choice.

According to Gove, "Adler then described the magnitude of the attack STURP wished to launch on the shroud -- [...]." Adler said there would probably be four or five conservation textile persons as well as other experts involved. Chagas suggested that Flury-Lemberg be involved in this aspect. He said he would like to see the dating done as soon as possible so when could STURP take this on? In response, Gove wrote, "This comment appalled me but I remained silent. As far as I was concerned, STURP would never take on the job." Dinegar said STURP could extract the sample as soon as they received permission and that Flury-Lemberg was probably acceptable but they would need a few hours to decide for sure.

Gonella asked Dinegar how much lead time was needed for STURP to organize. Dinegar replied that not much time was needed just to take the samples. Chagas asked whether they could be ready by March or April and Dinegar replied in the affirmative. Gove then noted in his recounting, "I could see that Lukasik, by this time, was really licking his lips." Gove then asked if STURP wouldn't want to know if the Shroud was only 600 years old -- wouldn't that change their approach for the battery of tests they wished to perform. Dinegar replied that at the moment, he was not sure. According to Gove, Lukasik added that it wouldn't make that much difference "if STURP just took samples or if they carried out all the tests they lusted to do. The samples would have to be characterized (whatever that meant) [...]. " *[The "(whatever that meant)" was a parenthetical comment by Gove.]*

Chagas said that the discussion had drifted away from the procedure for the C-14 dating. The dating could be done very soon and wouldn't have to wait for the additional STURP tests. Meacham spoke up and disagreed. He believed the STURP tests should not be set aside. He said an archaeologist would never stop a dig just because a date came out younger than expected. Gove wrote of Meacham, "He had such a boyish enthusiasm for archaeology and yet his attempted analogies between that field and carbon dating the shroud seemed farfetched to say the least. He should never have been allowed to play a role in the workshop."

Dinegar stated that STURP accepted responsibility for taking the Shroud samples. Gove wrote, "I thought to myself -- over my dead body they will." Dinegar asked if it was agreed that STURP, along with Flury-Lemberg, would take the samples. Chagas said that after conservation assessment, samples would be taken at the beginning of May under Flury-Lemberg's direction. Chagas then said he assumed that all participants were pledged to keep matters as secret as possible until the final results were announced. Gove then announced he was surprised that the sample taking would be done by STURP. Since the samples would eventually be turned over to the British Museum, Gove wanted a simple operation with Flury-Lemberg extracting samples witnessed by the C-14 consortium. Gove wrote, "I said I saw no reason at all, and I was being more forthright and less diplomatic than I should be, that our carbon dating enterprise had to be connected in any way with any of the enterprises that STURP wished to carry out. I said I felt so strongly about this that I would have to consider the question of our laboratory's participation if STURP played any role at all."

On the question whether each lab would receive an actual Shroud sample, Gove said that one of the reasons he had for wanting to be involved in the Shroud dating was that he wanted his lab to occasionally be involved in something mainstream that the public was interested in, since it was their tax dollars that funded the lab. If he went to his lab saying that he had three samples, one of which may or may not be the Shroud, he would have difficulty getting everyone enthused. Gove then added, "Whether one called such sentiments **political** *[my bolding]* and not scientific, they were very real."

Regarding blind testing, Gove said that if they did, in fact, receive three samples and one was the Shroud, they would do the measurements, which would not be affected by whatever prejudices they had about the Shroud's date. Gove conceded that in science, "[I]t was true that in making a measurement sometimes one involuntarily got the answer one wanted by somehow unconsciously manipulating the apparatus and I assumed this was the reason for making the measurement blind." But he felt that they did not need to go the extra step of not giving a Shroud sample to one lab only. Gove said that if that extra step was taken, Rochester would withdraw.

On the question of not keeping secrecy pertaining to the test results, Gove said that would risk being dishonored in the C-14 community forever. So members of each and every lab had high motivation to keep secrecy. More discussion ensued about blind testing and whether the Shroud samples given to labs should be unraveled or whole. Most seemed to favor "whole." Hall opined that he would like to see the samples unraveled, but then Tite could put them and the control samples in envelopes to give to the labs' representatives. Hall reasoned that if the sample was whole, it would easily be identified as the Shroud. It was unlikely to find a dummy Shroud sample, so blindness was impossible. But he said it was important to do blind testing because if it wasn't done, they would be open to criticism. Gove wrote that he found Hall's argument confusing insofar as he seemed to be saying there was no feasible way to do blind testing, but they must do so anyway. Tite said there clearly was no question of blind testing unless the samples were shredded to some extent, and that he wasn't the one pushing for the blindness.

Gove wrote that Gonella "gave another of his tiresome monologues. He was concerned that there was a significant lack of trust among the participants." He inquired whether he should be present for each step of the process in each lab. Gove then wrote, "At the time this struck me as an inventive way to dispose of Gonella." Gonella was critical of the decision to use seven labs. He also said he didn't understand why the labs had to be present at the sample taking. Didn't they trust the British Museum? Finally, he said that Turin had not requested the blind testing. If there were no blindness, there was no need for seven labs and a dummy sample. He ended with the remark that "we all know that Professor Gove doesn't like STURP," and added that Gove had "said that many others felt the same way." Gonella wanted to know specifics on this. Chagas censured Gonella, saying he didn't want let personal conflicts intrude. Gove then jumped in saying he was just accused of not liking the STURP people. Gove stated that this was untrue and irrelevant.

Chagas reported that the samples would be taken outside the image area of the Shroud. If the Shroud were younger than two-thousand years, there would be a declining interest in the Shroud -- at least among the general public. He indicated that samples would be taken in early May 1987, the results would be announced at Easter 1988, and anyone from the labs was welcome to come to witness the sample taking. Chagas then asked Giovanni Riggi, who would end up being the one to cut the sample, about the logistics of getting the Shroud out of storage. In his book, Gove was critical of Riggi for cutting the sample without even wearing gloves.

Gove then related that Lukasik followed. "To me his monologues were almost as sanctimonious, tiresome and tutorial as Gonella's." Lukasik stated that STURP's main concerns were authenticity, conservation, and image formation, but that a C-14 dating was a first priority. Gove then asked if all labs would be getting a Shroud sample. Chagas asked what the various labs thought about not getting a Shroud sample and added that since Dr. Gove came from a democratic country, he would abide by the majority decision. Members from the various labs went on the record giving their answers. Chagas stated that the Shroud C-14 dating would be blind and added, "I do not think Professor Gove was serious when he said Rochester would withdraw." Gove replied "that perhaps Professor Chagas knew me too well -- we would not withdraw."

Although Chagas had said only a few minutes before that anyone from the labs could witness the sample taking, he then announced only one person from each lab could observe. Gonella expressed concern about the number of people that would be in the small room where the sample would be removed. Gove relates in his book that Gonella later told him that he was even concerned that a representative from the lab might be seized by religious fervor and rush forward to touch the Shroud. Gove wrote, "This struck me as so ludicrous I laughed in his face."

Source: Gove, Harry. Relic, Icon or Hoax?: Carbon Dating the Turin Shroud, (Bristol and Philadelphia: Institute of Physics Publishing), 1996, pp. 150-173.

Comments: Gove clearly wouldn't have been satisfied participating if his lab hadn't received a Shroud sample. Regarding Flury-Lemberg's comment, as asked previously, how could she know without having examined the cloth if an area had been repaired or not? Regarding Gove's comment about Gonella's "incoherent teeth grinding sounds," the question has to be asked, "Does anyone ever make coherent teeth-grinding sounds?"

Gove painted Lukasik as a "true believer" but doesn't give any examples of why that was so. Gove seemingly painted STURP with one large brush. Again describing Lukasik, it's not clear what an "unreconstructed cold warrior" is, but one can be sure it's not a compliment.

Despite the fact that Woelfli acknowledged that small samples are not sufficiently cleaned, the labs would later proclaim their test results with a 95% confidence level. One can get the impression that the C-14 consortium believed that certain C-14 problems just weren't applicable to the Shroud. Gove's recounting of Hall swaying on the number of labs based on whichever one would benefit his lab seems suggests that scientific objectivity did not trump -- well, how did Bromley describe in the foreword to Gove's book some behavior that went on? -- "deception, outright lies, low cunning, misrepresentation, and a pathological hunger for publicity [...]." The idea that future tests on the Shroud depended on its age ignored two very important points: 1) STURP did its battery of tests in 1978 without knowing exactly the age of the cloth and 2) given that C-14 results are sometimes wrong, it would be unwise to deem other tests insignificant.

Chagas' suggestion that STURP might make additional measurements next century would mean that over twenty years would have elapsed between their twentieth-century expedition and the proposed twenty-first century expedition. Needless to say, that's a large gap of time to expect any real continuity. Notice Gove's inflammatory language when speaking of other scientists, especially those from STURP.

When Adler talked of conservation measures, Gove described it as an "attack." When talking about the tests that STURP proposed, he says they "lusted." What exactly was the difference between Gove wanting to perform a C-14 test on the Shroud and STURP wanting to perform other tests (which actually included C-14)? The answer is none, other than perspective. Lukasik is described as "licking his lips" in response to a proposed STURP involvement. Gove questioned Lukasik saying the samples should be "characterized (whatever that meant) [...]." Presumably he meant they should be chemically analyzed, something the labs should have done but didn't. Lukasik's comments are termed "sanctimonious" by Gove. Even though the Shroud is an archaeological object, Gove saw no need for the archaeologist Meacham, with his "boyish enthusiasm," to be at the meeting. Once again, Gove seems to be overly critical of non-C-14 scientists. Gove seemingly admits that his motives for wanting to date the Shroud were more **political** *[my bolding]* than scientific. In the case of a highly emotional topic like the Shroud, one cannot underestimate the unconscious manipulation factor that Gove conceded occurs in science.

Regarding secrets being kept until the final results were announced, it was not done, as we shall see later. At the very least, the labs allowed unauthorized persons to view the datings, which was strictly prohibited in the agreement they signed. Given that, it doesn't matter much whether or not the leak came directly from a member of one of the labs. The dishonor Gove feared from early disclosure did not prevent most people from accepting the labs' results and, indeed, from lauding the supposedly wonderful job the labs did. Gove's critique of Hall's argument about the blind testing shows how convoluted the thinking was during the workshop.

When Gonella pronounced that Gove didn't like STURP, Gove responded that he was accused of not liking STURP people. Gove added the "people." Gonella wasn't talking about STURP as individuals -- he was talking about them as a group. But any rational person, based on Gove's own statements, would conclude that Gove didn't like STURP, either as a group or as individuals. And it was relevant, despite Gove's claim that it wasn't -- it played a major role in the discussions leading up to and throughout the workshop.

Note: sometimes Prof. Wolfli's name is spelled by some as "Woelfli." Also, the German authors Kersten and Gruber sometimes use the British spellings of words and sometimes the American spelling, sometimes even on the same word, e.g., "practise" and "practice."

1986 *(September and October).*

Adler recounted his recollections of the meeting and the subsequent results: "I was at that meeting, too, as the chemical advisor. It was written up in *Archaeological Chemistry IV*. Bill *[Meacham]* pointed out all the things that you could screw up if you didn't have an archaeologist involved in the sampling, to advise you what to do and what not to do. And a chemist, to tell you what to do and what not to do, before you start sampling. That was all in the original protocol. *They didn't follow it [italics in original]*.

They wrote a different protocol. *They didn't even follow that [italics in original]*. When asked why they took the sample where they took it, the answer was: 'Well, it was cut there before.' Now that is the stupidest argument in the world for taking one sample from the place where they took it. Because they know that area is an area that's been repaired; they know it's by a water stain; they know it's by a scorch; and they know that people have found previous chemical evidence that that area is peculiar. But nevertheless, that's what they did. And that's why we have a date that all sorts of people don't believe. Because they don't believe the accuracy of the thing."

Source: Case, T.W. The Shroud of Turin And The C-14 Dating Fiasco: A Scientific Detective Story, (Cincinnati: White Horse Press), 1996, pp. 77-78.

Comments: Having advisors at meetings don't do much good if their advice is not taken.

1986 *(September and October)*. Gove's summary of the protocol finally decided on:

1. This is the time for radiocarbon dating of the Shroud of Turin.

2. A minimum amount of cloth will be removed, which is sufficient (a) to ensure a result that is scientifically rigorous and (b) to maximize the credibility of the enterprise to the public. For these reasons, the decision was made that seven laboratories will carry out the experiment: five accelerator-mass spectrometer laboratories and two small counter laboratories.

3. The samples should be taken from an unobtrusive part of the shroud, and from a portion which is not likely to yield other useful information. The samples should not include charred material. They should be prepared in a form, not too small, so as to allow reasonable pre-treatment processes. In addition to the shroud samples [and the dummy sample*], the British Museum will also prepare and provide two control samples for each laboratory.

4. For logistic reasons, samples for radiocarbon dating will be taken from the shroud immediately prior to a series of other experiments planned by other groups. Selection of the material to be removed and the actual removal will be the responsibility of Mrs. Flury-Lemberg (Abegg-Stiftung, Bern, Switzerland).

5. Seven samples containing a total of 50 milligrams of carbon will be taken from the shroud. In addition, a single dummy sample will be prepared by the British Museum. These shroud samples and the dummy sample will be distributed to the seven laboratories in such a way as to ensure that the seven laboratories are not aware of the identification of their individual sample. This distribution will be the responsibility of the following three, certifying institutions: the Pontifical Academy of Sciences (Professor C. Chagas), the British Museum, (Dr. M.S. Tite) and the Archbishopric of Turin (Professor L. Gonella).

6. The taking of the samples will be done so that representatives from the seven laboratories, will have complete knowledge of the process. Samples will be delivered by the three certifying institutions (see 5 above) directly and immediately to the representatives of the seven laboratories who will thereafter be responsible for the samples.

7. At this time, a date will be chosen for submission of experimental results from the seven laboratories to the following three analyzing institutions: the Pontifical Academy of Sciences, the British Museum, and the Metrological Institute of Turin, "G. Colonetti". These institutions will keep the results in sealed envelopes until an agreed upon date, at which time they will be opened, for statistical analysis.

8. After the analysis of the experimental result by the three analyzing institutions, a meeting will be held in Turin between the three analyzing institutions and representatives of the seven laboratories to discuss the results of the statistical analysis with the objective of deciding the final result of the measurement program.

9. The radiocarbon groups will, through correspondence, establish a common format for presenting the experimental results to the analyzing institutions.

10. The cost of the experiments and the analyses will be borne by the participating institutions.

11. Travel and living expenses entailed will be provided by the Pontifical Academy of Sciences unless other arrangements are made.

12. Samples from the shroud will be taken by May 1987. It is hoped that the final result will be available by Easter of 1988. This final result will be published in an appropriate scientific journal as a collaborative paper.

13. The Archbishop of Turin, Cardinal A. Ballestrero, will issue a press release concerning this Turin Workshop."

[*Parenthetical phrase added by Gove for clarity.]

Source: Gove, H.E. "Turin Workshop on Radiocarbon Dating the Turin Shroud." *Nuclear Instruments and Methods in Physics Research* **B29** (1987): 193-195.

Comment: See the entry under "1988 *(October)*" that has excerpts from a confidential letter to me from one of the major participants, with interesting comments about this meeting. Also, for documents given by Cardinal Ballestrero to the participants, see Appendix D in my book Wrapped Up in the Shroud: Chronicle of a Passion (Revised and Updated) (https://www.amazon.com/Wrapped-Up-Shroud-Chronicle-Passion/dp/1734813024/ref=sr_1_8?dchild=1&keywords=shroud+of+turin&qid=1596519943&s=books&sr=1-8).

1986 *(October)*. On the 4th, the Church sent out a press release regarding the meeting. It stated that Ballestrero had submitted the request to date the Shroud to the Pope, the recently-concluded workshop was presided over by Chagas, and the procedures agreed upon "now awaits approval from the authorized superiors." It also said that the methodology must be in the context of preservation and that results were likely to be released by Easter 1988.

Lukasik represented STURP at the workshop and offered options if Gonella's wish of only 50 mg. total was to be taken for the Shroud was implemented, which would involved three labs only, or if seven laboratories were chosen. Also Harbottle had a plan to use charred material in a secondary role, addressing one of Meacham's concerns.

Source: Bracaglia, Giorgio. Uncovering the Paradox within the Archives of the Holy Shroud Guild, (Honeoye, N.Y.: Holy Shroud Guild), 2019, pp. 217-219.

1986 *(October)*. On October 5th, the Italian newspaper *La Stampa* announced that the Pope had given approval for the Shroud to be dated by seven labs. When Gove discussed the story with Canuto, he said that the Pope, in fact, had not approved the testing. The protocol was being prepared in Rome and copies would be sent to everyone, include the Pope and Cardinal Ballestrero. Chagas started having second thoughts about the tests that STURP proposed after the removal of the samples. He was concerned about floodlights and X-ray and UV irradiation that the Shroud would be exposed to. Chagas wanted to get advice from the other labs about this. Canuto said that he could call Harbottle and Donahue and suggested that Gove contact the other four.

The next day Hall said he would write to Chagas that he was less worried about X-rays than UV light. Wolfli, Harbottle and Donahue said they would write to Chagas. Duplessy and Otlet declined to offer advice. Harbottle claimed there was nothing new in STURP's proposed tests. Gove wrote to Chagas on October 7th saying that if the C-14 date did come out first century, no further testing should be allowed "until they had been approved by some reputable and dispassionate international group of scientists."

Source: Gove, Harry. Relic, Icon or Hoax?: Carbon Dating the Turin Shroud, (Bristol and Philadelphia: Institute of Physics Publishing), 1996, pp. 177-178.

Comments: Harbottle's remark about nothing new in the proposed new round of tests was incorrect -- STURP specifically wanted to do additional testing that they were unable to do from 1978. Gove's comment about additional tests being "approved by some reputable and dispassionate international group of scientists," was obviously meant to exclude STURP's participation.

1986 *(October)*. Italian medical doctor Baima Bollone cited a quote from Gonella, (taken from Benedetto P. P., "Carbon test for the Shroud," *La Stampa*, 5 October 1986): "The dating will not be precise because a tolerance oscillating between 100 B.C. and 400 A.D. must be considered, which with the appropriate corrections drops to half. It would be enough, however, that the ^{14}C gave such a dating to keep those who claim that the Shroud prints are the work of a skilled forger in silence". Bollone then commented: "Frankly, these declarations seem to be marked by a euphoric lightness, as will be bitterly ascertained to prove the facts. It is likely that Gonella's unwary optimism has infected the ecclesiastical environment since in Fossati's *[Fr. Luigi Fossati]* article, also dated 5 October, we read: 'It is presumable -- it is said in a low voice in the circles of the Turin Curia -- that if the results are positive, that is to say if it is established that the cloth can be brought back to the time of Christ, Turin will have to prepare for another great ostentation'. Personally, since then I thought I would distance myself and I concluded an article aimed at illustrating the method of analysis in the following skeptical and cautious terms: 'Dating with radiocarbon will certainly be the result of prominent scientific investigations on the Shroud of the second half of the 80s, could be otherwise, also due to the exceptional qualification of the researchers, among the greatest in the world. We look forward to knowing the results, to know how they have been obtained and how the various difficulties of the method and in particular of its application on the Shroud have been remedied so as to verify the reliability of the results in the light of other scientific knowledge' [...]."

"I have never had, and from the first day, the slightest confidence in the radiodating of the Shroud, on the management of the investigation and its results. I will immediately explain the reasons for such a cut and dried statement. I have always known the position on the origins of the Shroud of Cardinal Anastasio Ballestrero: an attitude of detachment, just from those who believe that a purified faith does not need scientific supports. [...] I have already reported what he had to tell me about it in chapter 6. More or less at the same time as this episode, an interview was published with the following statement: 'The Shroud is just an image. Out of respect for science, I never went so far as to say that the Shroud is a relic in the strict sense of the word: in fact, I believe that science will never be able to tell me that that blood is the blood of Jesus.' It was all too evident that the Cardinal Ballestrero would not have become a partisan of the Shroud and therefore could not have defended it from the so-called *science* when it decided to condemn it. The priest-journalist Pietro Accornero echoed it with the following comment: *Baima Bollone's statements are of great interest, but scientists do not expect the blessing of the Church*.

In fact, I had recently managed to show that the sheet contains traces of blood and that this blood belonged to a man. Clearly, such an attitude created my apprehensions and anxieties for the future.

When I learned of the Curia's consent to the request by the carbonists to exclude all the experts and scholars of the Shroud from the design and execution of the ^{14}C radiodating samples, I realized that my concerns were materializing. In any case, my small surviving hopes of an attitude of defense of the Shroud were definitively stifled the day following that of the withdrawals from a piece of news obtained from a Turin newspaper."

Source: Bollone, Pier Luigi Baima. Sindone O No. (Torino: Societa Editrice Internazionale), 1990, pp. 276 and 279.

Comments: Certainly the labs had no concern to "verify the reliability of the results in the light of other scientific knowledge."

The Accornero quote was put in italics to cut down the number of (potentially confusing) quotation marks from multiple individuals.

1986 *(October)*. Meacham wrote, "a few weeks after the Turin conference, things began to unravel. According to Gove's own account, in October 1986 'Chagas was having second thoughts about the tests STURP planned to carry out after the removal of the Shroud material for [C-14]. He was concerned about the floodlights and the X-rays and the ultraviolet radiation [...]. Chagas would like the advice from the laboratories on the possible danger that these tests might pose to the Shroud.' And if accurate, this account puts Chagas in an extremely bad light. And he was asking the radiocarbon labs for advice about the conservation impact of tests that STURP was planning on a textile? These were the very same labs whose directors mostly could not see the relevance of these tests for the carbon dating; needless to say, they would not have a clue about such a conservation matter. And yet, amazingly, they jumped in where angels fear to tread, and Gove, Harbottle, Hall and Donahue wrote to Chagas expressing the objections to the STURP tests. This was probably also a **political** *[my bolding]* move, to finish the job of shoving the STURP aside completely. The whole enterprise was shameful."

"Gonella would write later about this move on the part of Gove and his cohorts: 'at the beginning [...] they [the radiocarbon labs] had guaranteed us the utmost seriousness and completeness in the analysis, as well as promising to collaborate with the custodian of the Shroud, the archbishop of Turin and with his scientific advisor, the undersigned. Seized however by a feverish desire for celebrity, they began to renege on their promises: no further interdisciplinary investigations; just the carbon 14 test. They even badgered Rome, bringing pressure to bear so that Turin would have to accept their conditions [...]. Scientifically, I would have been happier and have my mind at ease if the dating operation had been carried out in the context of comprehensive, wider ranging and thorough chemical and physical investigation of the Shroud as originally planned. The carbon 14 laboratories preferred to work independently and they did not wish to collaborate with other scientists, something that, from the point of view of scientific methodology, left me greatly puzzled and certainly not satisfied'."

Source: Meacham, William. The Rape of the Turin Shroud: How Christianity's most precious relic was wrongly condemned and violated, (Lulu.com), 2005, pp. 76-77.

1986 *(October)*. Italian researcher Franco Faia stated, "*[Gove]* involved the head of the British Museum, getting him to sign something without being aware of what it was. From all over the world the Church was blackmailed, being told, 'either you let us examine it or it means you're scared and it's false.' Gove even put pressure on the three labs chosen to refuse the dating of it if it were carried out only by three laboratories."

Source: *The Night of the Shroud (La Notte de la Sindone)*, documentary directed by Francesca Saracino, 2011. In 2016, it was revised and retitled "Cold Case: The Shroud of Turin," which is available at amazon.com. I have a review copy of the original version, which has an English voiceover. The revised version has English subtitles. (The material cited here can be found between approximately the fifteen- minute and thirty-five-minute range on the original version review-DVD.)

1986 *(November)*. On the 4th, Wolfli wrote to Meacham, "Dear William, thank you very much for your letter of October 27 and the interesting documentation. I am not surprised about your strange ^{13}C. Measuring ^{14}C and ^{13}C on about 1,000 samples per year we observe quite often similar high ^{13}C values for C3- and C4-systems. Unfortunately, so far no systematic research on this particular problem has been made, as far as I know."

"Concerning ^{14}C-dating I share your opinion on the contamination problem. As a matter of fact it is the problem in small sample dating. I was quite surprised to learn in Turin that most of my colleagues obviously are not yet fully aware of the problem. Otherwise they would have voted for more sample material."

"Concerning the influence of the 11 year solar cycle on the production rate on ^{14}C I fully agree with Prof. Baxter's opinion. From our own ^{10}Be concentration in Greenland ice (see enclosed reprints) *[reprints not reproduced in this book]* we know that in this century the production rate in the upper atmosphere varied by about 30% due to this effect. Using the CO_2-model and this result as input, it is possible to calculate the ^{14}C-variation in tree rings. According to this model these variations should be not more than 2%o *[sic]*, variations which are just at the present day detection limits. I know about the Russians' results, but have no reasonable explanation for it. It should be noted however that much stronger ^{14}C variations can be seen in tree rings for two reasons: First of all, irregular sun activities particularly observed in the 17th century are responsible for a 5% increase of ^{14}C and a factor of about 2 in ^{10}Be production rates. These spike-like structures can be seen best in ^{10}Be and in the most precise calibration curve published most recently by Pearson in *Radiocarbon* Vol. **28**, 1986. In addition, a long term variation is observed (de Vries effect), which most probably is the result of changes in the geomagnetic field."

"In order to convert a Radiocarbon age (corrected for ^{13}C) into a calendar age, these effects have to be taken into account. Using this most recent Pearson curve, as well as some sophisticated computer programs, it is possible now to reduce the uncertainty of the convertion *[sic]* procedure considerably, as you will see soon. Our pyramid paper is not yet ready for distribution. I will sent *[sic]* it to you as soon as possible. Looking forward to meet you again in Turin."

Sincerely,

[signed]
Willy Wolfli

Source: Letter of November 4, 1986 from Willy Wolfli to William Meacham. Luigi Gonella archives, of which the author has a copy.

1986 *(November)*. "On the 5th, Chagas wrote to Casaroli and Ballestrero that he had received three letters from C-14 specialists (but didn't identify them), citing short excerpts that were denouncing the dangers of the proposed STURP tests. Gonella was angry that anonymous letters were being pushed for an agenda, and that these specialists likely did not even have the expertise to evaluate the STURP tests."

Source: Petrosillo, Orazio and Emanuela Marinelli. La Sindone: Storia Di Un Enigma, (Milan: Rizzoli, 3rd edition, 1998), pg. 142.

Comments: To continue with this thread, go to the Petrosillo/Marinelli entry for "1987 *(Spring)*."

1986 *(November)*. According to Gonella, on November 5th, Canuto informed Gove that Chagas had lunch with the Pope and had informed him of the Turin meeting, and that he would make recommendations against the STURP exams when he reviewed it during the week. On that same day, Chagas wrote to the Secretary of State that at the Turin meeting it was found that STURP intended to make examinations that the radiocarbon experts judged to be dangerous -- that he had received in this regard information from someone who supported this judgment and believed that the examinations had no value; on the same day he wrote to Cardinal Ballestrero in more succinct terms that there had been remarks by "some scientist" (without names or quotations) on the dangers that the Shroud would run following some of the other proposed exams and therefore suggested postponing every other examination until the dating was completed. On November 11th, the Secretary of State wrote to Cardinal Ballestrero transmitting the letter of Chagas and asking him for comments and information on the dangers reported for the Shroud.

In Gonella's eyes, the labs and Chagas, under the pretext of being concerned for the Shroud's safety, expressed generic concerns about some of STURP's proposed tests and then jumped to requesting a block of all of them. Gonella believed the action of the laboratories and of Chagas was clearly specious and constituted a serious violation of scientific ethics. Gonella elaborated further. He said it was unheard of that scientists accuse colleagues of dangerous procedures without knowing the technical data (which the radiocarbon laboratories did not know at all); in any case, the correct procedure would have been to express one's concerns to the proponents of the examinations, perhaps, by sending a copy of the letter to the person responsible for security of the object deemed to be at risk, i.e., to the Custodian of the Shroud or to his scientific advisor. Chagas, receiving said letters would have had to inform the senders of the correct procedure and transmit the letters to the Custodian of the Shroud, perhaps with his comment -- instead the letters were never transmitted to Turin, nor the names of their authors. By now it had fallen to the level of anonymous complaints.

Chagas had the first written proposal of STURP in hand for over eighteen months but, despite having given a negative opinion for "uselessness", he had made no mention of dangers; they were asked to do a study on the conservation of the Shroud before proceeding to the examinations while a good part of these was aimed at the study of the service itself that was difficult to face without the essential data of some of them. On a **political** [*my bolding*] level this action and the way in which it was conducted showed (even without direct knowledge of the background revealed by Gove) that there was a stronger bond than ever between Chagas and Gove, which their objectives were to prevent at any cost any examination other than dating and to remove the control of the procedures from Turin, and that does not take into account the outcomes of the Turin meeting; Chagas behaved like he believed he was responsible for the Shroud and didn't even need the ratification of Ballestrero; Casaroli seemed to accept this state of affairs and seemed to have lost sight of the multidisciplinary nature of the research that had also been recommended in the original protocol.

Source: Gonella, Luigi and Riggi di Numana, Giovanni. Sindone: il mistero continua, (Milan: Fondazione 3M), 2005, pp. 56-58.

Comments: Clearly there was a lot of worldwide interest in seeing multi-disciplinary tests on the Shroud. It is clearly a shame that the Chagas/Gove opinion won out.

1986 (*November*). Gove talked with Canuto on November 5th. Chagas and Monsignor Rovasenda had gotten together with the Pope for lunch the week of October 20th. The Pope had been "pleased at how the workshop had gone but had expressed concern about STURP." Canuto related that Chagas had received letters from Gove, Donahue, Hall and Harbottle regarding the proposed STURP tests. Canuto summarized them for Chagas to deliver in a letter to the Pope the week of November 5th. Canuto remarked that if the dating did not come out to first century, interest in STURP's tests would diminish.

Source: Gove, Harry. Relic, Icon or Hoax?: Carbon Dating the Turin Shroud (Bristol and Philadelphia: Institute of Physics Publishing), 1996, pp. 178-179.

Comments: Canuto's prediction that if the Shroud dating turned out not to be first century that interest in STURP's proposed testing would diminish did not come true. In fact, given that STURP's 1978 tests were the impetus for a C-14 dating, the C-14 results were doubted by many from the very beginning, and that Cardinal Ratzinger had in 1985 actually approved STURP to have two full weeks with the cloth, the importance of the proposed STURP tests can hardly be overestimated. Sadly, though, the Church never gave STURP permission again to do more testing.

Gove did not elaborate in his book exactly why the Pope had expressed concern about STURP. Perhaps Chagas' concerns impacted him. Perhaps the Pope would not have had concerns had he been able to meet with a scheduled STURP contingent on May 13, 1981. Unfortunately, that was the day of the assassination attempt on the Pope, and the contingent never did meet with him again.

1986 *(November)*. Gonella wrote in a report to Ballestrero,

"**1.4 The Political Context**"

"Research on the Holy Shroud arouses great public interest and seems to excite emotional reactions even more among 'agnostics' than among believers. One of its major problems is that of maintaining a strictly scientific imposition in the face of the tensions that tend to transform it into a publicity event, in which judgment is delegated to the general public (that is, to the mass media), while I take due account of the reaction of the public. It is above all a matter of perspective, since in the long term any deviation from scientific practice is pernicious, even if at the moment it may seem appropriate to give space to other considerations."

"The scientific works published following the '78 examinations, substantially to the detriment of the American and Italian scientists of

STURP, have made the thesis that the Holy Shroud is a medieval artifact. This thesis (which was, moreover, entirely legitimate in the context of pre-78 knowledge and had authoritative supporters in the ecclesiastical field), and also supported by some scientists who worked on the Holy Shroud, in particular by Dr. McCrone, the American microscopist-criminologist who made a preliminary optical examination of the adhesive tapes placed by STURP on the Shroud, and who enjoys wide publicity and esteem in the Anglo-Saxon world."

"The thesis of the 'medieval fake' is particularly dear to those 'secularist' environments which seem to be bothered by the idea that the Holy Shroud can be authentic and who like to classify, more or less explicitly, as 'superstitious,' 'unscientific,' 'fideistic,' etc. any assertion favorable to authenticity. These environments naturally welcomed McCrone's conclusions, but were puzzled when the publication of more than 20 papers in peer-reviewed journals by STURP '79 and 1981 refuted claims based only on partial evidence. Since then, the rumor was systematically fueled that STURP scientists are not 'credible' or 'objective' because their judgment is obscured by an 'emotional attachment' to the Shroud or religion. The controversy emerged from the scientific field to transform itself, from a 'secularlist' point, into a process of intentions: nobody was able to counter STURP's theses on the scientific plane, and therefore attempts were made to disqualify the authors *a priori*. STURP scientists are considered 'not credible' because they present unacceptable theses as they are favorable to religion[10]."

"In this climate, C-14 dating has been repeatedly claimed as the 'decisive proof', and the Church has also been accused of having so far denied it for 'fear of the truth'. Not only that, but we proceed to proclaim that the C-14 is the only measure to be done, and any other research program must be suspended until we have the results of this 'decisive test.' This corollary clearly shows its **political** *[my bolding]* nature: from a scientific point of view I also know (by hypothesis) the Holy Shroud, should it be of medieval origin, both its conservation problems and those relating to the formation of an image of such extraordinary characteristics would remain to be solved. But by focusing entirely on the topic of dating, the rest of the work and the other arguments in favor of authenticity are automatically disqualified (such as, for example, the demonstration of the presence of coagulated whole human blood); furthermore, scientists who have already worked on the Holy Shroud developing these topics, to entrust the task to other scientists 'more objectives' (i.e., favorable to the thesis of the medieval forgery); and if since the date turns out to be of the first century nothing has been lost because, as has already been asserted, this would only mean that someone had carefully preserved an ancient cloth to manufacture it as a false relic."

"This line was obviously supported by McCrone after rejecting the scientific debate with STURP. Gove also evidently joined it, who in the informal meeting in Trondheim tried to put it at the basis of the joint dating enterprise apparently of the six laboratories concerned: some passages of Gove's letters almost literally repeat McCrone's arguments. The line has followed quite a bit in Britain (to which a certain English-style anti-Americanism towards STURP is not extraneous) -- Hall seems to be inclined to do so."

"The disparaging attitude towards STURP and the '78 exams automatically reflects on the Turin Authorities, responsible for their management. The protocol followed on that occasion, with its careful division of responsibility, and generally ignored, and therefore the need for external controls as 'guarantee of objectivity.' With this in mind, responsibilities are neglected and generally ignored, and the need for external controls to 'guarantee objectivity' appears. With this in mind, the responsibilities of the Custodian of the Holy Shroud with regard to his safety and integrity, and his right to be informed in turn on the works of the scientists, are being neglected. While there is nothing contrary to any *collaboration* with Turin aimed at guaranteeing objectivity, it can certainly not be ascertained that Turin is disqualified *a priori* in terms of control of the procedures (which in 1978 proved to be perfectly capable of exercising) in terms of considering the foreseeable future activity on the Holy Shroud. In this situation, adherence to 'reservations on STURP' by people in the ecclesiastical sphere does not constitute a demonstration of 'objectivity' by the Church, but supports the unjustified suspicions of a 'domesticated research': it is not in fact, a question of a scientific discussion of results, but of prejudices in the literal sense of the word."

"The intrusion of this **political** *[my bolding]* component into Shroud research naturally influenced the attitude of the C-14 laboratories. They know very little about the research on the Holy Shroud and they did not care to deepen their knowledge (which is obvious, since for them it is an archaeological object like all the others). They feel uncomfortable for obvious publicity reasons in being associated with a scientific group that it rumored to be 'non-credible', without going into the validity of the claim. They suspect that the ecclesiastical authorities will try, more or less subtly, to push the results in a certain direction[11] (also because the 'lay people' attribute to the authenticity of the Holy Shroud a disproportionate value in relation to the Truths of Faith) and therefore seek safeguards in external controls. They are also flattered to see their work considered 'decisive' for a problem which at the media level has been called 'the mystery of the century.' Being brought into the limelight has made public relations considerations dominant over strictly scientific considerations in defining operating procedures. For this reason they come to a procedure so different from the usual practice such as blind dating performed in parallel by many laboratories."

"The trouble for us is that this situation has come to entail a request for much more scientifically- necessary material, and consequently we too face the problem of assessing which Shroud we must sacrifice to unjustified suspicions of 'poor objectivity' of a part of public opinion."

"It should be borne in mind that the labs are certainly not indifferent to the advertising that derives from the company, especially those of the University Physics Departments which are increasingly difficult to obtain the funds necessary for the maintenance of the expensive AMS equipment whose use in other fields is decreasing. The seven laboratories in question are not the only ones, not necessarily all and seven of the best. At the availability surveys conducted by STURP from '82 onwards, some laboratories replied that they did not want to be involved in investigations of religious objects; The Vancouver laboratory, among the most quoted and active in the world, made it clear that they would participate willingly but did not want to be involved with Gove."

[10]"The accusation of 'non-credibility' against STURP has never been supported by any documentation. When I asked Gove on what basis he declared his fellow scientists not credible, he replied, 'but everyone has seen the photographs of Jackson examining the Shroud with a cross hanging on his neck!' I protested that I did not consider it admissible to judge the validity of a scientist by his mode of dressing, and asked him if he had found anything of little purpose in Jackson's articles; he replied it was a big cross and that Jackson's articles are extremely objectionable because it is evident that makes an effort of objectivity to hide his true feelings. It was clear to you why it is evident that he makes an effort of objectivity to hide his true feelings. It was clear, I told him, that he was more emotionally touched by the Shroud than Jackson was. Although whispered in a low voice in some scientific circles, and supported by a certain media opinion, the assertions of the 'non-credibility' of STURP have not been openly supported by any scientist except McCrone and Gove, and even these have never been brought to the field of scientific publications."

[11]"When I had informally described the work of 1978 to a group of colleagues, one of them heartily recommended 'be careful when working with priests because sooner or later they make them say what they want'!"

"2 The advice of the Pontifical Academy of Sciences"

"It is difficult to describe the action of the Pontifical Academy of Sciences on the occasion of the consultancy requested by the Custodian of the Holy Shroud through the Secretary of State."

"The Academy has many illustrious physicists and chemists who could have provided a self-reliance on the Custodian of the Holy Shroud for the problem described above[1]." But its President, Prof. Chagas, preferred to deal personally with the mater, relying only on his personal friend, Dr. Canuto a theoretical astrophysicist (from a NASA office in New York) who provides scientific information to the Vatican Observer at the United Nations, no academician was involved -- only at the end Prof. Leprince-

Ringue was invited to participate in the Turin meeting, but declined the invitation (understandably)."

"Neither Chagas nor Canuto contacted Turin to inquire and discuss the situation and problems. Instead Canuto made contact with Gove in July 1985 and in October Gove conferred with him and Chagas in New York at the Vatican Observer at *[the]* U.N^2."

"In August 1985, Chagas sent his report to the Secretary of State, enclosing a Canuto report and Gove documents. Report and reaction did not even take into consideration the problems mentioned above in §1: they merely subscribed to Gove's approach uncritically, recommending that his proposals be followed. Canuto's report decidedly embraced Gove-McCrone's thesis that no other exams had to be done except the C-14 and that STURP was not 'credible,' without any documentation supporting the peremptory statements expressed by Gove and McCrone in their letters to other people. At the end of his report, Chagas proposed a meeting of the six laboratories, of STURP and of representatives of Turin and the Academy to be held in Rome at the Academy or in Turin 'under the responsibility of Cardinal Ballestrero' following the desire of the Holy Father."

"Turin ignored the comments about the other exams, not being pertinent to the consultancy requested, and those about STURP, not wanting to enter on an evident personal rivalry. It was clear that the Academy did not intend to provide any effective advice and that the problems should have been addressed in the proposed meeting, which was established to be held in Turin under the responsibility of the Custodian of the Holy Shroud. The date was set by Chagas for June '86."

"The first direct contact between Turin and the Academy occurred in January '86 on the initiative of Cardinal Ballestrero, who spoke to Chagas and asked him to discuss the various problems with me. I spoke with Chagas at the Academy on 17/1/86, explaining to him the problems outlined above and the concerns about the amount of withdrawal requested and the attitude towards STURP, which apart from any other consideration, and in fact and the formal presenter of the dating proposal is entitled. He told me that had reservations about STURP without specifying the nature, and that he thought it would be necessary to give a sample to each lab if they did not ask for more than 1 cm^2 each, without however motivating the judgment."

"At the request of Cardinal Ballestrero, and in agreement with Chagas, I also went and found Gove (who I already knew as a colleague) to try to understand the reasons for his position and make him understand our problems. I asked him to visit his lab in Rochester (I also wanted to see his equipment), but instead I set an appointment in New York at the L'Osservatore Vaticano headquarters at the U.N., where Canuto would also have attended (who then did not come because of influenza). It was a cordial conversation and we left as good friends, but with poor results because Gove was evasive on all technical problems and seemed obsessed with the idea of excluding STURP from the scene as 'not credible' (see note 10 §1.4). I told him that it was my duty to justify in strictly scientific terms every milligram removed from the Holy Shroud, and he replied that he would follow Chagas' indications. I immediately reported this interview to Chagas in Brazil by phone and letter."

"Immediately after (24/2) Chagas wrote to Cardinal Ballestrero that given the difficulties of holding the meeting in Turin (?), he proposed that it be held at the Academy headquarters in the Vatican 'in neutral territory ... under your Presidency of Honor and my direction.' Naturally, the Cardinal reiterated that the meeting had to be held within the terms already established, and to cut off further disputes, I set the location in the Archdiocese of Turin."

"Chagas telephoned me from Rio on 4/4 to tell me: the meeting was to be considered an event of the Academy[3] and by the rules of the Academy itself it should therefore have been held at its headquarters in the Vatican, but for exceptional consideration it was accepted to held in the Archdiocese of Turin, if it was agreed that the samples were taken on the last day -- also mentioned was which other people to invite. The same day he phoned me at a STURP meeting in the U.S. to warn me that Gove had told colleagues that he would obtain samples at the meeting. I reported to the Cardinal that I replied to Chagas, by telephone and telex, that the meeting was essentially procedural, to be conducted in already-established terms, only after its conclusion and outcomes could specific decisions on the follow-up of the tests and any withdrawals be made, the work would have to be opened by the Cardinal, the agenda items were specified, placing in the foreground the question of the number of the samples, and the names of some people to invite."

"On 14/4 Chagas sent the summons for the meeting from Rio, which was defined as organized by the Pontifical Academy of Sciences and to be held 'under the other patronage' of the Archdiocese of Turin. It was specified that the Academy would cover the expenses of those who could not be paid by his institution. In the short list of topics to be discussed, no mention was made of the problem of the number of samples and other points requested by Turin, while it was asserted that the meeting would decide, with the approval of the Cardinal, on the responsibility for the execution of the agreements entered into. The presence of the STURP chemistry expert whose presence was requested was not invited[4]. The invitation letter was not even sent to Turin as an alert: we only came to know via the courtesy of one of the guests who sent me a copy."

"Meanwhile, an interview with Chagas appeared in an American Catholic weekly in which they gave details on the procedures that would have been followed 'if the Vatican allowed the exam', assuming that six laboratory samples would be given and that the operation would be coordinated by the British Museum, without a word on the responsibility of Turin[5]."

"The Custodian of the Holy Shroud expressed to the Secretary of State his doubts about developments so different from the agreements confirmed several times and from his responsibilities, and the Secretary of State invited the Academy to reconvene the meeting within the terms established at the time, also asking that the invitations be limited to strictly necessary people (of three laboratories two representatives had been invited against one of the others, and it was not clear in what capacity Canuto intervened). On 16/5 a telegram from the Academy informed the guests that the meeting was postponed for organizational reasons, announcing a letter from Chagas (which was never written)."

"The postponement aroused a vibrant protest by Gove who on 19/5 sent a telegram in unheard-of terms to Cardinal Casaroli and Cardinal Ballestrero, speaking of reasons, 'not credible' and requiring immediate revocation. The telegram also bore the signatures of Donahue, Hall and Harbottle, who said later that they had joined Gove's initiative to protest the postponement but that they were unaware of the text he drafted, and that of the British Museum Director, Sir David Wilson, whom I learned later was not even aware of it."

"The meeting was reconvened by Cardinal Ballestrero's initiative which, having received only one interlocutory and recriminatory letter from Chagas without proposals on the matter, I asked him to meet with me, as your scientific trustee, to concretize the meeting sometime in September."

"At the appointment (24/6) I also found Canuto present. It was a difficult because it was Turin's intention to get the jobs done correctly but without breaking with the Academy. Chagas had an offensive weapon in Canuto to lash out against Turin in such a way that I explicitly asked Chagas in what role intervened in the discussion. Chagas told me that Canuto was his representative in the USA and 'scientific consultant at the Nunciatures at the U.N.' and continued the recriminations, asserting that the provisions of the Secretary of State gave him *carte blanche* in organizing the meeting. He wanted to show me the letter of assignment and was speechless when he had reread it he found that he was asked instead to organize the meeting in close collaboration with the Archbishop of Turin who was responsible for it. We then proceeded to draw up a convocation in the joint name of the Academy and the Cardinal, a text that required discussions on every single sentence. Since the idea of reimbursement of expenses to participants was unusual (unusual in scientific conferences), the Cardinal had established that if there was expenditure to be addressed, they would be at your expense as responsible for the meeting and commissioner of the Academy's consultancy (which, moreover, would have rebounded the costs on the Secretary of State), but Chagas refused to explain in the letter that the costs were borne by Turin, slipping to 'let the Academy know' if a refund was requested with the agreement which then would have communicated to me in Turin who needed to be reimbursed[6]. I also refused to limit the number participants, 'because it would have been said that the meeting had been postponed only to exclude certain people'. The Cardinal asked that a clear agenda be drawn up, but Chagas decided that the list of topics in the invitation letter was sufficient and that it was better to discuss the agenda at the first meeting when rescheduled. Also I told him that I thought it appropriate to send the attendees a memo on the state of the problem, so that they could be documented on the subject, but he told me not to do it, because was better to make it the object of the opening speech. However, we arrived at an acceptable text, which starts from Rome on 2/7 (after further modifications requested by Chagas and accepted by the Cardinal) without being accompanied by a list of guests."

"All the prospective participants accepted the invitation (or asked to send a substitute) except, as already said, the Academician Leprince-Ringuet."

"Shortly before the meeting a letter from Chagas sent from Rio on 8/9 with a detailed agenda prepared by Gove, and a thank-you from Chagas to Gove himself for organizing the meeting upon request from the Pontifical Academy of Sciences (!) It was, of course, quite different from what had been discussed in June, and in particular it obscured any reference to the problem of the number of samples to be taken. Faced with the protest of the Cardinal for this umpteenth violation of the Chagas agreements, he telegraphed that it was only a draft to be discussed at the opening of the meeting."

"At the meeting, the Pontifical Academy of Sciences finished speaking only with its full administrative structure -- President, Director, and Deputy Director -- but not with its structure and scientific expertise, replaced by Dr. Canuto, invited without formal justification as Chagas' personal advisor."

"It is easy to understand how the 'collaboration' of the Pontifical Academy of Sciences was a particularly frustrating experience for Turin. No scientific help was provided, indeed a scientific discussion could not even be set up, while a disproportionate expenditure of energy had to be faced, wasting a year of time just to launch a simple meeting of people already identified at the start."

"Throughout this period Chagas has shown with the facts (and in the end also with formal thanks) that he had closer and more cordial contacts with Gove than with his clients, despite Gove being one of the parties involved with what we had to face -- it seems that Chagas does not remember from one time to the next what has been established, and it is clear that for him personal and prestige reasons appear dominant. After this experience I see with considerable apprehension the expected intervention of the Pontifical Academy of Sciences in the future development of the works."

[1] "Physicists 21 with 9 Nobel Prizes, chemists 25 with 15 Nobel Prizes (including 14 biochemists with 10 Nobel Prizes)."

[2] "I learned of these meetings only at the end of '85, because Gove wrote to STURP to tell them that he already agreed with Chagas on dating procedures, and STURP sent me a copy of the letters."

[3] "For an agreement between the Pontifical Academy of Sciences and the National Science Foundation of the USA, NSF pays the expenses to a U.S. scientist who takes part in meetings of the Academy. In this way, representatives of the U.S. labs were able to get the trip subsidized by the NSF. Gove wrote to STURP that he gave NSF the funds for participation in the meeting of a representative for the laboratory, but not for representatives of STURP as he did not believe that STURP already participated in the renouncement. STURP representatives came at their expense."

[4] "Prof. Adler, who had thoroughly studied the chemical characteristics of the Holy Shroud, refuting McCrone's thesis and demonstrating that the bloodstains are actually of coagulated whole blood."

[5] "*Our Sunday Visitor* 13/4/1986, authored by P. Jennings."

[6]"They all thanked the Academy profusely for his generosity and were embarrassed when they learned they were guests instead of the Archbishop of Turin. It should be noted that this was the first time that the Church had to spend money on research on the Holy Shroud: on previous occasions scientists had not even thought about asking for reimbursements. The Academy explicitly asked us to lodge all the guests in a first class hotel, and to host Mrs. Chagas together with her husband."

"4. Meeting results and ongoing"

"The meeting was a success from the point of view of personal relationships, but not so much from the point of view of the integrity of the Holy Shroud."

"The situation has been settled, a direct relationship with all the parties has been established, and an acceptable protocol in principle has been reached, and still 'negotiable', insofar as there are inserted in the minutes such ideas to allow the Custodian of the Holy Shroud and to the Higher Authority to lighten withdrawals and procedures if they deem it appropriate."

"The problem is now to fix the protocol in detail, solving the environmental points (especially on the number of samples to be taken from the H.S.) and clearly setting the responsibilities of the various actors in relation to the operations to be performed."

"The outline delineator in the final report of Canuto (§3.12) and somewhat cumbersome, provides for a withdrawal of Shroud fabric which is in my opinion exaggerated, and is rather lacking in dividing tasks and responsibilities among the various agents."

"I am not at all sure that I have fulfilled my duties toward the Holy Shroud on the one hand, and on the other, to maintain good relations with scientists and the Pontifical Academy of Sciences, in the presence of considerable personal susceptibilities and under the pressure of Dardozzi's calls that implied my responsibility in case of 'disagreements',

I did not know until that point to insist on certain positions, all the more considering that I could not commit *a priori* the decisions to be taken by the competent ecclesiastical authority, and that there was a certain interest in exploiting the alleged divergences between Rome and Turin. I therefore tried above all to define the best in technical points and explicitly separate them from the **political** *[my bolding]* ones -- the competent Authority will certainly know better than I how to find the right balance between the two fields."

"After having analyzed in the light of the meeting the position of the bodies involved -- the laboratories, STURP, the Pontifical Academy of Sciences and the certifying and analytical bodies -- I list in §4.6 the points on which it is now necessary to take a position."

"4.1 The laboratories"

"A good relationship has been established with the C-14 labs (except perhaps Rochester) that promises reasonable collaboration."

"Of course, the dating labs don't approach the Holy Shroud with that spirit of enthusiasm and respect that animated STURP and other groups in '78. For them the dating of the Holy Shroud and above all a company of great notoriety -- not only for the individual participants but also (perhaps more importantly) for their common scientific discipline which can present itself to the public as capable of 'resolving a mystery'." I have reported in some detail the progress of the discussion so that we can understand how the reaction of the public opinion (i.e., of the mass media) to them is dominated by their work to better the result they will present."

"Their representatives came to Turin expecting to have to overcome general resistance of the 'clerical' stamp and counting on the support of the Pontifical Academy of Sciences (through Gove) to impose a scientific vision. Instead, they had to face a strictly scientific-technical discourse which had not prepared[1], and which exposed the gap between their publicity needs and a strictly scientific approach -- a gap they mostly explicitly admitted."

"The same publicity motivations that led to such a cumbersome protocol make the laboratories position towards us weak, if we wish to lighten it. The laboratories are well aware that the request for 6-7 samples is not scientifically reliable: the justification they have brought, that is to make the measurement 'credible' in the public eye, exposes them to easy criticisms of other scientists, who may very well ask them since when has science delegated to the press the task of judging the validity of its measures. That's why they had to say the Holy Shroud is not an object like the others due to its religious importance (it is necessary to clarify which one) and we have seen the strange spectacle that some of the scientists (and Chagas) who talked about relations between science and faith in the context of this dating after the Pontifical Custodian had clearly said at the opening of the works that the Church wanted them to be held strictly on the scientific level since this research did not touch any problem of Faith[2]."

"In reality, the fact is that no wants to stay out of such a highly-publicized enterprise, whatever the conditions for remanaging it. When Gove tried to force his thesis solemnly declaring that Rochester would have withdrawn if they had not been followed, he did not receive the least support (nobody said 'Much better -- one less to share the cake', but there was no need), and his stature certainly does not increase for not having kept his promise. They admitted (p.21) to be too many, but as Gove said, no one would voluntarily withdraw to save the Shroud fabric (and this shows clearly that their part in the enterprise amounts to more than the Holy Shroud), and therefore they do not separate as to reduce the number, especially considering that the press (and Chagas) regards them as 'the best laboratories in the world' precisely because they are engaged in this venture. The use of controls, which reduces the withdrawal without reducing the number of participants, offers a good way out if you want to minimize the impact on the Holy Shroud. I therefore believe that if the responsible Authorities decided to use more controls and less samples they would not encounter resistance."

"A point that has not been clarified at all and the formal modality with which the 7 labs are put to work -- the 'who calls who?' that I asked in vain to put on the agenda, or in technical terms, who is the submitter for the dating of the Holy Shroud. It is not convenient for the enterprise to appear as commissioned by the Church: there would be all the disadvantages mentioned in §1.3 without any of the advantages of a client on the choice of the laboratories and the procedures. We cannot say that we accept a proposal from the laboratories themselves, because there is no document proposed by them signed and presented. The same applies to the Pontifical Academy of Sciences, which also explicitly said that its role is only to help Turin to settle the matter. All in all, I think the best solution is to treat it as it actually took place, that is, as an evolution of the STURP proposal of 1984, integrated by the French one of 1986, agreed at this meeting specifically convened between the involved labs, STURP, the British Museum and Dr. Flury-Lemberg[3], which the Pontifical Custodian of the Holy Shroud accepts *juxta modum*, making the necessary corrections to better safeguard the integrity of the Shroud itself, meaning this acceptance with corrections in a letter to the various bodies involved."

"It should be remembered that this letter specifies that the subject of the research and exclusively the measurement of the radiocarbon date of the Shroud cloth and *not* the 'authenticity of the Holy Shroud,' given that some of the participants seemed to assume that they were called to 'authenticate a relic[4]'."

"4.2 The Pontifical Academy of Sciences"

"It is clear from the transcript how the Pontifical Academy of Sciences did not bring any scientific contribution to the meeting, and how in the Chagas presidential office he did not support the requests of Turin, behaving almost as if he were the consultant of the laboratories instead of the Custodian of the Holy Shroud. The meeting was for us a continual struggle to rationalize the withdrawal and to pinpoint control over the operations by the Custodian of the Holy Shroud, a struggle that often had to be conducted even more against Chagas than against the principal laboratories[5]." Chagas' action seemed systematically aimed at ensuring that all the laboratories had all the Shroud material that made him comfortable with a limiting criterion, in stark contrast to the philosophy followed so far by Turin for the Holy Shroud exams, which required timely justification of each withdrawal (§1.3)."

"Chagas' incomprehensible bond with Gove appeared clearly on many occasions. I begin by proposing an agenda elaborated by Gove after refusing to have it processed by Turin, and ends, on the last day, by discussing again at Gove's request what was decided the day before, blatantly failing to the practice followed by any president of a meeting. He never intervened on the attacks brought by Gove on other participants in the meeting, but I block my requests for clarification defining them a personal matter. He took a stand against him only at the end, when it became clear that Gove was completely isolated, and he also asked to modify the timetable set by Chagas himself for convenience. I was forced to formally ask to record my dissent from the procedure followed to be able to discuss what Turin had been asking for months to discuss, and only in this way was it finally possible to demonstrate, by explicit admission of the labs themselves, that the basis of their requests had no scientific justifications but only **political** *[my bolding]* ones ('credibility in front of the public') -- a fundamental point that allows the Authorities responsible for the Holy Shroud to take a thoughtful and responsible position on the extent of the withdrawal without exposing the accusation of entering into the merits of scientific problems."

"This link is probably at the root of Chagas' position on the problem of the number of samples to be taken from the Holy Shroud, which from the beginning had made him present as the thorniest given the exceptional situation that saw so many laboratories involved in the affair. If we had talked about reducing the participants, it was obvious to exclude the least qualified labs, and that of Gove and in fact the least qualified of the group, as clearly appeared from the comparison of §3.6 (p.26). For Chagas I try in every way not to discuss the number of samples and take it for granted that a Shroud sample was given to each requesting laboratory. After having made sure that the question was not put on the agenda (§2), he did not even consider my official proposal of discussion as a first point, accepting instead a proposal by Canuto (not even openly advancing it) that discussed before the sample size (p.22). Thus a global figure for the withdrawal came out, and Chagas immediately accepted it as small enough not to deserve further discussion, without specifying any criterion of judgment (but without discussing the criterion of metrological *[physics]* need that he proposed) and without noting that was almost double than in January he had judged a higher limit."

"Once accepted, on the proposal of the laboratories themselves, the idea of working with a control, Chagas' decision to take in any case 7 Shroud samples to give one to the British Museum was very strange, but it becomes understandable in light of the subsequent request by Gove to discuss again, all so that everyone could be sure they have a Shroud sample in their hand. And the fact that, despite the repeated decision to work with a control, the proposal to take 7 samples was kept in the final report of Canuto suggests the intention to return to the question at the time of the collection ('since we took 7 samples, why not give one each?'). Without this forced piloting of Chagas, the discussion would have easily moved to the level of the 4 samples. While I obviously tend to have as much material as possible (as Hall frankly said -- p.21), the labs were well aware that they could not justify 7 samples, and in fact they started the general discussion talking about a 5-sample job (normal bargaining moves in the front of the reference at 3), and they spoke again in these terms the next day, but both times Chagas ignored the idea and carried over the discussion to 7 samples. Likewise I ignored the cue advanced by the British Museum, which with the controls the number of samples was less than that of the laboratories, continuing to present them as one."

"The second line Chagas pursued in presiding over the meeting was to ignore or compress Turin's role in controlling operations and to arrange his personal intervention at this stage. Together with Gove and Damon, he continued to speak of the British Museum as the body destined to preside over the certification collection of the samples, and it must have been Hall of Oxford together with the British *[Museum]* to explicitly recall the role that Turin had to play there, at which point he took to a vote for a suggestion from Damon to propose the Academy (in the person of himself) as a 'third certification body.' Throughout the meeting I have taken for granted that the works would have been carried out under the supervision and control of Turin, but they failed to understand if this point was actually implemented by Chagas, and consequently by someone else present. At the end of the meeting Chagas delegated the task to summarize the results to Canuto, who had not contributed to the discussion and has no role in the Academy and these promptly proceeded to give the interpretation as close as possible to the Gove line (§3.12). I wonder with concern whether Chagas intends to be admitted, if the Pontifical Academy of Sciences intervenes, it must do so with its official members."

"4.3 Certificators and Analyzers"

"The meeting proposed that there be three 'certification' bodies and three 'analyzer' bodies. The certifiers, with the task of identifying and documenting the samples and distributing them to the laboratories in conditions of blind examination together with the controls, will be the Archdiocese of Turin (in the person of Gonella), the British Museum (in the person of Tite) and the Pontifical Academy of Sciences (in the person of Chagas). An excellent relationship has been established with the British Museum: Tite has shown good understanding and willingness to collaborate. We all agree on the opportunity that an institution such as the British Museum collaborates with Turin on the certification of samples provides, and Tite is the first to recognize Turin's primary responsibility in this regard. The British Museum should also provide inspection and control samples."

"I doubt much about the usefulness and opportunity of a true certifier. The proposal started from Damon with the evident intent to open a role for Chagas in the operational phase, and Chagas took it on the fly. But it does not seem to me that a third entity can add anything to the 'credibility' -- indeed it can detract from it. From the point of the guarantee towards collusion, having multiple verifiers only reinforces the aforementioned that there is someone who wanted this; from that of secrecy in the identification of the samples (for the blindness of the exam), the more people there are, the more likely there will be indiscretion. The presence of a 'secular-cultural' entity such as the British Museum alongside the scientific representative of the Custodian of the Holy Shroud has its meaning of reciprocal coverage in the face of the suspicions of the two adverse parties, but what role does a third entity linked to the ecclesiastical environment have? Secular critics can only interpret it as an ecclesiastical desire for preponderance (the Pontifical Academy is certainly not seen as neutral, and would not intervene with a scientist of his own renown and competence), others would see it as the Vatican's distrust of Turin."

"Considering Chagas' manifest link with Gove, and his behavior in setting up and presiding over a meeting for which the Archdiocese was to be responsible, I see with learning his presence in the certification and distribution of the sample. I don't think I can take responsibility for the security and integrity of the Holy Shroud if the possible role of Chagas in the operations connected with the collection of the samples is not defined and limited very precisely. In October I spoke with Prof. Anthos Bray, Director of the 'G. Colonetti' Metrological Institute, about the proposal that his Institute should be called to analyze the results of the analyses together with the British Museum, and he assured me of their full availability. IMGC is one of the largest metrology institutes in the world, and can guarantee us a correct assessment of the uncertainty of dating."

"Also for the phase of analysis of the results I do not see the need or the opportunity of a third body, I specify a statistical institute not identified by the scientific group responsible for the research but chosen specifically by the Pontifical Academy of Sciences. The analysis of the results of a measurement is not a matter of pure statistics, but is a matter of metrological evaluations where statistics is only a tool. The general and highly qualified metrological competence of the IMGC suitably complements the specific competence the specific competence of the British Museum in archaeology and technique of the C-14 and does not understand what further contribution a pure statistic can give you. Here too the work of many chefs could ruin the sauce. I do not know of any case in which a statistical institute has been called to analyze physical measurement data -- the proposal came out casually, in a perspective of 'triumvirates' by analogy with the idea of the three certifying bodies, but I have had the clear impression that there were many perplexities about it that they did not want to express in front of the first-person position taken by the President of the Pontifical Academy, who had initially seen the Academy itself in this role. For the reasons explained above in §3.12 (p.38) I consider completely inappropriate that the Pontifical Academy of Sciences directly enters on the analysis of the results, and these reasons can also be applied, albeit more weakly because they are indirect, to an institution specifically chosen by the Academy."

"4.4 STURP"

"STURP brought to the meeting a very positive contribution, both **political** *[my bolding]* and scientific, systematically supporting the ideas aimed at minimizing withdrawal from the Holy Shroud[6] and ensuring its control by the Custodian. Harbottle, who had been invited as a representative of one of the C-14 laboratories and not as a member of STURP, had a somewhat swaying demeanor between the STURP theses and that of his fellow radiocarbonists[7]. Having actually promoted the dating request, with many years of work, STURP obviously cares about recognizing this role, but has proven to be completely available to any evolution of its original program, and to keep the C-14 dating completely separate from the rest of its research program."

"Given his previous work experience on the Holy Shroud -- in particular that of Riggi (which joined STURP after carrying out the investigation on the back of the Shroud cloth in '78) -- STURP constitutes the obvious choice for the group to which to entrust the material execution of the sampling, and therefore Chagas asked him in the meeting (§3.9, p.31). STURP accepted without reservations the modalities that would result from the protocol approved by the Custodian of the Holy Shroud, and with the collaboration of Flury-Lemberg, proposed by Chagas as a trusted textile expert of the Pontifical Academy of Sciences (which in turn explicitly accepted this collaboration). Only Gove opposed this choice, and consequently Canuto removed any reference to STURP in his report (§3.12). However, I do not believe that there can be doubts about the execution of the withdrawal by STURP. It would be absurd to renounce their experience and their proven reliability only for the opposition of a person with implausible motivation. The execution of the sampling by STURP immediately before the start of other examinations programmed by it will allow using sampling equipment otherwise difficult to find and put in place, and in particular to proceed with preliminary measures on the fabric suitable to avoid, as far as possible, the danger of taking 'young' carbon-contaminated material that could result in a more recent date."

"4.5 The size of the withdrawal"

"The somewhat confused way in which the discussion was conducted has meant that at the end of the meeting, considerable perplexity remains regarding the precise amount of material to be taken from the Holy Shroud. In the discussion (§§ 3.5 and 3.7) there was a tendency to highlight that the required material was not so much so as to need to precisely define the quantity. But for the actual withdrawal we will have to know how much to cut -- it would not be appropriate to cut a few more centimeters more just for lack of precise data."

"The situation can be summarized as follows:"

- Laboratories with small counters require more carbon than those with AMS, but I do not know where more fabric corresponds because the conversion yield is better for small counters -- from the data offered it seems that the difference is not significant.

- AMS laboratories use different pre-treatment and conversion methods with different yields. Furthermore, it is clear that their requests related to different measurement modes (different number of separate measurements). Only Wolfli provided his conversion yield.

- It has been proposed to make 5 mg available to the individual AMS labs and 25 mg of carbon to the two small-counter labs, but it is unknown how much these numbers actually correspond.

- The figure of 12.5 cm^2 of fabric corresponding to 50 mg of carbon available at 7 labs came out of a proportion with a Wolfli datum on a cloth with a density other than the Shroud, and without checking which measurement methods were referred to. With a different interpretation of Wolfli's words resulting from the transcript, I can reach 16 cm² or 11.8.

* It was clearly decided to use a control, which implies the use of only 6 Shroud samples. The effects on the withdrawal could be different using the control with an AMS or a small-counter.

* However Chagas and the Canuto report talk about taking 7 samples in any case, one of which could be left unused.

* The specification of the Canuto ratio that the sample must contain 50 mg of carbon does not take into account the conversion yield, and would lead to a sample certainly insufficient for 7 samples, and probably also for 6.

"In this situation, the only thing to do is to write a very technical letter to the 7 laboratories asking for thoughtful and precise data. As this is purely technical information, I can also write it while it is still in the interlocutory phase, but it may be preferable to wait until the competent Authority has made its decisions on the procedures to be followed. In the absence of these specifications, I can assume that one sample for AMS (with 5 mg available in the measure) corresponds to 1.7 – 2.5 cm² and the two samples for small-counter to 3.5 – 4 cm². The 5 mg figure for AMS is leveled to the maximum, to allow a number of separate measurements to provide sufficient statistics for a dating performed by a single lab. In this case, where several labs are at work, there is no scientific motivation for many repetitions within each one, having a collective statistic, and it would therefore be more reasonable to speak of 4-3 mg of carbon per laboratory[8] -- then the sample size for AMS may drop to 1.4 or 1.0 cm². Using a control (for AMS) the sampling of 6 samples could therefore be from 7.5 to 14 cm². With two controls (one for AMS and one for small-counter) from 5.8 to 12 cm². With three controls (two for AMS and one for small-counter) from 4.8 to 9.5 cm²."

"It is likely that the sample must be taken for the edge of the sheet, but it is also possible that enough non-carbonized material can be found under the patches, to limit the collection from the visible parts. This could only be decided at that moment, relying on textile experts to evaluate where the withdrawals bring the least structural damage and the least disfigurement."

"**4.6 Decisions to be made**"

"They take for granted of course, the intention to proceed with dating on the general lines set up in the meeting, they present themselves at the Custodian of the Holy Shroud the following points on which it is necessary to take a stand."

*1) Role of the Pontifical Academy of Sciences in the continuation of the works, in particular in their executive phase. Given the nature of the consultancy that the intervention requested from the Academy, the way in which he carried it out by reducing it to a meeting with the interested parties, and the repeated affirmations of Chagas that his sole purpose was to help Turin *set up* the enterprise of dating, I think it would be completely obvious to the Custodian of the Holy Shroud and the Secretary of State to thank the Academy for its valuable contribution and to tell you that its task ended with the meeting, which allows Turin to move on to the executive phase of its competence. This would be the best solution to allow us to work with the serenity and precision, without having to continually look behind us and confining Gove in his place of one of 7 measurers. In anticipation of this probability, Chagas has, however, preordained his role in the execution phase for the certification of the samples, which would give him access to the sampling operations with the consequent opportunity to condition them, and also an entry, at least by proxy, on the analysis of the results. For the reasons already explained (p.44) I consider these hypotheses of intervention by the Academy to be inappropriate and rather perilous. Chagas could perhaps be allowed, as a diplomatic exit route, to have a personal presence at the withdrawal operations, without any official function that would allow it to intervene. In any case it should be clear that there is absolutely no further role for Canuto.

*2) Supervision of withdrawal operations. It must be established without any misunderstanding that the operations are carried out under the control of the scientific-technical commission designated for this purpose by the Custodian of the Holy Shroud, responsible only to him, and in an appropriate location for all present. The experience of '78 has shown how many unexpected events can be encountered in the operations on the Shroud cloth, which must be resolved on the spot and on the spot with a clear idea of the priorities to be respected and the scientific implication of the different options. It also showed that not all scientists have the same material care, the same respect for the Holy Shroud, and the same scrupulousness in sticking to work plans[9]. The sampling supervision should not be confused with the certification of the samples: for the latter, two or more 'certifiers' of equal authority can be appropriately obtained, but a single authority is indispensable for supervision.

*3) Participation in the withdrawal operation. To make effective supervision and avoid possible accidents, only technically and scientifically necessary people for the collection and official certifiers are present in the work rooms. At the meeting, some representatives of the laboratories expressed their wish to attend the sampling, and Chagas promised them the financial support of the Pontifical Academy of Sciences for the trip to Turin[10]. It was noticed that the request was unusual (the analysis laboratories are never present for archaeological sampling) and there was a distrust towards the certifiers[11], and I pointed out, with the consent of Chagas, that at most they could observe from several meters from the Holy Shroud, based on the experience of '78. The presence of laboratory representatives in Turin would certainly simplify the distribution of samples, which otherwise would require a trip by the certifiers[12], but this is not a problem. In the Canuto report it was written that the sample "will be made so that the representatives of the 7 laboratories have complete knowledge of the process", which can also be accomplished by sending them a copy of the video recording of the sample-taking, which will in any case be made for documentation. It can therefore very well be said to the representatives of the labs that they will not be admitted to the premises (also because this would be antithetical with a blind examination) and therefore their presence in Turin at the time of sampling is useless. Given the terminology of the Canuto report, it will be good to specify that the withdrawal operations will be carried out by STURP with the collaboration of Flury-Lemberg, under the supervision of the scientific-technical commission nominated by the Custodian of the Holy Shroud; they certainly cannot be called the "responsibility of Flury-Lemberg," which I do not doubt the lady fully agrees with.

*4) <u>Certifiers</u>. It is necessary to decide whether to accompany the representative of the Custodian of the Holy Shroud in the certification work of the sample[13], the British Museum is sufficient or the Pontifical Academy of Sciences is added in the person of its President. I am convinced that two certifiers would also be undesirable for its link with one of the laboratories, highlighted in the preparation and conduct of the meeting in Turin.

*5) <u>Analyzers</u>. I am very perplexed on the modalities for the call in place of the "analytical bodies" referred to in §3.8, 4.3. The analysis of the results is obviously an integral part of the actual scientific research, on which we have never entered. However, the point of having the raw results analyzed by bodies external to the measurement laboratories has been set by the laboratories themselves as an integral part of the Holy Shroud, and someone must call these bodies to work. This can be easily done for the British Museum and IMGC, explicitly proposed with scientific motivation during the meeting. The problem is constituted by the possible third entity, which has no scientific reason to be added to it. I said in §4.3 because it does not seem to me to be able to add the Pontifical Academy of Sciences or an institution chosen by you. I believe that the competent Authority could very well indicate in the protocol only the names of the British *[Museum]* and IMGC as analytical bodies explicitly requested by the participants in the meeting -- the absence of a third name would have been motivated (if deemed appropriate to account for it) by the simple fact that this name was not expressed, and one could certainly not assign a task to a blank entity.

*6) <u>Number of samples</u>. I consider unacceptable the idea of taking 7 samples once it has been decided to an exam with a control. With 6 laboratories at work, there is no need for an "accident reserve[14]". The competent authority must decide whether to use the cues provided to use multiple controls in order to limit the withdrawal from the Holy Shroud. It is a **political** *[my bolding]* decision having been clearly admitted that the reason for using more than three samples was only of public credibility (§3.7). In this political evaluation, the opportunity to underline the right of the Custodian of the Holy Shroud to have a scientific justification for the required exams, and the repercussions that the position taken in this case will have regarding future requests for other exams.

*7) <u>Call to the work of various operators</u>. The above specifications must be communicated to the 7 labs, STURP, Dr. Flury-Lemberg, the British Museum, the Metrological Institute "Colonetti", as well as the Pontifical Academy of Sciences. While for the laboratories and STURP the communication has the character of acceptance with amendments of a proposal[15], for the others it is a request for work as a result of availability offered for it (the British *[Museum]* must also be asked for the contextual supply of the control samples.)"

"This call in place should be made by the Custodian of the Holy Shroud, so that it is clear that we must refer to him in the next operational phase. It would be opportune for him to clarify incidentally how he has the full approval of the Superior Authority, to block any further attempt to circumvent the limitations of the protocol in the name of pretexts of no precise indications and Vatican sources."

[1]"In fact, they did not have the necessary punctual documentation, except for Harbottle and Wolfli who expected a technical speech (the first via STURP and the other for having occasionally called me the week before.)"

[2]"I remember once Cardinal Ballestrero rightly pointed out to me how Shroud research did not concern the relationship between science

and faith, but rather that between the scientific world and the ecclesial world."

[3]"This approach is also shared by Chagas in his latest speech in response to Gove -- cf. §3.10, p.36."

[4]"We also remember Hall's note on the repercussion that this dating will have for the other relics (clients in sight ...)."

[5]"With all the difficulties due to the continued imperative from Dardozzi to avoid disagreements -- our task would have been easier in the face of an openly-hostile presidency."

[6]"Lukasik also presented an articulated proposal to optimize the collection, with 3 samples of intact material and the other of charred material, which Chagas did not question (p.28)."

[7]"It should be borne in mind that STURP is not an institution, but a free association of scientists who have a research program on the Holy Shroud and intend to collaborate with each other while maintaining their freedom of research."

[8]"I was not allowed to set this up during the meeting."

[9]"In 1978 for example, I had to intervene heavily on Frei so that he would not apply his adhesive tape on the image instead of only on the edges as he has asked, and *[also had to intervene heavily]* on *[photographer]* Ghio so that he would not take unauthorized photographs with a close flash."

[10]"It would be appropriate to check whether it intends to upload it to the Academy's ordinary budget or relegate it to the Secretary of State as an expense related to the consultancy requested from the Academy."

[11]"And it is also useless for the purpose of 'authenticity of the samples' in the context of a blind examination, in which certifiers must then mix

the Shroud samples and controls with a code known only to them, obviously carrying out the operation in secret."

[12]"Dispatch by mail or third party is to be excluded."

[13]"I suppose Cardinal Ballestrero does not to intend to certify the samples in person but delegates a member of the supervisory scientific-technical commission for this purpose."

[14]"In addition, the idea itself is unethical with the blind exam because the laboratory that asks for it to make up for an accident would be the only one to be sure of having a Shroud sample. Note that nobody mentioned having backup samples."

[15]"For the laboratories, remember the point referred to at the end of §4.1 (p.41) regarding the authentication of relics."

Turin, 20 November 1986

[signed]
Prof. Luigi Gonella

Source: Report of November 20, 1986 by Luigi Gonella titled "The C-14 Dating of the Holy Shroud." Luigi Gonella archives, of which the author has a copy.

Comments: This report was sent to Ballestrero. It was fifty-two pages in total. The above excerpt I have reproduced were the concluding sections. The two sections that Gonella refers to in the above excerpt are in another report, reproduced in the entry immediately below. That report was apparently sent to the Pope. Most of the material in the above report to Ballestrero that has not been reproduced here would seem to have been covered by other sources in this book, including excerpts from Gonella's own 2005 book co-authored with Riggi. There is some overlap with the material from these two reports and the entries based on the Gonella/Riggi book, as well as a gap in the chronology of the entries. For example, Gonella, in his book, related events that occurred in February

1986, and the entry for that source is listed for that month. This report to Ballestrero was dated November 20, 1986, so the entry for this one falls in this time frame.

One might get the impression from Gonella that Gove and McCrone were very closely aligned. While it was true that both Gove and McCrone had issues with STURP, Gove, not surprisingly, had some negative things to say about McCrone. In his book, Gove wrote (pg. 19), "I sometimes think that McCrone dreamed of becoming history's greatest iconoclast. Having in his view, demolished the authenticity of the Vinland Map, he saw the chance to do the same to the Turin Shroud!" And referring to the latter, Gove wrote (pg. 20), "It remained, however, for others to settle the question and to do so with somewhat greater objectivity and with a great deal more credibility." Gove's self-assessment of his own objectivity and credibility appeared to be off the charts.

Regarding the postponement of the Turin workshop, Gonella revealed the mind-blowing facts that Gove, in his protest letter to Casaroli and Ballestrero, added Donahue, Hall and Harbottle's names as signatories to the letter even though they had not seen the text and also added the name Sir David Wilson, who wasn't even aware of the letter!!! Then Chagas believed he had the power to organize the meeting however he wanted, until Gonella had him reread the Secretary of State's letter, at which time he discovered the small detail that he was, in fact, supposed to organize it with Ballestrero! Both Gove and Chagas seemed to be adept at what today we would call "fake news."

Gonella had to correct Chagas' perception from a letter from Secretary of State that Chagas had *carte blanche* to run the Workshop. Dinegar had to correct Gove's perception from a document about one of the planning aspects of the dating (see Dinegar's entry for "1985 *(June 26th)*." It seems that both Chagas "read into" documents "facts" that were not there.

It's rather surprising that Gonella opined that the meeting was a success from the point of view of personal relationships, when in fact, it was those relationships that caused so many of the problems.

Gonella mentioned the importance of doing chemical analysis of the chosen samples to make sure that the samples taken were representative of the main cloth, one of the most crucial points that were ignored. STURP was to have taken the samples. That was not done. Gonella also did not want to cut more material than what the labs would use for their dating, but that was done. Gonella did not want the labs' representatives to be present for the sample-taking but that happened as well. Basically, for every criterion that Gonella wanted, the opposite happened.

It is striking that Gonella asserted "Chagas had an offensive weapon in Canuto to lash out against Turin [...]." Although Gonella downplayed Chagas' role in the dating and wanted to see Canuto eliminated completely, I state later in comments to another entry that a highly-reliable source told me that the Academy continued to call the shots and even let Gonella be the fall-guy for their bad decisions. Even more striking was Gonella, when discussing Gove wanting to keep STURP from cutting the sample and Canuto subsequently cutting STURP from his report about Trondheim, said, "It would be absurd to renounce their experience and their proven reliability only for the opposition of a person with implausible motivation." But sadly, that's exactly what ended up happening! Several people quoted throughout this book have alluded to the relationship between Rome and Turin. Obviously, Rome won out in this situation.

1986 (November). Gonella[1] wrote a report, seemingly for the Pope's eyes. Gonella wrote,
   "1)"

"In two letters of 5/11/86 *[American format]* to S.E. Cardinal Ballestrero and to the Secretariat of State[2] the President of the Pontifical Academy of Sciences, Prof. Chagas, reports and makes his own concerns on presumed risks to the Holy Shroud deriving from the examinations proposed by STURP, expressed by unnamed persons who would participate in the meeting of Turin of 29/9/86 *[European format -- Gonella was not consistent in his use of formats]* on the dating of the Holy Shroud, and takes its cue to recommend that these tests not be performed at all until a study has been carried out on the conservation of the Holy Shroud itself."

"Whereas:"

* The concerns are expressed in terms so vague and imprecise under the technical profile that they could apply to any operation on the Holy Shroud;

* The concerns are communicated anonymously through a third party, contrary to normal scientific practice. Most of the exams in question are precisely aimed at studying the conservation of the Holy Shroud. Prof. Chagas knows how the Guardian of the Holy Shroud is well aware of the problem of risk associated with scientific tests, and has in fact established a protocol for these tests which provides for the examination of their technical plan by university experts of his trust;

* The Pontifical Academy of Sciences had already known the STURP proposal for a year and a half, the relative program having been delivered to it on the occasion of the consultancy requested for the C-14, but in all this time it has never mentioned any danger of the proposed exams;

* The Pontifical Academy of Sciences declared until 1/10/86 *[European format]* that it was completely extraneous to any other examination proposed for the Shroud beyond dating to the C-14;

* The execution or otherwise of other tests does not in any way prejudice the procedures proposed at the Turin meeting for the collection of samples for dating at the C-14, which provide for the collection before any other examination, at the time established by Chagas at its convenience, after consultation with STURP;

* It is suggested to suspend all the exams proposed by STURP, not only those for which presumed risks are reported, it is clear that this sudden preoccupation and the resulting proposal are not presented in the framework of a positive collaboration with the Custodian of the Holy Shroud for the sake of scientific research on it, but soon within the framework of the links already highlighted between Prof. Chagas and people, such as Prof. Gove, who oppose

on principle the best scientific awareness of the Holy Shroud (cf. Confidential report on the dating of the Holy Shroud, §1.4 and 2).

"2) The technical-scientific aspect"

"The complaint by Chagas and his anonymous suggestion of the risks that the Holy Shroud would run for the exams proposed by STURP and based only on arbitrary adjectives, without any quantitative or at least documentary support. There is talk of 'serious dangers' due to 'very strong lighting' or 'concentrated lighting' various wavelengths, of 'degradation of the polymeric structure of the fibers',

'degradation.' No data is provided on the irradiation that STURP is supposed to use, nor on the relationship between irradiation and alteration of the cloth."

"This speaker professionally deals with risk assessment[3] (as well as the interaction of radiation with the material) and has now learned to distinguish between scientific analyses, based on numbers intended to define safety limits, and generic complaints in which the risk is used as a pretext to block an initiative -- this complaint belongs to the dark category, characterized by the use of adjectives instead of numbers[4]."

"It should be clarified that obviously the Holy Shroud takes risks for whatever you do or don't do. You have a finite probability that a process of 'degradation of the polymeric structure of the fibers' will be triggered for each day of permanence in your cassette as it is without protections or instruments due to the thermal agitation in the atmosphere with intriguing acids and microbiological attacks. The bloodstains are somewhat altered, in principle, each time it is unrolled and rewound. In the long run, the image may slowly fade due to the progressive darkening of the background due to the anti-oxidation/dehydration of the cellulose. It could very well happen that by opening the box after several years (or decades), there is the cloth damaged by mold -- the excellent conservation that has taken place so far is not a guarantee for the future, given the rapid changes of the last decade in the composition of the atmosphere[5]. It is for all these reasons that STURP and other groups have proposed a series of tests aimed at measuring the physicochemical parameters relevant for the conservation of the cloth and the image, in order to have objective data, and to determine the safety procedures for future ostensions[6]. And these are precisely the tests that Chagas proposes not to do."

"The Guardian of the Holy Shroud has always had these dangers well in mind, and consequently has systematically recommended that scientists prioritize tests relevant to conservation in their research programs. The Turin scientific environment has always shared these concerns with positive contributions. Shortly before the letters of Chagas in question, a colleague of mine at the Polytechnic of Turin, Prof. Ferro of the Department of Energetics, wrote to the Custodian of the Holy Shroud to offer him to install, by and at the expense of his Department, a monitoring instrumentation in the altar of the Holy Shroud -- an approach to the problem somewhat different from the purely negative one of the Pontifical Academy of Sciences."

"The STURP group, author of the offending proposals, includes 5 textile experts and 5 other specialists in organic chemistry and biochemistry. Chagas' suggestion to have the proposal examined by textile experts has already been fulfilled at the start. The experience of '78 had also shown the attention paid by STURP to the safety of the Holy Shroud with regard to the examinations to be carried out -- the details of the preventive verification measures on samples are provided in the memoirs published on the various examinations, memoirs which obviously Chagas and his advisor have taken care to read. On the other hand, the protocol envisaged for the exams by the Custodian of the Holy Shroud provides that the proposals are examined by a scientific-technical commission of university experts of your trust, with the task of verifying the scientific qualification of the proponent and any danger of the proposed exams, of which a precise technical project is required[7]. This commission, of which a textile chemist is expected to participate, has not yet been established for the new examination proposals (also presented by English and French groups in addition to STURP) owing to deference to the Pontifical Academy of Sciences: we wanted to wait for the outcome of the C-14 test before going into any other exam, in order to avoid an issue in which the Pontifical Academy of Sciences was involved. Chagas is aware of this procedure which I had described to him in January and which I have amply illustrated in the opening of the Turin meeting on 29/9 (saying precisely for the C-14 this meeting covered the role that for other exams it was and will be entrusted to the aforementioned commission of the Custodian of the Holy Shroud), but I have no intention of making the slightest mention of it in his letters[8]."

"For its part, STURP certainly did not present its proposal in terms of 'take it or leave it,' but respectively submitted it to the Custodian of the Holy Shroud, asking for it to be examined in merit by a scientific-technical commission of your appointment and waits to discuss the details with this commission. The Guardian of the Holy Shroud already last year had previously mentioned to STURP that it seemed appropriate to limit the number of exams to those that can be performed in a shorter period of time, for general safety reasons, asking to establish an order of priority that would prioritize those most relevant for the conservation of the Shroud[9]."

"As you can see, the safety considerations have been and are well presented, both in terms of and in the organizational procedures, to the Custodian of the Holy Shroud, to the Turin scientists, and to the proponents of STURP."

"The anonymous Chagas advisor instead dealt with the question in a very superficial way, without even bothering to give the two words a minimum of technical plausibility: he simple selected the 12 work packages[10] in the description of which the words, 'photography,' 'spectroscopy,' or 'X-rays,' assuming that the times indicated are exposure times (while instead they are the total operating times of the working group on the Holy Shroud, inclusive of all adjustments and preliminary measurements) and makes it possible to offer any assessment of the intensity of this exposure, which is *a priori* judge of the Holy Shroud. He says that two of the exams proposed by STURP concern conservation and therefore 'deserve serious consideration, perhaps from a commission of qualified textile museum experts,' but he overlooks the fact that (a) these two work packages are proposed by qualified textile museum experts; (b) they explicitly request data collected from three other listed work packages as 'dangerous;' (c) 18 of the 25 work packages beyond C-14 more or less directly concern the conservation problem."

"His words do not make any contribution to safety, but are measured in order to excite the legitimate apprehensions of the ecclesiastical authorities in order to block further examinations on the Holy Shroud and discredit to STURP, which is presented as a group of incompetent enthusiasts undermining their qualification in the field in question and the scientific work already published by them. The style in which he speaks of STURP makes his identification with Gove probable (similar phrases are found in his other letters.)"

"It should be noted that X-ray fluorescence and infrared spectrometry are techniques normally used in investigations of works of art or archaeological finds, and are universally considered as the least intrusive possible. I would be very curious to see which short or long-term effects on cellulose, other than the normal existence at room temperature, can be attributed to an infrared exposure to the irradiation values used in these cases. Ultraviolet photography and fluorescence are also normally used for non-destructive tests on artistic or archaeological objects, and the ultraviolet was in fact used in the 1969 and '73 by the commission appointed by S.E. Cardinal Pellegrino, which included three museum-qualified textile experts, who did not even feel they had to monitor the irradiation as STURP physicists carefully did in 1978."

"In short, the concerns expressed, formulated by people who remained anonymous and declaredly not competent in the matter, are completely generic; they have long been held in due consideration, and with due consideration, by the Custodian of the Holy Shroud; they are not such as to justify the suppression or delay of the proposed exams. Indeed, it should be underlined that there would be serious reasons of concern for the safety of the Holy Shroud if it were not allowed to have such tests: if this were the case, it would be my duty of professional conscience to advise the Custodian of the Holy Shroud of the danger for the conservation of the Holy Shroud itself."

"3) The Procedural Aspect"

"The question was presented by Chagas and his correspondents in a manner so alien to normal scientific practice that it legitimately obscures his real intentions."

"Scientists criticize and quarrel among themselves as (and perhaps even more) than others, but they do so following modules of behavior that never contemplate anonymity. If a scientist has something to object to in a colleague's research proposal, he usually addresses him directly in the style, 'why didn't you do that ...?' Leaving aside the courtesy, wanting instead to be very formal in pointing out the risks of an operation, scientific ethics wants us to openly address the official safety managers of the operation. The participants in the Turin meeting had been invited by Cardinal Ballestrero in his capacity as Pontifical Custodian of the Shroud and they were personally welcomed and educated by him in his address on the opening of his responsibility. Therefore, if they had concerns about the security of the Holy Shroud they would logically have to expose them to him, with a signed letter. If they preferred to send to Chagas instead, it means that they had reason not to respect the hierarchical framework officially proposed to them by the competent authority. Not knowing the text nor the authors of the communications mentioned by Chagas, I cannot make other hypotheses."

"Chagas' behavior was questionable, both towards the Custodian of the Holy Shroud and towards the scientists who would have turned to him. The expressed concerns by Chagas were transformed into an anonymous complaint that no authority can take into account. Any person in his position would have simply sent the letters to the Custodian of the Holy Shroud, with his eventual comments, for an appropriate knowledge of the technical-scientific commission, and informed the writers of this transmission."

"The fact is that with this action Chagas has once again shown that he does not take into account the position and responsibilities of the Holy Shroud Pontifical Custodian, treating it as if his was a sinecure with a courtesy title. Addressing primarily the Secretary of State[11] Chagas shows that he considers the Custodian of the Holy Shroud only as a material executor of decisions taken elsewhere. This is perfectly in line with the behavior that he has followed since the beginning in his action regarding the Holy Shroud. As I have already documented in the previous dating report, Chagas has systematically disregarded the indications of the Custodian of the Holy Shroud, despite repeated interventions of the Secretary of State, instead becoming an explicit spokesman for Gove's theses. He seemed to have understood reason with the reconvening of the meeting in Turin, but evidently not so, and the latter episode, which confirms several negative aspects of his conduct of the meeting itself, seriously affects the possibility of further collaboration with him for the work on the Holy Shroud."

"The exams judged to be perilous are presented by Chagas to the Secretary of State as something we learned during the Turin meeting, following my presentation of the STURP exams of '78 [12]. But instead the Pontifical Academy of Sciences was aware of it from the first moment it was called into question, about a year and a half ago, the full text of the formal STURP proposal having been sent to it, and in all this time neither Chagas nor its collaborators have ever done the slightest mention of their possible danger. Canuto's report sent by Chagas to the Secretary of State in August '85 (see report on dating §2) claims that no other tests should be done except the C-14, according to Gove and others, the reason given and that they would be useless, repetitions of things already done, and that STURP is not 'credible:' there is absolutely no danger to the Holy Shroud. Why is the Pontifical Academy of Sciences only now aware of the 'serious danger' that would bring? What kind of advice did Chagas provide if he didn't think about the Holy Shroud (well, does it take care of the Custodian and the authors of the proposals?) until the point was raised by unnamed people? I don't know how Chagas could answer these questions (which an industrial client would certainly ask him without having to admit to negligence on his part, or citing very different reasons from security. In reality, this point has been raised as an extreme attempt to block research on the Holy Shroud, and to cover the exaggerated demand for fabric for dating[13]."

"The style and procedure used by Chagas, in such a clear violation of a reliable practice in such circumstances, sound open mistrust in the capacity of the Guardian of the Holy Shroud to manage the research and conservation activities that concern you. They also put the Secretary of State in an embarrassing position, in that they exploit it to give the impression of sharing this distrust. A clear position of the Secretary of State on the role of the Pontifical Academy of Sciences towards the Custodian of the Holy Shroud and the higher authorities would be appropriate, to avoid further misunderstandings and allow the Archdiocese of Turin to fulfill its role as Pontifical Custodian without conditioning interferences. Chagas' attitude has aroused dangerous confusions in the mind of the scientists participating in the dating operations, attempting them to 'play Rome against Turin' to favor their own interests in the face of the supervision of the samples. The continuation of his intervention on this line -- and it is now clear that there is no hope that changes -- makes it difficult for the Custodian of the Holy Shroud your prerogatives with the necessary authority in front of the scientists involved in the actual moment of the operation."

"Regarding the people who would have taken the initiative to report the alleged Chagas dangers, he is very vague. To Cardinal Ballestrero he only mentions 'some scientist who attended the Turin meeting;' to Cardinal Martinez he talks about 'those who deal with dating' (saying that they are not experts of the effects of radiation on fabrics) and then of three letters received, citing a section of one of them. This vagueness cannot be justified in a situation which scientific ethics would have required to resolve with a simple transmission of letters. Do we have to think that these letters also contained some unscientific elements that Chagas did not consider appropriate to make known? While the concerns are generally attributed to the daters attending the Turin meeting, the three letters are not necessarily from them -- if they came only from Canuto, Gove and his assistant Brignall, for example, it would have a well-defined appearance."

"Another serious misconduct shines through from Chagas' letter to the Secretary of State. His anonymous suggestion had suggested the formal proposal of STURP, which explicitly mentions the order number and working hours of the various work packages, which have not been published elsewhere. Now this formal proposal is a confidential document, which bears at the head the explicit declaration that it is the property of STURP and cannot be reproduced (even partially) or distributed without the express written consent of STURP or at the discretion of the Archdiocese of Turin. This confidentiality was desired, and closely maintained, by STURP, in agreement with Turin, because it was not intended that advertising the details of the proposal could become an instrument of pressure regarding those who had to decide on the approval of the individual exams. One wonders how the anonymous Chagas prompter got a copy of it."

"4) The Political Aspect"

"The Chagas letters explicitly propose to block all exams on the Holy Shroud other than dating, and also appear to be intended to transfer control of any future examinations to the Pontifical Academy of Sciences. Given the evident link between Chagas and Gove, we cannot fail to consider the possibility that his action is not impromptu but is part of a plan. In any case, it actually promotes a well-defined design."

"As already illustrated in the dating report (§1.4), in recent years certain 'secular' circles have carried forward the idea that the only exam to be done on the Holy Shroud is the dating of the cloth, and that the other investigations are conducted by non-'credible' people such as STURP scientists. This thesis is supported by Gove (who attempted to run the Holy Shroud dating protocol) and was delivered by Canuto, the 'personal representative' of Chagas in the USA on which Chagas has exclusively relied to deal with advice on the Holy Shroud."

"Discrediting all other investigations into the Holy Shroud (identification of the blood, three-dimensional image, etc.) and by preventing its continuation to focus everything on the C-14 dating, you can get the result of sinking the scientific meaning of the Holy Shroud and throw a bad light on the Church whatever the results of the dating. If in fact the date was later than the time of Christ, the Holy Shroud would be proclaimed a fake and any possibility of investigating any cause of error in the dating itself due to the particular nature of the fabric would be lacking. If the date was compatible with the era of Christ, it would be said, indeed it has already been said, that this does not prove authenticity, but only proves that someone has 'made a relic' using an older cloth. Given that dating has an uncertainty of many hundreds of years, it would be easy to suggest that it is only by chance that the assigned interval covers the beginning of the first century. In this eventuality it would be ideal that it was the Church itself to block the continuation of research on the nature of the image, because this would be interpreted as an implicit acknowledgment that their results were not such as to contest the claim that it is an artifact: McCrone's assertion of the ochre painting would definitely triumph in the newspapers and the demonstrations to the contrary would remain labeled as the work of religious fanatics."

"It is indeed very strange, quite unusual in the scientific field, the fury of some scientists so that certain searches are <u>not</u> done. It happens, alas, too often, that scientific groups try to block lines of research that are competitive with them for the distribution of funds or other resources, but this is not the case: none of the scientists active in Shroud research proposes to do either dating or other research -- and instead only a small group (Gove, McCrone, supported by certain circles) that supports dating only but no other research, not at all competitive with dating itself[14]."

"We do not understand the nature of the influence that Gove has on Chagas, but the link between the, also through Canuto and now evident (and also publicly recognized): Chagas faithfully believed Gove's propositions and theses even when they were in clear contrast to the requests that came from the Custodian of the Holy Shroud. At the Turin meeting, Gove did everything possible to exclude STURP and that their other researches (which were not the subject of the meeting) were somehow eliminated, without however obtaining support from his colleagues. Now try the pretext of security and Chagas will try to push it through. Of course he does not mention it, but the anonymity in which he leaves his suggestions, the style of the passage quoted in his letter, and the historical precedents easily allow the supposition that it is him."

"It is especially noteworthy how Chagas pushes his proposal to block research much further than implied by the short passage quoted. In this step, 12 work packages are indicated as 'dangerous' because Chagas also proposes to postpone the other 13 for which no danger is reported?

He takes over a motivation that does not have to do with security but follows the theses expressed by Gove and McCrone, 'I think dating should precede all other research,' 'such research could be performed subsequently once the age of the Holy Shroud has been determined,' adding they 'would lose their value if dating showed that the Holy Shroud came back to a very distant date' (far from what?)."

"In all this we completely lose sight of the point that is most dear to the Custodian of the Holy Shroud, because the need to continue research aimed at producing a reasonable conservation project, therefore a better knowledge of the physical-chemical nature of the image. It is very strange that Chagas comes to talk about research 'without great interest for the Church'!"

"Another point systematically carried out by Gove and that research on the Holy Shroud: it should only be managed by the Pontifical Academy of Sciences, putting aside the Archbishop of Turin. This supported Trondheim, and his attitude towards the Turin meeting was in this direction. The passage quoted in the Chagas letter to the Secretary of State offers a starting point in this regard, where there is a talk of a 'commission of textile experts that (Chagas) could convene'."

"The intent of Chagas is purely negative towards the STURP researches (and of the English whose proposal is systematically forgotten) because their execution or not has nothing to do with the C-14 dating, for which it was asked additional Pontifical Academy of Sciences advice. It was clearly established at the Turin meeting that the sampling for dating must precede any other research."

"It is entrusted to STURP, with the collaboration of Flury-Lemberg, to exploit the experience of STURP on the Shroud cloth, its equipment present for other research, and to avoid interference with these. But nothing would change if these were postponed, except for a possible lesser availability of equipment by STURP. Chagas is well aware of this, but he is careful not to remember the role of STURP in the withdrawal, since, if he request for postponement is accepted, it can be said that then it is not a reason that STURP participates in the withdrawals and achieve the objective by Gove to eliminate it completely from the research framework, making us forget that he was the actual initiator of the same dating proposal[15]."

"In the whole discussion there is no mention of discussing things with STURP or the English group, whose scientific qualification is systematically solved. Chagas praises their 'enthusiasm' a lot but doesn't treat them at all like colleagues worthy of respect on a scientific level, here too he follows Gove's approach. Now, the objective facing the scientific world is the following: we have a set of proposals put forward by a large laboratory researcher. The President of the Pontifical Academy of Sciences asks for a mouthpiece based on the undocumented hypothesis of dangerousness of the exams, put forward by three people declaredly not competing in matters that remain anonymous. If this were to happen, when the thing was known it could not fail to reflect very negatively on the prestige of the Pontifical Academy of Sciences and on the relationship between the scientific world and the ecclesiastical world. It would appear quite evident from the bare facts that the Pontifical Academy of Sciences has lent its name to well-known and publicized coverage of personal revaluations."

Turin, 28 November 1986

[signed]
Luigi Gonella

[1] "Luigi Gonella, 28 November 1986."

[2] "Prot. n° 1227 and 1229 of the Pontifical Academy of Sciences."

[3] "In these days he has been called to be part of the preliminary investigation commission for the National Conference on Energy for the safety and environment sector. "

[4] "Typical in this regard is Chagas' phrase about establishing 'with absolute certainty' that the planned research absolutely cannot damage the

Holy Shroud. 'Absolute' certainties do not exist in science, but only evaluations to the best of our knowledge."

[5] "Sulphurous smog in the Shroud cloth was detected in '78 by STURP scientists."

[6] "The microscopic investigation conducted by STURP in '78 showed how the blood was almost completely abraded on the crown of the fibres and was transported up and down the whole cloth by repetitive handling throughout the centuries."

[7] "A.D. Adler, chemistry professor at Western Connecticut State University, G.W. Carriveau, professor at Wayne State University and consultant at Detroit Inst. of Arts, Metropolitan Museum of Art and Center for Science Applied to the Archeology of the Pennsylvania Museum; W.J. Cairney, biochemist, deputy director of the Air Force Academy Biology Department; J.R. Druzik, J.P. Getty Conservation Inst; J.K Hutchins, curator of the Merrimack textile museum; N. Kajitani, curator of the Metropolitan Museum of Arts; C.C. McLean, Los Angeles Museum of Arts textile coordinator; R.J. McNeal, biochemist, Bioenvironmental Research Group, Perkin-Elmer Corp; J. Peisach, biochemist, professor at Albert Einstein College of Medicine; G. Riggi, curator of postal documents in Italy."

[8] "For example, the measurements with infrared performed with mini-irradiation of the irradiation, after proving the absence of damage on samples subjected to irradiation 100 times higher, the withdrawals with adhesive tape were accompanied by a spectrophotometric beat before and after to verify that there was no trace of the adhesive on the cloth (a special "market" adhesive was used specifically for this verification), etc."

[9] "Five of the criticized exams are considered to be of STURP's lowest priority."

[10] "The STURP Proposal is divided into 26 'work packages' or work segments of which constitute a separate examination (such as no. 6 --

C-14 dating) and others a data-collection operation with a particular technique, to be used for several interlaced exams."

[11]"Note that Chagas sends a copy of the letter sent to the Custodian of the Holy Shroud to the Secretary of State, but not vice versa (do you think that the Secretary of State shares his appreciation of the position of Cardinal Ballestrero), and is much more detailed and explicit in his letter to the Secretary of State -- it is evident that the letter to Cardinal Ballestrero is an executive summary, even if registered first."

[12]"In reality it was a presentation of the documentation available on the Shroud fabric. Obviously through it we have seen the STURP work of '78, because it is the only scientific work from which this documentation could be obtained. Most of the participants were very impressed because they weren't aware that such a thorough and conscientious work had been done -- and this evidently disturbed those who have an interest in scientifically discrediting the Holy Shroud investigation."

[13]"Knowing very well that the amount of the withdrawal depends only on the number of laboratories who wish to actually have a piece of the Holy Shroud, having explicitly declared at the Turin meeting that no one would withdraw only to save the Shroud fabric, fearing that the ecclesiastical authorities require a reduction of the withdrawal (perhaps suggesting using more 'controls'), Gove went on the counterattack trying with the bogeyman of security to carry out his plan to eliminate STURP from the research framework and transfer its control to Chagas who supports his game."

[14]"There is perhaps competition in the face of advertising. Do not exclude that some of the defectors, even indifferent to the secular design of which above, do not see favorably other research that can take away part of the publicity resonance for the 'solution of the mystery'."

[15]"Also to fuel the legend that scientists who expressed themselves in favor of authenticity oppose dating."

Source: Report of November 26, 1986 by Luigi Gonella titled "The Intervention of Prof. Chagas on the STURP research program." Luigi Gonella archives, of which the author has a copy.

Comments: It's notable that the Archdiocese was offered, at no cost to them, monitoring instrumentation, to make sure that any of STURP's tests would not be harmful, which would have negated the concerns of Chagas, Gove and Canuto, which Gonella clearly found suspicious. Sadly, the Archdiocese and/or the Vatican did not accept the offer and the objections were allowed to stand, despite Gonella's mistrust of the motives of Chagas, Gove and Canuto.

It is clear that STURP was much more considerate to the Turin authorities than the C-14 scientists were.

Based on the report reproduced earlier under "1985 *(July)*," there is little doubt that Chagas' "anonymous advisor" was Vittorio Canuto. Since that report comes from the Gonella archive, it's a puzzle why Gonella didn't name Canuto. Gonella's observations seem to suggest that Canuto had not read the STURP document carefully. The reader should also recall that in the letter from Dinegar to Gove reproduced under "1985 *(September)*," Dinegar had to remind Gove, who complained that STURP didn't consider C-14 important enough because it was only number 6 in their proposal, that STURP had simply listed the tests <u>alphabetically</u>. Dinegar chastised Gove for not understanding that -- because it was discussed in previous correspondence AND Dinegar had also explained it to Canuto. Since Canuto was Chagas' right-hand man and Chagas was close with Gove, STURP's intentions should have been clear, but Gove and Chagas acted as if they had become totally biased toward STURP, and apparently could not objectively evaluate the latter's test proposals.

Regarding footnote 10, Gonella is recounting the same exchange with Gove that was described in the entry from the second "1986 *(July)*" entry above with the Gonella/Riggi book as the source. But there is a major discrepancy. In his book Gonella wrote, "Asked him if he found anything to object in Jackson's <u>papers</u> on the Shroud – 'no' he said, because it was quite evident in those papers that Jackson wrote with absolute objectivity because he was leaning over backward to correct for his feelings." In this report from November 1986, Gonella said that Jackson was not even capable of writing with absolute objectivity. But one thing was clear, as has been shown throughout the book: Gove was antagonistic toward anything or anyone STURP-related.

1986 *(December)*. On December 14th, Cardinal Ballestrero wrote a letter to Vatican Secretary of State, Cardinal Casaroli, regarding the strange behavior of Chagas. Gonella said "The actions of Chagas over the laboratories were clearly a pretext. It was a serious violation of scientific ethics. Those involved have now sunk to the letter of anonymous letters."

Source: *The Night of the Shroud (La Notte de la Sindone)*, documentary directed by Francesca Saracino, 2011. In 2016, it was revised and retitled "Cold Case: The Shroud of Turin," which is available at amazon.com. I have a review copy of the original version, which has an English voiceover. The revised version has English subtitles. (The material cited here can be found between approximately the fifteen- minute and thirty-five-minute range on the original version review-DVD.)

Comments: The narrator in the *Night of the Shroud* documentary posed the question, "What lies behind these alliances?" The most likely answer is that people will work with whoever can help them with their respective agendas.

1986 *(December)*. By December, Gonella had heard that Gove and his colleagues had written to Chagas with their concerns about STURP's proposed non-destructive tests. Chagas forwarded the letter to Cardinal Casaroli. When Gonella heard this, he contacted Lukasik to warn STURP that some changes might be needed for their proposal.

Source: Bracaglia, Giorgio. Uncovering the Paradox within the Archives of the Holy Shroud Guild, (Honeoye, N.Y.: Holy Shroud Guild), 2019, pg. 225.

1986 *(December)*. On December 17th, Gove asked Canuto if there were any new developments. He said that the Chagas letter to the Pope had also been sent to Cardinal Ballestrero. Gonella had been shown the letter and apparently showed no agitation. Gove wrote "One would have expected him to be enraged by the thought that STURP might be slighted. Canuto was suspicious of his calm reaction."

Source: Gove, Harry. Relic, Icon or Hoax: Carbon Dating the Turin Shroud, (Bristol and Philadelphia: Institute of Physics Publishing), pg. 179.

Comments: Here, Gove skimped on elaboration. Canuto said he was suspicious of Gonella's calm reaction but goes no further. I sense that Gove told this story simply because he did not like Gonella.

1987. Cardinal Ballestrero wrote an official letter to all participants of the 1986 Turin conference and said, "[S]ome participants in the Workshop [...] stepped out of the radiocarbon field to oppose research in other fields, with implications for the freedom of research of other scientists and on our own programs for the Shroud conservation that asked for thorough deliberation. Besides, when the competent Authorities advised me they deemed we ought to proceed with three samples, a concerted initiative was taken to counter the decision, with the outcome of a telegram sent to H.E. *[His Eminence]* the Cardinal Secretary of State and myself by some participants in the Workshop, a telegram where the meaning of my introductory words at the Workshop was heavily misinterpreted."

Source: Meacham, William. The Rape of the Turin Shroud: How Christianity's most precious relic was wrongly condemned and violated, (Lulu.com), 2005, pg. 84.

1987 *(January)*. Meacham sent a letter to Gonella advising that he eliminate both Gove and Harbottle from the C-14 testing. Then on the 5th, Meacham sent a letter to Chagas criticizing him and the Academy for not keeping to the agreement reached at the Turin workshop. That letter was not well-received at the Vatican or in Turin. Chagas represented the Pope, and that was emphasized in a stern reply by Dardozzi to Meacham.

Dardozzi concluded his letter by saying that Meacham was free to withdraw from the dating process.

Source: Bracaglia, Giorgio. <u>Uncovering the Paradox within the Archives of the Holy Shroud Guild</u>, (Honeoye, N.Y.: Holy Shroud Guild), 2019, pg. 226.

Comments: Meacham sent a reply back to Dardozzi in February, which will be reproduced further below.

1987 *(January)*. Here is Meacham's letter to Chagas on the 5th: "Dear Prof. Chagas, I was most distressed to learn recently that certain participants in the Turin C-14 Workshop had taken it upon themselves to raise anew the question of other planned testing of the Shroud and its supposed 'impact' on the relic. Aside from the fact that this question is totally outside their area of expertise, this action reflects an attitude hostile to the proper scientific study of the relic, as indicated in the provocative and improper remark made at the meeting (regrettably allowed to pass without reprimand by the chair) that 'STURP is on some kind of crusade in their zeal to study the Shroud, in the full glare of publicity'."

"The consensus arrived at in the Turin Workshop ('Conclusions and Procedural Steps') represents, in my view, a program with several major weaknesses, constructed by committee and motivated not only by scientific considerations but also by **'politics'** *[my bolding]*, the desire to cater to the various personalities involved, and the ultimate appearance of the project in the eyes of the general public."

"In spite of its shortcomings, however, the 'Turin protocol' was a consensus elaborated through negotiation. The conclusions were agreed upon in gentlemanly fashion 'in a spirit of amity' and, more importantly, in an effort to give due consideration and weight to the various interests and perspectives represented. The timing and relevance of the other testing programs were discussed in Turin at length, and the agreement on this point (on which I have serious reservations) is expressed in item 4 (first sentence) of the Conclusions paper. An attempt now to alter this agreement through backroom dealings appears to me to be subversive and under-handed, and I would not hesitate to withdraw my support of the 'Turin protocol' if this very weak provision for the securing of contextual information relevant to the C-14 samples is not respected."

"I sincerely hope that you and the Academy, as chairman and co-organizer of the Turin Workshop and in the spirit of our accord laboriously hammered out, will endeavor to insure that the consensus, as expressed in the Conclusions paper will be strictly adhered to. No faction should be allowed to maneuver, through whatever connections they may have, their priorities to the fore or erect obstacles to the procedural steps recommended after having failed to obtain majority support for their position at the Workshop. Either we have a basic agreement and protocol or we have none, in which latter case the Workshop will have been little more than an academic debate, and the authorities responsible for the Shroud would be justified in re-considering the entire program for radiocarbon dating."

"With every hope that the C-14 and other testing of the Shroud will proceed smoothly [...]."

Source: Meacham, William. The Rape of the Turin Shroud: How Christianity's most precious relic was wrongly condemned and violated, (Lulu.com), 2005, pp 77-78.

1987 *(January)*. Ballestrero wrote in January to Casaroli expressing his surprise and indignation for Chagas' action against other Shroud examinations based on generic and anonymous accusations of unjustified dangers; he was perplexed over the personal conduct of Chagas, who had not leaned on the advice of the Pontifical Academy of Sciences, and for his solidarity with Gove, whose origins and aims were not understood. He was further perplexed (as was Casaroli) about the proposals that emerged from the Turin meeting, noting that there was no reason to grant more than three samples, to involve representatives of the laboratories to draw, nor to involve Chagas in the certification of the samples, and even less in the analysis of the results. Ballestrero also noted at that time that Canuto's summary report of the meeting didn't correspond with the actual discussions. *[!]*

Meanwhile, Meacham filed a formal protest against the push by Gove *et al.* to eliminate STURP's other tests. Meacham wrote on January 5th to Chagas (sending a copy to Ballestrero, Tite and Dardozzi) that the blocking of such examinations, besides being scientifically harmful, was an open violation of the proposal that painstakingly emerged from the Turin meeting. Chagas was very offended, and on January 27th dictated to Dardozzi a violent response in which, among other things, he claimed to have been commissioned by the Vatican to control all research on the Shroud; on the same day he sent a copy of this correspondence to Cardinal Ballestrero and also protested the intervention of Meacham (who he mistakenly believed belonged to STURP).

Source: Gonella, Luigi and Riggi di Numana, Giovanni. Sindone: il mistero continua, (Milan: Fondazione 3M), 2005, pg. 58.

Comments: For those who are familiar with the American political phrase "alternative facts," which popped up in 2016, it appears to me, based on Ballestrero's assessment of the report of Chagas' right-hand man, that Canuto was already engaged in that in 1987. As far as who was in charge in the Church regarding the Shroud dating, it appears to have been a classic case of "the left hand not knowing what the right hand was doing" (Mt. 6:3).

1987 *(January)*. Meacham sent a letter on the 5th to Ballestrero, "The Workshop was, under your patronage, a successful meeting in that a consensus was obtained. However, as you will note from my letter to Prof. Chagas, it was very disappointing for me to learn of the actions of certain participants in the Workshop to re-open issues covered by the consensus."

"These actions having been taken, if it is deemed necessary by Your Eminence to examine further the details of the C-14 program, I should wish most humbly to re-submit to you my original proposal of May 1985 for a panel of experts from relevant fields to consider and advise on such details. Even after the deliberations of the Turin Workshop, it does still seem that a co-ordinating panel with representation from the radiocarbon profession, textile studies, chemistry and archaeology would be most beneficial for the smooth implementation of the C-14 dating project."

"In addition, may I convey to Your Eminence the indication I have had from responsible persons in the Istituo Italiano per il Medio ed Estremo Orienta (IsMeo) of their interest in the scientific study of the Shroud and their willingness to be of any assistance they can offer. IsMeo is a state research organization conducting historical, archaeological and scientific studies in the Middle East, and it is held in the highest esteem by scholars around the world [...]."

Yours in Christ,

[signed]
William Meacham

Source: Luigi Gonella archives, of which the author has a copy.

Comments: Meacham' letter to Ballestrero must have struck a chord because as seen in the previous entry, the latter did write to the Vatican Secretary of State with complaints about various actions of Chagas and Canuto.

Sadly, Ballestrero did not take Meacham's advice to involve IsMeo.

1987 *(January)*. On the 25th, Cardinal Ballestrero wrote to Cardinal Somalo, Vatican Undersecretary of State, "Most Reverend Excellence, in response to Office No. 184.214 of 16/10/86, in which Your Excellency expressed my concerns about the protocol proposed for the dating of the Holy Shroud at the end of the Turin meeting, and following the same number of 11/11/86, in which he communicated Prof. Chagas' observations and proposals regarding the investigations scheduled on the Holy Shroud, after careful reflection I believe I must observe the following."

1. I cannot fail to express my surprise at Prof. Chagas' way of proceeding, who makes very serious statements referring to documents of which he does not think it appropriate to attach a

copy or communicate the authors *[names]*. It seems to me that it would be correct if those documents were exhibited to Your Excellency and me.

2. I find it offensive that Prof. Chagas thinks that I would have authorized the analyses without paying due attention to the factors he is now concerned with. In fact, I have not yet given consensus, although many of the analyses proposed are of great interest for the conservation of the Holy Shroud: the analysis proposals are being studied not by the Pontifical Academy (which, although knowing them for some time, had so far shown no interest for them) but rather those scientific skills that have shown so much concern and so much specific competence in the previous analyses.

3. I assure you that any other exams will take place only after the sampling for C-14, with a program that takes account of every need or even for this reason it would be interesting to have knowledge of the letters mentioned by Prof. Chagas, in order to be able to take into account specific scientific observations that may be contained therein.

4. As for the Turin meeting on the C-14, I can assure you that the attached report *[not reproduced in this book]* by Prof. Gonella is objective and responds accurately to the atmosphere of that meeting, which ended with a report, as for the common goodwill to arrive at the dating with the C-14 which was the subject of the meeting itself. On the details of the proposal, I share the perplexities mentioned by the Most Reverend Excellence.

5. Personally I am not convinced by the choice to have the analysis even carried out from 6-7 labs, which seems to me quite disrespectful of the seriousness of science: in fact I do not really know how to judge that speech on the opportunity to increase the credibility of research among the public multiplying the procedure in a quantitative and non-qualitative way: this is indeed the only reason given for this proposal, as the agreed scientists explicitly

recognized that there were no scientific reasons to suggest such an unusual procedure, never adopted in such enterprises. The taking of three samples would be the only scientifically sufficient one, as can be seen from the discussion held in Turin. Since with this scenario not all the involved laboratories would have a Shroud sample, it could be possible to complement the Shroud samples with a suitable number of control samples, as already suggested during the same Turin meeting, or to draw the distinctions by lot, possibly after exclusion of the less-skilled laboratories. Indeed, it turned out in the Turin meeting that the laboratories of Prof. Gove and Dr. Harbottle only occasionally perform dating activities, and other sources suggest that they do not seem to enjoy the tactical trust of colleagues.

6. Another reason for perplexity for me is that from this protocol it is not at all clear the full extent of the samples removal to be carried out for analysis, because this data is subject to the efficiency of the conversion treatment on which sufficient information has not been provided; the data mentioned in the final report presented by Prof. Chagas are contradictory and do not correspond to the terms of the discussion held.

7. A further reason for my perplexity and in my opinion the excessive interference of the owners of the analysis laboratories in the sample collection; too many people would be involved without skills or responsibility for this delicate phase of operations. The responsibility of the analysis labs begins in fact only when it has received from the client the sample on which to carry out the analysis itself. They do not understand what the President of the Pontifical Academy of Sciences has to do with this operational phase, given that he has not specific competence in this regard and that it is a particular technical operation, not a ceremony.

8. Not least we can understand the participation of the Pontifical Academy of Sciences in the results/analysis phase, especially considering that the Academy has no statistical or metrological competences as explicitly declared during the meeting by its own

President. It would be the first time that a body directly linked to the Church takes part as an actor in the scientific investigation of the Holy Shroud, and this would certainly not make a good impression, and without any gain in terms of scientific knowledge.

9. I would add another observation to propose the analysis of the results, and that, as a layman, I do not understand why almost a year must pass from the delivery of the samples to the communication of the results, unless there are implied intentions to agree to manipulate them, or generally considerations that longer belong to the scientific discourse.

10. I totally miss the reason why, by arranging the Pontifical Academy of Sciences with specific skills in the matter, none of them was officiated and instead only the President intervened, who specific expertise is not in the matters in question, which then made use of only the advice of a personal friend, even without specific expertise in the matter. One would say that an interpretation of honor rather than rigor and scientific method was given to the question. Finally, such solidarity between Prof. Chagas, Prof. Gove and Prof. Canuto clearly emerges from the story, that I can't help but wonder how it was born, because it is so tenacious; and I must confess that it makes me suspicious!

"I remain at your Excellency's disposal for any opportunity to seek a convenient solution to prevent the exam in question from being compromised and at the same time we can guarantee a less-complicated procedure and less exposed to personal or partisan interests. I attach to the present a complete transcript of what was recorded in the Turin meeting and a scientific report from my consultant, Prof. Luigi Gonella of the Polytechnic of Turin [...]" *[transcript and report not reproduced in this book.*

[signed]
Anastasio Card. Ballestrero
Archbishop of Turin

Source: Luigi Gonella archives, of which the author has a copy.

Comments: Notice that in point 9, Ballestrero expressed a concern about the manipulation of the results. Despite Ballestrero's fears and the aforementioned Gonella concerns, the Chagas/Gove/Canuto coalition, which Ballestrero terms "tenacious" and of which he was very suspicious, were able to get their partisan interests to rule.

1987 *(January)*. Dardozzi wrote to Meacham on the 27th, "Dear Prof. Meacham, This is in reply to your letter of January 5, 1987, to Professor Carlos Chagas, who has asked me to answer on his behalf. Notwithstanding our dismay at noticing that the central motivation of your letter are topics and arguments that were part of a private correspondence (between the President, the Cardinal of Turin and the Vatican Authorities) that was intended to remain private, I must point out that the choice of words like 'backroom dealing', 'subversive' and 'underhanded', can only reflect your lack of appreciation of some basic facts:"

1. Your "reprimand" of Prof. Chagas in the discharge of his duties as Chairman of the meeting is at the same time disrespectful and naive: disrespectful to the President of the Pontifical Academy of Sciences who has accumulated an unprecedented forty years of experience in chairing international meetings, and naive for it misses entirely the main point. In fact, giving the floor to a speaker is not tantamount to agreeing with what is going to be said.

2. The duty of Prof. Chagas regarding the Shroud (as requested of him by the Authorities of the Vatican) entails his presenting the whole spectrum of potential pitfalls, doubts and shortcomings of all intended experiments, no matter when and by whom the former are brought to light, so long as they are scientifically sound.

3. Prof. Chagas himself brought up the very same points during his final remarks in Turin, as our review of the Records of the Meeting has confirmed.

4. Prof. Chagas is a biophysicist with a time-honored experience regarding damage induced by radiation, the Shroud being just one such example.

"Your further allegation of 'factions maneuvering to bring their priorities to the fore after they have failed in their attempt in Turin', is particularly unpleasant for it implicitly puts into question the ability of Professor Chagas to arrive at a balanced decision once he has given equal opportunity to all the parts involved to express their points of view."

"Much as the Carbon-14 dating process was scrutinized by a panel of experts, so should future experiments of any kind concerning the Shroud. This is, and will remain, the firm position of the Pontifical Academy of Sciences in discharging its duties as an unbiased, non-partisan advisory body to the Pontiff on scientific matters. As for your possible withdrawal from the Turin protocol, you may feel free to take whatever action you deem necessary."

Sincerely yours,

Renato Dardozzi

Source: Letter of January 27, 1987 from Msgr. Dardozzi to William Meacham. Luigi Gonella archives, of which the author has a copy.

Comments: Although Chagas was not a clergy member, Msgr. Dardozzi ostensibly reacts as if Meacham had criticized a bishop or Cardinal. As I have compiled information for this book, I've been struck by the flowery titles and salutations such clergy are given. It seems that human nature dictates that, often the more titles one has, whether clerical or scientific, the less one is inclined to admit fault.

1987 *(January)*. Cardinal Ballestrero sent a letter to the Pope requesting that he authorize revisions to the protocol. Gove stated that he believed that the revisions had been suggested by Gonella and "were just the ones Gonella wanted." Gove said that Canuto related that Chagas was not sent a copy of this letter.

Meacham wrote a letter to Adler, the late Paul Damon, Donahue, Hedges, Lukasik, Otlet and Wolfli. Meacham did not include Gove, but someone sent him a copy. Meacham related "that he had received very disturbing news from Gonella that cause him to write to Chagas." In the letter, which was also sent to Cardinal Ballestrero and Michael Tite, Meacham expressed alarm that some participants at the Turin workshop had voiced concerns about the proposed STURP tests and the negative impact those tests might have. He asserted that the tests were outside their areas of expertise and that questioning them exhibited a hostile attitude toward appropriate study of the Shroud. Meacham said that this attitude had been "reinforced by the provocative and improper remark made at the meeting, regrettably allowed to pass without reprimand by the chair, that STURP was on a crusade to study the shroud in the full glare of publicity."

Meacham believed that the consensus reached at the workshop had several significant weaknesses. But he acknowledged that it had been reached through negotiations. The timing and relevance of the proposed STURP tests were discussed at length and the agreement on them was spelled out in Item four of the conclusion paper, specifying that the samples would be taken right before other tests. Meacham felt that "any attempt now to alter this agreement through back-room dealings appeared to him to be subversive and underhanded." He hoped that Chagas would ensure that the agreed-upon consensus would be adhered to. Meacham further stated that no faction should be allowed to maneuver through whatever connections they might have to get their priorities to the fore or to block steps that had been agreed upon at the workshop. He concluded by saying either there was a basic agreement on the protocol or not. If there wasn't, the workshop was simply an academic debate and the authorities responsible for the Shroud would be justified in reconsidering the whole issue of C-14 dating for it.

Source: Gove, Harry. Relic, Icon or Hoax: Carbon Dating the Turin Shroud (Bristol and Philadelphia: Institute of 1996), pg. 195 (re: Ballestrero letter to Pope); pp. 180-181 (re: Meacham letter).

Comments: Gove guessed that Meacham's allusion to the "provocative and improper remark" at the Turin workshop was "probably his version of some quote from me." Gove seemed to have a hard time entertaining the idea that such negative thoughts could have been expressed by him. And there was a double standard regarding publicity. Although the C-14 consortium, based on various quotes reproduced in this book, definitely thought of the publicity aspect for their labs, Gove, at least, apparently accused STURP of having publicity uppermost in their minds. Meacham, it should be pointed out, was neither a member of STURP nor of the C-14 consortium, so probably could be considered one of the more objective participants. See the entry for "1988 *(October),*" which has excerpts from a confidential letter to me from one of the major participants, who had an interesting comment regarding this letter from the Cardinal to the Pope.

1987 *(Early February).* Canuto returned from Rome. He said that the authorities in Turin "were still dragging their feet." Cardinal Ballestrero had visited the Vatican recently and had been invited to also visit the Pontifical Academy of Sciences but declined. While Cardinal Ballestrero was in Rome, Chagas had dinner with the Pope. Chagas repeated to the Pope his concern about the STURP tests and the Pope said he was going to meet with Ballestrero the next day and would remind the Cardinal that some action must be taken with regards to the C-14 testing. Canuto believed that the STURP testing was the big obstacle and that Gonella was still pushing for them. Canuto suggested that Chagas send Monsignor Rovasenda to Turin to meet one on one with Cardinal Ballestrero, as one clergyman (and friend) to another. In that way, Gonella could be excluded. Chagas obliged. Canuto told Chagas that Cardinals were powerful individuals. Gove said that he began to realize that in some matters, the Pope was just a brother Cardinal. Gove saw Ballestrero and Gonella as "the key players in this shroud drama." Gove added that Canuto remarked, tongue in cheek, "How is the Pope to know that Gonella is a second rate scientist and that the rest of us are super scientists?"

Source: Gove, Harry. Relic, Icon or Hoax?: Carbon Dating the Turin Shroud, (Bristol and Philadelphia Institute of Physics Publishing), 1996, pg. 181.

Comments: It's worth remembering here again that Cardinal Ratzinger had officially approved in 1985 STURP getting two full weeks to do additional testing on the Shroud. Regarding Canuto's remark about Gonella, given all the previous comments by Gove and the closeness of Gove and Canuto, one can wonder how firmly Canuto's tongue was in his cheek. [See e.g., "1987 *(October)*" entry further below in which Kersten and Gruber quote Gove as seriously asserting that Gonella was "a second-class scientist."]

1987 *(February)*. After Chagas' angry letter to Dardozzi in late January that was copied to Ballestrero, the latter replied on February 18th by refusing to go into the divergence and expressing his perplexity at the statements that seemed to imply a role of the Academy parallel to the task of the Pontifical Custodian. Casaroli wasn't sure what to do and didn't take action until May.

Source: Gonella, Luigi and Riggi di Numana, Giovanni. Sindone: il mistero continua, (Milan: Fondazione 3M), 2005, pp. 58, 62.

1987 *(February)*. Meacham sent on the 20th a reply to Dardozzi, after Meacham's original letter in January. Meacham clarified the intent and assured Dardozzi than no disrespect was intended. However, Meacham was frustrated because he had believed that the Turin workshop had produced the final protocol; he expressed his outrage in a letter sent to Rinaldi.

Meanwhile, STURP was having various problems. The biggest was that they had no money to finance their proposal. They tried to solicit donations but were unsuccessful. They proposed letting British director/producer David Rolfe film the whole procedure to raise the needed revenue, but Ballestrero would not allow it, citing the inappropriateness of the exploitation of a relic.

Source: Bracaglia, Giorgio. Uncovering the Paradox within the Archives of the Holy Shroud Guild, (Honeoye, N.Y.: Holy Shroud Guild), 2019, pp. 226-227.

1987 (February). Meacham replied on the 20th to Dardozzi's letter of January 27th: "Dear Ing. Dardozzi, I have your letter of Jan. 27, and clearly we have very different attitudes, perspectives and concepts of protocol. Nevertheless, one should always aim at being constructive; this was the original intent of my letter of Jan. 5, in spite of the strong language employed."

"Let me assure you and Prof. Chagas that no disrespect to him was intended. My criticisms were directed at those who I believed to be attempting to subvert the consensus hammered out at Turin. I had no knowledge of the 'topics and arguments' in the private correspondence you mentioned, and it is disappointing to glean from your letter that the Academy is not pressing for strict adherence to the conclusions and procedural steps drawn up in the Turin Workshop. As stated in the fifth paragraph of my Jan. 5 letter, I hoped that Prof. Chagas and the Academy would work to uphold the provisions of the Turin protocol."

"If my criticisms apply to Prof. Chagas as well as to certain individuals involved in the C-14 controversy, let me stress again that I certainly do not intend any disrespect, unless it is considered that criticism itself constitutes a disrespect, as your letter seems to imply. In fact, I have great respect for Prof. Chagas, deriving from the fine and dedicated manner in which he chaired the Turin meeting. You will recall that I stated this in rather lavish terms at the end of that meeting. This was not an attempt to flatter or curry favor, but a sincere expression of praise. By the same token, criticism should be taken as an equally sincere communication of views. I trust that you would agree that every one of us is on occasion capable of error or of having subjective bias influence our decision-making. Yet you find it 'disrespectful' that I point out what I perceived to be a minor error in the chairing of the meeting, and 'particularly unpleasant' that I implicitly put into question the ability of Prof. Chagas to always arrive at a properly balanced decision. Allow me to elaborate further on these aspects:"

1. I believe that most independent observers would agree that the proper intervention of the chair should have occurred after Prof. Gove's inflammatory and gratuitous remarks casting aspersions on the motivations of STURP members, rather than after Prof. Gonella's simple observation that "Prof. Gove apparently does

not like the STURP people." My comment was certainly not naive; that an inflammatory remark like Gove's did not draw any rebuke from the chair was a simple error, compounded by subsequent events seen in retrospect.

2. I believe however that Prof. Chagas did an excellent job overall in chairing the meeting, insuring that all participants had equal opportunity to express their points of view. The Turin protocol represents a balanced consensus obtained at the meeting. What I objected to was a private tampering with the consensus at a later stage, initiated by a certain faction, during which time other participants in the workshop had no opportunity to comment further on the issue raised. Nor did the scientists whose proposed experiments were in question have the opportunity to respond. Nor were views sought from other, dedicated scientists who have devoted considerable time and energy to the proper study of the Shroud. It can only appear to the outsider that the Academy has taken a position on the promptings of a few influential individuals, without input from other scientific circles, and thus not in the manner of an unbiased, non-partisan advisory body.

"Again, I wish to emphasize that these critical views are offered in good faith, in the belief that subjectivity and bias affect us all, and that procedures need constantly to be reviewed in order for an organization to be as unbiased as humanly possible."

"I must confess to being completely unaware of the basic fact in item 2 of your letter, and I feel certain that other scientists interested in the study of the Shroud were also unaware of it. Would it not have been a more non-partisan approach to have raised the question of possible pitfalls or doubts of the intended experiments directly with the scientists concerned, so as to allow them to respond to or clarify any problems or critique? Such an inter-play between proponents and critics of a proposed project is the normal procedure for an outside body to determine whether the project is designed on a scientifically sound basis."

"I do not recall any mention during the Turin meeting of the Academy's intention to review and approve all intended experiments on the Shroud. The date of May 1987 for C-14 sampling and other testing was fixed primarily as the earliest date possible for the organization of the other testing. My own position and that expressed by several other participants is that the C-14 test must take place in the context of the other testing. This sentiment was incorporated in a compromise form in item 4 of the procedural steps and it constitutes part of the scrutiny of the C-14 dating process by the panel of experts. Rather than altering one of the crucial points of consensus arising from the Workshop, would it not be better to consider amending item 11 instead, if necessary? I trust that the Academy will give due consideration to this suggestion, and strive to promote harmony among those who have a genuine interest in the scientific study of the Turin Shroud."

Sincerely,

[signed]
William Meacham

Source: Letter of February 20, 1987 from William Meacham to Msgr. Dardozzi. Luigi Gonella archives, of which the author has a copy.

1987 (February). Meacham wrote to Gonella on the 22nd, "Dear Luigi, You were right. It was indeed useless to write to Chagas. Except maybe to have on record some protest from a participant in the Turin workshop against the post-workshop tampering with the agreed upon procedure."

"I was surprised at the tone of Dardozzi's letter, as if Prof. Chagas should somehow be above criticism. Of course I employed some fairly strong language, but it was directed against those attempting to undermine the Turin protocol, and I did not specifically name Chagas in that regard. Dardozzi's reply is a case of 'thou dost protest overmuch, milord.' Anyway, I thought you would like to see it."

"I trust that by now you have been to the U.S. and spurred things on there. Are things taking shape for testing in May? Have any new problems come up?"

"Drop me a line when you get a chance. I shall let you know as soon as we get things organized for a Japan exhibition symposium, hopefully in early 1988."

As ever,

Bill

"P.S. I will do nothing more on the 'Chagas-Pontifical Academy of Sciences' issue. Dardozzi may not reply to my letter, and if he does I probably won't make any further reply to him. By the way, do you find it strange that Chagas himself did not reply? Do you suppose this was so that someone else could sing his praises, or do the dirty work? Or is this supposed to indicate, Curia-style, his contempt for my criticism?"

"However, if things really get sticky and the Academy really does succeed in obstructing the other testing, I could take the whole matter to the press, but only if advised by you that such a course of action might be of some use. I think the Academy ... *[here the page ends and I did not find the 2nd page in the archive]*.

Source: Letter of February 22, 1987 from William Meacham to Luigi Gonella. Luigi Gonella archives, of which the author has a copy.

1987 *(March)*. On March 3rd, Tite sent a circular to the laboratories, copying Chagas and Gonella, proposing a standard format to report the results.

Source: Gonella, Luigi and Riggi di Numana, Giovanni. Sindone: il mistero continua, (Milan: Fondazione 3M), 2005, pg. 62.

1987 *(March)*. Heeding Gonella's advice from December that they might have to make adjustments in STURP's proposal, D'Muhala and Lukasik called a meeting for the 7th in Rye, New York. The purpose was to look at the original 1983 proposal *[formalized in 1984]* and to elicit new suggestions from members.

Source: Bracaglia, Giorgio. Uncovering the Paradox within the Archives of the Holy Shroud Guild, (Honeoye, N.Y.: Holy Shroud Guild), 2019, pg. 225.

Comments: A revised proposal was submitted in late June, but not without controversy among the STURP members, which will be recounted further below.

1987 (March). Sox called Gove on March 8th. Sox had recently been in Italy and told Gove he got indications while there that "all was not smooth sailing with the shroud dating project." He also "had no doubts that the problem lay with STURP. Sox learned that Fr. Coero-Borga, who had been head of the International Sindonology Centre in Turin, had died in December. Gove noted that Coero-Borga had always been opposed to the Shroud being dated, "so we had one less adversary. Plenty remained."

In mid-March, Canuto called Chagas. There still was no definitive word yet on the Shroud dating. Gove then wrote Chagas and said he was sure that the Vatican and Turin were not having any doubts about the efficacy of the AMS method, so "it must be because STURP was insisting that its battery of high technology, invasive and possibly deleterious measurements must immediately follow the removal of a small sample for carbon dating." Gove then added, "STURP's desire to conduct further measurements on the shroud seemed to have rendered them bereft of reason."

On March 19th, Gove received a letter from Mark Plummer, executive director of the *Skeptical Inquirer*, "a journal that investigated claims of the paranormal." Plummer inquired if "scientific observers" from his group could be present at every stage of the testing. Plummer also enclosed a copy of a letter written by Fr. Rinaldi of the Holy Shroud Guild written to Guild members that the Shroud would soon be undergoing the most thorough scientific examination ever attempted, including the C-14 dating. Fr. Rinaldi said it was the Pope's wish "that everything be done to solve the mystery of the shroud." The examination would involve many experts and would involve a lot of expenses. Donations were requested. Plummer wanted to know from Gove if the Guild was directly or indirectly funding Gove's lab's involvement. Gove responded to Plummer that his lab was not being funded by the Guild. Gove said that many groups would like to be present for every stage of the testing but that would be impossible. But he assured Plummer "that the radiocarbon measurements would be carried out in a manner that would satisfy even the most skeptical inquirer."

On March 28th, Gove informed Chagas about the letter from Fr. Rinaldi. Chagas was surprised as he had not been kept informed about STURP's proposal. Chagas then asked Gove to send him a copy of the Rinaldi letter.

Source: Gove, Harry. Relic, Icon or Hoax?: Carbon Dating the Turin Shroud, (Bristol and Philadelphia: Institute of Physics Publishing), 1996, pp. 181-183.

Comments: Gove saw Coero-Borga, and anyone else who didn't agree with Gove, as adversaries. Gove's attitude, needless to say, was not conducive to a cooperative effort in dating the Shroud. Gove made an educated guess, but practically bordering on fact in his eyes, that STURP was responsible for the delay in the dating. It's not clear why STURP's desire to do additional testing to complement C-14 dating, a test that is sometimes in error, left them "bereft of reason."

The *Skeptical Inquirer* is not a journal, but is rather, a magazine. They're "devoted to the cause of advancing science over pseudoscience, media literacy over conspiracy theories, and critical thinking over magical thinking." Had they been allowed to be present for the testing, and the test had come out first century, very few people, given the magazine's history of producing negative opinions about the Shroud, would have been surprised if they would have found a reason or reasons why the testing would not have been deemed valid. Gove was in error in his belief that the dating would be carried out in a manner that would satisfy even the most skeptical.

Regarding the Church's pronouncements via various individuals regarding particulars of the dating procedures, it was again obvious that the left hand did not know what the right hand was doing.

1987 (April). On April 1st, Chagas replied to Tite's circular of March 5th, saying that he was still waiting for a final word from Ballestrero and then he would communicate to him "the names of the three laboratories that have been chosen." Tite sent Gonella on April 13th a copy of this letter asking politely if there was something new, and Gonella told him that instructions were being awaited.

Source: Gonella, Luigi and Riggi di Numana, Giovanni. Sindone: il mistero continua, (Milan: Fondazione 3M), 2005, pg. 62.

1987 (April). Gove phoned Gonella in Turin on April 1st and asked if there were any new developments. He said that Turin was still waiting to hear from Rome. Gove told him he had heard that Rome was waiting to hear from Turin. Gove then read to Gonella a portion of Fr. Rinaldi's letter, and said the only way he could interpret it was that STURP had some big plans, which concerned Gove regarding the impact on C-14 dating. Gove suggested to Gonella that Cardinal Ballestrero call the Pope and pin down who was waiting for whom. Gonella said he would pass along the suggestion when he met with the Cardinal the following Wednesday.

On April 10th, Chagas wrote a short letter to Gove saying, "I think something has happened which will find a solution for the work of dating carbon-14." On April 15th, Gove sent a progress report to the six labs and to Tite at the British Museum. Six months had passed since the Turin workshop. At a C-14 conference held April 27-30, Gove met with representatives from five of the AMS labs involved in the Turin workshop to discuss the delay in the Shroud C-14 dating. Gove was requested by the others to write a letter to Chagas reaffirming their support for the protocol that had been agreed to and to press for action.

Source: Gove, Harry. Relic, Icon or Hoax?: Carbon Dating the Turin Shroud, (Bristol and Philadelphia: Institute of Physics Publishing), 1996, pp. 183-184.

Comments: According to the Gonella/Riggi book (pg. 65, footnote 43), Gove had organized the C-14 conference held on April 27-30. He presented a paper on the Turin meeting, citing the Canuto report as an "agreement" signed in the meeting and specifying which of the speakers would participate in the enterprise and which would not. Turin or the labs had not been informed of this; Gove would only on October 13, 1987 inform Gonella of the gist of the paper presented. At the same conference, Professor Dale (of the Department of Visual Arts of Western Ontario University) presented a Shroud paper. Dale cited only McCrone's researches, and concluded that the Shroud was a Byzantine icon of the eleventh century.

1987 *(April)*. On April 27th, Gonella was quoted in *La Stampa* saying "Only two to three laboratories would be involved with carbon dating. No more were needed." Canuto let Gove know that the Pope was ready to allow three samples and three labs and that a communication from the Vatican was being prepared. The labs couldn't reveal that they had learned that information from Canuto so Gove persuaded them to protest the statement of Gonella in *La Stampa*.

Source: Sox, David. The Shroud Unmasked (Basingstoke, Hampshire: The Lamp Press), 1988, pp. 114-115.

Comments: Gonella claimed in his book (pg. 62) that he had actually said "several" labs but that the writer changed it to "two or three." The subterfuge on the part of Gove and labs was simply unbecoming.

1987 *(Spring)*. Ballestrero had a meeting with the Pope, who agreed with Ballestrero's decision in February not to allow director David Rolfe to film the C-14 proceedings in order to be able to raise money for STURP. It was at this meeting that the Pope stipulated what was supposed to be the final protocol. The three labs were to be Arizona, Oxford and Zurich. It was Gonella's understanding at that time that although the Vatican was in constant contact with the Pontifical Academy of Sciences, Ballestrero was given discretion about procedures, which needed final approval of the Vatican.

Source: Bracaglia, Giorgio. Uncovering the Paradox within the Archives of the Holy Shroud Guild, (Honeoye, N.Y.: Holy Shroud Guild), 2019, pg. 227.

Comments: There is a proverb: "Too many cooks spoil the broth," which means if too many people are involved in a task, it will not be done well. The Shroud C-14 dating was a prime example of this.

1987 *(Spring)*. Petrosillo and Marinelli wrote, "In the spring, STURP proposed the first definitive-analysis protocol that was envisaged as being done at the same time as the sampling for the C-14. Gove and his camp, of course, wanted C-14 alone and there was still a question of how many labs would participate in the test. But Gonella dropped a bombshell in an April 27th interview in *La Stampa*, in which he said only two or three labs would take part. Canuto confirmed to Gove that the Vatican was leaning toward only three, something that was supposed to be confidential. Thus, the labs can't reveal that they knew that via Canuto, and Gove convinces them to protest Gonella's *La Stampa* comments with a telegram to Ballestrero and the Pope. The Turin workshop had settled on seven labs and they hoped that the target of three by the Vatican wasn't set in stone. Gonella telephoned Wolfli in Zurich to find out the genesis of the protest telegram. Wolfli replied, 'Physicists are never good diplomats'."

"It was clear that some of the seven labs would be eliminated, but which ones? Gove's conflicts with Gonella made Rochester an easy choice. Harbottle from Brookhaven had angered Gonella by disclosing at an American Chemical Society meeting the preliminary agreement in 1986 at the Turin workshop, so Brookhaven was out. Harwell lab had negative publicity for a significant error on the dating of the Lindow Man. Zurich was in danger due to a one-thousand-year error in one of their recent tests. An issue with Oxford was that Hall was on the Board of Directors of the British Museum. But the seven labs try to convince the authorities that using seven labs would give the results more public credibility. Gonella expected that only an English lab and an American lab would be chosen, with one using proportional-counter and the other using AMS. New Zealand art historian Dennis Dutton, a friend of Sox, sent a letter to *Nature*, full of suspicions regarding the chain of custody of the samples, i.e., leaving open the chance of a sample switch." Dutton's full letter can be found in the next entry and Tite's reply can be found under "1987 *(June)*."

"Gonella was upset about the insinuations that the Church would switch samples and that the British Museum didn't deny them -- Tite himself vouched for preventing it -- as if the word of the Cardinal was not enough." Gove also replied to Dutton in a letter to *Nature* [also under '1987 *(June)*']."

Source: Petrosillo, Orazio and Emanuela Marinelli. La Sindone: Storia Di Un Enigma, (Milan: Rizzoli, 3rd edition, 1998), pp. 142-144.

Comments: To continue this thread, go to the Petrosillo/Marinelli entry for "1987 *(May)*."

Gove would later say in a press conference in January 1988 that he warned Ballestrero that if the number of labs were reduced from seven to two or three, "the whole exercise could be meaningless." The number of labs was reduced from seven to three and yet the labs in their report stated that the dates were accurate with at least 95% confidence, which Gove would accept. The question needs to be asked: how does one go so easily from "could be meaningless" to "with at least 95% confidence?" My speculation: they got what they needed with an apparent medieval date close to the time of the exhibition around A.D. 1357 in Lirey, France *[when the Shroud definitely enters the historical record]*, so why question it any further? For the labs, an apparent accurate date was sufficient -- they came out looking pretty good, so never mind if the cloth might actually be the burial cloth of Jesus -- to pursue that would just be opening up a can of worms and would be more trouble than what it was worth.

1987 *(May)*. In the May 7th issue of *Nature*, New Zealand researcher Dennis Dutton wrote to the journal, "SIR-Like most observers keen to know the historical provenance of the Shroud of Turin, I welcome the decision to subject the relic to radiocarbon dating. However, clouds loom on the horizon, in the form of confusions about the protocols for the tests. The procedures as so far understood involve a number of samples of the shroud which are to be divided among as many as seven laboratories. These laboratories will be asked to date dummy samples along with the shroud, and none will even know which of their samples are from the Turin relic. This blind procedure will avoid any possible taint of prejudice on the part of the testing laboratories."

"However, such a protocol leaves serious unanswered questions about the possibility of tampering with the samples themselves. How are independent observers to know whether any of the samples which testing laboratories receive are in fact actual linen fragments from the shroud? Are we simply to take the Vatican's word for it? Repeated enquiries in this matter made by me and by the U.S. Committee for the Scientific Investigation of Claims of the Paranormal have so far elicited no satisfactory answers. One prominent shroud authority, Father Peter Rinaldi, has given assurances that the British Museum is acting as 'guarantor' of the tests. But the relevant person in the British Museum, who was in fact present at the meeting in Turin last autumn which recommended the testing procedure, has declined to divulge any information about testing protocols because of 'confidentiality.' He has referred correspondents to Cardinal Ballestrero in the Vatican and to the Pontifical Academy of Sciences. Inquiries there have so far gone unanswered."

"The situation as it now stands is most disturbing. After years of discussion, there is agreement to go forward with ^{14}C tests on the Shroud of Turin, but apparently so far without due regard for an open disclosure of procedures for taking the samples. Evidence for or against the authenticity of a relic of such widespread veneration involves deep religious passions: for some people there is a great deal potentially to be lost. So there must be no hint that, for example, fibres of mummy linen might have been supplied to the laboratories, rather than actual shroud samples. If those conducting the tests wish the results to be taken seriously, they must offer their procedures to open inspection by independent observers. 'Confidentiality' is out of the question."

DENIS DUTTON, School of Fine Arts, University of Canterbury, ChristChurch 1, New Zealand."

Source: "Still shrouded in mystery." [Letter], *Nature*, **327**, 10 (7 May 1987).

Comments: Tite would send a reply to *Nature* in response in June. See further below. Despite Tite's assurances in that letter, the sloppiness of the process and major discrepancies by several principle individuals involved facilitated the very thing that Dutton had feared. Dutton would write two more letters-to-the-editor, which would be published in 1988. See further below for full texts of those letters.

1987 (May). On May 7th, Gove sent a telegram to senior representatives of the six other labs and to the British Museum. Gove informed them he planned to hand-deliver it to Chagas in New York on May 16th or 17th. Gove said that in his view, "Gonella and STURP are being deliberately mischievous concerning carbon dating." He further stated that if Turin workshop is not followed to the letter, he would no longer be willing to be involved. He then asked the recipients to approve the letter. In particular, they affirmed 1) all seven labs must be directly involved; 2) Mechthild Flury-Lemberg must be responsible for the selection and removal of the samples; 3) representatives of all seven laboratories should be in attendance when the sample is removed; 4) representatives from the Pontifical Academy of Sciences, the British Museum, and the Archbishopric of Turin will supervise the removal of the samples and their transfer to the representatives of the seven labs.

Hall from Oxford called Gove that day and expressed his concern that the letter might actually make the situation more complicated. He believed Chagas preferred only two or three labs. Gove replied that it was his impression that the Turin authorities was bypassing Chagas and that this letter could strengthen his hand. Hall also said that if the number of labs was reduced it would mean starting again. Although it was clear to Gove that Hall opposed the reduction of the number of labs, Hall would later change his mind.

On May 8th, Gove received a message from Woelfli in Zurich saying that he fully agreed with all the points made in the letter and would sign it. He also stated categorically that he would withdraw if the Turin workshop agreement was not followed to the letter. Gove noted that Woelfli would also change his mind later on.

On May 11th, Gove sent the letter to Chagas; it was signed by the heads of the five AMS labs that gave their approval.

Source: Gove, Harry. Relic, Icon or Hoax?: Carbon Dating the Turin Shroud, (Bristol and Philadelphia: Institute of Physics Publishing), 1996, pp. 187-189.

Comments: Gove did not elaborate on why he felt Gonella and STURP were being "mischievous" regarding the C-14. Although Gove and Woelfli both had stated they would not be involved if the Turin workshop protocols weren't followed, both stayed involved: Gove, by attending Arizona's dating (even though the labs had agreed not to invite outsiders) and Woelfli with his Zurich lab being one of the three labs ultimately chosen. Hall, although not threatening to withdraw, did not complain when the number of labs was reduced from seven to three, with Hall's lab also being one of the three labs finally picked. There seemed to be a wave of scientists "not sticking to their guns." As will be noted later, Gonella believed the three labs would not refuse being involved because "the prize was too great."

1987 (May). After waiting several months after Chagas' angry letter to Dardozzi and copied to Ballestrero in late January, Casaroli sent on May 21st the following directives: the operation was under the total responsibility of Ballestrero, there would only be three samples given out, the participation of the British Museum and the Colonetti Institute was approved, and the presence of representatives of the laboratories to the collection was not necessary. Chagas could be invited by Ballestrero to attend as his personal guest. At the same time Casaroli himself wrote to Chagas thanking him for the work he had performed and sent him a copy of the instructions sent to Ballestrero.

Source: Gonella, Luigi and Riggi di Numana, Giovanni. Sindone: il mistero continua, (Milan: Fondazione 3M), 2005, pg. 62.

1987 (May). "On the 21st, Casaroli sent Ballestrero the following provisions: three labs would participate, the participation of the British Museum and Colonetti Institute was confirmed, and the presence of representatives at the sample-taking was not necessary. There were furious reactions from the labs, which protested to Casaroli by telex (some later apologized privately); Chagas protested in person to Casaroli. According to Ballestrero, those protests, as well as the pressure of other Shroud groups both pro and con were ignored. He asserted that three labs, Oxford, Tucson and Zurich, based on the Turin workshop meeting, were chosen based on the greatest experience of archaeological data and the smallest amount of Shroud material requested."

Source: Petrosillo, Orazio and Emanuela Marinelli. La Sindone: Storia Di Un Enigma, (Milan: Rizzoli, 3rd edition, 1998), pp. 144-145.

Comments: To continue this thread, go to Petrosillo/Marinelli book entry "Enigma…" entry for "1987 *(September)*."

Gove's lab was not one of the three chosen and furious protests ensued. Some labs claimed that the AMS method wasn't ready yet, mainly because of the significant number of incorrect readings from small samples. Harbottle claimed that the chances of a measurement being incorrect were one in five. Cardinal Casaroli did not allude to the other tests (STURP's proposed twenty-six), which ultimately would never be done. Petrosillo/Marinelli, quoted Gonella, "These laboratories […] are a very closed group. I have become aware of this; they are much more concerned about the good relations between themselves not to offend their individual susceptibility, even more than about their relationship with the outside world" ["Enigma …" book (pg. 38)].

1987 *(June)*. In response to Casaroli's directive to Ballestrero that Chagas could attend the sample-taking, Chagas said on June 3rd he was still waiting for Ballestrero's invitation. Chagas had learned that the Vatican was going to settle on three samples and leaked the news. On June 17th, Chagas wrote to Casaroli that it should have been foreseen that there would be complaints if the number of samples were less than seven as he had given indication that seven labs would be chosen, and that there was a scientific article in *Radiocarbon* that showed that more than three samples were needed to obtain consistent results.

Source: Gonella, Luigi and Riggi di Numana, Giovanni. Sindone: il mistero continua, (Milan: Fondazione 3M), 2005, pp. 62-63.

1987 *(June)*. Tite sent a letter to *Nature* that was published in its June 11th issue responding to a previous letter that was concerned about the possibility of a sample-switch taking place. Tite wrote, "Sir-I first wish to assure Denis Dutton (*Nature* **327**, 10; 1987) that all the institutions involved in the proposed radio-carbon dating of the Shroud of Turin are fully aware of the crucial need to ensure that the 'chain of evidence' remains unbroken. It was to meet this need that the British Museum accepted the invitation to act as 'guarantor' and independent observer. The purpose of the meeting in Turin last autumn was to devise procedures for every step of the sampling and testing, procedures which could and would be monitored at every stage by the three certifying institutions, the British Museum, the Pontifical Academy of Sciences and the Archbishopric of Turin, to preclude any possibility of tampering with the samples. These procedural steps have yet to be finally agreed by the Pontifical Academy of Sciences and the Archbishopric of Turin so I am not at liberty to divulge their details. But, I can reassure Dutton that should the proposed procedures be amended to introduce a possibility of tampering with the samples, the British Museum would decline to act as a certifying institution. Nor would the radiocarbon dating laboratories then necessarily be willing to participate in the project."

M.S. TITE Research Laboratory, British Museum, London WCI B 3DG, UK"

Source: "Turin shroud." [Letter], *Nature* **327**, 456; (11 June 1987).

Comments: Almost every major facet of the 1986 Turin planning meeting was ultimately ignored when the sample was taken on April 21, 1988. There is no way that the chosen labs would have withdrawn from the project due to any ethical considerations. There wouldn't be the enormous amount of material for this book if that were the case. The C-14 field is now a several-billion-dollars-per-year industry, due largely to the publicity it received from the Shroud dating. Italian scientist Paolo Di Lazzaro commented in an Italian article (https://www.aboutartonline.

com/sindone-dipinta-o-si-e-scientificamente-incompetenti-o-si-e-in-malafede/), "It amazes me how the British Museum statistician who worked on the data didn't realize there was something wrong. It must be considered, that in 1988 the mass spectrometer accelerator technique was the newest technique, it was in its childhood. It was still being learned how to use it. At this point the laboratories had two possibilities: either to request another sample, admitting at that point that the technique had not succeeded in intending and to affirm the failure of the technique itself, or to choose the simplest way, that is, to publish the data hoping that nobody noticed the inconsistencies. One can imagine what would have happened to admit that the technology was not suitable."

1987 (June). Gove, like Tite, sent a letter-to-the-editor to *Nature* addressing Dutton's concerns. "SIR-I can assure Denis Dutton (*Nature* **327**, 10; 1987) that all the participants in the workshop on 'Radiocarbon Dating of the Turin Shroud' are acutely aware that the operation must be completely credible."

"The workshop at Turin from 29 September to 1 October last year, chaired by Professor Carlos Chagas in his capacity as president of the Pontifical Academy of Sciences, involved representatives of the seven laboratories that will make the measurements, the British Museum, the Archbishopric of Turin and a representative of the Abegg-Stiftung in Bern, who will remove the sample from the shroud."

"I presented the conclusions and procedural steps agreed to at the workshop as a poster at the International Symposium on Accelerator Mass Spectrometry at Niagara-on-the-Lake, Ontario, on 27-30 April. It will be followed by a paper to appear in *Nuclear Instruments and Methods*."

"The procedures recommended are clearcut and straightforward. Although the testing laboratories will follow blind carbon-dating procedures, there will be no possibility of 'tampering' with the shroud samples except as a result of collusion by a number of organizations including the British Museum, the Pontifical Academy of Sciences and the Archbishopric of Turin. The removal of the shroud sample by a noted textile expert from the Abegg-Stiftung in Bern, Switzerland, will be witnessed by representatives of the seven carbon-dating laboratories.

A representative of the Pontifical Academy, the British Museum and the Archbishopric of Turin will supervise the shroud samples from their removal to their delivery, together with a dummy sample and control samples, to each representative of the seven laboratories. Equally careful procedures will attend the final analysis of the results from the seven laboratories. Six of the seven laboratories have already participated in blind interlaboratory comparison measurements supervised by the British Museum."

"It is clearly important that the most significant scientific test on the Shroud of Turin, radiocarbon dating of the cloth, should be carried out in a manner that will convince people like Dutton that the results, whatever they may be, are believable. The only interest of the participating carbon-dating laboratories in 'confidentiality' is that they be able to carry out the measurements under reasonably serene conditions."

HARRY E. GOVE, Nuclear Structure Research Laboratory, University of Rochester, Rochester, New York 14627, USA"

Source: "Turin Shroud." [Letter], *Nature*, **327**, 652 (25 June 1987).

Comments: This was less than one year before the sample-taking. Not only were procedures changed from this, but they also changed after a meeting in London in January 1988, just several months before the excision of the samples. Nothing about the dating process could be described as "serene." In his book, Gonella (pg. 65, footnote 43), would say there was an unhealthy atmosphere of suspicion regarding the dating procedures. Gove claimed, "The procedures recommended are clearcut and straightforward." What ended up happening was anything but that.

1987 *(June)*. Gove wrote a letter to Chagas sharply critical of the way that the Turin authorities had handled Shroud matters in the past and were continuing to mishandle matters. Gove stated that the Shroud "had been subjected to a number of scientific tests of dubious value carried out in ill conceived ways by scientists of unknown reputation." He first cited the 1973 investigation of textile expert Prof. Gilbert Raes, who analyzed some threads. Gove claimed that Raes' discoveries had minimal significance, and that his and Turin's control of the samples were so careless that they were judged to be not suitable for carbon dating. He then recounted the 1978 STURP study, "carried out by people who were already convinced they were dealing with Christ's shroud." He asserted that STURP's findings yielded "negligibly significant results" and that STURP "subjected the shroud to a number of intrusive stresses." He further stated "almost every aspect of the STURP organization was distasteful to many other scientists. This included their clear religious zeal, their questionable sources of support, their military mind set, and last, but not least, their assumption that the Turin Shroud was their property as self-appointed investigators of its origins and properties."

Gove's key points to Chagas were: 1) If Chagas was not allowed to continue his leadership of the C-14 dating process, the C-14 "consortium would probably become disenchanted and withdraw their participation." He added that if STURP was involved in any way in the C-14 dating, the consortium withdrawal would be guaranteed. 2) If the C-14 dating was delayed due to conservation considerations, "conservation experts should be contacted by the Pontifical Academy and not by STURP." 3) Gove noted that STURP had proposed that Shroud samples would be removed from behind the patches in order to measure stable isotope ratios to try to determine the geographical origin of the cloth. Gove stated he "described this, quite charitably, as outrageous nonsense and asked whether there was nothing that could be done to hold STURP in check." 4) He reminded Chagas of the pressure that he claimed STURP had put on Flury-Lemberg during the 1986 Turin workshop "and, exaggeratedly, compared it to the Spanish Inquisition."

Gove had hoped that Pontifical Academy under Chagas would bring "a proper degree of international dispassion and integrity to the scientific endeavours to solve the mystery of the Turin Shroud." Gove said that "So far it had not because, clearly, he was unable to control the antics of STURP." He further stated that "One would be amused by the whole farce if one did not feel so saddened by the consequences STURP's activities would have [...]."

In his conclusion, Gove says that those directly involved in the C-14 dating hoped that the Shroud would "be subjected only to sensible and prudent" testing. He acknowledged that STURP might be allowed to do other tests, but "what is in our power, however, is to ensure that STURP plays no role in carbon dating." Gove related that he received no reply from Chagas, nor any indication of his reaction. He thought that Chagas may have realized, as Gove did not at the time, that the Pope had nixed the Academy's involvement.

Source: Gove, Harry. Relic, Icon or Hoax?: Carbon Dating the Turin Shroud (Bristol and Philadelphia: Institute of Physics Publishing), 1996, pp. 191-193.

Comments: It doesn't seem to have occurred to Gove that if the previous scientific tests had not been allowed, the C-14 test probably wouldn't have been performed. Regarding Raes, he was from the Ghent Institute of Textile Technology in Belgium. A Dutch web site says, "The Belgian textile industry is one of the most advanced and most successful industries in the world." That does not fit with Gove's assertion of "scientists of unknown reputation." His characterization of STURP, also relegated to that group by him, is laughable. The only point on which he was remotely accurate was the military mind-set. STURP did have various members who had been in the military; perhaps the group could have used a little less of that mind-set. But they definitely were not "scientists of unknown reputation." As noted previously, many of them worked in the U.S.'s space and nuclear programs. (See www.shroud.com/78team.htm for a list of STURP members and their organizations.) While the group included several devout Catholics and some believed that it could be Jesus' Shroud, their main objective had been to try and discover how the image got on the cloth, which they actually were unable to do. Far from producing "negligibly significant results," they produced an enormous amount of data, which took three years to analyze. STURP had agreed to do only non-destructive testing and even constructed a special table at the cost of $20,000 (in 1978 dollars! -- about $80,000 today) to be as careful with the Shroud as possible. Some money and equipment for the project were donated by corporations and some STURP members even spent some of their own money to get there so it's not clear what Gove had in mind regarding "questionable sources of support." Gove's claim that they assumed "that the Turin Shroud was their property as self-appointed investigators of its origins and properties" is just absurd. They worked with various Church authorities, who gave them permission to work on the cloth, and the results were openly reported in peer-reviewed scientific journals. Gove clearly unfairly demonized STURP and seemed to severely criticize any scientists outside the C-14 consortium (for whom Gove also leveled some criticisms).

Regarding the numbered points Gove made in the letter: 1) even though the Pontifical Academy of Sciences ended up basically having no role in the C-14 dating, the C-14 consortium did not withdraw. 2) Once again, Gove exhibited his fanatical antipathy toward STURP. 3) Gove revealed his bias that only a C-14 test had any significance, which is absurd given all the scientific disciplines that had been involved with the Shroud. 4) With all the pressures that Gove had been putting on various individuals to try and get his own way, it's laughable that Gove was complaining about STURP putting pressure on an individual. He even admitted his analogy about the Spanish Inquisition was exaggerated!

Gove judged that STURP had lacked objectivity and integrity and was presently up to "antics." And by implication, he was saying that none of that applied to him. The reader can judge if Gove was blameless in those areas. In the letter's conclusion, Gove again stated his intense (and obsessive) desire to at least keep STURP from being involved in the C-14 dating.

1987 *(June)*. Gove told the others they should have a position before Gonella sprung the announcement about the reduction of the number of labs. He thought the C-14 consortium would stick together and just inform Gonella that they refused to accept the decision. He thought, "That might bring Turin to its senses. Harbottle told Gove he would say "no." Donahue said he wouldn't do anything without more discussion but would want to stick by the original protocol. Gove thought it might be good to get the input from the other four labs. Gove decided he would inform all seven labs that as a result of Cardinal Ballestrero's request to the Pope, 1) The Pontifical Academy of Sciences would not be involved from this point onward, 2) Gonella was in charge of the whole enterprise and 3) fewer than seven labs, probably only two or three would be chosen to do the dating.

On June 30th, Gove wrote Woelfli suggesting that a group letter be sent to Cardinal Ballestrero requesting that another planning meeting be convened "or we would just say to hell with it." Woelfli said he would agree to such a letter. Gove then talked to Hall. Hall said he suggested that Gove not do anything. Hall told Gove, "If you do anything to your enemies in Turin it will be curtains." Hall thought Gove's wording of the letter was threatening, and that Gove should just request another meeting and not threaten to withdraw. Gove actually took Hall's advice. Gove followed by talking with Donahue, who expressed his thought that what had happened was almost certainly the work of STURP. He thought it would be tough to get it reversed. Gove "replied that I could not help thinking that it was STURP's enmity toward me that was causing the problem." Donahue told Gove he didn't think Gove was being singled out, but that STURP didn't like their meddling. Donahue agreed with Hall regarding not including the threat to withdraw. Harbottle suggested to Gove that they make some mention of the Cardinal's opening address at the Turin workshop, in which he said it was important that the procedure be carried out properly. The original protocol was drawn up on that basis, and the changes were not in the best interest of carrying it out. Gove then phoned Tite, who seemed to Gove "a bit cagey as to what he knew." Gove felt Tite knew much more than he was letting on. He said he would sign Gove's proposed letter. Tite also said if the number of labs were reduced to three, the labs themselves should decide which ones would be involved.

Source: Gove, Harry. Relic, Icon or Hoax?: Carbon Dating the Turin Shroud, (Bristol and Philadelphia: Institute of Physics Publishing), 1996, pp. 197-200.

Comments: Tite's comment that if the number of labs were reduced to three, the labs themselves should decide which of the seven would participate, is most bizarre. First of all, why would he assume that the Vatican and Turin authorities would be content with that? Secondly, given the existing chaos and human nature, how could he think that the labs could make an objective decision?? Note that Gove believed that Tite was keeping some information from him.

1987 *(June)*. Regarding the exclusion of the Pontifical Academy of Sciences, "Its exclusion was perceived as a **political** *[my bolding]* gain for Turin: it would be clear that they were to decide, without Rome interfering. But they were not completely pleased either with the choice of these institutions, and particularly of the British Museum: they felt that Dr. Tite was willing to act as a guarantor for the laboratories, but not for them. There were some hard feelings left after both Tite and Gove had published a reply to a letter from Dutton, in *Nature*, where the latter had warned against possible fraud. Indeed they answered by saying more or less that they would take any precaution 'not to be tricked'. This was interpreted in Turin as 'not being tricked by the Turin authorities'."

Source: Laverdiere, H. "The Socio-Politic of a Relic: Carbon Dating of the Turin Shroud," (Doctoral Thesis) 1989, pg. 85. Accessible via free download at http://ethos.bl.uk/OrderDetails.do?did=1&uin=uk.bl.ethos.531916.

Comments: Laverdiere noted (pg. 254) that Dutton [...] "did not complain about the protocol after the results indicating a medieval date."

1987 (June). After STURP's meeting on March 7th, a revised proposal was submitted to Gonella and Riggi on June 30th. According to Bracaglia, D'Muhala and Lukasik did not share with other members the news they had received from Gonella in December. This was brought to light when physicist Larry Schwalbe saw the draft proposal; his six-work package, X-ray fluorescence, and radiography had all been eliminated. Schwalbe demanded another meeting, and there was talk of replacing Lukasik as the coordinator.

Source: Bracaglia, Giorgio. Uncovering the Paradox within the Archives of the Holy Shroud Guild, (Honeoye, N.Y.: Holy Shroud Guild), 2019, pp. 225-226.

1987 (June). On the 30th, STURP's Lukasik wrote, "[...] [Q]uestions of the accuracy of radiocarbon have been raised by Meacham, especially in the case of an object like the Shroud that has been exposed to diverse conditions and potential contamination during its long life, both certifiable and unknown. These are not raised here because the issue is one of test priority, not technical detail. Nevertheless, there are significant problems in radiocarbon dating objects of unknown provenance. More importantly, the results, if unrecognizably incorrect, could actually put a true Shroud at risk by apparently establishing that it is not authentic.

Source: Lukasik, S.J. "Draft Protocol For The Next Examination Of The Shroud of Turin." http://freepages.rootsweb.com/~wmeacham/religions/sturp87b.pdf, pg. 13.

1987 (July). Representatives of seven different C-14 labs being considered to perform the Shroud dating test sent on the 1st a letter to Ballestrero: "Your Eminence:"

In your opening statement to the delegates at the Turin Workshop on Radiocarbon Dating of the Turin Shroud you charged us with designing "a valid and acceptable project for at last carrying out the radiocarbon dating of the Shroud cloth. You reminded us that, owing to "the unique and singular character of the object," such measurements "certainly could not easily be repeated." You asked us to devise "concrete operative proposals."

Your Eminence, it was to satisfy this desire on your part for a convincing and valid proposal for radiocarbon dating the Shroud that we drew up the protocol. A most important article in the protocol concerned the minimum number of independent measurements required to fulfill your charge of achieving a credible result. In our judgement that number should involve measurements by seven different laboratories.

We were therefore alarmed to read in the April 27, 1987 issue of *La Stampa* a statement attributed to Professor Luigi Gonella, your scientific advisor on matters concerning the Shroud of Turin that only two or three laboratories will be involved in the measurement. If that is indeed the case you are risking the possibility that what may be the first and only chance to date the Shroud cloth will fail. The material removed from this precious object will have been wasted.

We urge Your Eminence, before making a final decision on this question, to reconvene a meeting of the seven carbon dating laboratories and the British Museum with your science advisor Professor Gonella to more fully apprise him of the dangers of modifying the Turin Workshop protocol in this fundamental way.

The protocol was carefully crafted to meet your charge that the results of the measurements be credible to the general public and to knowledgeable scientists alike. As participants in the workshop who devoted considerable effort to achieve your goal we would be irresponsible if we were not to advise you that this fundamental modification in the proposed procedures may lead to failure.

Yours sincerely,

[signed by]
[representatives of the seven radiocarbon labs]

Source: Letter of July 1, 1987 from seven labs to Cardinal Ballestrero. Luigi Gonella archives, of which the author has a copy.

1987. According to Meacham, "I heard from several sources that the number of labs was going to be cut to three, and that STURP and other groups would not be allowed to run any of its planned testing until after the C-14 dates were announced. This was a sad state of affairs, as STURP was the main group studying the relic. Clearly Michael Tite of the British Museum was only in the picture as a referee of sorts; he would not be directly involved in the sample-taking or in the interpretation of the results. Sensing an opportunity for a small group to play a role in the project, I contacted two Italian archaeologists I knew -- Roberto Ciarla and Maurizio Tosi -- both of whom had worked in the Middle East. Together we formulated a proposal to be involved in the sampling and in the final interpretation of the results. Unfortunately Gonella did not take up this offer, and in the end chose his colleague Riggi, plus two textile experts who knew nothing about the Shroud, to assist in selecting the sampling site. This was a terrible decision on the part of Gonella, matched only by his equally appalling handling of the announcement of the results."

Source: Meacham, William. The Rape of the Turin Shroud: How Christianity's most precious relic was wrongly condemned and violated, (Lulu.com), 2005, pg. 83.

Comments: The shocking exclusion of STURP from the process -- and failing even to ask STURP's advice on a suitable C-14 sample Shroud location -- were errors as serious as the decision to choose only one sample region for the carbon dating. Gonella ignored a lot of valuable advice from Meacham (and others).

1987 (July). Canuto phoned Gove on July 1st and informed him that Cardinal Casaroli would be travelling to the United States. Gove discussed the revised letter he had recently penned and wanted to send to Cardinal Ballestrero. Gove wanted to send Canuto the final version and have him translate it into Italian for the Cardinal. According to Gove, "This would bypass the need for Gonella to translate it with all the attendant potential for mischief." That day, Gove sent the completed letter to Canuto, representatives of the seven labs, and the British Museum and told everyone he hoped they would all sign it, even though he actually feared that it was already too late. Gove called Canuto the next day, and it was decided that a copy of the letter would also be sent to Cardinal Casaroli and the Pontifical Academy of Sciences. Donahue and Damon from Arizona called Gove that day and said they would sign. Damon was appalled at the change in the protocol and said that if his lab was chosen they would probably refuse.

On July 3rd, Otlet and Tite indicated they would sign the letter. Hall said he was not pleased with the letter. He was open to having seven labs involved but thought even additional labs should be used. Gove informed Hall who had agreed to sign the letter and then asked Gove if he had talked with Tite recently.

Gove replied that they had talked within the past several hours, and Tite said he would sign. Hall then made the cryptic remark, "It didn't change anything, then, hmmm." This led Gove to believe that Hall, like Tite, knew more than he was willing to admit. Gove thought Hall probably knew that Tite had some information -- probably from Gonella -- about which labs had been chosen and that Hall's Oxford was one of them. Gove and Hall then had a discussion about seven labs versus three. Gove said that using four additional labs wouldn't use that much more material, so Gonella's main concern couldn't be saving cloth. Hall told Gove he suspected that Gonella did have other motives. Hall finally agreed to have his name added to the letter. Gove reached Duplessy in France on July 6th, and he agreed to have his name added as well. That day, Gove sent off Canuto's Italian translation by cable. Canuto then called Gove on July 9th. He had lunch with Cardinal Casaroli and had driven him to the airport, but they had no chance to discuss Shroud matters. Gove was disappointed that Canuto could not find the time to discuss with the Cardinal such an important matter.

Source: Gove, Harry. Relic, Icon or Hoax?: Carbon Dating the Turin Shroud, (Bristol and Philadelphia: Institute of Physics Publishing), 1996, pp. 201-205.

Comments: Gove did not even trust Gonella enough to assume he wouldn't do a straightforward translation of the letter. Damon's threat to withdraw, like Gove's multiple ones, was not carried out after Arizona was chosen to be one of the three to date the Shroud. It appears from comments and Gove's perceptions that even some of the individuals of the C-14 consortium were trying to keep secrets from the others. It would be interesting to know what Cardinal Casaroli and Canuto discussed in their time together instead of the Shroud.

1987 (July). On the 6th, a telegram signed by the seven laboratories and by Tite protesting the reduction to three of the number of samples was sent to Ballestrero (and to Chagas and to Casaroli). It was claimed to be a violation of the "protocol," which Ballestrero himself had wanted to credible to the public. One telegram arrived in Turin while the official letters were starting with which Ballestrero communicated the provisions received for the withdrawal (drawn up in accordance with the letter of Casaroli of May 21st and subsequent talks), and naturally blocked the operation. On July 18th, Casaroli transmitted to Ballestrero a copy of the telegram he received and Chagas' letters dated June 3rd and June 17th. Meanwhile Gonella was calling from colleague to colleague to Tite, Wolfli, and Hall asking them how they had come to sign such an offensive text, which treated as if it were an "agreement," a proposal that had clearly been said would be examined by the superior authorities and distorted the words of Ballestrero. Tite told Gonella that it was an initiative of Gove; he did not know that he would also be sent to Casaroli, he had signed so as to be "not losing credibility" and he had already regretted it, he wrote Gonella July 16th to apologize to the Cardinal for his unwise signature. Wolfli said he had signed only to please Gove, and that scientists are known to be bad diplomats. Hall said he had proposed radical changes, but Gove had told him that others had already signed and personally thought that Gove and Harbottle's labs should not participate in the dating because they were poorly equipped.

Source: Gonella, Luigi and Riggi di Numana, Giovanni. Sindone: il mistero continua, (Milan: Fondazione 3M), 2005, pp. 63-64.

1987 (July). Rinaldi from Turin wrote a letter to Otterbein on 12th, saying, "A letter from the Cardinal, accompanied by one from Gonella, was mailed last week to all the participants of last fall's carbon 14 test workshop held here in Turin. In it, the Cardinal makes it clear that only three laboratories will be called on to perform the test, this decision having been made by the Holy Father himself. The laboratories are Arizona, Oxford and Zurich."

Source: Email of January 17, 2020 to author from Giorgio Bracaglia, overseer of the Holy Shroud Guild archives.

1987 (July). Ballestrero mailed notices to the various labs to inform them that only three facilities would take part. One of the striking elements about the notice was that STURP was not mentioned in connection with the C-14 dating. In addition, nothing firm was stated about STURP doing their other proposed tests. STURP members were devastated and called for a Board meeting to discuss the possibility of removing D'Muhala as president on the grounds of his performance deficiencies and lack of communication from him. The seven labs, with Gove taking the lead, contacted both Ballestrero and the Pontifical Academy directly, urging to keep to the seven-lab protocol -- or they would not participate.

Gonella met with Rinaldi and told him that the Vatican and Turin would definitely stick to having only three labs. Tite originally signed the letter from Gove demanding that seven labs be kept but within a week, Tite apologized for his action and agreed not to boycott the test. In a separate letter sent to Gonella, Tite assured the scientific advisor that Oxford and Zurich would cooperate and agree to the decision established by the Vatican and Turin. Only the American labs seemed unwilling to agree to the revised protocol.

Source: Bracaglia, Giorgio. Uncovering the Paradox within the Archives of the Holy Shroud Guild, (Honeoye, N.Y.: Holy Shroud Guild), 2019, pp. 227-228.

1987 *(August)*. Rinaldi called Otterbein on the 1st to update him on various matters. Here is Otterbein's memo regarding the call:

"Rinaldi had meeting with Gonella yesterday. Card. *[Ballestrero]* has been to Vatican again and they agreed to hold their position: No more than 3 Labs for C-14 test. Prof. Tite of Oxford Lab signed Gove's letter of protest and had agreed to boycott the test. Gove's letter was sent to Vat. Sec. of State and to the Cardinal of Turin. However, recently Tite wrote to the Cardinal in Turin and apologized and said he made a mistake in signing the letter. He is now willing and anxious to make amends and is willing to cooperate. In fact, he is pretty sure that the three labs picked by the Vatican: Arizona, Switzerland and Oxford will all go along. Gonella will write to the three labs this week in the name of the Cardinal. This seems to be a good compromise and answer to my problem. When Gonella said they would get other

Labs and go ahead, I feared that the results of a test by three small, unknown Labs would be ignored, criticized, and of little value -- would not be accepted and hence would be a waste of time and money. However, if TITE and the three well known Labs are involved, there will be criticism by Gove and perhaps Chagas, but they will not be able to just brush aside the results."

"Peter *[Rinaldi]* thought the samples might be taken in October. I raised several questions. 1. Gonella seems favorable to have a representative of STURP at the sample taking -- ADLER would be the man. 2. MONEY -- Gonella said the Pontifical Academy paid travel for meeting in Sept. 86. Cardinal might do it. Card. thinks he might be able to raise some money in Italy. I told Peter, not to talk NOW, but if everything is set for samples, and only problem is money, we will get the money. Peter mentioned that Arizona had told him usual price for test was $300, but they waived it for the Oviedo cloth. 3. SEPARATION OF C-14 and STURP TESTS -- this seems definite. Possibility of C-14 in October & of STURP Tests in December 1987. 4. STURP TABLE and PHOTO RECORDING of C-14 Sample Taking -- I mentioned it and Peter will take it up with Gonella. I pointed out advantages but also the problems e.g., to have table ready for Oct. and shipping. I had mentioned to Kevin Moran yesterday the importance of getting table ready as soon as possible, JUST IN CASE we get word of green light for ANY test. He said STURP decided: NO TEST until summer of 1988."

"I asked about calling Bob Dinegar, but Peter thought this is all too confidential and delicate at the moment. Leave Gonella to contact the three Labs. Peter would call Mike M*[inor, STURP lawyer]* and give him the above info."

Source: Email of January 17, 2020 to author from Giorgio Bracaglia, who has access to the Holy Shroud Guild archives.

Comments: Bracaglia commented further, "The summation of months of research based on the Guild's documents on this topic, the underlying fact that I believe can be substantiated is: Turin/Gonella was nothing more than a figure head representing the Shroud. It was clear that the Holy See was in full control and always was [...]. I surmise the issue was the amount of material the Pontifical Academy of Sciences was willing to give up 50 mg of material or there about based on the minimum amount of carbon required. Every AMS lab stated they only needed 4 mg of carbon. So when the Pontifical Academy of Sciences realized the request jumped to 200 mg of material, that was not what they agreed on. It did not matter what Gonella said or agreed on in 1986. He was not in the position to speak for the Holy See. In fact, neither was Ballestrero [...]. The actual operation of C14 took as much as 700 mg of material of which 240 mg was usable. (NOT EXACT MEASUREMENT. ONLY USING THESE NUMBERS FOR THE SAKE OF THIS DISCUSSION) only half was distributed. Since the 3 labs received around 40 mg of material they should have been able to conduct 4 tests. Did they? I doubt it [...]."

Fr. Otterbein mentioned a C-14 dating of the "Oviedo cloth." That is a cloth believed by many to have been the cloth that the Gospel of John mentions having been on the face of Jesus and was found in a place by itself in the tomb. (John 20:6-7). Some scientists and researchers also believe there is evidence that the Shroud and the Oviedo cloth wrapped the same man. The Oviedo cloth has a solid history of having been in Spain since the early seventh century. If, in fact, the Oviedo cloth is authentic, the 1988 C-14 dating is off by <u>at least six-hundred years</u>. For more information on the Sudarium of Oviedo, one can use the search engine at www.shroud.com.

1987 (August). At the beginning of August, Ballestrero wrote Casaroli that the latest events had once again demonstrated "that a coalition is operating between Chagas, Gove and Canuto with objectives very different from the best interest of the Church;" the protest had been a personal initiative by Gove who realized that only if seven labs received Shroud samples would his ill-equipped lab be included in the project.

In the meantime, a copy of the progress report that Gove had sent on April 15th to Tite and the six labs arrived in Turin. Hall wrote to Gove on August 12th that as he had predicted "your last broadside against Turin" had only caused trouble, and that he would not sign other documents. He waited for the Italian decisions, hoping for the best, and he sent Gonella a copy of the letter expressing regret for having signed the telegram. *[Hall was referring to the telegram that Gove and the seven labs had sent on July 6th to Ballestrero, Chagas and Casaroli, and not the* *progress report from Gove sent on April 15th to Tite and six labs.]* Gonella felt that Chagas' objection to STURP's tests were based on false data and constituted "heavy interference in the freedom of research of other scientists."

Source: Gonella, Luigi and Riggi di Numana, Giovanni. Sindone: il mistero continua, (Milan: Fondazione 3M), 2005, pg. 64.

Comments: Note that Ballestrero reemphasized a perceived coalition between Chagas, Gove and Canuto "with objectives very different from the best interest of the Church." That statement is followed by a semi-colon that indicates that the different objectives were tied to Gove's desire to make sure his lab would be included. So why were Chagas and Canuto, who were employed by the Church as part of the Pontifical Academy of Sciences, so closely tied to Gove's objectives?????? It is currently a question without a known answer.

1987 (August). On the 19th, Gove received a letter from Hall, who brought up the conversation they had concerning Gove's proposed letter (in Italian) to Ballestrero. Hall said he considered the letter unwise but had agreed to sign it as a friendly gesture. Hall learned that the Archbishop had been displeased (and told Gove that the Cardinal "was now probably laying various mystical punishments" on their heads), so he and his colleague Robert Hedges intended to distance themselves between the Gove camp and the Gonella camp. He felt "that any further hectoring" would only delay things. Hall informed Gove that he intended to keep a low profile and just await developments from Italy. Gove commented in his book, "[H]e was breaking ranks -- he was letting the side down." Two days later Gove sent a response to Hall. Gove said he couldn't understand why his letter why the letter would have displeased the Cardinal.

Gove went on at length: "Let me hasten to assure you that my 'hectoring' as you call it is directed toward STURP and only peripherally toward Professor Gonella to the extent that he champions STURP's cause. It is certainly not directed to him in his capacity as the Cardinal's science advisor. By all accounts, however, he is not held in particularly high regard in that capacity outside Turin. He is unfortunately still *the* power in Turin as far as the shroud is concerned. I have had almost ten years experience with STURP and regard them as a pack of religious zealots, who could really queer the pitch for carbon dating unless they are held at bay. I fear the cold and malevolent eye of Mr. Lukasik much more than your suggested mystical imprecations of the Cardinal. I have received recent information that STURP's influence is on the wane and high bloody time I would say."

Hall was told by Gove that a paper he had written about the protocol was going to be published in the November issue of *Nuclear Instruments and Methods in Physics Research, section B* and that it was refereed by Paul Damon of Arizona. They went on to discuss the question of seven labs versus three and the hope that the British Museum and the Oxford lab would be able to participate. Gove wrote, "Why the hell should Turin be permitted to put us in this 'Sophie's Choice' situation? Do any of us so lust to have our laboratory involved that we are willing to behave in unseemly ways? I suggest we continue to stick together. There is no earthly reason why three laboratories are better than seven. They cannot date it without us and who of us really cares if it is ever dated? I suppose I should care the most since I have put so much time and effort into the enterprise [...]." Gove related in his book that he was still annoyed with Hall breaking ranks for what he suspected were "self-serving reasons."

Gove proceeded on August 24th to send a letter to all of the other labs. Gove summarized Hall's letter and his own response. Gove advised, "[R]epresentatives of the seven labs must also be present during the sample taking in Turin and personally receive the samples and controls for delivery to their labs. If Turin intends to modify any of these protocol provisions I believe none of us should accede. I suggest in such circumstances Turin be invited to find three other carbon dating laboratories which would be willing to take on the task. Such a stance should not be taken as a threat to Turin but rather as an act of prudence and responsibility."

Source: Gove, Harry. Relic, Icon or Hoax?: Carbon Dating the Turin Shroud, (Bristol and Philadelphia: Institute of Physics Publishing), 1996, pp. 205-209.

Comments: Gove and Hall's discussion about the letter that Gove had sent Ballestrero clearly brings out the cliques and courting of favors, elements which should not be present in a pure scientific exercise. Even after it was revealed that the letter displeased the Cardinal, Gove couldn't understand why -- another example of him seeing through "Gove-colored" glasses. And once again, Gove took the opportunity to bash STURP and Gonella. Given that Gove feared "the cold and malevolent eye of Mr. Lukasik," it again confirms that Gove demonized STURP.

Since Gove and Damon were close colleagues, what were the chances that Damon, in his role of referee for Gove's paper, would suggest radical changes or even reject it? Regarding Gove's question "Do any of us so lust to have our laboratory involved that we are willing to behave in unseemly ways?," he apparently answered it with his comment about Hall's "self-serving reasons." Gove seems to have contradicted himself: he asks Hall who of us really cares if it is ever dated, but all of his statements actions clearly indicate that he did. Previously, Gove had several times threatened to pull his lab from the enterprise and acknowledged he wouldn't when pressed by Chagas; now he's suggesting that the other six labs pull out if their demands weren't met. Does anybody think that there was any realistic chance of that happening???

1987 *(September)*. According to Petrosillo and Marinelli, "Despite the directive by Casaroli to Ballestrero back in May, things were still uncertain. Turin and Rome did not agree on major points. Ballestrero and Gonella were anxious to proceed, while Chagas' failure to be on the same page with Gonella caused delays. The number of samples still had not been specified and STURP was having some financial issues. Ian Wilson of the BSTS expressed the hope that the dating won't be done without the multidisciplinary tests proposed by STURP, ASSIST, and the British *[BSTS]*. Meacham vehemently asserted that the ^{14}C should not be the supreme arbiter of the authenticity of the Shroud. He stated there could be hidden problems. Dating is based on various assumptions that cannot be subjected to lab tests. Regarding the number of samples needed, the request had been made not only to date cloth, but also to take them from image and bloodstain areas. In 1986, the first ^{14}C of blood residues on prehistoric stone tools was carried out in Canada.

The technique needed further fine-tuning, but the results obtained corresponded to the expected age, thus showing that blood could be dated. The amount of carbon needed was less than one milligram. Naturally, it had to be uncontaminated carbon and care had to be taken that purification techniques themselves didn't inadvertently introduce contaminants during the preparation process."

Source: Petrosillo, Orazio and Emanuela Marinelli. La Sindone: Storia Di Un Enigma. (Milan: Rizzoli, 3rd edition, 1998), pg. 145.

Comments: To continue this thread, go to the Petrosillo/Marinelli entry for "1987 *(October)*."

1987 *(September)*. Gove received a reply from Hall on September 3rd regarding the August 21st letter. He said his main point was that Cardinal Ballestrero would be the only making decisions. The C-14 consortium made their wishes known but he wasn't obliged to follow them. Hall said any lab was free to withdraw. Hall said he would "have to think hard if less than four labs were involved." Gove was then quick to point out, "He clearly did not have to think too hard when it was decreed that only three labs would be involved -- one being Oxford. He accepted with alacrity." Hall mentioned that the C-14 consortium had all agreed they would all pull out if STURP "ran the show."

Woelfli from Zurich also replied on September 3rd, regarding Gove's August 24th letter. Woelfli informed Gove that he had received from Gonella a phone call saying that the cable to Ballestrero was upsetting to him and even might put the C-14 dating in danger. According to Woelfli, Gonella was agitated when Woelfli told him that he still favored the procedures as outlined in the Turin Workshop, and that any change would necessitate another meeting of the C-14 consortium. Gonella told Woelfli that the final decision on procedures would now definitely be delayed.

Gove received a letter on September 3rd from Otlet from Harwell. He told Gove he was sorry to hear of the exchange with Hall. Otlet heard from a colleague that Hall had given a talk near Harwell and had ridiculed Harwell's proportional-counter method, the time it would take a measurement, and the sample size needed, which he described as handkerchief-size. But Gove wrote that Hall knew full well that due to advancements, Harwell would only need a sample slightly larger than what Oxford needed with its AMS method. Otlet told Gove that he was uncomfortable with the latter's approach to the Cardinal. But he said that any significant change in the protocol wouldn't be acceptable to him, and he wrote Chagas directly to tell him that.

Tite wrote Gove on September 11th regarding the latter's letter of August 24th. Tite told Gove he agreed with many points in the letter but didn't want to comment further until something more specific came from Turin. Gove also received a letter from Damon of Arizona (with copy to his colleague Donahue) saying that the consortium should push for the original protocol or not participate and that Donahue agreed.

Source: Gove, Harry. <u>Relic, Icon or Hoax?: Carbon Dating the Turin Shroud,</u> (Bristol and Philadelphia: Institute of Physics Publishing), 1996, pp. 205-209.

Comments: For Hall to have said that Harwell would need a handkerchief-size sample to date the Shroud when he knew it was actually smaller was unethical, to say the least. Once again, the labs made threats about withdrawing if the original protocol was changed, but when it actually did, not one of them did withdraw.

1987 *(October)*. On the 10th, Ballestrero sent the following letter to the representatives of the seven labs:

> "To all participants in the Turin Workshop on the radiocarbon dating of the Shroud of Turin"

Dear Sirs,

At the end of May I received positive instructions from the Holy See, personally signed by the Cardinal Secretary of State *[Casaroli]*, on how to proceed to the radiocarbon dating of the Shroud of Turin.

The instructions agree to the main line of the proposal put forward at the Turin Workshop of last year, but do not accept a few items. In particular, they direct that no more than three samples be taken, to be used for measurement by different laboratories. As for the measurement, the instructions agree to the suggested procedure, i.e., to use the method of blind testing with control samples, to apply to the competence of the Shroud samples, and to entrust to the competence of the same British Museum and of the Institute of Metrology "G. Colonetti" the statistical analysis of the measurement results.

As a consequence, in the first place I wish to express my thanks to all who participated in the Turin Workshop with generous availability, even though I find myself unable to take advantage of the competence of all participants, as it was in the wishes of the meeting.

The choice of the three Laboratories among the seven which offered their services was made, after long deliberation and careful consultations, on a criterion of internationality and consideration for the specific experience in the field of archaeological radiocarbon dating, taking also into account the required sample size. On this criterion the following Laboratories are selected:

Radiocarbon Laboratory	University of Arizona
Research Laboratory for Archaeology	Oxford University
Radiocarbon Laboratory	ETH, Zurich

The operations for taking the samples have to be presided by myself, in my capacity as Pontifical Custodian of the Shroud. H.E. Professor Carlos Chagas, President of the Pontifical Academy of Sciences, will be invited to be present at the operation, as well as at the eventual final meeting, as my personal guest, in consideration of the collaboration he gave in working out the project. The instructions from the Holy See do not deem it necessary for representatives of the measurement Laboratories to attend the sample-taking operations.

The decisions took more time to be worked out than originally wished, owing to the situation without precedents created by a number of competing offers tied into a rather rigid proposal, and also by initiatives of some participants in the Workshop who stepped out of the radiocarbon field to oppose research proposals in other fields, with implications on the freedom of research of other scientists and on our own research programs for the Shroud conservation that asked for thorough deliberations.

Besides, when the competent Authorities advised me they deemed we ought to proceed with three samples, a concerted initiative was taken to counter the decision, with the outcome of a telegram sent to H.E. the Cardinal Secretary of State and myself by some participants in the Workshop, a telegram where the meaning of my introductory words at the Workshop was heavily misinterpreted.

After further deliberation and scrutiny of the situation with the Cardinal Secretary of State we are now proceeding on the already-decided terms, which I was just going to write you when I received the above quoted telegram.

In consideration of the great attention from the public and the press that all of us know this measurement is attracting, it seems to me worthwhile to stress again what I said in my opening address at the Turin Workshop, about the purely scientific character of this enterprise, which does not mean to, nor could, address any issue of faith related to the death and resurrection of Jesus Christ. Nor do I mean with this analysis to charge the Laboratories that have been selected with the task of "authenticating" the Shroud of Turin: the analysis is strictly meant to ascertain the radiocarbon date of its cloth, as an objective datum to the scientific quest that has long been growing on the illustrious image entrusted to my stewardship.

[signed]
Anastasio Card. Ballestrero
Archbishop of Turin
Pontifical Custodian of the Shroud of Turin

Here is the final draft of a letter sent by Donahue, Hall and Wolfli:

"Anastasio Cardinal Ballestrero, Archbishop of Turin, Pontifical Custodian of the Shroud of Turin."

Your Eminence:

We have received your letter of 10 October 1987, and we are honored to have been selected to participate in the determination of the age of the cloth from the Shroud of Turin. However, we are concerned to learn that a decision has been made to limit the number of participating laboratories to three. We are in agreement with the conclusion reached at the workshop held in Turin in September/October 1986. That is: "A minimum amount of cloth will be removed, which is sufficient (a) to ensure a result that is scientifically rigorous and (b) to maximize the credibility of the enterprise to the public. For these reasons, the decision was made that seven laboratories will carry out the experiment."

We believe that reducing the number of laboratories to three will seriously reduce "the credibility of the enterprise" which we are all so anxious to achieve. As you are aware, there are many critics in the world who will scrutinize these measurements in great detail. The abandonment of the original protocol, and the decision to proceed with only three laboratories will certainly enhance the skepticism of these critics.

While we understand your desire to use a minimum amount of material from the Shroud, we believe that the increased confidence in the measurements which would result from the inclusion of more than three laboratories in the program, would justify the additional expenditure of material. Although improvements in statistical errors resulting from including more measurements might not be great, the possibility of the occurrence of unrecognized, non-statistical errors would be substantially reduced. For example, if only three laboratories participate, and one of them obtains a divergent, non-understandable result, the entire project could be jeopardized. But if results from a larger number of laboratories are available, a divergent result could be more easily recognized as such, and could be treated appropriately, in a statistically accepted manner. Clearly, it is the reduction of the possibility of unrecognized non-statistical errors in measurements that leads to increased confidence in the final result.

We would very much like to take part in the program to determine the age of the cloth in the Shroud, but we are hesitant to proceed under the arrangement in which only three laboratories would participate in the measurements. We urge that the decision to change the protocol of the Turin workshop and the decision to limit participation to only three laboratories be given further consideration.

Respectfully,

[signed by]
Professors Douglas Donahue, Edward Hall and Willy Wolfli

Source: Copy of letter from Ballestrero to the labs, and draft of letter from representatives of the three chosen labs, both in possession of author.

1987 (October). After having conferred with Casaroli on September 7th, Ballestrero sent on October 10th prepared letters to the labs. It was decided that only three labs would do the testing. Arizona, Zurich and Oxford were chosen; Harwell was eliminated because it needed more material *[using the older C-14 method]*. Ballestrero sent a letter to all the participants of the Turin meeting in which he communicated and gave motivation for the decision, a letter to the three chosen laboratories in which their proposal was accepted, one to Tite to invite him to perform the actions required of him in the Turin meeting, and one to the Director of the Colonetti Institute to ask him to work on analyzing the results together with the British Museum. For all the letters (except the last one) the English translation was provided at the same time, the pronouncement provoked a predictable reaction on the part of some of the excluded. Gove pressured the chosen laboratories to refuse, prompted negative articles in his hometown's newspaper, and even convinced the Catholic bishop of Rochester to write a letter to Ballestrero. On October 27th, the Director of the Colonetti Institute had replied to Ballestrero accepting the invitation to analyze the dating statistics.

Source: Gonella, Luigi and Riggi di Numana, Giovanni. Sindone: il mistero continua, (Milan: Fondazione 3M), 2005, pp. 64-66.

1987 *(October)*. Kersten and Gruber wrote "On 10 October 1987 Cardinal Ballestrero wrote to the seven laboratories and told them that he had received the go-ahead from the Holy See for the experiment. But now suddenly the procedure looked completely different. All that was left of the agreement made a year before was the role of the British Museum as procurer of control samples. The most interesting and certainly the most unexpected change was the total exclusion of the Pontifical Academy of Sciences. Chagas was now only to be admitted to the sampling as personal guest of Cardinal Ballestrero. There was not a word about precautions to prevent the specimens being swapped; and this even though this precise danger had been addressed in an article which appeared shortly before, which had excited a lot of attention -- accusations, discussion and apologia -- and must still have been fresh in the organizers' memory. No mention was made either of the Swiss textiles expert Flury-Lemberg, who was originally to have taken the specimen from the relic. Only Tite of the British Museum was left as guarantor of the correctness of the procedure. Neither the Pontifical Academy nor the IMGC *[G. Colonetti Institute]* would take part in the analysis of the final data. This too was to pass exclusively into the hands of Tite. Only three laboratories -- Tucson, Oxford and Zurich -- were selected; a decision which had apparently been reached as far back as May. All three worked with the newer AMS technique."

"Harry Gove, who had assumed the role of sole leader of the radiocarbon investigation, was furious. He fired off letters to the Pope, the journal *Nature* and the British Museum. His colleague Harbottle at the Brookhaven laboratory also aired his disappointment. They had both worked to develop the classical technique further as a method for small samples. The exclusion of the Harwell laboratory baffled the others, because it had the most experience of them all, and was renowned for the most precise datings. In their joint letter to the Pope, Gove and Harbottle classed this decision as scientifically short-sighted. The original Turin protocol with the seven laboratories would eradicate errors like those at the Zurich laboratory during the 'test run.' It would be better to do nothing at all, they added, rather than dare to go ahead with such a truncated experiment. In another letter of protest Robert Otlet of the Harwell laboratory voiced the suspicion that someone in Italy wanted to obstruct the course of science. Reducing the number of laboratories to three would lead to a scientific catastrophe. 'It is,' he wrote, 'like ordering a bulldozer to run over an archaeological dig site before you had examined it.' The rumour spread that the Vatican had restricted the test to three laboratories in the hope of obtaining contradictory results. Gove accused the Archbishop of having a false advisor. He said Gonella was not qualified for the post, he was nothing but a 'second-class scientist'. Gonella countered that they were dealing with a real 'radiocarbon Mafia', who were seeking their own advantage."

Source: Kersten, Holger and Elmar R. Gruber. Jesus Conspiracy: The Turin Shroud And The Truth About The Resurrection, (Rockport, MA: Element), 1994, pp. 44-45.

Comments: Kersten & Gruber added "One thing emerges from these reactions. The excluded parties felt deeply offended, and did not hesitate to speak out against these decisions, taken by anonymous backroom men at the Vatican, and to attack their 'adversary'."

1987 *(October)*. Not knowing that on the 10th, Ballestrero mailed out letters to the seven labs saying that only three would be chosen, Gove sent to Gonella on the 13th his summary report of the workshop from the previous year. He wrote, "[A] minimum amount of cloth will be removed, which is sufficient (A) to ensure a result scientifically rigorous and (B) to maximize the credibility of the enterprise to the public. For these reasons, the decision was made that seven laboratories will carry out the experiment: five accelerators-mass spectrometer laboratories and two small counter laboratories." He also stated that a total of 50 mg. would be taken and a single dummy sample would be prepared by the British Museum. The Shroud samples and dummy sample would be distributed in a way that all the labs would not be able to identify their samples. The responsibility for the distribution of the samples to the labs would be divided among Chagas, Tite, and Gonella.

As previously noted, if only 50 mg. total were used, it would not be possible to have seven labs involved. Gove had not specified in his report that some labs would receive only dummy samples. It's clear from Meacham's book that dummy samples were included in the testing procedures. Even if only one small-counter lab were used, only two other labs would receive actual Shroud material to stay within Gonella's fifty-mg. maximum. Apparently, Gonella was unaware that the amount of carbon that can be extracted from material necessitated the piece to be larger than the target figure -- Lukasik estimated that two-hundred mg. would actually need to be excised.

Another issue was that both Chagas and Coero-Borga did not want the Shroud to be dated. Coero had some radiocarbon-experts advisors, who told him the Shroud was not a suitable object for C-14 dating. Many in that circle believed that the Shroud would end up dating only to the fifth or sixth century. Rinaldi, however believed that Coero-Borga was using his information to try and eliminate STURP altogether. In a letter to another sindonologist (an Episcopal priest), Rinaldi wrote, "I know you agree with me that the carbon-14 test must be approached with caution. Not to do so would be to provide the Centro and others, too, with arrows -- some real, some fancied -- to aim at the 'ugly Americans.' I could not agree more with you when you state that 'the Centre's proprietary relationship to the Shroud' is behind Don Coero's efforts to decide how and by whom it will be tested, but the very least we can do is again, to forestall some of the Centre's guerilla tactics."

The coordination of the chain of custody of extracted Shroud samples to the chosen labs was assigned to the British Museum. Dr. Mechthild Flury-Lemberg, the Swiss textile expert, was chosen to remove the sample at the assigned time. Bracaglia believed that at this point, any hopes that STURP had about participating in the C-14 dating had disappeared.

Donahue of Arizona, in one last-gasp attempt to the original protocol, drafted a letter that all seven labs would sign. Their concern was that if one of the three labs obtained a result that was an outlier, it could very well jeopardize the success of the project. The letter was sent to Ballestrero. The effectiveness of the letter was lessened since Tite had already hinted to Gonella that Oxford and Zurich had several months earlier agreed to the revised plan.

One big problem with the letter was that it didn't stress the importance of retaining utilization of the proportional-counter method. If the Vatican and Turin decided to return to the original protocol with seven labs, that method would have been used. However, since in Donahue's draft there was no emphasis on it, AMS would be the only method. Maloney was concerned about this, and convinced Rinaldi to relay the matter to Gonella. Maloney furnished to Rinaldi a document from Dr. Stuart Fleming, Director of the Museum of Applied Science Center for Archaeology. That document supported the original protocol and detailed five conditions that can significantly alter the accuracy of C-14 dating. Fleming asserted that extraneous contamination could be introduced in the necessary pre-testing process used for AMS. Fleming believed that to properly date the Shroud, a minimum of twenty samples and multiple locations from the cloth would be needed. Rinaldi thought it best if Maloney met directly with Gonella, so a meeting was planned in Rye, New York for November 21st. (See further below.)

Source: Bracaglia, Giorgio. Uncovering the Paradox within the Archives of the Holy Shroud Guild, (Honeoye, N.Y.: Holy Shroud Guild), 2019, pp. 219-221, 228-229.

Comments: In the days when snail mail was used for most overseas communications, it would normally take about eight days for a letter from Italy to reach the United States. When Gove sent on the 13th his letter to Gonella, still believing (or possibly just hoping) that seven labs would be chosen, he wouldn't receive the Cardinal's letter for at least another five days. One might ask why Gove in the summary proceeded to refer to a seven-lab protocol when there were strong rumors since May that there would only be three. There's no doubt that Gove would have been in contact with Chagas and Canuto; it could be they had told him the three-lab information had not been official.

Why Coera-Borga's advisors believed the Shroud would be dated to the fifth or sixth century is not known. The Episcopal priest refers to "Don Coero." "Don" is an Italian term for a priest and the second part of his name "-Borga" is often left off when referring to him. It's rather striking that two priests, discussing another priest, used terms like "guerilla tactics" when describing events surround the supposed burial cloth of Jesus.

Regarding Rinaldi's reference to "ugly Americans," Bracaglia relates in his book (pg. 225) that Coera-Borga, shortly before he died, bellowed to Rinaldi (who was a dual citizen of Italy and the United States), "You Americans will destroy the Shroud!"

In 1988, the AMS method had been in use for only about eleven years. The way the test was conducted was a far cry from the conditions suggested by Fleming.

1987 *(October)*. On October 19th, Otlet called Gove and informed him he received a letter from Cardinal Ballestrero that three labs chosen to date the Shroud were Arizona, Oxford and Zurich. Gove wrote, "My first reaction was to say that the exclusion of Rochester meant I should probably take a course in diplomacy." He added, "I admitted that I was most disappointed and he said that he understood." Otlet faxed Gove a copy of the letter. Various points stood out to Gove: 1) The Pontifical Academy of Sciences would not be directly involved; 2) the labs were told that representatives of the labs need not be present at the sample taking, which Gove found "outrageous;" 3) it was clear that the cable that was sent to Ballestrero had intensely agitated him. Gove added that the Cardinal's "thinly veiled accusation that we were attempting to prevent STURP from carrying out its scientific investigation was quite accurate."

Gove phoned Donahue later in the day. The latter was taken aback by the letter, and said Arizona probably wouldn't go along with it. He suggested a discussion at a C-14 conference to be held in Yugoslavia the following year. Gove said he would try to contact Woelfli. Harbottle called Gove the next day to say he was going to contact Dutton, the New Zealand art historian, the *Skeptical Inquirer*, National Public Radio, and whoever else he could think of to complain about the decision. He said he would call Canuto and possibly Chagas, whom Harbottle thought could get through to the Pope. When Gove talked with Harbottle the next morning, the latter said he talked with Canuto, who agreed that

Chagas should try one more time to have the Pope advise Cardinal Ballestrero that he was receiving bad scientific advice. Gove suggested a joint press release involving himself, Harbottle, Damon and Donahue. Canuto phoned Gove shortly after and suggested they send an open letter to the Pope via Chagas that the seven lab protocol should be followed.

On October 26th, Gove phoned Woelfli in Zurich. Woelfli also was upset that the Cardinal indicated that the lab representatives need not be there for the sample taking. He was not in favor of Gove's press release, but did want to see both a draft of the letter to the Pope and the press release, and then would decide if he would participate. On October 29th, Gove prepared a memo with both drafts that was sent to the other six labs. Gove indicated he would only send the letter to the Pope if at least six of the seven labs signed it. It was decided to resort to a press release only if the proposed letter to the Pope failed. Duplessy called from France the next day and said he had recently received a call from Gonella, whom he told that he didn't want any changes from the Turin Workshop protocol. Duplessy said he would sign the letter to the Pope if the others did but, like Woelfli, did not like the idea of a press release.

Below are some excerpts from the drafts of the letter to the Pope and also the press release. (Keep in mind while reading them that Gove did not complain about the final results after only three labs participated.)

Draft to the Pope: "It is our collective impression that Cardinal Ballestrero as received very unwise scientific advice. The proposed modifications will confirm the suspicion of many people around the world that the Church either does not want the Shroud dated or it wants to have it done in an ambiguous way. The procedure that the Cardinal of Turin is suggesting is bound to produce a result that will be questioned in strictly scientific terms by many scientists around the world who will be very skeptical of the arbitrarily small statistical basis when it is well known that a better procedure was recommended. Since there is great world expectation for the date of the Shroud, the publicity resulting from a scientifically dubious result will do great harm to the Church."

"We respectfully urge Your Holiness to persuade the Cardinal of Turin that the scientific advice being given to him is not shared by the world experts in this field. He should be urged to seek the advice of the eminent scientific organization expressly created to advise you, namely the

Pontifical Academy of Sciences that enjoys the respect of the scientific world at large."

"Rather than following an ill advised procedure that will not generate a reliable date but will rather give rise to world controversy, we suggest that it would be better not to date the Shroud at all."

Concluding paragraph of completed Press Release: "The new procedures suggested to the Cardinal of Turin and that he has now embraced, will, if implemented, yield a result for the date of the Shroud that will certainly be vigorously challenged by the world scientific community for their flimsy statistical basis. We urge the Cardinal of Turin to seek scientific advice from an unimpeachable source that was available to him from the very beginning, but that he chose to ignore, namely the Pontifical Academy of Sciences, which enjoys worldwide respect in the world scientific community. Only with the best advice of world experts on carbon-14 dating can a scientifically credible date for the Shroud of Turin be arrived at."

Source: Gove, Harry. Relic, Icon or Hoax?: Carbon Dating the Turin Shroud, (Bristol and Philadelphia: Institute of Physics Publishing), 1996, pp. 213-219.

Comments: Very few people would probably disagree that Gove could have used some lessons in diplomacy. Although Gove in August had written to Hall, "who of us really cares if it is ever dated," his response did not reflect his earlier statement. Regarding the Pontifical Academy of Sciences non-involvement, a highly-reliable source with connections to Turin told me that the Academy continued to call the shots, and even let Gonella be the fall-guy for their bad decisions. Although this is not proven beyond a reasonable doubt, it would be consistent with the other skullduggery reported throughout this book. Regarding the lab representatives supposedly not being present at the sample taking, they were allowed to be there when the samples were taken on April 21, 1988.

The fact that Otlet faxed a copy of the letter to Gove indicates that Gove did not even actually receive the official letter from Ballestrero. Considering that for at least one point in time, Gove was in a leadership position for the dating project, it's rather shocking that he wasn't even sent the letter. I believe it's an excellent example of the lack of communication among some of the principal figures. Anyone who reads this book from beginning to end will probably find at least one example of each major player criticizing in some manner another major player, and in some cases, even vowing to keep that other player out of the process.

1987 *(October)*. Petrosillo and Marinelli wrote, "On the 10th, which was a little over a year since the Turin workshop, Ballestrero wrote to the seven labs and participants. He declared that he received the instructions from the Holy See on how to proceed with dating. He asserted that the instructions were in line with the main points from the workshop, but Hall opined that details had changed. In fact, the only thing in common is the inclusion of the British Museum which will help with the certification of the samples provided and the statistical analysis of the results. The Pontifical Academy is eliminated from the guaranteeing and coordinating institutions. It's involved only insofar as Chagas is to be invited to the sample-taking as a guest of Ballestrero. There was no mention of procedures to ensure that the Shroud samples are not replaced during the project. Flury-Lemberg was not mentioned. Tite remains the sole 'guarantor' master of everything. The Holy See had authorized the taking of no more than three samples, therefore three laboratories were chosen after reflections and careful consultations with advisors from the Pontifical Academy. Taking seven samples was deemed to be too destructive, so the proportional-counter method was excluded. The AMS labs -- Oxford, Tucson and Zurich -- only needed about one-third of the material that the proportional-counter labs, about 20 centimeters total. Other criteria specified for the labs were internationality and accumulated experience in the field of archaeological dating. Those, according to Petrosillo and Marinelli, were just pretexts. They believed each of the seven labs was equally qualified. Gove himself recognized this at the Dubrovnik radiocarbon conference held in June 1988. The protests of the supporters of the proportional-counter and of the excluded laboratories came fast and furious. Some said the AMS method was not yet reliable, especially given the frequency of false results with small samples and also because it didn't have much experience dating fabrics."

"Gove (AMS) and Harbottle (proportional-counter), both pioneers of radiocarbon dating, were particularly offended. Gove penned letters, first, and indignant one to the Pope, then to *Nature*, and also the director of the British Museum. Why were the Pontifical Academy of Sciences and Flury-Lemberg removed from the project? The authorization that limited the number of labs to three had already been signed in May by Casaroli. So, the operation has been pruned from seven labs to three, from two C-14 methods to one and from three guaranteeing institutions to two. Ballestrero explained that the decision was slowed by a number of proposals in the competition and also by the initiatives of some who wanted the C-14 test to the exclusion of other tests, which both prevented the freedom of research of other scientist and blocked programs relating to the conservation of the Shroud. The maneuver had the air of "**political**" *[my bolding]* action to eliminate those who deal with the problem with modern means in various disciplines favoring access to the Shroud only to a very limited circle of people willing to provide an uncontrollable result and therefore unreliable."

"The other work plans had been given the go-ahead by Church authorities, but the controversies delayed them. Ballestrero harshly criticized the labs' decision to send a telegram to the Vatican complaining about him. The complaints were obviously based in Gove's attempt to have the ^{14}C test done exclusively. In fact, in his communiqué, Ballestrero didn't indicate that the other tests would be done, nor did he indicate they would be refused; STURP was dismayed."

Source: Petrosillo, Orazio and Emanuela Marinelli. La Sindone: Storia Di Un Enigma, (Milan: Rizzoli, 3rd edition), 1998), pp. 146-148.

Comments: It seems strange in the extreme that Ballestrero complained about the Gove *et al.* effort to have the C-14 test only, but then seems to give in by not announcing at the time that the other tests were planned to done (which unfortunately, never ended up happening).

1987 (November). "Due to the non-committal by Ballestrero regarding multidisciplinary tests in his communiqué of October 13th, thirteen STURP members met with Gonella in Port Chester, New York on the 20th for three days of restricted talks in which they take stock of the situation and agree to wait for a truly final decision."

Source: Petrosillo, Orazio and Emanuela Marinelli. La Sindone: Storia Di Un Enigma, (Milan: Rizzoli, 3rd edition, 1998), pp. 146-148.

Comments: To continue this thread, see the Petrosillo/Marinelli entry for "1987 *(December)*."

1987 *(November).* Gove's book related that on November 2nd, he received a call from Canuto, who said he had talked to Chagas. Canuto was told by Chagas that he had felt he had been kicked in the stomach when he read the October 10th letter from Cardinal Ballestrero. Canuto told him about the letter and news release that Gove and others were proposing. Chagas thought the letter was a good idea and that he would likely endorse and deliver it. Chagas was particularly interested in what Woelfli felt. Canuto said that he thought that Donahue in Arizona was the key person because if that lab pulled out, the United States would be out of the loop, which could have some impact on Turin. Gove then called Woelfli, who said he would be agreeable to sign the letter to the Pope as long as both Oxford and Arizona signed it. He said he had now received a letter from Ballestrero specifically inviting his lab to participate -- Arizona and Oxford had received similar letters. Woelfli said he felt that they should jointly answer the invitation. He thought Oxford would be prepared to go and wanted to hear from Donahue. Woelfli wanted to coordinate the reply to Ballestrero; he thought it best for all three labs to agree not to participate, and also to sign the letter to the Pope. Donahue called later in the day saying he was in favor of the three labs trying to negotiate with the Cardinal and if that was unsuccessful, they should withdraw.

The next day, Hall told Gove that he would not sign the letter to the Pope, feeling that Ballestrero would perceive it as blackmail. Hall said that he didn't know how Arizona and Oxford would respond to

Ballestrero's letter, but he assured Gove that he (Hall) would not go it alone. He also said that everything would be fine if all three labs got the same date, although there was some risk. Hall told Gove that he thought that Gonella reduced the number of labs from seven to three to get back at Gove. Hall believed that the representatives from the labs must be at least in the next room from where Tite would supervise the sample extraction and should be given the samples immediately to protect against a possible allegation that Tite substituted the samples. In fact, Tite did package the samples instead of handing over the samples right away, and that allegation was later made.

Gove was not surprised that Hall wouldn't sign the letter to the Pope. Hall believed that any protest and threat to withdraw would be seen by both Rome and Turin as blackmail and could lead to the cancellation of the dating. Gove knew that Hall wanted Oxford to be involved and as things currently stood, they would be so why take any risks? Gove then talked to Donahue, who said he was composing a letter that hopefully would be signed by all three labs saying they were upset with the protocol changes, and would also likely request that five or six labs be involved. In any event, the letter would say that the three labs did not want to go along with just that number. Donahue told Gove that if Arizona was chosen and Gove's lab was not, that Gove could be involved with Arizona as a consultant or some other capacity.

On November 4th, Canuto informed Gove that he spent an hour on the phone with Fr. Rinaldi, who was very unhappy about the cable that had been sent to Ballestrero. Rinaldi, who was unaware about the article that had been in *La Stampa*, said that the decision to use only three labs had actually been made long before the Turin workshop. Gonella's statements before the workshop reflected that, and Gove believed that the number wasn't reduced because of Gonella trying to get back at him. Gove called Donahue later in the day and said he thought that Donahue should tell the Cardinal that Arizona wanted to keep the original protocol. Donahue said he didn't want to sign the letter to the Pope, but suggested that Gove call Damon. Gove made the call and Damon also said he didn't want to sign the letter. Arizona didn't want to be preparing a reply to Ballestrero's letter and also signing a letter to the Pope. Damon said that Donahue's original draft to the Cardinal would say that Arizona preferred the original protocol. If the Cardinal refused to go back to that, then Damon said they didn't know what they should do.

Gove received a message on November 5th from the Arizona lab that contained a draft of the letter that the three labs were planning to send to Ballestrero. It emphasized that the reduction of labs from seven to three would threaten "the credibility of the enterprise." It informed the Cardinal that they were "hesitant to proceed" if only three labs participated. The following day, Gove read the draft to Canuto, who thought the letter was good but should be specific on what they would like the Cardinal to do, e.g., hold another meeting. On November 17th, Donahue told Gove that the translation of their letter by Canuto into Italian would be expressed mailed to Hall the next day, then to Woelfli and the Cardinal. Canuto called Gove shortly after the Donahue call, and said that Fr. Rinaldi said that Gonella had talked with STURP's president, Tom D'Muhala, and said he would actually not push the three labs decision if the three labs stood firm in their opposition.

On November 25th, Rochester's paper *Democrat & Chronicle* carried an editorial that talked about Gove desire to be involved in the dating of the Shroud. It stated "his hopes gave way to one bureaucratic hurdle after another." When plans were put in place to date the Shroud, he was optimistic again that he would be involved, but when the decision was made to use only three, his lab was excluded. The author of that editorial, a friend of Gove's, wrote "After so many years spent thinking about the image of that bearded man on the shroud, he deserves to prevail." Gove sent a copy to Donahue, Woelfli, Hall and Canuto with a little note: "Does this not bring tears to your eyes. I would enclose a small piece of linen to dab them, but you know how it is -- Harry."

The day the editorial came out, Gove called Canuto. Fr. Rinaldi told him that Dinegar of STURP had obtained a copy of the letter that the three labs had sent to Ballestrero, but no one knew how Dinegar got it. STURP had recently held a meeting in Rye, New York, which Gonella attended. STURP was very angry at Gonella. Dinegar asked Gonella what would happen if the three labs withdrew. Gonella replied he would go outside the original seven labs. Gove then called Fr. Rinaldi, who said his position was delicate. He thought that only three labs was a mistake, but Gonella did not. Gove asked Rinaldi what labs he thought Gonella could go to, because the heads of all the other labs were close colleagues, and

Gove felt that he would have trouble getting other labs to be willing to take the place of the three. Gove urged Rinaldi to phone Ballestrero, and he said he was considering that.

Source: Gove, Harry. <u>Relic, Icon or Hoax: Carbon Dating the Turin Shroud,</u> (Bristol and Philadelphia: Institute of Physics Publishing), 1996, pp. 220-225.

Comments: Hall had predicted that everything would be fine if all three labs got the same date but because of the fact that the samples had only been taken from one disputed area and because of all the politics being documented in this book, a limited number of people ultimately accepted the labs' results. Given the secrecy agreement that the labs signed, it seemed inappropriate that Donahue would offer Gove a consultant position. Gove, in fact, was invited by Arizona to observe their testing. Considering Gove's own actions, it would have been particularly galling to many people if he had been allowed "to prevail." The many back and forth decisions and behind-the-scenes actions after various planned protocols did not bode well for a rigorous dating test.

The paper in Gove's hometown asserted that he deserved "to prevail" after "so many years spent thinking about the image of that bearded man on the shroud." First of all, Gove was, at best, a once-a-year Episcopalian, so I doubt the image itself was a big concern to him. Secondly, why didn't STURP deserve to prevail, since most, if not all of their members had spent their own money to get to Turin in 1978 and spent countless thousands of hours collecting and analyzing the data, and writing over twenty articles in peer-reviewed scientific literature?

1987 *(November)*. Harbottle on November 3rd called Dinegar of STURP, telling him that Gove's attempt to exclude STURP and Turin by entrusting everything to Chagas had been a tragic mistake, asking for the support of STURP to maintain seven labs. If Gove's Rochester lab was one of the seven to be chosen, they would have to let McMaster University in Ontario actually test the sample, but Gove would be an advisor. Dinegar replied that the decisions belonged only to Ballestrero. In an interview in *Science News* on November 7th, Gove said that the reduction of labs to three would compromise the statistical data. Canuto let it be known that he would go to see Casaroli to try to put things right. Tite wrote to Gonella on the 18th asking him to communicate his acceptance to be overseer of the project. On the 21st, Casaroli asked Ballestrero by letter to confirm with the three labs that they accepted the invitation to date the Shroud. Hall phoned Gonella to probe what the reactions of Turin would have been to a possible refusal of the labs to proceed if only three were designated; Gonella told him that they would refuse. Hall called back on the 26th to say that they had combined on a letter recommending returning to seven labs only for diplomatic reasons, so as not to be accused of prior agreements with the Church, ending with the classic phrase "this call was never made." This letter, in which the heads of the three labs chosen politely asked Ballestrero to reconsider a seven-labs operation, arrived in Turin on November 30th (before the aforementioned letter from Casaroli of November 21st).

Source: Gonella, Luigi and Riggi di Numana, Giovanni. Sindone: il mistero continua, (Milan: Fondazione 3M), 2005, pp. 65-66.

Comments: It's rather ironic to discover that Gove's own lab would not have been able to test the Shroud had it been chosen. Recall Gonella's previous statement: "At the request of Cardinal Ballestrero, and in agreement with Chagas, I also went and found Gove (who I already knew as a colleague) to try to understand the reasons for his position and make him understand our problems. I asked him to visit his lab in Rochester (I also wanted to see his equipment), but instead I set an appointment in New York at the L'Osservatore Vaticano headquarters at the U.N., where Canuto would also have attended (who then did not come because of influenza). It was a cordial conversation and we left as good friends, but with poor results because Gove was evasive on all technical problems and seemed obsessed with the idea of excluding STURP from the scene as 'not credible'." Harbottle was probably too closely aligned to Gove to have had Brookhaven being chosen to be one of the participants.

1987 (November). Archaeologist Paul Maloney wrote to Rinaldi on the 16th: "Dear Fr. Peter, During our conversation last week I mentioned to you some of the possible objections the scientific community might raise if the planning for the carbon dating moves in the direction of only one or two samples as now appears to be the case. I had reflected a conversation I had had several years ago with Dr. Stewart Fleming, Director of MASCA (Museum Applied Science Center for Archaeology) at the University Museum of the University of Pennsylvania. You will recall that you requested me to set these comments in a letter."

"I drafted a preliminary copy of this letter after discussing it at length with Dr. Fleming last week and sent it to him. I have just discussed this draft with him by phone this morning and now want to get this to you immediately so that you'll have it before Prof. Gonella arrives from Turin this week."

"Given the current state of development of AMS (Accelerator Mass Spectrometry) Dr. Fleming would agree that the original protocol for carbon dating the Shroud is the best one -- i.e., that two small proportional counters and five accelerators be involved in the testing; that all these labs each receive one or more samples from the Shroud; and that blind samples be included for each of these labs. When completed we would have at least 7 dates of material from the Shroud. The average from these results would be a sounder base than those resulting from any other alternative proposed to date, which, as you know, would comprise fewer labs, drop the use of the small proportional counter, or reduce the number of samples removed from the Shroud to one or two."

"There are good reasons why the original protocol is best. First, Dr. Fleming points out that all labs have worked with 'Rogue samples.' These are samples which deviate wildly from the expected date. Dr. Fleming referred to a case where the date produced from the shell of a hazelnut was drastically different from material from the same tree! He suspected that perhaps one in ten samples handled by laboratories were "Rogue samples! In fact this could easily be checked out if one were to go to the published data on carbon dates (in *RADIOCARBON*, for example) and check on the actual material dated by the very labs which will be a part of the protocol -- Brookhaven, Harwell, Lyon, Oxford, Rochester, Tucson, and Zurich -- to discover how prevalent this is. Who is to say that the Shroud will not have such rogue samples? So many factors may be at work here: methods of handling the cloth in ancient times, extraneous sources of contamination throughout its history, and the lack of complete knowledge about whether or not or how these various factors will affect the date which is finally derived. The greater the number of labs, the greater the variation in technique (i.e., both small proportional counter and AMS), and the greater the number of samples, all contribute to a sounder base for drawing conclusions. In this light Harwell's request for eight samples is not at all out of line. In fact, Dr. Fleming says that any program should use 20 or more samples of all types of linen. He believes that if each lab received at least one sample from the Shroud this should be acceptable if all labs -- small proportional counter and AMS -- are involved as planned in the original protocol."

"A second line of reasoning is this: Dr. Fleming is not certain that linen has been adequately tested using AMS -- it may require a special chemistry. By this I interpret that the existing data base for linen in general is not enough to provide a proper evaluation of the results from testing the Shroud. Therefore, it would be important to do -- as was originally suggested -- tests on many different samples from many different types of areas on the Shroud. This would include material from beneath the patches where the Shroud has been burned. And in order to get enough of the non-burned material for the small proportional counter one could recommend removing linen from the same area from which the 'Raes Sample' was taken in 1973."

"Dr. Erle Nelson, Director of the Simon Frazer University AMS laboratory, referred to a study done some years ago where the ratio of carbon 12 and 13 to carbon 14 was different than, say, that found in cotton. He suggested that if, as Bill Meacham claims, fractionation had drastically affected the linen in the burn areas, tests done by the same lab on samples taken from both burn and non-burn areas could identify this change. This would then provide a scientific accounting for any resulting deviation. In other words, the peculiar signature of the 'normal' linen ratio would act as a check on the final results. This alone argues strongly for labs to be able to have more than only one sample from the Shroud to test. Science likes to account for whatever problems it encounters. This would put it in a better position to do so."

"A third factor makes Dr. Fleming reluctant to see only AMS labs the sole testing agent for the Shroud. He pointed out that with the accelerator technique there is a lot of pre-testing chemistry. With each step in the process there is the possibility that extraneous contamination could be introduced. For example, as you will recall, the Zurich lab -- so well known for being one of the best accelerator labs in the world -- produced an outlier of 1,000 years on one of the linen samples in the pilot test supervised by the British Museum. They later admitted that the sample was improperly prepared -- another way of saying the outlier was a <u>lab</u>-induced result."

"Dr. Fleming pointed out that if only a few AMS labs are used and, say, only <u>one</u> of those labs dates a sample from the Shroud, how can the results be evaluated if a radical outlier is produced from the testing? Dr. Fleming's analogy was, 'It's like playing Russian Roulette!' The fact that one of the best AMS labs in the world has already produced such an outlier ought to make us very cautious about the protocol currently being promoted. He noted that labs can always rationalize ways to account for their outliers but if it is <u>after the fact</u> it has already muddied the picture. Those who are scientifically oriented, those who are avowed skeptics, or even those who are committed to the view that the Shroud is probably 2000 years old, would be able to point out to such unclear results in support of their own case. In essence, nothing would have been accomplished by the carbon dating and, technically speaking, it would have to be done over again -- <u>correctly</u> -- in order to obtain results which could be properly evaluated. In the long run this would require <u>more</u> samples than would be the case if done properly now. But it might have a devastating psychological impact on any future carbon dating plans. It would therefore be better not to do any carbon dating at all than to do the job halfway."

"Dr. Fleming notes a fourth factor. When dealing with cultural items one likes to have as tight a data as possible. If a lab could take their testing to one standard deviation -- say +/- 35 years, this would provide a relatively tight 'window'; but characteristically, labs may regard 2 standard deviations-- +/- 70 years --to be a proper measure of precision. Unfortunately no AMS facility is currently achieving that precision; +/- 150 years is more common. And to this one must add a certain amount for calibration. If <u>only</u> AMS is involved, if only a few of such labs are testing, or worse, if only <u>one</u> of those labs gets an actual sample from the Shroud -- if those final results turn out to be plus or minus 300 years, or 450 years or more, the window is so wide as to be capable of many different interpretations. True, one may be able to produce a date in, say, the 3rd or 4th century AD and successfully remove the date from the

14th century or the 1st century -- but with so wide a window one has only transplanted that debate from one time frame to a different one. IF the goal is to test the case for authenticity it is most desirable to produce as small a window as possible."

"As you are already aware there are scholars who have promulgated views which would place the Shroud within a few centuries of the time of Christ. Dr. Robert Drews of Vanderbilt University would find a carbon date based on such a wide window as supportive of the case he has built. James R. Foye, of San Diego, has taken Dr. Drew's position a step further by suggesting that the Carpocratians made the Shroud. He too would find a 'wide window' as supportive."

"Dr. Fleming states, 'The whole purpose for testing is to cut out ambiguity.' He believes that anything less than what has already been proposed in the original protocol could well spell disaster for the entire effort."

"One issue which has been raised by Turin is, 'What precedent will we set if we allow a larger number of samples to be taken?' My answer is: 'No precedent at all!' The reason for removing these samples is to obtain the best scientific results achievable under the current state of scientific ability. If 7 labs are required for checks and balances, and multiple samples are needed to create an average date from each lab and to minimize the possibility of 'rogue samples', then this is a one time procedure that should have nothing whatsoever to do with samples needed for testing unrelated to carbon dating either now or in the future."

"Finally, it is important to point out here that when the testing is finished and the results are published, it will be the radiocarbon scientific community -- those uninvolved directly in the testing -- that will take the hardest look at the manner in which the testing was conducted. If it does not approve, you can be quite sure that it will become one of the most publicized critiques of a carbon test ever to become available to the general public. If the current plans proceed as now projected -- <u>any</u> resulting date will raise a storm of protest."

"Dr. Fleming would be happy to discuss this matter with you..."

"Please feel free to use this letter or its contents in whatever manner you see fit. The issues at stake are so terribly important that we must apprise the authorities of the ramifications of the current planning [...]."

With warmest regards,

[signed]
Paul C. Maloney
General Projects Director, ASSIST

cc: Dr. Stewart Fleming, Dir. MASCA, Philadelphia, PA; Rev. Albert R. Dreisbach, Dir., Atlanta International Center, Atlanta, GA; Fr. Adam J. Otterbein, C.SS.R., Pres., Holy Shroud Guild, Esopus, NY

Source: Copy of letter from Paul Maloney to Fr. Peter Rinaldi, sent to author by Maloney.

1987 *(November)*. Paul Maloney revealed in a conference presentation, "I met with Luigi Gonella on Saturday evening, Nov. 21, 1987 and discussed with him the possible sites on the Shroud from which a sample might be taken. Gonella placed great emphasis on the conservation of the Shroud.
Although he would not openly admit that the 'Raes' Corner' would be the site from which a sample would be taken, all of the implications in his conversation with me were that, the Raes' Corner would indeed be the 'best' candidate site simply because it was thought the most appropriate place since a sample had been excised previously from this corner for Raes and since the royalty of Italy had stitched repairs on the patches, and they wanted to preserve that history, therefore no samples should be removed from anywhere else on the Shroud. By the end of that meeting my growing conviction was that Gonella was leaning very strongly toward taking a single sample only from a single site, the 'Raes' Corner'. He was perhaps following the wishes of Cardinal Ballestrero who had appointed him science advisor for the project."
"Following that dialogue I conducted a series of technical phone interviews around the world with leading specialists in the field of radiocarbon dating and from that series I developed a 'white paper' which summarized the collection of data. On March 22, 1988, nearly a month before the April 21 sample removal session, I sent a copy of this paper to Pope John Paul II (actually to his then secretary, Cardinal

Cassaroli) via his Papal Nuncio, Cardinal Pio Laghi in Washington, D.C. in a diplomatic pouch. I sent a second copy to His Eminence Anastasio Cardinal Ballestrero, and one to Gonella himself urging the need for convening a new Turin Workshop wherein specialists could analyze the fresh data. One of the most important points in this paper was made by Marian Scott of the International Radiocarbon Calibration Program, headquartered in Glasgow, Scotland: she asserted that a minimum of three samples must be taken from three different areas on the Shroud so that the results could be compared with all other results. Without this we would not know if the date obtained from the 'Raes' Corner' represented the date for the main body of the Shroud. I also suggested in that paper a method that could be used to circumvent Gonella's argument that royalty had helped stitch the patches: single yarns could be teased out from under the many burn patches without interfering with any of the stitches known to have been placed there by members of the royal family. But the final decision, on April 21 of that year, was to take a single large sample only from the 'Raes' Corner'."

Source: Maloney, Paul. "What Went Wrong With the Shroud's Radiocarbon Date? Setting it all in Context" (pp. 4-5), presented August 16, 2008 at "The Shroud of Turin: Perspectives on a Multifaceted Enigma" Shroud conference in Columbus, Ohio. Maloney's paper is accessible at https://shroudofturin.files.wordpress.com/2014/12/maloneywhatwentwrongwiththeshroudversionfive2014.pdf.

Comments: Giorgio Bracaglia of the Holy Shroud Guild told me he remembers seeing a copy of Maloney's "white paper" in their archives, but has not been able to actually locate it. If it can be found, he has given me permission to reproduce it. If and when it is found, I will do that in my web site's dedicated page for this book, mentioned in the Preface.

1987 *(November)*. In his meeting with Maloney, also attended by some STURP members, Gonella confirmed that three AMS labs would be used, which spelled the end for the proportional-counter method. Gonella also firmly countered Gove's claim that a firm protocol agreement had been reached at the Turin workshop meeting. Maloney could tell that Gonella was concerned about preservation; Gonella knew that to make the Turin ecclesiastical figures content, the best place to take the C-14 sample was right next to the Raes sample area.

Dinegar wanted to raise with Gonella some points that had come up in a phone conversation between Dinegar and Harbottle. Harbottle had informed Dinegar that not a single lab would accept the reduction from seven to three and neither would the British Museum. Harbottle also admitted that Gove was determined to try to eliminate STURP from participation by campaigning for the Pontifical Academy. Apparently, the seven labs were hoping that, given STURP's close association with Gonella, it could help persuade Gonella regarding the number of labs.

Bracaglia believes, based on the aforementioned and other Guild archives, all the parties had their particular agendas. By this time, Tite had apologized for his actions and on behalf of the European labs assured Gonella of their willingness to participate. Lukasik had adapted STURP's tests proposal, eliminating X-ray fluorescence and radiography. During the Rye meeting, Harbottle had received a letter from D'Muhala, who reprimanded him for stating in an interview in *Science News* that he was a STURP member. Harbottle countered by saying that STURP had assigned him the title of "Co-investigator." (See exchange in following two entries.)

Support for the reduction from seven labs to three came from a surprising source: D'Muhala and Lukasik, who supported Gonella's position. Gonella would have known about the STURP discussion about their participation since the 1984 proposal had been submitted. Gonella had been keeping D'Muhala and Lukasik informed of the C-14 discourse, which included Ballestrero's conversations with the Vatican. According to Bracaglia, that explained why Lukasik eliminated the X-ray fluorescence and radiography from the revised draft proposal. Gonella told the Rye participants that Ballestrero "was forced to distance himself from STURP for **political** *[my bolding]* reasons." Ballestrero was actually sympathetic and supportive of STURP, but he had to concede to the wishes of the Vatican.

Source: Bracaglia, Giorgio. Uncovering the Paradox within the Archives of the Holy Shroud Guild. (Honeoye, N.Y.: Holy Shroud Guild), 2019, pp. 229-230.

Comments: Once again, Harbottle's role in the big picture is intriguing. He had close connections with both Gove and STURP, both of whom opposed each other vigorously.

The Vatican, at that point, was heavily being influenced by Chagas and Gove.

1987 (November). D'Muhala wrote to Harbottle: "Dear Dr. Harbottle, A copy of the November 7, 1987 issue of Science News containing an article entitled 'Shroud dating isn't ironed out' has been brought to our attention. In this article, a copy of which is attached herewith *[not reproduced here]*, you are quoted as saying you are a 'Co-coordinator' of the Shroud of Turin Research Project, Inc. I am directed by the Board of Directors of the Shroud of Turin Research Project, Inc. to advise you that you are not presently and have at no time in the past been a member of STURP and we request that you not so represent yourself in the future."

Sincerely,

[signed]
Thomas F. D'Muhala
President

cc: Anastasio Cardinal Ballestrero
Prof. Dr. Luigi Gonella
The Editor, Science News
Chitty and Minor Law Firm

Enclosure: Science News, Vol. **132** #19 November 7, 1987

Source: Letter of November 27, 1987 from Tom D'Muhala to Garman Harbottle. Luigi Gonella archives, of which the author has a copy.

1987 (December). Harbottle replied to D'Muhala a few days later, "Dear Tom, I am in receipt of your rather insulting letter of

November 22. A single telephone call could have cleared the matter up. What I actually said to the reporter from Science News, Steffi Weisberg, was that I was 'co-coordinator of the Shroud of Turin Carbon 14 Dating Research Project' or words to that effect. I was, after all, so designated by you people (see your enclosed Formal Proposal of 1984) *[accessible at http://www.mondosindone.com/Site/documenti/DSS001_02%20-%20Sturp%201984.pdf]*. I would point out that your own designation of me, 'co-investigator' implies an even stronger connection than I was willing to have, but I let it stand because of my respect and long-standing friendship with Bob Dinegar, and for no other reason. 'Co-investigator' certainly implies some kind of relationship to STURP, and that is what I tried to give the reporter for Science News."

"You are right about one thing, and that is, that I am not presently and have never been a member of STURP. Nor could I be, since I disagree so totally with STURP's approach, goals, attitudes, and methodology. I think that there is a distinct possibility that the 1978 physical examination of the Shroud by the STURP team damaged it, and I am 99% certain that the proposed new examinations would do so if carried out. That is why I have pressed to have the Church authorities in Turin and at the Vatican forbid any further physical testing until a full conservation examination of the Shroud, viewing it as a museum object, be undertaken. It would be my hope that such an examination could be made by a team of textile conservators of established reputation -- experts who have no connection whatever with the STURP organization. After they have made their report, another panel of experts, again independent of your organization should meet to decide whether the Shroud can tolerate any further assaults by x-rays, laser light, high-intensity illumination, scotch tape, etc. and whether the information on the Shroud image obtained by these museologically-questionable means is worth the additional stresses placed on the Shroud. In short, the Church should demand that your proposals be subject to peer-review. This second panel should certainly include members of the Clergy, distinguished museum conservators and perhaps a few philosophers and scientists. When this second panel -- the preservation panel -- meets, they should have the results of the carbon-14 dating in their hands. For as I pointed out in my 1984 paper, if the linen cloth of the Shroud dates only to the 1300's, then STURP's work becomes largely irrelevant. Hence carbon 14 should take precedence over all other physical tests: the outcome of the carbon 14 measurement may save a lot of wear and tear on what may be, after all, an important example of medieval religious iconography."

"I often think that, in the whole sorry struggle over carbon 14 dating and the STURP examination and testing, the feelings of ordinary Church-goers and the physical durability of the Shroud and its image have been somehow overlooked. I hope that the Church authorities will see to it that this is no longer the case, and that the Archbishop of Turin H.E. Cardinal Ballestrero, will act vigorously in his office of 'Custodian' of the Shroud, to preserve it from further 'examinations' like those proposed by STURP."

"I am not at all sure that STURP has, or ought to have, a future, but if it does, Tom, I can assure you that I will not 'represent' myself as having any connection with it. I hereby dissolve any role I may have had as liaison, 'co-investigator' or 'co-coordinator' between the carbon 14 laboratory group and STURP."

Sincerely,

[signed]
Garman Harbottle

enclosure: page 44 from the 'Formal proposal' of STURP, 1984 *[not reproduced here]*

CC: H.E. Anastasio Cardinal Ballestrero
      H.E. Augustino Cardinal Casaroli
      The Editor, <u>Science News</u>
      Reverend Robert Dinegar
      Professor Carlos Chagas
      Father Peter Rinaldi
      Professor Luigi Gonella
      Professor Vittorio Canuto

Source: Letter of December 2, 1987 from Garman Harbottle to Tom D'Muhala. Luigi Gonella archives, of which the author has a copy.

Comments: Harbottle included "scotch tape" in the dangerous elements he enumerated that he believed STURP used in 1978. In fact, it was not a member of STURP, but Swiss criminologist Dr. Max Frei, who was going to put scotch tape on the facial image, when STURP member John Jackson started to physically restrain Frei from doing so; Gonella had to intervene.

1987 *(December)*. According to Gove, on December 1st, Harbottle called him and said that he had called Dinegar at the STURP meeting in Rye held on November 21st. Dinegar had told Harbottle there was no need for the latter to drive there because Gonella had announced there that even if the three chosen labs proceeded with the testing, STURP would not be allowed to do additional testing afterwards. STURP was prepared for the testing, but they would not be allowed to proceed, which was apparently why D'Muhala was angry towards Gonella. Gove was told by Harbottle that he was having his own dispute with STURP. He had received a letter from D'Muhala with a copy of an interview Harbottle had given to *Science News*. Harbottle was described as a STURP spokesman. D'Muhala told Harbottle he had no right to be speaking for STURP and should cease and desist. Harbottle was furious; he called Dinegar and said that D'Muhala should be kicked out.

Fr. Rinaldi called Gove on December 3rd, and said that Gonella had called him the day before saying the Cardinal had received the letter from the three labs, but didn't reveal his or the Cardinal's reactions to it. Rinaldi thought that Ballestrero would contact the Vatican; the former felt that it was concerned about its image and might reconsider. Rinaldi emphasized to Gonella that there would be bad publicity if the Turin authorities kept making bad decisions. On December 7th, Canuto told Gove that Gonella had advised Ballestrero to stick to the three-labs decision. Someone called from England to tell Gove that Hall had made an offer to a person from the BBC to give them an exclusive on the dating in Oxford "for a fancy price."

On the 28th Gove talked again with Rinaldi. Rinaldi said that on December 18th, the Cardinal had written to the three labs trying to convince them to agree that there was no chance to return to the original protocol. Gove then called Donahue, who said he had just received the letter from Ballestrero. The Cardinal (Gonella in Gove's eyes) said that the labs had claimed that seven labs would provide a higher accuracy to the measurements but claimed that wasn't valid. Donahue said the Cardinal didn't understand the difference between statistical and non-statistical errors. Donahue said he had talked with Hall, who had received the letter from Ballestrero the week before. Hall was going to agree to the three-labs decision, and that Woelfli would do the same. Donahue had not yet talked to Woelfli, but surmised that the three labs would proceed with the dating. Gove related, "So despite all the high-minded statements he, Damon, Woelfli and even Hall had made to me in writing that they would stick by the protocol, it all went down the drain as soon as their bluff was called."

Canuto called Gove and informed him that he had tried to persuade Donahue to stand firm. Canuto asked Donahue whether Ballestrero's letter had really answered the joint letter sent by the three labs and Donahue replied it had not. Canuto asked, "why cave in now?" Canuto asserted that would make the whole workshop a farce, and that Chagas had felt he and the Academy had been kicked in the stomach. But Donahue seemed unmoved. Gove told Canuto that they should have a press conference. Gove called Harbottle and told him about the Cardinal's reply and his idea about the press conference, which should be held before the three labs officially agreed to proceed with the dating. Harbottle agreed; Gove said he would call Woelfli the next day.

Gove reached Woelfli on December 30th. Woelfli asserted that Turin wanted to play the leading role. According to him, there was friction between Turin and Rome as to who controlled the Shroud. He said he would probably say "yes" to his involvement, but with restrictions. Woelfli informed Gove that Hall had talked with Gonella and had suggested that Tite and the representatives from the three labs should meet with Gonella in London for a final pre-dating meeting. Woelfli felt that if the three labs didn't participate, other labs would. Gove asked Woelfli which labs specifically, and he said Otlet, Harbottle or Duplessy. Gove wrote that he was surprised that Woelfli didn't include Rochester. That same day, Donahue told Gove that he had talked with Woelfli and was surprised that the latter was going to go along with the new protocol. Donahue believe that Woelfli felt that if he did not go along with it, Gonella would find another lab besides Zurich. Donahue said he would go to the meeting in London, and also that he would still like to have Gove involved in the Arizona dating. However, he would not mention that at the meeting because he was afraid it would "infuriate Gonella."

Source: Gove, Harry. Relic, Icon or Hoax?: Carbon Dating the Turin Shroud (Bristol and Philadelphia: Institute of Physics Publishing), 1996, pp. 225-228.

Comments: According to an article by Ian Anderson in *New Scientist* (Jan 21, 1988) the three labs' letter to Ballestrero included the following statement, "As you are aware, there are many critics in the world who will scrutinize these measurements in great detail. The abandonment of the original protocol and the decision to proceed with only three laboratories will certainly enhance the skepticism of these critics." Hall's offer to the BBC is a clear example of a lab trying to take financial advantage of its participation in the Shroud dating. Gove criticized Damon, Hall and Woelfli for backing down but makes no mention of his own previous multiple unfulfilled threats to withdraw. The fact that Woelfli's recommendations for substitute labs didn't include Gove's speaks volumes. The various exchanges detailed in this entry show that the labs were only concerned with their own advantages.

Harbottle complained to Dinegar that D'Muhala should have been kicked out of STURP for having sent the letter to Harbottle, but D'Muhala's letter stated clearly he was sending it at the direction of the STURP board of directors.

1987 *(December)*. According to Petrosillo and Marinelli, "On December 12th, the Cardinal wrote to the directors of the three chosen labs that there were no reasons to change the guidelines. The recipients responded by stating that they preferred the original Turin protocol. Three of them were hesitant to proceed. They argue that if one of the three achieves a divergent result, the whole project could be in danger. 'As you know,' they wrote, 'there are many critics around the world who will examine these measurements thoroughly and in great detail. The abandonment of the original protocol and the decision to proceed only with three laboratories will certainly reinforce the skepticism of these critics!' Paul Damon of Tucson, one of the chosen laboratories, expressed his doubts: 'When doing research, there is always something wrong. What did we do wrong? Maybe we mixed the samples? Was there an error in the laboratory? You cannot proceed in such a way with an artifact like the Shroud. You got to do it right the first time. I'd prefer seven labs to three for several reasons.' Nonetheless, Gonella thought it better to proceed."

"Someone insinuated that the Church hoped to obtain an ambiguous result by reducing the number of labs. The three chosen are among the most capable, but do not have the same experience as Harwell and Brookhaven in the pretreatment of the samples. As for radiocarbon dating, then, Harwell had more experience than all the other six put together. 'Harwell probably makes the most accurate dating of all of us,' admitted Gove, while Hall said he believed Harwell would have been a good addition. Gonella replied, however, that the three chosen had more specific experience in archaeological dating. The pioneers of both methods, Brookhaven and Rochester, and the most experienced laboratory in the field were therefore eliminated: Harwell. Gonella was totally oriented towards the new and fast AMS method, without considering that scientists and the public would have felt more relaxed with the inclusion of a laboratory that used the proportional-counter method, older, slower, but more hopefully tried and tested. With this method, counts on the same sample can be repeated, whereas with AMS it can be done only once because the sample is destroyed."

Source: Petrosillo, Orazio and Emanuela Marinelli. La Sindone: Storia Di Un Enigma, (Milan: Rizzoli, 3rd edition, 1998), pp. 148-149.

Comments: To continue this thread, go to the Petrosillo/Marinelli entry for "1988 *(January)*."

1987 *(December)*. Paul Maloney phoned Dr. Marian Scott, statistician at the International Radiocarbon Calibration Programme headquartered at Glasgow University in Scotland and asked, "What is the minimum sample one should take from the Shroud in order to have a reliable date?" She responded: "At minimum, for a project as important as this, one should take no less than THREE samples (preferably more) from disparate and separate places on the Shroud. If you take only one and you get a result that is not commensurate with other pieces of evidence, you have no way to compare or question if the result is historically viable. If you take two samples and they differ in date from each other, you have no way to determine which of the two date results is the correct one. Only if you take three disparate samples from different areas will you have a working basis for questioning outliers." Therefore, I do not believe that we can yet -- scientifically -- make the second assumption. I am personally resoundingly in favor of a new radiocarbon dating project."

Source: Email from Paul Maloney to various Shroud researchers, including author, on May 20, 2016.

Comments: Maloney had sent this information to Gonella. Sadly, this would be additional expert advice that the Turin authorities would ignore.

1987 (December). Ballestrero replied on the 11th to the three labs' letter that arrived in Turin on November 30th, observing that if the three labs thought it better to have seven labs involved, the three labs must not be confident of their own results, and advised Casaroli of the current situation. Hall immediately telephoned Gonella to accept and advised others to do the same, proposing a meeting in London on January 22, 1988 to define the outstanding technical points. Gonella indicated he would go to London as long as the three labs agreed, which they did.

Source: Gonella, Luigi and Riggi di Numana, Giovanni. Sindone: il mistero continua, (Milan: Fondazione 3M), 2005, pg. 66.

Comments: While Ballestrero made an excellent point about the three labs' confidence in their results, one has to wonder why he didn't take the logical step of allowing seven labs to do the dating. Hall's speed in accepting the three-lab proposal confirms there was no way Gove's suggestion that the three labs withdraw if seven labs weren't involved would be implemented.

1987 *(December)*. Meacham wrote to Lukasik with some ideas in the event that STURP would be involved in some manner: "Now two suggestions:"

1. That a micro-sample (say 3 to 5%) be removed from each lifted sample before they are dispatched to the C-14 labs, and these tiny pieces be either sent to STURP, or held in Turin, for detailed testing by STURP for contaminants – something which the C-14 labs will probably not do. They will more likely conduct the standard pretreatment, and even this might need to be severely limited because of the critical sample size involved. Such testing on the micro-samples would also provide corroboration or otherwise, of pretreatment performance re contamination.

2. That STURP run a specific set of examinations at the points where the C-14 samples were lifted. I am thinking particularly of micro-Raman spectroscopy and/or an analytical spectrometer (assuming that both exist in portable units and can operate directly on the Shroud fabric) for any alien organic or inorganic compounds; uv florescence *[sic]* for scorch damage not visible to the naked eye; and other types of testing which might reveal any other anomalies in the area sampled, e.g., higher than average proportion of carboxyls. Neither of these suggestions has anything to do with conservation, but would contribute much to put the C-14 measurements in a better perspective, and provide data for the assessment of the C-14 results on the samples lifted *vis-à-vis* the entire cloth.

Source: Meacham, William. Letter of December 15, 1987 to Steven Lukasik. http://freepages.rootsweb.com/~wmeacham/religions/wmslcorr.pdf.

1987. Gove remarked, "I am, unfortunately, I think, in many ways better known because of my involvment *[sic]* in the Shroud having never measured it […]. [T]he best thing that can happen to us is to always be on the verge of measuring it but never measure it."

Source: Laverdiere, H. "The Socio-Politic of a Relic: Carbon Dating of the Turin Shroud," (Doctoral Thesis) 1989, pg. 78. Accessible via free download at http://ethos.bl.uk/OrderDetails.do?did=1&uin=uk.bl.ethos.531916.

Comments: It's hard to interpret this quote any other way than Gove would have liked to milk the association with the Shroud dating as much as he could, independent of any result.

1988 *(January).* At the beginning of the month, well-known Shroud scholar Ian Wilson from England called Gonella and asked if there was any room for compromise. Wilson felt that Harwell should be one of the labs chosen. Gonella basically said "no." Wilson called Gonella back on the 12th and said that Gove had told him he would withdraw from a planned press conference if he had an assurance that there were reasonable prospects of compromise at the London meeting. Once again Gonella declined.

Source: Gonella, Luigi and Riggi di Numana, Giovanni. Sindone: il mistero continua, (Milan: Fondazione 3M), 2005, pg. 67.

1988 *(January).* "Gonella assured the sindonologists that other tests will be allowed shortly after taking the samples for dating with the ^{14}C and and before the results of the latter are disclosed, which would have more authority than the isolated announcement of the ^{14}C test. Two months later, however, Gonella himself announced that Ballestrero decreed that the ^{14}C would be done before any other tests."

Source: Petrosillo, Orazio and Emanuela Marinelli. La Sindone: Storia Di Un Enigma, (Milan: Rizzoli, 3rd edition, 1998), pg. 149.

1988 *(January)*. Gove wrote: "Rinaldi had told me that Ian Wilson, author of the most authoritative and, in some ways, the most fanciful, book on the Turin Shroud, was opposed to the use of only three labs and he might have some influence with Gonella, so I phoned Wilson on Monday 11 January 1988. He said he had spoken to Gonella and had raised the question of there being no representatives of the small-counter labs involved in the dating. He tried to persuade Luigi to include the small-counter labs -- specifically Otlet's. Gonella told him that no more than three samples could be taken from the shroud. Wilson said that Hall was certainly going to agree to do it. The publicity he would receive from dating the shroud would be too tempting for Hall to resist [...]."

Gove stated: "Damon returned my call *[evening of the 13th]*. He said he would be in London on Wednesday 20 January and that Donahue would arrive the next day. He said that they would make demands of Gonella that would make the whole affair workable. I, of course, had continued to hope that perhaps at least Arizona would decide not to go along with Gonella's dictate but it was pretty clear from talking to Damon that they were going to proceed. I knew Damon to be a person of considerable rectitude and decency. Apparently the lure of dating the Turin Shroud was so great it overcame his previously expressed reservations. He had said he was opposed to limiting the number of laboratories to be involved to three, but he now seemed neither remorseful nor contrite about changing his mind. I was surprised and saddened."

Source: Gove's book: Relic, Icon or Hoax: Carbon Dating the Turin Shroud, (Bristol and Philadelphia: Institute of Physics Publishing, 1996), pp. 229, 231-232.

Comments: In Gove's eyes, this was obviously an ethical compromise by Damon.

1988 *(January)*. STURP member Larry Schwalbe sent on the 13th a memo to various STURP members and several others:

To:
Alan Adler John Jackson
Rudy Dichtl Eric Jumper
Bob Dinegar Tom D'Muhala
Bill Ercoline Giovanni Riggi
Jane Hutchins Richard Speck

"Last November, STURP met with Luigi Gonella in Rye, NY to discuss the status of the project's research proposal. Those who were present remember that most of the news as quite disappointing: Luigi told us of his decision to have the radiocarbon work well underway before we could even begin our planning. At that time, his idea was to bring the Shroud out sometime in late February to remove the C-14 samples. However, when I asked if he would consider giving STURP short-term access at that time (to assist with the sample removal and to begin some preliminary studies), he agreed."

"Since November, the planned time for removing the C-14 samples has slipped from late February to sometime in the range between mid-March and mid-April. As far as I know, Luigi is still willing to give us access, and so I am enclosing some material for you to consider. The first document, entitled 'Proposal and Protocol for Testing on the Shroud of Turin,' is a draft proposal for work I believe we can usefully do in a 12-h period preceding the removal of the radiocarbon samples. I am sending you this because you are either a board member or one of the listed principal investigators. Please read it and return your comments to me before January 25, either by mail or phone [...]. The second proposal, 'Conservation Testing on the Shroud of Turin-B' is included for your information. *[This proposal not reproduced in this book.]* You may comment on it or not, as you wish. In any event, I would like the board's approval either for this plan or for some higher-evolved one by February 1, so that we can have a 'go-ahead' by the middle of February. If I hear nothing from you by the 25th, I'll take the lack of response as tacit approval."

[signed]
Larry Schwalbe

"PROPOSAL AND PROTOCOL FOR TESTING ON THE SHROUD OF TURIN"

PRINCIPAL INVESTIGATORS:

J.P. Jackson, Ph.D.
Kaman Sciences Corporation

L.A. Schwalbe, Ph.D.
Los Alamos National Laboratory

A.D. Adler, Ph.D.
Western Connecticut State University

R.H. Dinegar, Ph.D.
Los Alamos National Laboratory

G. Riggi
Turin, Italy

R. Speck
Spectron Engineering

J.K. Hutchins
Merrimack Valley Textile Museum

"I. OBJECTIVE"

"We propose to conduct a brief (12-h) study on the Shroud of Turin at the time the cloth is brought out for removing C-14 samples. Our workplan has the following threefold objective: 1) to prepare the Shroud for removing the radiocarbon samples and to help decide where these samples will be taken, 2) to inspect the cloth, particularly in the area tested in 1978, to assess any damage that may have occurred from the earlier procedures, 3) to perform preliminary tests and to begin preparations for more intensive future experiments. The latter activities primarily include unstitching operations and the location and marking of permanent test points on cloth. Concurrent analytical tests and observations will involve no more than hand-held equipment."

"II. STATEMENT OF THE PROBLEM"

"Radiocarbon dating has been redefined as a separate research activity from the main proposal of the Shroud of Turin Research Project (STURP). The Turin authorities are now planning to remove the Shroud from its repository sometime in the near term for C-14 sampling and to conduct the remaining conservation and analytical tests at an unspecified future date. The near-term plans provide STURP with an opportunity to assist in preparations for removing the C-14 samples and, at the same time collect information necessary to plan more extensive testing when a longer period of access becomes available."

"This workpackage serves a dual need. On the one hand, STURP has gained considerable experience and specific expertise in Shroud research and can well serve the Turin authorities in their job of selecting, accessing and excising the best C-14 samples. On the other hand, by allowing STURP a temporary and limited access to the Shroud, Turin can provide the project with much of the information it needs to plan a fruitful and efficient follow-on expedition. At the same time, the opportunity for another direct examination will help the project reestablish its perspectives and rejuvinate [sic] its activities. Even brief inspections are sufficient to answer some of the questions that have lingered for years; many of the proposed observations are designed to support specific, ongoing physical and chemical studies."

"III. METHODS AND MATERIALS"

"We have identified three objectives for our limited study."

- A. <u>Radiocarbon Samples.</u> (Riggi, Hutchins, Adler, Jackson, and Speck) The primary reason for removing the Shroud at this time is to select and remove samples for radiocarbon dating. The exact locations for sampling are still undetermined, but current plans are to remove three specific sewn patches. If the amount of usable sample in the exposed areas is sufficient, G. Riggi will remove approximately 50 mg for the test. If the amount or quality of the material is unsatisfactory, the sample will be taken from some other location. We propose that Giovanni Riggi be primarily responsible for unstitching the patches with Jane Hutchins and Al Adler assisting. Both during and after this operation, John Jackson and Richard Speck will collect spectrophotometric manufactured by Spectron Engineering. The device gathers reflectance spectra in the wavelength range 240-1100 nm, and the data should be useful in evaluating the degree of cellulose degradation and contamination before the sample is removed. In addition to these specific tasks, all of the principal investigators will participate in discussions concerning the sample selection. When the sample location is determined, the team will complete the documentation with photographs and spectrophotometric scans for permanent record.

B. Inspection of 1978 Test Points. (Schwalbe, Dinegar, Speck, and Hutchins) A vital consideration of any test applied direct to the Shroud is whether it will cause some irreversable [sic] damage.

C. In planning their experiments thus far, STURP researchers have done as much as possible to ensure safety, but at this point, further refinements can only follow new, direct information about how earlier procedures may have affected the cloth, blood, and image. In 1978, STURP carefully documented all of its test procedures and the locations where they were applied [3]. We therefore propose that a reinspection of these areas be undertaken by both a qualified conservator and a team of analytical researchers. The equipment required for the observations will include a set of hand-held color-comparison charts and the portable optical spectrophotometer described above.

D. Conservation Testing. (Jackson, Schwalbe, and Dinegar) A revised version of the proposal entitled 'Conservation Testing of the Shroud of Turin-B' accompanies this document. *[This particular proposal is not reproduced in this book.]* The major objective of the workpackage is to characterize the physical and chemical state of the Shroud and its components (image, blood, etc.) by applying a battery of precise measurements at a number of well-defined, permanent test points. We propose to locate these test points and begin some of the experiments discussed in the proposal. The latter will include measurements of the surface pH and optical reflectance, for example. The latter measurements on both front and back surfaces of the Shroud are additionally necessary as benchmark data to support ongoing studies of the kinetics of cellulose degradation [4]. The equipment required for this work is all hand-held and portable.

"IV. SCHEDULE AND PROCEDURE"

"The operation we propose should take 12 h to complete. A reasonable schedule may begin in the morning of the testing with the Shroud being taken from its repository at, let us say, 7:00 AM. One hour may be needed to move the cloth to an appropriate testing area and to unroll it on a flat, horizontal inspection table. We will begin at 8:00 with a half-hour period of visual inspection and discussion to finalize plans for the day. At 8:30 AM, Riggi, Hutchens [sic] and Adler will begin unstitching the patches that have been selected for removing the underlying radiocarbon test material. After the patches have been removed (ca. 14:00), Riggi will continue unstitiching [sic] operations along the edge of the cloth to expose the rear surface for later testing, and Adler will begin removing microsamples for his proposed program of chemical testing [1]. These activities will continue until 19:00, the time scheduled for removing the radiocarbon samples."

"The remaining work described in the previous section will proceed concurrently. At 8:30, Jackson and Speck will begin their spectrometric studies, and Dinegar and Schwalbe will start relocating and marking (temporarily) the 1978 test points. The latter work will begin at the most critical locations, those analyzed by x-ray fluorescence and those where the sticky-tape samples were taken. We have allocated 2 h for this investigation. If time permits, we will extend the inspection to remaining test locations; that is, those where infrared and optical reflectance measurements were made. When this phase is completed (ca. 10:30), Jackson, Speck and Schwalbe will begin locating and marking the permanent test points, as they have described in the accompanying proposal [2]. Dinegar and Speck will follow at each point by applying the surface pH and spectral reflectance measurements described in the preceding section. These activities will continue, with intermittent periods of discussion concerning plans for C-14 sample removal, until 19:00."

"REFERENCES"

1. Collection of Materials from the Surface of the Turin Shroud of Chemical and Physical Testing --Adler, A.D. and McNeil, R.J. (Workpackage) 1984 Proposal.

2. Conservation Testing on the Shroud of Turin-B – Schwalbe, L.A. and Jackson, J.P. (Workpackage) 1984 Proposal, Revised January 1988.

3. Schwortz, B.M., Mapping of Research Test-Point Areas on the Shroud of Turin, *IEEE Proceedings of the International Conference on Cybernetics and Society*, 538-547 (1982).

4. Jackson, J.P., Arthurs, E., Schwalbe, L.A., Sega, R.M., Windish, D., Long, W.H., and Stappaerts, E.A., Accelerated Aging of Cellulose by Laser irradiation, accepted for publication in the Proceeding of the 1988 Spring Meeting of the Materials Research Society, Symposium L: *Materials Issues in Art and Archaeology*, April 5-8, Reno, Nevada.

Source: Memo of January 13, 1988 and Proposal by STURP member Larry Schwalbe. Luigi Gonella archives, of which the author has a copy.

Comments: This document shows the elaborate planning skills and precision that allowed STURP to conduct the 1978 examination. It included a plan to go over some of the details they had obtained in 1978 and to even see if any of their tests had caused any damage. Compare that to the helter-skelter planning and execution of the C-14 labs and Church authorities in preparation for the Shroud C-14 dating.

The link for references [1] and [2] is http://www.mondosindone.com/Site/documenti/DSS001_02%20-%20Sturp%201984.pdf.

1988 *(January)*. Dutton's second letter to *Nature* was published in the January 14th issue: "SIR-One of the essential points of my previous remarks about the confidentiality surrounding the dating protocols for the Turin shroud (*Nature* **327**, 10; 1987) is that the matter unavoidably involves religious passions. No better demonstration could be found than P.R. Smith's denunciation (*Nature* **328**, 11; 1987) of my letter as a 'gross insult' to the experts whose job it has been to design the procedures for the carbon tests. Smith suggests I am hostile to religion; but anyone who has been critical of attempts in recent years to give scientific legitimacy to the shroud will be accustomed to such accusations."

"The point is, however, that there has been unacceptable secrecy and confusion about how the tests are to be conducted, beginning with an early report in *La Stampa* (Turin, 5 October 1986), at the conclusion of the closed conference of experts, indicating that the 'timetable and methods of investigation are secret'. In this respect, the open response from Harry Gove (*Nature*, **327**, 652; 1987) is welcome, but it stands in stark contrast to a 'Vatican spokesman' whose reaction to my previous letter in *Nature* confirmed the suspicions it expressed. According to a wire service report, the spokesman told the *London Daily Telegraph* that it was indeed likely that 'only the results' of the tests would be made available to scrutiny by independent observers -- precisely the issue that is so unsettling -- adding that they 'would be made available in a couple of years'. This last assertion also curiously contradicts the initial announcement about the tests, which stated that the results were to be made public at Easter 1988."

"I hope and expect that there will be a detailed and satisfactory disclosure of the test protocols before the samples are taken. An important first step would be for the person who is in a position to speak authoritatively about the conduct of the tests to identify himself or herself. I also trust that M.S. Tite is correct in suggesting (*Nature* **327**, 456; 1987) that neither the British Museum nor the seven testing laboratories will be party to the tests unless each of them is confident that the protocols absolutely preclude tampering with the samples by the introduction into the chain of evidence of ^{14}C-depleted linen, such as mummy linen."

"There is no specific reason to believe that any of the scientific devotees of the shroud or its Vatican owners would seek to rig the tests. Nevertheless, besides the obvious involvement of religious sensibilities, all of us must face the fact that a veritable industry has been built up around the shroud. In this respect, the Vatican can rightly be seen as having a vested interest in keeping alive at least the possibility that the shroud is the actual burial cloth of Jesus. As a negative test result is clearly going to spoil the fun, it is imperative that the protocols for the sample handing be spelled out in advance and that they be seen to be beyond reproach."

DENIS DUTTON, School of Fine Arts, University of Canterbury, ChristChurch, New Zealand

Source: "Protocols for Turin Shroud." [Letter], *Nature* **331**, 108, (1988).

Comments: Dutton raises the point about the Vatican having a vested interest in the results, but the labs also had a vested interest in making sure that the public was convinced that the testing was reliable to show that C-14 was a reliable tool.

1988 *(January)*. Gove tried again an intimidating maneuver, with a press conference held in New York on the 15th with Harbottle along with the support of Canuto. Gove and Harbottle protested against the "arbitrary" decisions of the Archbishop of Turin saying that a dating made by only three laboratories was not valid and it was necessary to return to the project of seven laboratories proposed and studied by the Pontifical Academy of Sciences, and in particular contesting the competence scientific report by Gonella. A few days before the *Chicago Tribune* interviewed Gonella by phone and told him that Gove had announced to the press his intention to attack Gonella's scientific qualifications and asked what he had to say in his defense. Gonella told him that for his qualifications, he was answering to the Faculty of the Polytechnic of Turin and not to a professor of Rochester. Canuto actually wanted to have the press conference under the auspices of the "Permanent Observer of the Holy See to the United Nations," but Casaroli denied permission. So Canuto held it at Columbia University, but didn't actually speak himself.

Source: Gonella, Luigi and Riggi di Numana, Giovanni. Sindone: il mistero continua, (Milan: Fondazione 3M), 2005, pg. 67.

Comments: Although Gove's press conference did not result in his lab being one of the labs chosen, he could find consolation that he was the main force behind the Church's decision to perform a C-14 test only, and not the additional twenty-five tests that STURP proposed also be done. Plus, he was invited (against the agreement signed by all the labs) to Arizona to witness their dating process. Gonella also said, "but aside from this, using a press conference to broadcast their protests is certainly not a laudable procedure." We hope that the serious business of researching the Turin Shroud will not end by becoming a race as to who can get there first." (Paglia, Guido. "'Open Letter' to Pope Accuses Cardinal Ballestrero." *La Stampa*, January 13, 1988.)

It's pretty amazing that Canuto, who was an advisor to the president of the Pontifical Academy of Sciences, the Pope's scientific advisors, had wanted permission from the Pope's Secretary of State to have a press conference in which two American scientists were protesting the decision of the Turin Cardinal, obviously also under the Pope's authority. The biblical verse about a house divided against itself cannot stand comes to mind.

1988 *(January)*. Here is the text that Gove and Harbottle used: "Press Release for Press Conference -- Friday, January 15, 1988 -- Professor H. Gove* -- Dept. of Physics and Astronomy, Univ. of Rochester, Rochester, NY 14627 / Dr. G. Harbottle* -- Chemistry Department, Brookhaven National Laboratory, Upton, NY 11973

*Affiliations for identification purposes. The views expressed are the personal views of the two scientists concerned."

"The Turin Conference"

"At the opening of the Turin Conference, Sept. 29, 1986, whose goal was to prepare a concrete plan for carbon-dating the Shroud, His Eminence Anastasio Cardinal Ballestrero, Archbishop of Turin, under specific instructions from the Vatican, charged the conference to prepared [sic] 'concrete operative proposals' for 'a valid and acceptable project for at last carrying out the radiocarbon dating of the Shroud cloth, a test that, owing to the uniqueness and singular character of the object, certainly could not be easily repeated'."

"With this charge in mind, the conferees worked for three days, literally in a cloister, and prepared a proposal for action, or protocol, for carrying out the dating experiment. The protocol was carefully designed to (a) ensure a result that is scientifically rigorous, and (b) maximize the credibility of the enterprise to the public. Provision was made for blind measurement, for the laboratories to have knowledge of the actual sampling procedure, and for the distribution of non-Shroud (i.e., dummy) samples of identical appearance. All samples would be encoded so that no scientist would know the identity (Shroud or dummy) of the cloth being datd [sic]. The British Museum would encode and distribute samples, and collect and process the results, along with representatives of the Pope (Professor C. Chagas) and the Archbishop (Professor L. Gonella)."

"Most importantly, the protocol specified explicitly that all seven laboratories represented at the Turin Conference should take part. There was a good reason for this."

"The Intercomparison Test"

"To establish the degree to which different laboratories having miniature-sample dating capability could get results that agreed, an intercomparison test was held in 1983-84. This test was arranged by the British Museum: samples of a 4000 year old Egyptian linen mummy wrapping were distributed to six of seven laboratories. The age and identity of the cloth was unknown to the labs; they just dated it and sent in their results. When the results were examined, all agreed and were on target except one, which was 1,000 years off!"

"This result demonstrated that, because of the complexity of the dating procedure, occasional very large 'non-statistical' errors can creep in -- errors that are much larger than ordinary statistical errors. The only way you can recognize one of these big errors is by having enough labs to establish a result in agreement, and hence identify the lab that has had a problem with its procedure. If one lab does the dating, and is in error, you will never know it. If two labs disagree, you can surely recognized the maverick date and throw it out. That was the basis of the protocol."

"The Archbishop's New Plan"
"In April 1987 the Archbishop's science advisor, Professor Gonella announced in an interview with 'La Stampa', of Turin that only 'two or three laboratories would participate in the Shroud dating experiment.' All of the laboratories then sent a joint telegram in Italian to the Archbishop, expressing great concern that if the protocol was breached, the whole exercise could be meaningless. Credibility before the public could certainly be in jeopardy if only two or three did the job."

"The Archbishop's reply, based on Professor Gonella's advice, came in a letter dated October 10, 1987, naming only the three laboratories as University of Arizona, Oxford, and Zurich. It is interesting that the Archbishop stated 'experience in the field of archaeological radiocarbon dating' was a criterion: in fact the Harwell laboratory, left out, has had more experience than the three chosen laboratories put together. Professor Damon of the University of Arizona, however, has much experience with conventional carbon dating. No reason excluding four laboratories was given."

"The Archbishop's plan throws the protocol and its safeguards out the window. Although the Zurich laboratory was the same one with the 1000 year error in the earlier intercomparison, all three labs are highly competent. The Zurich laboratory afterwards identified the source of their error. They are all, however, using the same 'accelerator' method. They do not include the most experienced lab, Harwell, nor the two labs where the two new methods were developed, Rochester and Brookhaven. It is interesting that, in a joint letter to the Archbishop in November 1987 the three laboratories themselves pointed out the problem of non-statistical errors, saying, 'if only three laboratories participate, and one of them obtains a divergent, non-understandable result, the entire project could be jeopardized.' This was all the more significant since of the three, Zurich, had in fact obtained just such a result in the earlier test as mentioned above."

"Public Credibility"
"Dating the Shroud of Turin cloth presents the world with the most significant interaction of the Church and Science since Galileo. It can probably be done only once as the Archbishop intimates. No matter what result is obtained, what date, a large group of people will be disappointed. Those who true believers or skeptics, will attack a disappointing result with any means at hand. The Archbishop's plan, disregarding the protocol, does not seem capable of producing a result that will meet the test of credibility and scientific rigor. The protocol was designed to convince all people, for all time, as to the correct date for the Shroud."

"The Position of the Vatican"
"The Holy See, not the Archbishop of Turin, owns the Shroud. In a broader sense, however, the Shroud is a part of the heritage of all mankind. For this reason it cannot be dated like an ordinary historic artifact or a lump of charcoal from a dig. It was given to the Church in the will of Umberto II, last of the Savoys. The Vatican participated through the Pontifical Academy of Sciences in convening the Turin Conference and drawing up the protocol."

"What Needs To Be Done"
"It is our position that, given the importance of obtaining a valid and convincing date for the cloth, it is probably better to do nothing than to proceed with a scaled-down experiment. Whenever science has wished to determine a particular number with the utmost reliability for example the velocity of light, many laboratories have participated. The same is true here. It is also of vital importance, as stressed in the protocol, that the chain of evidence from the Shroud cloth to the scientific output, be ironclad and unbroken, for all laboratories. Thus, it is clear that the best course is to return to the original protocol. An alternative plan, better than wasting the precious Shroud cloth in an ill-conceived and unconvincing exercise, would be to reconvene in about a year's time the Turin conference, this time with the participation of outside observers

from both the high religious and highly skeptical sides. Distinguished scientists from the Pontifical Academy would again be involved. And in the cold light of open discussion the merits of the different proposals, and new proposals, could be dispassionately examined once again, to finally achieve the goal originally set forth."

Source: Transcript of Gove/Harbottle press conference on January 15, 1988. Luigi Gonella archives, of which the author has a copy.

Comments: Notice that Gove, who had complained vehemently when the 1986 Turin workshop was postponed from June to September, now advised the Pope to wait another year before reconvening the workshop participants!

Gove and Harbottle mentioned Galileo. It took the Church three-hundred and fifty years to admit that he was right and they had been wrong. Let's hope it doesn't take the Church that long to right its wrongs in the C-14 dating of the Shroud.

1988 *(January)*. Petrosillo and Marinelli wrote, "On January 15th, Gove and Harbottle publicly protested the choice of only three laboratories press at a press conference at Columbia University. They wrote a letter to the Pope. The Turin protocol, according to their judgment, would have ensured a scientifically rigorous result and with the maximum credibility of the initiative for the public. To recognize errors like that of Zurich, which was one of the chosen ones, you need to have enough labs to have agreed results and identify the one that possibly had process problems. Better to do nothing than have a diminished experiment. There will always be doubts about results from limited research. Damon, also one of the chosen ones, expressed his disappointment. Hall, however, defended their right to make the decisions. 'All decisions concerning the relic must be made by them; why should they take into account other points of view? Who are the representatives of the labs to decide what should be done. Even if there was an agreed protocol, they would be in a position to change it.' Gove asserted that the Pope and Cardinal of Turin were being badly advised. He claimed that Gonella didn't have the needed competence for being advisor. Gove was direct, 'Gonella is simply the wrong man; he is a second-class scientist, a man no one has ever heard of. I would like to know how he became a scientific advisor'."

"As one might expect, Ballestrero was not happy with Gove's remarks. He declared, 'the Americans are assuming choices yet to be verified have already been made. We are still in a delicate phase of confrontation.' Gonella announced that Gove and Harbottle had no reason to complain. An agreement had not yet been signed. In mid-January the executive work plan for the withdrawal is finally set. The operation must be carried out without the knowledge of the press with a few operating people, in the least possible and with the maximum safety for the object, with the minimum expense and the maximum documentation. The place of the collection will be the sacristy of the cathedral of Turin. Two textile experts, the late Franco Testore and the late Gabriel Vial, were chosen to ascertain the best area from which to take the sample. To avoid doubts about the object and about a possible replacement of the samples, the invited witnesses will be able to follow all the operations live and without interruption. The crucial moments of the operation not possible to be shown live would be shown on closed-circuit television. Finally, a working protocol is drawn up."

Source: Petrosillo, Orazio and Emanuela Marinelli. La Sindone: Storia Di Un Enigma, (Milan: Rizzoli, 3rd edition, 1998), pp. 149-151.

1988 *(January)*. Gonella stated that the original protocol at the Turin meeting was only a suggestion and never an official agreement. He also observed that he wasn't aware of any archaeological dating for which more than two independent labs were used (see reference below under "Jennings"). However, one C-14 project (see reference below for "International Study Group") had twenty different labs each date eight samples from one tree. And, interestingly, the results revealed an "existence of systematic bias and unexplained variability." According to C-14 expert Reidar Nydal (see reference below), who wrote about the study, "the main reason for using more than one sample [...] is generally to get some idea of the magnitude of other sources of error rather than purely statistical ones."

Harbottle wrote regarding the new proposal, "Professor Gove and I have deep reservations about this. We're concerned that we may be opening the door to enormous controversy and endless, endless bickering and recriminations that could go on and on" (see reference below under "Clark"). Harbottle told another journalist, "The original protocol was pretty fail-safe. I think this way will be chancy. As an experiment goes, it is not very well drawn up. If you do an experiment like this, you should do it right" [...]. With only three pieces of data, the project is fraught with danger. Even if it goes well, skeptics will have a field day. They can say the Church had a chance to rig the results. If there is a problem, the whole thing will be a fiasco. They can lose more reputation than they gain. It's a wild scientific problem these fellows have gotten into, a win-or-lose proposition" (see reference below under "Kava").

Paul Damon of Arizona stated, "When you're doing research, you always have that thing that doesn't fit. What did we do wrong? Could we have mixed the samples? Was there an error in the lab? You can't do that with something like the Shroud. You've got to get it right the first time. I would prefer seven labs to three for a number of reasons [...]" (see reference below under "Clark").

Otlet of Harwell remarked, "I think it's as much a catastrophe as it would be if you allowed bulldozers to go over an archaeological site before you examined it" (see reference below under "Glass.") He went on to say that the changes were made by "someone in Italy obstructing the true path of science" (see reference below under "Wright").

Archaeologist Paul Maloney, General Projects Director of the "Association of Scientists and Scholars International for the Shroud of Turin" (ASSIST), sent a report to Cardinal Ballestrero after discussing which C-14 method would be best with many eminent C-14 scientists, most of whom were not even connected with the Shroud project. He says he found many felt a "grave concern [...] that accelerator technology is not yet ready to do what the Church wants it to do" -- mainly because "small samples often give fallacious results" (see reference below under "Raloff").

Another archaeologist, William Meacham, who was concerned about contamination that might have been present in the Shroud due to the fire it had been in 1532, was told by one of the directors of the three labs chosen to date the Shroud that "I share your opinion on the contamination problem. As a matter of fact, it is *the problem [italics in original]* in small sample dating. I was quite surprised to learn in [the] Turin [conference] that most of my colleagues are not yet fully aware of the problem]" (see reference below under "Meacham").

Sources:
Clark, Kenneth R. "Shroud of Turin Controversy Resumes." *Chicago Tribune*, October 14, 1988, pp. 1, 4, on pg. 4.

Glass, Robert. "Modern Technology May Finally Fix Age of the Shroud of Turin." *Chicago Sun Times*, April 8, 1988, pg. 4.

International Study Group. "An Inter-Laboratory Comparison of Radiocarbon Measurements in Tree Rings." *Nature*, **298**:619-623 (1982), on pg. 619.

Jennings, Peter. "Shroud of Turin to Undergo Radiocarbon Testing." *Our Sunday Visitor*, February 14, 1988, pg. 3.

Kava, Brad. "Scientist Protests Vatican Changes in Shroud Testing." *Corpus Christi Caller-Times*, April 23, 1988, pp. 14A-15A, on pg. 14A.

Meacham, William. "Turin Shroud Carbon Dating." Unpublished manuscript. Copy in possession of author.

Nydal, Reidar. "Optimal Number of Samples and Accuracy in Dating Problems." *PACT* **8**:107-121 (1983), on pg. 107.

Raloff, J. "Controversy Builds as Shroud Tests Near." *Science News*, April 16, 1988 pg. 345.

Wright, Pearce. "New Dispute on Dating Tests." *London Times*, January 16, 1988, pg. 3.

Comments: Not surprisingly, the protocol that would be adapted at the meeting in London was also substantially changed. Regarding "unexplained variability," one has to wonder why it's permissible to acknowledge that in other cases, but apparently not in the case of the Shroud, especially considering how much other evidence is known about it. The enormous controversy and endless bickering that Harbottle feared ensued. It is significant that C-14 people involved in the Shroud testing found so many grounds for criticisms -- and yet, so many people have not critically questioned the final results. Most people are familiar with the computer phrase "garbage in -- garbage out," which means that end results are affected by the quality of the data one starts with. Harbottle was spot on with: "Even if it goes well, skeptics will have a field day."

1988 *(January).* Gove and Harbottle's press conference lasted nearly one hour, blaming Ballestrero and Gonella for abandoning the 1986 Turin workshop protocol and reducing the number of labs from seven to three. Gove also mentioned his open letter to the Pope that complained bitterly about the change. Gove and Harbottle spent most of the press conference explaining with technical details why the three-lab protocol would be discredited. A reporter asked them why they thought they had to right to question the Church's decisions. Gove replied, "We do not question the right of the Pope and of the Archbishop to decide what they think best for the Shroud. What we question is the expertise of the man who leads them to make those decisions. The Archbishop's reply to the carbon-14 experts, based on Prof. Gonella's advice, stated that 'experience in the field of archaeological radiocarbon dating' was the criterion that decided the choice of the three laboratories. But, in fact, the Harwell laboratory has had more experience than the three chosen laboratories put together. This is but one of several reasons, and not the most important, why we feel that the Cardinal was badly advised." What Gove didn't mention about Harwell was that it would have required a minimum of 60 mg. of material, which was a lot more than the Vatican was willing to allow for one lab, to be taken from the Shroud.

Source: Bracaglia, Giorgio. Uncovering the Paradox within the Archives of the Holy Shroud Guild. (Honeoye, N.Y.: Holy Shroud Guild), 2019, pg. 231.

Comments: But Gove never complained about the results of the three labs after the testing

1988 *(January)*. On the same day as the Gove/Harbottle press conference, the organization known as "CSICOP" (Committee for the Scientific Investigation of Claims of the Paranormal, of which New Zealand researcher Denis Dutton, who wrote several letters to *Nature*, was a member) sent a circular letter to various scientific figures saying the Vatican's reduction from seven labs to three would render the examination inadequate and discredit those who accepted it. CSICOP also sent a letter to Turin requesting they be allowed a representative to attend the sample taking. Turin did not respond to the request.

Source: Gonella, Luigi and Riggi di Numana, Giovanni. Sindone: il mistero continua, (Milan: Fondazione 3M), 2005, pg. 67, footnote 49.

1988 *(January)*. On the 18th, Otlet called Gove and told him he heard from Harbottle that the press conference had gone well. Reuters and the BBC carried stories and the *London Times* was going to run a story on Saturday. Otlet mentioned to Gove that Edward Hall and Sir David Wilson (of the British Museum) "were members of the millionaire's club so that one had to be very careful in dealing with them." Otlet was "still worried about the possibility of collusion between the British Museum and Oxford."

The following day, Gove talked with Woelfli, who said that at the London meeting scheduled on the 22nd, they would make proposals for changes. Specifically, he would request 40 milligrams, larger than previously requested. He would also insist that at the sample-taking, two independent persons would be in attendance. He would voice these as non-negotiable and if Gonella would not accept them then he would withdraw.

On January 21st, Sox told Gove that Hall had gone to the producers of a BBC program called *The Chronicle*. He asked them to pay his lab to cover Oxford carrying out the Shroud dating. Hall was told that the BBC's *Timewatch* program was already involved in covering the dating so it was inappropriate for another program to get involved. Sox then told Gove "that Hall's stock in the BBC was now absolute zero."

Gove received a call from a reporter at the *Daily Telegraph* who said that the decision had been made to use only three labs, the sample taking would be videotaped and that Turin workshop protocol would be respected as far as possible. That evening, Harbottle informed Gove that he received a message from Otlet confirming that information, and that a textile expert would be present, although it wasn't known if it would be Flury-Lemberg. Harbottle agreed that Senator D'Amato should be pressured to help them with their cause.

Source: Gove, Harry. Relic, Icon or Hoax?: Carbon Dating the Turin Shroud, (Bristol and Philadelphia: Institute of Physics Publishing), 1996, pp. 234-236.

Comments: It's interesting that Otlet was concerned about the possibility of collusion between the British Museum and Oxford. That doesn't seem to say much about the perceived integrity of both institutions. Hall seemingly wanted to financially capitalize on having been chosen. Recall that after Hall retired, his lab was given a one-million-pound donation, and his position was taken over by Michael Tite of the British Museum, of which Hall was a trustee. In most people's views, those would qualify as conflicts of interest. For someone who thought there was no scientific justification for dating the Shroud, Gove sure put in a lot of time and effort trying to be involved, to the point of grasping for straws, no less.

1988 *(January)*. Gonella announced at dinner the night before the meeting on the 22nd that his Church superiors definitely would not revert to the old protocol. He also declared that if any lab protested the decision, a replacement lab would take their place. Gonella asserted that most archaeological testings use just one lab and a decision to use two would be exceptional. The total amount that the Vatican was willing to be excised just wasn't large enough to have seven labs involved. It was stated that the material would be taken from one corner. Each of the three labs would have enough material to do the test twice if necessary; that was done to address concerns about a lab possibly having an outlier.

Regarding the blind test, it was agreed that the Z-twist of the Shroud was so unique that it would be impossible to hide its identity. Also, unraveling the threads would not be done, since it would be not be compatible with the pretreatment process used in the AMS method.

Source: Bracaglia, Giorgio. Uncovering the Paradox within the Archives of the Holy Shroud Guild, (Honeoye, N.Y.: Holy Shroud Guild), 2019, pg. 232.

1988 *(January)*. On January 22nd representatives of the three chosen labs along with Dr. Tite along with Gonella met in London and agreed to the revised plans advocated by Cardinal Ballestrero as outlined below in a release by the British Museum.

* Representatives of the three radio carbon dating laboratories from Arizona, Oxford and Zurich, accepted by the Vatican to undertake the radiocarbon dating of the Shroud of Turin, met on 22 January 1988 at the British Museum together with Professor Luigi Gonella, Scientific Adviser to the Cardinal of Turin and Dr. Michael Tite of the British Museum, who has been invited to help in the certification of the operation.

* After discussion they accepted the decision of the Vatican to use no more than three samples in the interests of the conservation of the Shroud.

* The procedures for taking the samples from the Shroud and for the treatment of the results were discussed and the proposals on this will be submitted to the Archbishop of Turin for his agreement.

* It is proposed that, as far as possible, the spirit of the original protocol of the 1986 meeting will be retained. Each laboratory will be provided with control samples of known age. The Shroud sample will be taken from the main body of the Shroud away from any patches or charred areas, under the supervision of a qualified textile expert.

* Certification of the samples will be undertaken by Cardinal Anastasio Ballestrero, the archbishop of Turin and Pontifical Custodian of the Shroud of Turin; and by Dr. Michael Tite of the British Museum.

* Representatives of the three laboratories will be present in Turin to receive the samples.

* The overall proceedings will be fully recorded by video film and photography.

* The laboratories will submit their results for statistical analysis to the British Museum and to the Institute of Metrology 'G. Colonetti'.

* The timetable for the operation has yet to be established but it is hoped that a radiocarbon date for the Shroud of Turin will be released by the end of 1988.

* If these proposals are approved by the Cardinal, then a letter will be submitted to *Nature* giving further details of the procedure.

* The participants at this meeting wish to take this opportunity to record their appreciation of Professor Carlos Chagas, President of the Pontifical Academy of Sciences, who chaired the original meeting in Turin in October 1986, as well as the other participants who played a crucial role in moving the project forward.

"Present at meeting at British Museum on 22 January 1988:"

Professor Paul Damon	University of Arizona
Professor Douglas Donahue	University of Arizona
Professor Edward Hall	University of Oxford
Professor Robert Hedges	University of Oxford
Professor Willy Wolfli	University of Zurich

| Professor Luigi Gonella | Scientific Adviser to the Cardinal of Turin |
| Dr. Michael Tite | Keeper of the Research Laboratory—British Museum |

Source: "AGREEMENT ON DATING THE SHROUD OF TURIN (From the British Museum)." Official Press Release (Issued Jan. 22, 1988) by the British Museum. Copy in possession of author.

Comments: Two of the provisions stand out here: 1.) the proposal to retain "the spirit of the original protocol of the 1986 meeting." None of the major provisions of the 1986 meeting were kept, much less the spirit. 2.) "The overall proceedings will be fully recorded by video film and photography." Shockingly, the actual placing of the samples into the steel containers, the climax of the process, was not filmed!!!

1988 (January). A meeting in London on January 22nd was held. Tite presided. Arizona was represented by Damon and Donahue; by Hedges for Oxford; and by Wolfli for Zurich. On the amount of tissue needed for each sample, the laboratories admitted frankly that the data submitted seventeen months earlier in Turin were at fault, because then they were all too interested in showing how little cloth was needed; taking into account the presumable losses of material in the cleaning and pretreatment of the samples, 40 mg of cloth were required for each, that is about 2 cm^2 of fabric. They also admitted that it was impossible to carry out the examination really blindly because the Shroud fabric was easily recognizable having been minutely described; even pulverizing it could have been recognized by X-ray fluorescence with high strontium content. Moreover, shredding and shredding would have compromised cleaning operations and risked losing too much material. The laboratories, however, did not want to withdraw the idea of dating blindly, knowing that it was essentially a fiction -- Gonella reminded them that the proposal had been theirs, not of Turin, and it was up to them to decide, and it was decided that the requirements of good quality of the measure should prevail over the needs of "blindness."

The laboratories asked that the samples be taken from a single site in order to better guarantee the homogeneity of the results, and

Gonella found himself in agreement because this solution minimized the disfigurement of the cloth (and besides it was one of the few things on which everyone had been in agreement with at the Turin meeting). It was established that the site was chosen by a qualified textile expert chosen by Ballestrero and that other tests would be at the discretion of the Cardinal and not the three labs. The control samples were to be procured and certified by Tite, and would need to cover two age periods: the 1st century, and the Middle Ages, possibly around the 14th century. Gonella elaborated on the dates of the control samples. The primary purpose of the "control" samples was to guarantee against systematic errors in the measurement, not to ensure the "blindness" of the examination (even if for this they are obviously necessary), their age period was chosen on the 1st and 14th centuries not because they were the "most probable dates" for the Shroud, as it was said later, but because they were the extremes of the date field in which the Shroud could reasonably be placed: dated to the 1st century, but much more difficult to find than medieval ones, and for this reason greater latitude was left for the date range of the second sample.

The laboratories asked again if it was not possible for their representatives to be present at the sampling, declaring that in any case they would come to Turin to personally take the samples to ensure the chain of evidence; they were told that they would be rejected if their presence was in any way linked to the certification of the samples, but they could possibly be admitted as guests if they required it as a favor in a spirit of friendship. The laboratories undertook to complete the measures within three months of the reception of the samples and send the data for the statistical analysis to Tite and to the Colonetti Institute. At the end of the statistical analysis there was to be a joint meeting in Turin in which the identity of the samples would have been communicated to the laboratories and a scientific communication would have been prepared (at the level of letter to a magazine or preprint). At this point the results would have been communicated to Ballestrero, and the laboratories asked again whether it would be he who would make public the results. The laboratories undertook to maintain the strictest confidentiality on the measurements carried out until the final communication of the data, and also asked that the collection be conducted privately, without notifying the press so as not to be disturbed during their stay in Turin. At the end of the meeting, Tite released a brief press release. Ballestrero approved the proposals of the London meeting, leaving the point on the communication to the public of the results pending decisions by the Holy See.

After a few phone calls from Turin to document the various points laid out, a specification for the execution of the dating work was sent to the laboratories and to Tite for approval. Everyone gave their approval. However, although Hall accepted the specification, he wrote that it would have been "unwise" if members of STURP had been present at the sample-taking for textile consultancy. Ballestrero had decided to turn to an Italian expert, finding it rather ridiculous to look abroad for skills in which Italy and Piedmont have centuries-old traditions. The primary task of the textile expert was then to ensure the safety and better conservation of the Shroud in the operation, so he had to be directly responsible to Ballestrero, and certainly not to the radiocarbon laboratories. The task was entrusted to Prof. Franco Testore, Professor of Textile Technology at the Polytechnic University of Turin; to get additional advice on aspects concerning ancient fabrics, he enlisted the services of Gabriel Vial, technical general secretary of the Centre International d'Etude des Textiles Anciens of Lyon. The excision of the sampling was entrusted to Riggi, the Turin technician who, having worked in 1978 for eight hours on the Shroud was the person who had the best practical knowledge of it, and had shown on that occasion to have the talent and reliability required by the need. He also undertook to ensure the logistics of the operation and his complete video documentation.

Source: Gonella, Luigi and Riggi di Numana, Giovanni. Sindone: il mistero continua, (Milan: Fondazione 3M), 2005, pp. 70-72.

Comments: Regarding the information: "The laboratories asked that the samples be taken from a single site in order to better guarantee the homogeneity of the results […]," one has to ask: what was the goal of the test in their minds: that they all get similar results to show how good the AMS method was or to assist in finding out the actual date of the Shroud??

1988 *(January)*. With tensions high at the London meeting, the lab representatives requested 40 mg. each, which corresponds to about two centimeters of cloth. The reps admitted that the blind test (i.e., the Shroud sample being unidentified when dated) was impossible (because of the unique twill).

They pushed for sampling coming from one site only to better ensure the homogeneity of results. Gonella, keen to cause minimal defacement to the cloth, agreed. The sampling site would be designated by a qualified textile expert to be chosen by Cardinal Ballestrero; this expert would extract the sample. Tite would provide control samples, dated to the first and fourteenth centuries.

The representatives ask to attend the extraction of the sample to guarantee the chain of evidence. Gonella replied that their presence shouldn't be linked with sample verification, but they could be admitted as guests. The labs promised to complete the work in three months, to maintain strict confidentiality, and to send data to Tite as well as the G. Colonetti Institute of Turin for statistical analysis. Then there would be a meeting in Turin to coordinate communications, including to the official custodian of the Shroud, Cardinal Ballestrero, who would make the results public.

Source: Marinelli, Emanuela. "The Setting for the Radiocarbon Dating of the Shroud." Presented at 1st International Congress on the Holy Shroud in Spain – Valencia -- Centro Español de Sindonologia (CES), April 28-30, 2012, pg. 7. www.shroud.com/pdfs/marinelliv.pdf.

Comments: The point about the labs requesting only one site is crucial. Other experts had adamantly requested that multiple sites be used. By using only one site, the compilation and completion of the process would be much quicker. But this was going to be one of the most important C-14 tests ever, and expediency should not have been a consideration.

Although Mechthild Flury-Lemberg was the initial choice to take the sample, she would not perform the cutting, which was ultimately done by Italian scientists Giovanni Riggi di Numana and Franco Testore and the French scientist Gabriel Vial. During the sample taking, Testore would point to the blood from the side wound area of the image and inquire, "What is that?"

Gonella's comments about the representatives being at the sample-taking as guests as opposed to not being connected to certifying the sample authenticity, still ring hollow.

The "strict confidentially" of the labs would later be proved to have not been maintained. Gove was allowed to view the Tucson datings and even signed their non-disclosure agreement (*Night of the Shroud* documentary, original version; I personally saw it in the Gonella archive). Sox and a BBC film crew were allowed to view the Zurich datings (Sox book, pp. 135-140).

According to Petrosillo/Marinelli, Gonella hoped that "the serious research programme on the Shroud would not lead to petty attempts at stealing the limelight" ["Enigma ..." book (pg. 47)].

1988 (January). Petrosillo and Marinelli wrote, "A meeting took place at the British Museum on January 22nd between Tite, Gonella and the representatives of the three laboratories. The Cardinal's advisor warns that he will not attend the meeting if the representatives do not accept the procedure approved by the ecclesiastical hierarchy; they did agree. A week earlier, Damon and Donahue had threatened not to date if they were not satisfied with the protocol. Damon had lunch with Otlet and Wilson to discuss the matter. It was clear that he is ready to accept that there are three laboratories. Sox, Wilson and Rinaldi, animator of the U.S. research on the Shroud, tried to make Turin change their mind. In the meantime, Wilson and Tite listened to Gonella's complaints about Gove. Otlet believed the real problem is Gonella's aversion to Gove. According to the Turin scientist, a change of procedure at this time would have meant giving in to Gove."

"The procedures for taking and studying the results from the representatives of the three laboratories were discussed. Their recommendations will subsequently be approved by the archbishop of Turin and made known in Tite's letter to *Nature* of April 7, 1988. That paper was to be the only protocol of the experiment to which the twenty-one signatories of the final dating report would refer. Gonella made the point that it was the Holy See and not himself who decided that there was enough material to proceed with three labs. Then he wonders what could be the reason for the fuss over the story, given that the Shroud is not an article of faith for the Catholic Church. That there should be seven laboratories was only a proposal. In reality, therefore, there was not a decrease from seven to three, but an increase from one to three laboratories. According to Tite, this is the biggest change in the protocol. Meanwhile secret diplomatic attempts are being made to return to the original protocol with seven laboratories. The minimum quantity of tissue required for each for analysis is also established: 40 milligrams of fabric, about 1-2 square centimeters. Gonella agrees that the withdrawal is photographed and filmed under the supervision of two people. He didn't reveal, however, that the two likely, as Ballestrero had written, would have been he and Chagas. Analysts asked Gonella where the sampling site would be. He first remained vague, then mentioned the area near the lateral strip from which the 1973 Raes sample had been taken."

"All those present admitted the impossibility of carrying out a true blind test because, in any case, the Shroud sample would have been easily recognizable even if shredded. 'The blindness of the exam will be guaranteed by the good faith of the laboratories,' it was claimed during the London meeting. Gonella was explicit: 'The laboratories admitted that the 'blind' examination request was only for publicity reasons; in fact, the other samples (I knew it very well as a metrologist) served as an 'on-line' calibration control, and it was decided to privilege scientific safety over the 'blindness' of the examination in the treatment of the samples;' he ascertained the real quantity of material needed, but it was known that the collection was carried out in a confidential manner so as not to be tormented by the press, on which we all agreed. Damon has no problem that all three laboratories use the same AMS method. According to him, the comparisons between the techniques of the accelerator and the counts from which no difference arose resulted. Tite also adapts to the new agreement: in most archaeological studies only one laboratory is used; very rarely are three used. Hall was very understanding towards the excluded: 'I would have done the same. I would have been extremely jealous.' He will then say that, all things considered, it was better this way, with only three labs. With seven there would have been more accusations of indiscretions, and perhaps even the test would not have been done, at least for several years. Harbottle, however, pointed out that the tests will be conducted by those same scientists who had raised objections not long ago and had also signed the protest telegram. In addition, the American scientist makes a sensational accusation against two of the three chosen laboratories: 'The Oxford one is heavily supported by the British Science Research Engineering Council and the Tucson one in the same way by the National Science Foundation; it would be surprising if the dating of the Shroud was not clearly mentioned in their next request for grants.' Oxford has already done so. STURP was inclined to give their approval to the new protocol. The results of the ^{14}C would have been made public at the end of the year and STURP would have made its tests on the nature of the image after ^{14}C: they would not have known which date the three labs had discovered. In short, there was something for everyone, except for the four excluded laboratories. Instead, the laboratories still did not believe they had achieved complete victory. After the meeting in London, Gonella said he still received letters from some of the labs recommending that no other tests should take place with the ^{14}C."

"Sindonologists, for their part, protested the reduction to just three laboratories. But for other reasons. The most energetic was Paul Maloney, General Projects Director of ASSIST, who sent a twenty-seven-page document to Ballestrero and the Vatican in which he says that, according to many eminent ^{14}C scientists, accelerator technology was not yet ready, especially given the frequent false results with small samples. 'With the accelerator method,' noted the archaeologist 'the sample goes through more steps than with the proportional-counting method and, in each step, there is the possibility of introducing foreign carbon that will influence the date. The dual method is used as a point of reference when there may have been a divergence problem otherwise. Fabrics have been dated for years with the conventional method, while it seems that they have not been commonly dated with the accelerator method'."

"Gonella said that Ballestrero discussed with him in Turin the final details. It was decided to take the sample in the sacristy of the cathedral and to entrust its execution to Giovanni Riggi, the scientist who in 1978 had carried out the inspection of the back of the Shroud, because this work had made him the person with greater factual knowledge of the Shroud cloth and had made us appreciate his technical skills. The Cardinal asked Gonella to invite an Italian textile expert to follow the operations to guarantee the safety of the cloth; he wanted to emphasize his trust in Italian science and, given our national and regional traditions in the textile field, he did not see why we should have trusted people preferred by the ^{14}C laboratories instead of an expert who had to be responsible towards him. Professor Testore, the head of textile technologies of the Polytechnic of Turin, gladly accepted the assignment and we called to help him with his experience of ancient fabrics with the assistance of Gabriel Vial of the Lyon Textile Museum (with the clear understanding that the decisions were up to Testore only)."

Source: Petrosillo, Orazio and Emanuela Marinelli. La Sindone: Storia Di Un Enigma, (Milan: Rizzoli, 3rd edition, 1998), pp. 152-155.

Comments: To continue with this thread, go to the Petrosillo/Marinelli entry under "1988 *(January)*."

1988 *(January)*. "Most of the laboratories involved were very happy with this protocol. One of the most important aspect *[sic]* was the amount of cloth given to each laboratory. For some that was the main point, far more important than the number of laboratories. But for others this new amount of cloth was the proof that the preservation of the linen was not the true reason for reducing the number of laboratories to three."

"As for the textile expert, the opinions were divided. Some agreed with Turin's position that it was their responsibility 'because the main thing that a textile must do is to guarantee us that the minimum damage is made' (Gonella, interview). Others however saw it as a needless change which could further decrease the credibility of the whole enterprise."

Source: Laverdiere, H. "The Socio-Politic of a Relic: Carbon Dating of the Turin Shroud," (Doctoral Thesis) 1989, pp. 88-89. Accessible via free download at http://ethos.bl.uk/OrderDetails.do?did=1&uin=uk.bl.ethos.531916.

1988 *(January)*. Canuto informed Gove on January 23rd that Fr. Rinaldi had talked with author Ian Wilson, who "stated he was going to write a story deploring the situation." Canuto would later gather all news clippings to take to Rome with him. Fr. Rinaldi said that Gonella had told Wilson the samples would be taken at Easter. If there were any problems with the sampling, additional samples could be taken when STURP did its tests in June. Gove said this was the first he heard that the STURP tests were back on track. Canuto also said that Fr. Rinaldi suggested to Gove that the latter contact a man named Viki Weisskopf, who was a member of the Pontifical Academy of Sciences and long-standing friend of Gove, to try to intervene in some manner. Gove decided to talk directly with Rinaldi, who said that the press release from the January 22nd London meeting said that the new procedure still had to be approved by Cardinal Ballestrero. Maybe, Rinaldi thought, if Ballestrero was sufficiently pressured, he would compromise. Gove felt there was little chance of that. Rinaldi expressed the thought that this whole affair might discredit the Catholic Church.

Gove called Weisskopf at home. Weisskopf did not believe that the Shroud was genuine and preferred that the Shroud be left alone. Gove wrote, "I said there was absolutely no scientific justification for measuring the age of the shroud." If it were to be dated, Weisskopf wanted Gove to be involved. They left the matter there. Gove then resigned himself to the fact that Weisskopf probably would not be able to get any decisions changed. Gove wrote, "It was another case of my clutching at any straw."

Source: Gove, Harry. Relic, Icon or Hoax?: Carbon Dating the Turin Shroud, (Bristol and Philadelphia: Institute of Physics Publishing), 1996, pp. 236-237.

Comments: Gove had hoped that his friend Weisskopf could exert some influence on the Academy. There is another story of intrigue involving another member of the Academy, Monsignor Dardozzi. Dardozzi was the Chancellor of the Academy (i.e., Chagas' right-hand man). According to June 3, 2009 article in *The Guardian* (UK) (www.guardian.co.uk/commentisfree/belief/2009/jun/03/vatican-central-bank) Dardozzi smuggled out more than four-thousand documents pertaining to the scandal-racked Vatican Bank. The article said, "It is interesting to note that Dardozzi's motive for turning whistleblower was not unalloyed disapproval of the IOR's unethical conduct. His decision to smuggle his secret archive out of the Vatican was motivated, at least in part, by anger at the Institute's refusal to pay him a commission on the sale of a valuable real estate property near Florence. The unusual monsignor wanted to leave the money to his adoptive daughter, whose health condition required expensive treatment."

So, we have a financially-needy monsignor, who was the right-hand man to Chagas, alleged to be in Harry Gove's pocket. If there was a one-million-pound donation available for the Oxford lab after supposedly having proven the Shroud to be a fake, were there additional "donations" to pass around to key authorities to make sure that certain actions would take place for the C-14 dating of the Shroud? While there's no hard proof, it looks suspicious.

Petrosillo and Marinelli were also suspicious about the Chagas/Gove relationship. In their "La Sindone ..." book, they wrote (pg. 220): "As for Chagas, he left the presidency of the Pontifical Academy two weeks after the announcement of the results. A rumor immediately circulated that it was 'torpedoed' for the Shroud affair. Undoubtedly his attitude in the affair was disconcerting due to his disagreements with the engineer Gonella and with Cardinal Ballestrero (who complained to the Vatican), the double game and the agreements under the table with Gove, declared opponent of the authenticity and also interested for reasons of publicity related to the dating. The public tribute paid to the resigning Chagas by the Pope for the 16 years as president certainly does not exclude the reservations of the Secretary of State *[Casaroli]* on his behavior in the Shroud affair [...]."

1988 *(January).* On January 25th, Gove phoned Donahue, who filled him on details of the meeting on the 22nd. Donahue said that Gonella seemed agreeable to various requests they had made but he would have to check with the Cardinal. (Donahue said no "demands" were made.) Donahue said that Gonella still had anger toward both Gove and Chagas. Gove asked why Gonella was angry with Chagas. Donahue speculated it was because Gonella perceived that "Chagas was taking over the whole enterprise." Donahue read Gove the press release that was put out after the meeting. Gove asked if Flury-Lemberg would be the textile expert present. "Donahue said it would be some Italian expert

-- for purely **political** *[my bolding]* reasons." Donahue related to Gove that the Shroud samples would not be unraveled but that he didn't know what the Shroud actually looked like. He said "that Tite would make an effort to make the samples look as similar as possible but there was no pretence that it would be really blind because the shroud weave was so distinctive." Toward the end of the conversation, Gove told Donahue, "I can't disguise from you the fact that I envy the hell out of you." Donahue then told Gove he would try to have him in Arizona as an observer.

Gove called Canuto the next day and relayed the gist of his conversation with Donahue. Canuto asked Gove if he criticized Donahue "for knuckling under to Gonella." Gove said he did not because it wouldn't have done any good. Canuto suggested to Gove that he (along with Otlet and Harbottle) send a letter to the Pope via both the U.S. and Vatican ambassadors to the United Nations; Gove said he would think about it.

Source: Gove, Harry. Relic, Icon or Hoax?: Carbon Dating the Turin Shroud (Bristol and Philadelphia: Institute of Physics Publishing), 1996, pp. 239-242.

Comments: Gove told Donahue that he was unable to disguise his envy. Frankly, Gove did not seem able to disguise much of anything regarding his emotions.

It seems peculiar that Canuto, who worked for the Pope's scientific advisory group, asked Gove if he criticized Donahue "for knuckling under to Gonella." Why did Canuto apparently side with two C-14 scientists against Cardinal Ballestrero's scientific advisor? This once again points to the puzzling coalition of Chagas, Gove and Canuto that both Ballestrero and Gonella previously mentioned.

1988 *(January)*. On the 27th, Gove sent a letter to the seven laboratories, Tite, Chagas and Canuto, saying that the dating process had become a "shoddy enterprise." A copy of the letter was sent to Gonella by Hall (in which Gonella was described as suspect and of low-scientific standing). None of the laboratories responded to Gove's letter.

Chagas basically had always been against STURP's proposed tests. As far back as his first letter of August 7, 1985, Chagas had insinuated that Ballestrero was somehow compromised by STURP. However, during the Turin planning meeting in 1986, when all of the participants except Gove were open to STURP's tests, Chagas too briefly seemed open, but when Gove openly spoke out at the meeting, Chagas immediately sided with Gove's position. For some reason, Ballestrero, based on Chagas' opposition to the STURP tests, seemed concerned about a perceived collusion between himself and STURP. The insinuation seemed to be that if the dating results came out first-century, it was because STURP was biased in that direction. Casaroli could have ignored the positions of Chagas, who among other things had continued to speak out against the decisions taken even after they had been communicated to him, and not to fear the reactions of the scientific community. In this situation it was considered more prudent to perform the radiocarbon operation separately, without further complications, postponing other tests to a later time.

STURP, in fact, had not been the only group to propose tests. Wilson had presented on June 15, 1985 on behalf of the BSTS an operative proposal for some tests that supported and supplemented the STURP program. One of the British proposals included a collaboration with the Italian group ENEA, the Italian Agency for New Technologies, Energy and Sustainable Economic Development. The American group ASSIST had presented on March 3, 1986 a voluminous request for exams, which, however, according to Gonella, lacked effective operational indications. The French Professor Lucotte had asked for a large blood sampling for DNA studies. (The question, presented in May 1986 together with that of Duplessy dating that was treated separately, was in any case considered unacceptable by the Turin because it asked for the destruction of a couple of centimeters of a bloodstain in the hope of being able to determine if the blood belonged to a male or female person, and if it was affected or not by some diseases.) In October 1986, through the Turin Curia, Mr. Judica-Cordiglia of the Sindonological Center of Turin had sent a memorandum requesting to take photographs.

Source: Gonella, Luigi and Riggi di Numana, Giovanni. Sindone: il mistero continua, (Milan: Fondazione 3M), 2005, pp. 67-69.

1988 *(January)*. Although Gove had previously tried to persuade the three chosen labs to refuse to do the dating if more labs were added, the labs in Arizona, Oxford and Zurich decided to proceed with this limitation. According to Sox, had they refused to do the test, labs in Pisa and Udine in Italy would supposedly have been chosen to replace them. Apparently, Gonella believed the three labs would not refuse to do it because, as Sox put it, "the prize was too great."

Source: Sox, David. The Shroud Unmasked, (Basingstoke, Hampshire: The Lamp Press), 1988, pg. 117.

1988 *(January)*. On the 27th Gove sent a letter (that he called a "last-gasp effort") to the Director of the British Museum, Sir David Wilson, complaining about problems with the process up until that point: "Dear Sir David:"

I enclose a copy of a press release issued by the British Museum following a meeting called by Mike Tite. Both the certification of the shroud samples and the statistical analysis of the data will be the joint responsibility of Dr. Tite and Professor Gonella. Because Professor Gonella has little standing in the world scientific community, and because he has a vested interest in the shroud which renders him suspect, the entire burden of proof that the whole operation is credible rests on the shoulders of Dr. Tite.

I have no reservations whatever concerning Dr. Tite's honesty, integrity and credibility as a representative of the British Museum in this enterprise. However, there are many people who are overly suspicious of this entire operation. The situation is particularly exacerbated by the fact that the head of one of the three laboratories to be involved is Professor E.T. Hall of Oxford who is also on the Board of Directors of the British Museum.

The original protocol called for a third person to be involved in both the certification and data analysis namely the President of the Pontifical Academy of Sciences, Professor Carlos Chagas or his chosen representative. Dr. Chagas is such a distinguished scientist that, if both he and Dr. Tite had been involved and if the original seven laboratories had participated, the enterprise would have been as credible as possible.

I am astonished you would permit the British Museum to risk having its reputation called into question in what has become a somewhat shoddy enterprise. I am afraid that four of the participants at the meeting, namely Damon, Donahue, Wolfli, and Tite bowed to the dictates of Gonella, warmly supported by Hall.

I fear, sadly, that Mike Tite has taken on a responsibility which he, and the British Museum may live to regret.

Yours sincerely,

[signed] Harry Gove
H.E. Gove

Source: Copy of original letter shared among some sindonologists and in possession of the author.

Comments: There are several notable aspects in Gove's letter: 1) his intense dislike for Prof. Gonella; 2) he noted that many people are suspicious of the process; 3) a conflict of interest pertaining to Prof. Hall; 4) the breaking of the original protocol in terms of the number people involved and also the number of labs; 5) Gove termed it a "somewhat shoddy enterprise." Since Gove termed the dating as such, why did he so readily accept the A.D. 1260-1390 results??

Gove was not the only high-profile C-14 person to describe the process as "shoddy." STURP photographer, Barrie Schwortz, went to the University of Arizona in 2012 to photograph one of the two remaining Shroud samples they held. Dr. Timothy Jull, one of the scientists at Arizona involved in the dating, told Schwortz (recounted to me by Schwortz by phone on January 4, 2020) and a colleague Larry Schauf, a former federal prosecutor, that he believed the whole event "was a shoddy affair and should be redone." The fact that both Gove and Jull used the term "shoddy" to describe the Shroud dating speaks (= shoddy) volumes.

1988 *(January)*. According to Petrosillo and Marinelli, Gonella stated, "The managers of the laboratories had insisted on being present at the collection, declaring that in any case they would come to Turin to personally collect the samples that they did not want to be sent via the post office. The Cardinal therefore resolved to invite them to attend the operations as his personal guests, in consideration of their availability for the safety of the samples. Moreover, at this point, a refusal of their presence would certainly have been misinterpreted by the press. I drafted a letter in Italian and English with all the details of the case and sent it to the interested parties on 30 January." Petrosillo and Marinelli then commented, "Meanwhile the Cardinal had also come to what Gonella calls the painful observation that prudence dictated postponing other tests for the time being. The Cardinal's insistence at this point to include dating in a more complex research program would inevitably have been presented as an attempt to jeopardize the examination at ^{14}C and would have adhered to the thesis that STURP was 'enslaved to the Catholic Church.' 'It was a shame,' Gonella continued, 'to have to postpone the important exams for conservation and keep the scientists who had proved more correct and serious still waiting, but his diplomatic experience told him that it was better to take this action before proceeding: it was hoped that the rest of the research program could be carried out within a few months of dating. The STURP president accepted this decision willingly, saying by telephone that he certainly did not want to complicate the situation'."

"Meanwhile, in England, there were those who proposed to bet on the results. There were two possibilities that make the date of the cloth more successful."

Source: Petrosillo, Orazio and Emanuela Marinelli. La Sindone: Storia Di Un Enigma, Milan: Rizzoli, 3rd edition, 1998), pg. 155.

Comments: To continue this thread, go to the Petrosillo/Marinelli entry under "1988 *(March)*."

1988 *(February)*. On the 1st, Gove spoke to Canuto, who said it appeared that Chagas had lost influence in the Vatican. Chagas was disappointed because in his eyes, the three labs had caved in to Gonella. Canuto felt that a letter to the Pope "was very important, if only to give him some historical perspective on what was going on." Gove then composed a letter to Cardinal Ballestrero that he hoped would be signed by Harbottle and Otlet. Gove phoned a *La Stampa* representative in Washington, D.C. and asked him what he thought of the idea of publishing an open letter to the Cardinal in *La Stampa*. The rep said he thought it was an excellent idea and volunteered to try to make it happen. Gove faxed to Otlet and Harbottle on the 15th a copy of the proposed letter and asked if they would be willing to sign it. Gove also wanted Duplessy to sign but had problems contacting him.

Otlet replied on February 26th. Otlet said his hands were tied because such a letter would have to be vetted by their press officer who said they were not prepared to support any such contacts regarding the Shroud. Otlet didn't think that the open-letter would be effective. "It will be seen as a stunt and will be used as good evidence against you as the very reason your laboratory and the ones that sign with you were best left out of the dating exercise." Otlet suggested to Gove to write a direct letter to the Pope. Otlet was upset that their colleagues "have not the decency to say that the decision to leave out the founder laboratories was both scientifically and morally wrong."

In late February, Gove talked again with Harbottle, who reminded Gove that Senator Daniel Moynihan, was up for reelection in New York State. Harbottle suggested sending a joint letter to him with newspaper clippings, and a mention that Bishop Clark of Rochester was on their side. On February 26th, Gove put his thoughts down regarding the Shroud project to that point for his own records: "Far and away the saddest and most deplorable aspects of the whole shroud dating enterprise are, firstly, the fact that a group of estimable people comprising Canuto, Chagas, Damon, Donahue, Duplessy, Hall, Harbottle, Hedges, Otlet, Tite, and Woelfli, most of whom were colleagues and some of whom were personal friends, now have a distinctly different and more suspicious attitude toward one another and secondly, a scientific investigation that could have been exciting and challenging and, above all, a great deal of fun, has turned sour. None of us has come out of it whole and pure although some more than others. It is extraordinary to me that twelve people including me, each one of whom has a vastly greater scientific standing than Professor Gonella, should have allowed such a mean-spirited person to call the tune to which all of us danced in one way or another."

"As for me I will write a personal letter to the Pope that he will probably not even read and then let the matter rest. If I am invited to be a guest observer of the measurements at one of the three laboratories I will probably accept. Some day I hope to be involved in writing an account of the affair which has now spanned eleven years of my life and entailed considerable effort on my part. A lot of the effort was enormous fun however. All along, my chief motivation was great curiosity as to the shroud's age, the realization that it was a perfect artifact to demonstrate to the general public the power of AMS, and the desire that the measurement be made in the most credible fashion."

"Unfortunately it is not now going to be made in as credible fashion as it could have been and, what is worse, no reason has been given for the change in procedures. Obviously the reasons are so contemptible as to be embarrassing. However, maybe luck will prevail and the job will be done well enough. That, however, will not compensate for the fact that the affection and admiration some of us had for others in the group of twelve has lessened and, in some cases, even vanished. For that, I will never forgive Gonella. Knowing him as I do, I do not expect that will bother him in the least."

Fr. Rinaldi called Gove later that day. He said the Cardinal still had not officially approved the procedures agreed to in the January 22nd London meeting. Rinaldi felt that perhaps Gonella was rethinking his position. Gonella had been in touch with STURP's Lukasik, who believed that three labs were too few and who had Gonella's ear. Meanwhile, Gove contacted one of Senator Moynihan's staff, who took Gove's number and promised to get back with him. Rinaldi sent Gove a copy of a letter that author Ian Wilson had sent to several Shroud scholars. Wilson believed that seven labs were too excessive. He was in favor of doing a C-14 test but noted it was not infallible. Wilson's last paragraph was both insightful and humorous. Wilson mentioned how the Roman soldiers cast lots to decide who would receive Jesus' clothing. "Might there not even now someone looking down on the sorry scene and murmuring, 'Father, forgive them for they know not what do' […]."

Source: Gove, Harry. <u>Relic, Icon or Hoax?: Carbon Dating the Turin Shroud,</u> (Bristol and Philadelphia: Institute of Physics Publishing), 1996, pp. 242-247.

Comments: Chagas echoed Canuto's aforementioned concern about Donahue "knuckling under to Gonella," and expanded it into perceiving that all three labs had caved in to Gonella. Clearly all the power struggles among groups and individuals had a deleterious effect.

The giving of "historical perspective on what was going on" is exactly why I'm writing this book. Very few people understand how complicated and polluted the whole process became. Regarding the open letter to Ballestrero proposed by Gove, practically everyone knows that Church officials do not respond to open letters. It's amazing that Gove entertained the idea that it would work, considering that he knew that the Cardinal had been upset when the labs sent him a private cable the year before pertaining the Turin workshop. But Gove seemed to be a desperate man. He had already admitted he was grasping for any straw. Regarding Otlet's claim that leaving out the founders' laboratories was "morally wrong," since little or no morality had been seen up to that point, why did he think it would start at that stage? It is easy to imagine where Gove would have placed himself on his "whole and pure" spectrum. The fact that there was a loss of respect and admiration among individuals of the C-14 consortium points out how deplorable their behavior had been. If the change in protocol was contemptible and embarrassing, why were the final results basically unquestioned?

Gove was obviously a major player in all the chaos, but at least he did the world a favor by publishing his book, basically not having self-censored at all, and blatantly revealing disturbing thoughts and actions on his part, thus showing his negative and harmful effects on the whole process.

1988 (February). Meacham wrote to the *Chicago Tribune* on the 8th, "On Jan. 17 you reported the strong comments of two New York physicists 'outraged' at being dropped from the project to date the Turin Shroud by carbon-14. It is ironic, however, that these two scientists are now calling so stridently for adherence to the 'protocol' agreed at the Turin conference in September 1986."

"I attended the Turin meeting along with representatives of the C-14 labs and experts from other fields. In January 1987, the Director of the Pontifical Academy of Sciences in Rome revealed, in correspondence with me, that these same scientists now so outraged were working behind the scenes to change certain provisions of the protocol not to their liking. I wrote to several laboratories then involved to point out the dangers inherent in such manoevering [sic], but none reacted until they were dropped from the program."

"The 'protocol' agreed in Turin was in fact only a recommendation from the assembled experts to the Church authorities. There can be no claim that the Vatican or Turin have broken this agreement. If anyone violated the Turin protocol it is those who attempted to use their influence in the Pontifical Academy to have it amended."

"The dating of only three samples instead of the recommended seven will still yield a scientifically valid result, within the limitations of the C-14 method. It is alarmist (and possibly self-serving) to claim that seven laboratories must be involved to produce a reliable C-14 date on a piece of cloth."

"Much more important the number of samples is the provenance of each sample. The NY scientists had originally, in 1978, proposed to date a piece removed from the Shroud in 1973, even though this sample had been passed around and held in less than rigorous scientific conditions (in an old photo album at one stage). They then proposed to date the charred material from one of the burn marks, but this plan was roundly rejected at the Turin conference as unreliable."

"Finally, they wished to take a single piece and divide it into seven samples. To me, as an archaeologist with 17 years' experience in the application of C-14 dating to field contexts, this proposal seemed absurd. One should seize the opportunity to date samples from different parts of the cloth, avoiding a possibly anomalous (e.g., starched) area. This is the major scientific question now relevant."

"The dating of the Shroud is not, after all, a laboratory inter-comparison experiment. Three dates from reputable labs, hopefully from three different sites on the relic, should give a good indication of the radiocarbon age of the cloth, and whether or not random contamination or other problems exist which require more sophisticated testing techniques."

[signed]
William Meacham

Source: Letter-to-the-editor, *Chicago Tribune*, February 8, 1988 from William Meacham. Luigi Gonella archives, of which the author has a copy.

1988 February. On February 12th, Tite sent a request to the French C-14 expert, Jacques Evin, to procure a sample that resembles the Shroud. "I would certainly much welcome any assistance that you can give in obtaining a mediaeval control sample, which is as similar as possible in terms of weave and colour as the Shroud [...]. The material of the sample should be linen [...]. We are looking for a sample which dates from the 13th or the 14th century AD, preferably the latter [...]. The historical precision should obviouly *[sic]* be as good as possible, but one would certainly consider samples with an age range of fifty to a hundred years."

Source: Bonnet-Eymard Bruno (Brother). "The Holy Shroud Is As Old As The Risen Christ." *The Catholic Counter-Reformation in the XXth Century*, No. 330 (May 2000), pp. 27-28.

Comments: It's likely that Tite never thought this letter would be made public. One can understand him wanting a sample similar to the Shroud, as the idea of a double-blind study (labs not knowing which was the Shroud and which was a control), but his emphasis on the control sample being preferably from the 14th century instead of the 13th and with a tight precision window seemed suspicious to some. This sample, discrepancies about the actual number of control samples (as well as who was involved), along with discrepancies about the sizes and weights of the Shroud samples, would later lead some to theorize that other samples were substituted for Shroud samples. Piero Savarino, a chemistry professor from the University of Turin, who would eventually become the scientific advisor to Cardinal Poletto, a successor of Cardinal Ballestrero, stated,

"Unfortunately, a set of facts, or rather of deficiencies and carelessness, leaves the suspicion survive" (**Source**: Marinelli's Valencia presentation, cited in multiple places throughout the book, pg. 10). There will be more details further below and also in Parts II and III about the discrepancies regarding both the control samples and sizes and weights of the Shroud samples.

1988 *(March)*. Evin went to Saint Maximin's, and without even informing the parish priest, extracted a few threads from the cope of St. Louis d'Anjou (1274-1297). In the version recounted here (there are several versions), he wasn't able to send them to Tite because of a postal strike so asked Gabriel Vial to deliver them personally on the day of the sample taking.

In the meantime, Tite, not having heard from Evin, asked Ian Wilson of the BSTS to try to get a control sample.

Source: Bonnet-Eymard, Bruno. "The Holy Shroud, Silent Witness." *The Catholic Counter-Reformation In the XXth Century*, No. 295 (April 1997), pp. 19-20.

Comments: Other versions of the story regarding the threads from the cope will be detailed in Part III.

1988 *(March)*. Meacham wrote: "Elsewhere Gove commented on STURP's desire to 'characterize the sample' before it was dated by saying 'whatever that means.' This was the absolute nonsense! Clearly he had no concept of or interest in investigating the chemistry of a sample prior to running it through the standard pretreatment. The three selected labs were equally blinkered, they would conduct no research on where the sample should come from, and they planned to treat their prized Shroud fragment largely as they would any other archaeological specimen."

Source: Meacham, William. The Rape of the Turin Shroud: How Christianity's most precious relic was wrongly condemned and violated (Lulu.com), 2005, pp. 89-90.

Comments: Laverdiere cited (pg. 111) two experts regarding the pretreatment: (1) "'The thing is that the treatement [sic] that most of the laboratories who are going to date the shroud are using is just acid and alkaline and that does not remove the contamination. And the cloth is dirty, terribly dirty [...]' (Beukens interview). *[Refers to non-referenced interview of R.P. Beukens of IsoTrace lab in Toronto, whom Laverdiere claimed refused to take part in the Shroud C-14 test -- but doesn't specify why.]* So what is the right sort of pre-treatment? (2) 'Which particular methods are employed will to a large extent depend on the information provided by the user [...]. The user must be aware that no pretreatment procedure can garantee [sic] absolute decontamination'." *[Laverdiere's source for the second quote was R. Gillespie on pg. 12 of <u>Radiocarbon User's Handbook</u> printed by Oxford University Committee for Archaeology (1984).]*

1988 (March). Maloney sent a letter to Gonella on the 22nd: "Dear Professor Gonella,"

When we met on Saturday, Nov. 21, 1987 you shared with me your plans concerning the carbon dating of the Shroud. Although I personally felt very uneasy regarding those plans I felt it was inappropriate for me to debate the issues then without having had a chance to investigate them thoroughly to see what the scientific community actually felt about them.

But now I have just completed a series of interviews with the scientific community specializing in the field of carbon dating. I realize it is only a short time before the samples are to be removed from the Shroud for testing but I felt you should know what these scientists are saying about the current plans for dating the Shroud of Turin.

According to published information the plans would seem to be to take a sample from a single place on the Shroud, cut it into three pieces, and deliver them to the representatives of three accelerator laboratories for dating.

In the opinion of many of the most prestigious scientists who work with carbon dating, the accelerator cannot yet produce either a very precise or a very accurate date. I believe it would be a great shame to waste samples from the Shroud for a test whose current planning is being seriously questioned by these scientists.

It is true that under <u>normal</u> circumstances only a single lab would be used to date archaeological material as you noted to me during our last meeting. But whenever there has been a controversial date in archaeology, the specialists have relied upon conventional methods to verify that date. The Shroud fits this precisely because of the controversy surrounding it.

Scientists are specifically concerned about the following:

1. Conventional labs have been taken from the plans.

2. The sample size has been reduced to a point too small to make it feasible for the current labs to render a credible date.

3. There has been no external peer review of the three-lab protocol, especially the chemistry dealing with possible contamination of the samples.

Two scientists who know the radiocarbon field intimately are Dr. E. M. Scott, statistician, who is coordinator of the world's largest intercalibration study of carbon dating facilities, and her colleague, Dr. M. S. Baxter, a radiocarbon chemist who is member of the same intercalibration study team, both at the University of Glasgow, Scotland. They have co-authored a paper in which they have this to say about the accelerator:

> Despite the euphoria, enthusiasm and expense directed for a decade at the accelerator technique, it remains to prove fully its capabilities in terms of accuracy, precision, sample turnover rate and affinity for small samples. It is a highly expensive and technically advanced approach which deserves to succeed but there are those who believe its true value lies outside 14C dating in other areas of isotope science and who doubt its ability to give value for money in the 14C field. They would suggest that the relatively few very small samples requiring dating could be handled by low cost micro gas counting methods using existing

technology. [From a forthcoming paper on the subject of carbon dating.]

They suggest that it may be another 10 years before the accelerator by itself can do what the Church wants it to do.

I believe it would be very important for a conference, similar to that held during Sept. 29-30, Oct. 1, 1986, to be held to bring highly trained specialists together to discuss the nature of these tests with the three labs before any samples are taken. Such a group of specialists are suggested in the enclosed paper. I therefore urge you to reconsider the present planning by examing [sic] the enclosed paper [not reproduced here] and the recommendations which are offered at the end. The short delay may be well worth the time expended on behalf of a scientifically acceptable protocol for testing the Shroud.

I have sent a copy of this paper also to His Eminence Cardinal Ballestrero and His Eminence Cardinal Casaroli for their study. Thank you so much for your kind consideration.

Cordially yours,

[signed]
Paul C. Maloney
Vice President, General Projects Director, ASSIST

Source: Holy Shroud Guild archives. Sent by Giorgio Bracaglia via email on May 24, 2020.

Comments: The "forthcoming paper on the subject of carbon dating" is his "white paper" referred to previously.

Maloney's warning that it could be another ten years before C-14 could do what the Church wanted it to do is especially interesting in light of some of the later comments made by some of C-14 scientists (see the entries for each year under "2008-2010" further below).

Maloney sent Pope John Paul II the next day a letter with similar content. Maloney added in his letter to the Pope, "In view of your recent pronouncement authorizing the C-14 testing to be done on samples from the Shroud of Turin and in further light of the information contained herein, it would thus seem that the methods which are apparently going to be sued *[sic -- should be "used"]* may also cast reflection on the Church and Your Holiness."

1988 (March). Art historian Dutton sent yet another letter to *Nature*, published on March 24th: "SIR-Both the scientists involved and outside observers such as myself have been astonished at the recent decision of the Archbishop of Turin, Anastasio Ballestrero, to withdraw from four of the seven participating laboratories permission to carbon-test the Shroud of Turin."

"This action, supposedly made in the interests of conservation of the shroud linen, leaves in a shambles the carefully devised plans of the group of experts who met in the autumn of 1986 to draw up testing procedures for the cloth. As things now stand, only laboratories at the University of Arizona, the Technical University in Zurich and the University of Oxford will be given shroud samples. Shut out from the tests will be Dr Harry Gove of the University of Rochester and Dr Garman Harbottle of the Brookhaven National Laboratory, as well as the Saclay Laboratory in France and the Atomic Energy Research Authority in Harwell."

"Of equal importance is the fact that the Vatican officials in charge of the test have still not come forward with procedures to secure the authenticity of the samples themselves -- procedures, for example, to make it impossible for ancient mummy linen to be surreptitiously introduced into the chain of evidence. If the shroud linen is itself of ancient origin, but the tested samples are not provably from the shroud, then there will be no reason for anybody to take the test results seriously."

"I call on all the concerned laboratories to withdraw from the tests until such time as the Vatican decides to go back to the seven-laboratory plan, with strict, open procedures to ensure the authenticity of samples."

DENIS DUTTON School of Fine Arts, University of Canterbury, ChristChurch, New Zealand

Source: "The Shroud of Turin." [Letter], *Nature*, **332**, 300, (1988).

Comments: Dutton's statement "If the shroud linen is itself of ancient origin, but the tested samples are not provably from the shroud, then there will be no reason for anybody to take the test results seriously" also works the other way: "If it is of **recent** origin, but the tested samples are not provably from the cloth, then there will be no reason for anybody to take the test results seriously." There is no reason it should apply to one situation but not the other. There was no way the labs that were chosen would have withdrawn in protest of four labs being dropped.

1988 (March). Gove wrote, "Otlet said that Teddy Hall and Sir David Wilson *[Wilson was director of the British Museum]* were members of the millionaire's club so that one had to be very careful in dealing with them. (He may have to be but I do not.) He said he was still worried about the possibility of collusion between the British Museum and Oxford."

Fr. Rinaldi called Gove on March 17th to say the mystery had deepened. Turin was being totally silent and Gonella was "speaking in riddles." Apparently every move from that point on was going to be secret. STURP would play no role in the dating and samples would be taken in the next few days. Wilson found out from Gonella that STURP would not be allowed to do any testing until after the C-14 dating, which Wilson thought unsound since he believed that the C-14 dating should be done in conjunction with other tests. Gove asserted that those tests appropriately should not proceed because the approach would be different depending on the age.

Gove revealed, "On 18 March I talked to Ted Litherland. He had just come back from a trip to Oxford and he said that he had had a grand dinner at Teddy Hall's house but Hall seemed a bit ill at ease. He was defending his stand on the whole carbon dating enterprise on the grounds that Tite was involved. There was a four-year plan that was prepared for Hall's Oxford lab, that stated Oxford was 'chosen' to date the shroud and this was reason enough to support the lab. Ted said Hall would be retiring in the next year or so and Tite was the top contender to replace him. That was the first time I had heard of this possibility and it turned out to be true. No wonder Tite made no objection to only three labs being involved as long as one of them was Oxford!"

On March 24th, Dutton, the New Zealand art historian, had a letter-to-the-editor published in *Nature* expressing the concern that there were no procedures in place to guarantee the authenticity of the samples, so that for example, Egyptian mummy linen couldn't be substituted. Gove felt that Dutton was "snatching at straws."

That same day, Gove completed the final version of the letter he was planning to send to the Pope. Gove knew it was a lost cause but proceeded anyway because he didn't want to leave any stone unturned. The letter included the background of Gove's involvement, the summaries of all the meetings between 1977 and the London meeting of January 22, 1988, the changes in the Turin workshop demanded by Gonella; it encouraged the Pope to advise Ballestrero to revert to the original protocol. It also included a copy of the cable to the Cardinal warning him of the possible severe consequences of changes in the protocol as well as the letter from Ballestrero to all the workshop participants regarding the implementation of the changes. It was mailed on March 25th.

On that day, Gove sent a copy of the letter to Senator Moynihan. Because Gove was worried that his own letter wouldn't reach the Pope, Gove appealed to Moynihan to send a copy of the letter and materials to the U.S. Ambassador to the Vatican so that he could alert the Vatican Secretary of State.

On March 30th, Gove wrote to Donahue that it appeared certain that only Arizona, Oxford and Zurich would be permitted to date the Shroud. Gove indicated that he would accept Arizona's previous invitation to attend their dating process

Source: Gove's book: Relic, Icon or Hoax?: Carbon Dating the Turin Shroud, (Bristol and Philadelphia: Institute of Physics Publishing), 1996, pp. 234, 247-250.

Comments: Gove suggested here that Tite's "top contender" status to replace Hall at Oxford might have influenced Tite's acceptance (representing the British Museum, overseer of the testing) of the three-lab plan that included Oxford. This was a clear conflict of interest.

1988 (March). According to Petrosillo and Marinelli, "Cardinal Ballestrero had yet to give his approval to the procedures agreed on January 22. Gonella was having second thoughts and was faced with three choices: a) continue with the three laboratories; b) replace one of the three with another that used proportional-counter; c) add a laboratory that used a proportional-counter. But these last two possibilities would have required more material and Gonella had already granted twelve milligrams of more fabric for each laboratory than the Turin protocol. In fact, each laboratory would have a sample of forty milligrams (x three = one-hundred and twenty mg.) while with the Turin protocol everyone was to have received twenty-eight milligrams. In total (multiplying by seven) two-hundred milligrams of tissue would have been taken (= fifty mg. of carbon). Gonella remained firm in the decision to invite three laboratories. At the end of March the conditions for carrying out the sampling were verified and the work plan, established in January, was confirmed. Tite, at that point the only guarantor of the operation, communicated to the three laboratories the definitive protocol for radiocarbon analysis. The terms were disclosed in a letter published in *Nature*." *[See full text of letter under "1988 (April)."]*

"Tite had not set a deadline for the end of the operation. The reason is clear, given that the excision had not yet taken place. However, it was significant that the indeterminacy was interpreted in other ways. Sox felt that a Shroud dating could reveal an erroneous date as in the 1985 experiment. According to Bro. Bruno, however, this indeterminacy served to mask a well-regulated plan: the dating would not have occurred simultaneously but one after the other, after an exchange of information among laboratories. Gove then sent another letter to *Nature*. *[See full text of letter under "1988 (May)."]* Tite, meanwhile, was getting increasingly agitated over the way the preparations were going. The past fall, when Ballestrero expressed the opinion that the presence of the representatives of the labs for the sample-taking was not necessary, Tite reacted harshly. He asked if he and Chagas were expected to bring the Arizona sample to Tucson to ensure that the chain of evidence was not broken. Tite started to say that he has had enough of the Shroud and was looking forward to the whole matter being concluded. Apparently convinced that he was dealing with a medieval fake, Tite was surprised by the 'fanatic' enthusiasm that some Americans showed toward the relic while acknowledging that the date wouldn't affect someone's religious beliefs. About the time that Tite had written his most recent letter to *Nature*, he asked Evin for a medieval sample of about six square centimeters, equal to one-hundred and twenty milligrams, as similar as possible to the Shroud, ivory linen with a V-shape and between the thirteenth and fourteenth centuries, preferably the latter." *[Evin and Vial gave a different account, claiming that Gonella had asked for the sample. Additional details can be found in various other entries.]*

"Tite sent photographic enlargements of the Shroud as a model to Evin. They unsuccessfully asked the Cluny museum for the necessary amount of fabric to be divided into three 40 milligram fragments as foreseen by Tite in the letter published in *Nature*. Evin found the sample suitable for the requests received an English colleague in a cope preserved in the basilica of Saint-Maxim-la-Sainte-Baume in Provence. The liturgical cloak was worn by Saint Louis of Anjou, son of the King of Naples, Charles I, and nephew of the King of France, Saint Louis IX. Appointed bishop of Toulouse in 1296, Luigi died the following year at the age of 23. The cope is therefore perfectly dated: 1296-97. But it is not embroidered on an inverted herringbone linen, as Bruno Bonnet-Eymard wrote at first. The base of the cope is made up of a simple orthogonal canvas, completely embroidered up to the edges with evangelical scenes. The background of the figures consists of an inverted herringbone gold embroidery so worn that in some places the linen canvas is entirely uncovered. The fact remains, as Bonnet-Eymard points out, that his date was precisely that expected by Gove, McCrone and Sox."

"Tite, unable to go to France to personally collect the sample, asked Evin to send it to the British Museum by mail. For what purpose? For Bonnet-Eymard there was no doubt. Tite wanted to replace it with that of the Shroud when he retired to a room together with the Cardinal to package the samples. Fearing post-strike strikes, Evin *[in this version -- there are others]* gave it to Vial to hand-carry to Turin on April 21st. This upset Tite's plan. In reality, this sample was not suitable for replacement; however, the question remains as to the reasons for its appearance at the last moment among the control samples and a perfect coincidence of the dates shown for the Shroud and the Cope. For the ^{14}C they are two 'twins'. The removal of the threads from the cope was carried out in the absence and without the knowledge of the parish priest of Saint-

Maximin, who was surprised and upset. He discovered the removal while mending the old vestment. He asked what had been taken. He was shown a closed envelope, saying that only a few threads had been removed."

Source: Petrosillo, Orazio and Emanuela Marinelli. La Sindone: Storia Di Un Enigma, (Milan: Rizzoli, 3rd edition, 1998), pp. 156, 158-160.

Comments: Tite's attitude can certainly be criticized. Before the sample has even been taken, he was wishing the project was over. For such an important exercise, one would have hoped for a more committed overseer.

Much has been made of one of the control samples being 14th century, which led some people to believe that a sample-switch was involved. The fact that everything in the process was recorded on video EXCEPT the placing of the samples into the steel cylinders, did nothing to discourage such a belief.

1988 *(April)*. Petrosillo and Marinelli recounted, "The working protocol was approved by the scientific and religious supervisors in the latest version and started running from April 16. This protocol was dated: Turin, April 15, 1988. It consisted of twenty-six manuscript pages by Riggi. The role of each participant is meticulously established. Nothing was missing for the course of the day, apart from the most important thing: the indication of the site of the withdrawal. Riggi also thought of the emotions of the present. And, in fact, it dispensed recommendations to the participants: 'The impact with the Holy Shroud is not simple and can lead to internal upheavals and unforeseeable sensations. So please focus on the operations as much as possible. Let's reserve the time for the emotion and the conscience at work at the end of the operations.' Bonnet-Eymard found it surprising that Riggi, a member of STURP, made himself an instrument of Ballestrero and Gonella to discard the 1987 STURP protocol. In addition to the Riggi operations group, there was to be two control persons (Gonella and Tite), two textile experts (Testore and Vial) and four priests. Between April 16th and 20th about two tons of equipment were unloaded to the sacristy of the cathedral."

Source: Petrosillo, Orazio and Emanuela Marinelli. La Sindone: Storia Di Un Enigma, (Milan: Rizzoli, 3rd edition, 1998), pg. 163.

Comments: To continue this thread, go to the Petrosillo/Marinelli "La Sindone ..." book entry under "1988 *(April 21st)*."

1988 *(April)*. Robert Hedges, one of the Oxford C-14 scientists, was asked how confident he was of being able to establish the Shroud's age, admitted, "I wouldn't bet my life on it."

Source: Glass, Robert. "Modern Technology May Finally Fix Age of the Shroud of Turin." *Chicago Sun-Times*, April 8, 1988, pg. 4.

Comments: Despite Hedges' caution, when the dating results were released in *Nature*, with Hedges being one of twenty-one authors, it was claimed it was with a 95% confidence level.

It should also be noted that according to the Gonella/Riggi book (pg. 123), after the Shroud had been taken out for the testing, its container was treated with the chemical thymol to eliminate mites and other parasites. Some scientists that later heard about this were horrified, because they believe that action would render any possible future C-14 testing of the Shroud impossible. This was another example of the Church making a bad decision.

1988 *(April)*. Meacham wrote, "One certainly felt that an historic occasion was approaching; I doubted that a bulls-eye date of first century would be obtained, but was cautiously optimistic that a result would indicate some antiquity for the Shroud, perhaps back to the 4th or 5th centuries, owing to some intractable contamination. This could be taken as a good indication the Shroud was the genuine article. I made one last effort to persuade Gonella of the need for a small team of advisory archaeologists that I had suggested, particularly for the selection of sampling sites. Again, I put it to him in the strongest terms that a minimum of two sites was needed. By this stage he was not listening, and I did not receive a response. Fr. Rinaldi kept me informed what he learned of the developments, and it was clear that Gonella was proudly running the show. No one knew just how much he was going to ruin it, but there was a shadow in my mind of continuing nagging worry that he would take the sole sample from a bad location, and the Shroud could be assigned an incorrect age."

Source: Meacham's book. The Rape of the Turin Shroud: How Christianity's most precious relic was wrongly condemned and violated, (Lulu.com), 2005, pp. 87-88.

1988 (April). Text of Tite's letter to *Nature*: "Sir —Following my letter of 11 June 1987 (*Nature* **327**,456; April 7, 1987), I am now able to provide an outline of the procedures that have been finally agreed for the radiocarbon dating of the Shroud of Turin. Of the seven original offers to undertake the dating of the Shroud, three have been accepted by Cardinal Ballestrero, Archbishop of Turin and Pontifical Custodian of the Shroud. The radiocarbon laboratories concerned are at the University of Arizona, the University of Oxford and the Federal Institute of Technology in Zurich, and each has now agreed to proceed with the project."

"Each laboratory will be provided with a sample from the shroud, together with two known-age control samples, one of which will have been independently dated by conventional radiocarbon dating. The shroud samples will be taken from a single site on the main body of the shroud away from any patches or charred areas. In order to ensure that ample carbon for dating survives after pretreatment, the weight of each cloth sample (that is, shroud and controls) will be 40 mg. All the samples will be given to the laboratories as whole pieces of cloth without being unraveled or shredded. A blind test procedure will be adopted in that the three samples given to each laboratory will be labelled 1, 2 and 3 and the laboratories will not be told which sample comes from the shroud. Even if the samples were shredded, it would still be possible for a laboratory to distinguish the shroud sample from the others. It is therefore accepted that the blind test depends ultimately on the good faith of the laboratories."

"The removal of the samples from the shroud will be undertaken under the supervision of a qualified textile expert. These samples will be weighed, wrapped in aluminum foil and sealed in numbered stainless steel containers. The control samples will be similarly treated. All these operations will be watched over and certified by Cardinal Ballestrero in collaboration with myself. After they have been packaged, we will immediately hand over three samples (shroud and two controls) to representatives of each of the three radiocarbon dating laboratories who will be in Turin for this purpose. In addition, all stages will be fully documented by video film and photography."

"On completion of their measurements, the laboratories will send their data for the three samples to both myself at the British Museum and to the Institute of Metrology 'G. Colonetti' in Turin for preliminary statistical analysis. The laboratories have agreed not to discuss their results with each other until after they have deposited their data for statistical analysis. A final discussion of the measurement data will be made at a subsequent meeting in Turin between representatives of these two 'institutions' and representatives of the three laboratories at which the identity of the three samples will be revealed. The results as finalized at this meeting will form a basis for both a scientific paper and for communication to the public. The timetable for the operations has not yet been fully established but it is hoped that a radiocarbon date for the Shroud of Turin will be released by the end of 1988."

Source: M.S. Tite. "Turin Shroud." [Letter], **332,** 482 (7 April 1988).

Comments: Although there was anywhere from ten to sixteen hours *[why can't this be pinned down?]* of video of the various proceedings, the actual placing of the samples into the stainless-steel cases was NOT recorded, a fact that has never been explained and certainly opened the doors for a sample-switch belief.

Although it had been previously acknowledged that the labs would know which sample was the Shroud, Tite still stated that the labs would be informed which sample was the Shroud!

1988 *(April)*. On April 7th, a letter from Michael Tite was published in *Nature*. Tite stated that the samples would be given to the labs without being unraveled, since even if the samples were shredded, the labs would be able identify the Shroud sample. Tite was quoted in a newspaper article saying that if the date came out medieval, it couldn't be the burial cloth of Jesus. If it came out about two-thousand-years old, it would only show that it could be genuine. Gove wrote a letter to *Nature* in response to Tite's letter published April 7th. *[Gove's letter was published a few weeks later -- see further below.]*

Gonella was quoted in another paper as saying the authorities had never officially agreed to allow more than three labs to participate. They didn't see it as reducing the number from seven to three but rather increasing from one to three.

On April 16th, an article was published in *Science News*, which expressed concern about there being only three labs involved. Harbottle was quoted as saying there was a one in five chance in any given measurement that the answer would be wrong. With only three labs, it could be difficult to determine which lab had the bad reading.

Just about the time that the samples would be taken, Fr. Rinaldi called Gove. There was complete secrecy about the sample taking. The original April 25th date was cancelled, and it wasn't known if the removal would happen before or after that date. No one except Tite, Gonella, and an Italian textile expert would be involved in actually cutting the Shroud, which would be videotaped. Rinaldi indicated that Wilson had called Gonella around April 7th to inquire whether STURP could do some measurements before the C-14 date was announced. Gonella said there was no chance of that -- STURP would not be involved in the sample taking or in taking any measurements before the dating.

Source: Gove, Harry. Relic, Icon or Hoax?: Carbon Dating the Turin Shroud, (Bristol and Philadelphia: Institute of Physics Publishing), 1996, pp. 250-252.

Comments: Even though it was acknowledged that the labs would be able to identify the Shroud, procedures for the "blind testing" would be kept in place when the samples were taken on April 21st.

Tite's remarks about a medieval date disauthenticating the Shroud and a two-thousand-year-old date not authenticating the Shroud shows the dating was really a no-win situation for those who believed the Shroud to be authentic. Since the labs had total faith that C-14 couldn't be wrong and because no amount of scientific data, at least in some peoples' minds, could not prove the cloth genuine, Shroud advocates found themselves between a rock and a hard place.

Regarding the Church's stance on the number of labs, since so many changes were made, it seems pointless to designate what was official and what wasn't. Despite years of planning, the whole enterprise was a fiasco. And given Gove's statement that the results will be "vastly less credible" because of changes, one has to ask again how the labs could proclaim that the results they produced had a 95% confidence rate. If no changes had been made, would they have proclaimed a 100% confidence rate?

True to form with other numerous changes in procedures, the specifics of the sample-taking promulgated shortly before the actual event, did not actually happen. Unfortunately, Gonella's statement about STURP not being involved, did turn out to be true.

Part II
Day of the Sample Extraction, April 21, 1988

1988 *(April 21st).* The following account is from the Gonella/Riggi book. As agreed previously, Ballestrero personally invited Chagas to attend the sample-taking, and he sent the deputy director Dardozzi to represent him. Laboratories representatives were allowed to attend the event as guests of the Ballestrero, due to their willingness to come to Turin to take samples to guarantee the maximum documentary chain. Tite found it very difficult to find a medieval control sample and asked Evin if he could find one in France with the good offices of the Lyon textile center; Vial, through an official request of Turin, obtained permission to make a withdrawal from the St. Louis d'Anjou Chapel preserved in the Basilica of St. Maximin in Provence.

Another Italian group was involved in the process of taking the samples: the "Ministry of Cultural Heritage and Activities and Tourism," which under national regulations, all restoration work is subject to prior approval. The Shroud chapel at St. John the Baptist Cathedral was included under their jurisdiction. The superintendent and a colleague arrived very early on the day of the excision. The superintendent told Gonella that she might have some objections. When Cardinal Ballestrero arrived, Gonella informed him of the possible snafu. Ballestrero firmly reminded her that the Shroud was the property of the Holy See, and the operations took place with the supervision of the Pontifical Academy of Sciences. The superintendent and her colleague remained in the room. The former, in fact, insisted that the lighting in the room be reduced so as to lessen the danger to the cloth. No quantitative data were provided

-- it was only a subjective assessment by eye -- nor were they told how many lux were "safe" and how many were "dangerous". However, a disagreement at that time would have been inappropriate and the Cardinal ordered that some lights be turned off, which made the work of Riggi more difficult. The textile experts present agreed that the site of minimum damage and risk was on the corner of the main body of the sheet to the left of the frontal image, where in 1973 the Raes sample had been removed. Gonella noted the cloth was already disfigured by an ancient part of the side strip, and there was no selvedge whose cut could cause damage. The chosen area and the surrounding ones were carefully examined to verify the absence of foreign material or junctions.

It was planned to take a net strip of approximately 1×7 cm^2 for the laboratories to be able to supply the required forty mg of cloth, taking into account the possibility that the area mass in the sampling area was lower than the average, but a larger fragment was cut to keep a reserve sample available to Ballestrero for further exams in the event of disputes. The presence of a reserve sample, known to the labs, was also a further insurance against any temptations of someone to "adjust" the measurement data. The figure of 1×7 cm^2 has often been incorrectly reported as concerning the entire cut. Since the cut was made near the edge a wide margin had to be kept to eliminate the edge strip containing the seams, in order to avoid any possible contamination by possibly younger material; in fact the area mass of the sampling area was greater than the estimated average for the Shroud, and the fragment cut out was larger than originally envisioned.

As control samples, Tite had brought a fragment of 1st century linen and another dated to about the 11th century, which he had managed to find in the last days. Not having informed the French that he had also found a medieval control sample, Vial had taken the one taken from St. Maximin and decided to use them both; the French sample was intrinsically dated with better precision, but it was frayed and therefore immediately recognizable in front of the others presented as fabric; given that they had prepared, as per the specifications, three cylindrical containers signed for each laboratory, the Shroud sample and the two supplied by Tite were placed in the containers, and the French sample was added to the box in a separate envelope. As a concession to the laboratories' desire to maintain the "blindness" of the examination at least on the formal level, following the specifications, the samples were placed in the containers arranged in a secluded room (the Chapter Hall adjacent to the sacristy) in the presence only of Ballestrero, Gonella and Tite. Gonella and Tite each kept separate and unique notes on which numbers corresponded to the Shroud and which to the two control samples. The containers were then packaged in the sacristy sealed with the seal of Ballestrero and delivered by him personally to the representatives of the three laboratories, together with a certificate of origin that the said representatives signed for to confirm receipt. Later, some researchers, because of the lack of a presence of a notary, would assert that the dating was thus invalid. Gonella countered that it was a scientific operation and not a legal act -- nor was any report requested by anyone for the operations carried out in the laboratories. From the documentary point of view, Gonella felt that the continuous video of the operation was much more informative and reliable than any "report." On that occasion, Ballestrero wanted to avoid any impression of pressure or checks on the examiners. The fragment taken from the reserve was entrusted by Ballestrero to Riggi and Gonella to keep it at his disposal; given the attitude of the Superintendency of Cultural Heritage, it was not considered prudent to leave it in the Chapel of the Shroud, which falls under its jurisdiction. It was immediately sealed in wax with their seals and placed in a safe in the bank.

Source: Gonella, Luigi and Riggi di Numana, Giovanni. La Sindone: Storia Di Un Enigma, (Milan: Fondazione 3M), 2005, pp. 73-76.

Comments: The episode of having to reduce the lighting is another example of poor planning. I asked Bill Meacham, who attended the 1986 workshop in Turin, if the lighting issue had been discussed. He told me in an email of April 30, 2016 that it had not. He also said that during the 2002 restoration of the Shroud, lamps were left for hours on the Shroud, even when no one was actually working on it!! Meacham also said that in an interview for Italian TV "Ghiberti is talking about the 'conservation of the Shroud' while in the background a lamp shines right on the cloth, just around 12 inches away from it!" Riggi described the 1988 cutting as being, "in a generalized semi-darkness." Riggi mentioned in the Gonella/Riggi book (pg. 123) that the change in lighting put Testore, Vial and himself in "great difficulty." Riggi would note after the dating that he excised some foreign fibers, and the reduced lighting made it harder for him to detect those.

Over the years during interviews, Tite consistently only mentioned himself and the Cardinal being present for the placing of the samples in the containers. Gonella had consistently maintained he was present for it. The discrepancy is difficult to reconcile. Interestingly, Riggi only mentions the Cardinal and Tite in one account, pg. 110 of the Gonella/Riggi book, but on pg. 141 he says that "the hall was then cleared of strangers, the door was closed and what happened inside is known only to Prof. Tite, to the Cardinal and to Prof. Gonella."

As previously noted, there are reasons to doubt that the examiners could have determined just with the naked eye that there was no foreign material in the sample area. Gonella talked about having supposedly verifying the absence of foreign material but then acknowledged the possibility of younger material -- possibly alluding to recent repairs made in the area. The fact that Gonella said that the area mass of the sampling area was greater than estimated average for the Shroud suggests that the sampling area did have foreign fibers. In addition, Riggi (co-author of the book with Gonella!) stated specifically that he removed foreign fibers from the excised sample!

Although Gonella believed that the video recording of events was more reliable than a report, it must be pointed out that this did not prevent numerous discrepancies in the data, and it must also be pointed out again that the placing of the samples into the containers was not recorded on video.

1988 *(April 21st)*. Petrosillo and Marinelli wrote, "The sample-taking was carried out with a lot of improvisation. Over thirty people were involved, including eleven operators, two consultants, the supervisor, five clergy, the technical referee, five representatives from the three laboratories (who in theory were supposed to have carried out the test blindly), and envoys of the Ministry of Cultural Heritage. Chagas declined Ballestrero's invitation to attend; Chagas sent Msgr. Dardozzi to represent the Pontifical Academy. Shortly after six a.m., Ballestrero, Tite, Hall, Hedges, Wölfli, Damon and Donahue arrived. Hall reported that upon arrival they were informed of the arrival of a '**political**' *[my bolding]* obstacle and that the excision could not take place. Gonella completely denied these claims. Scientists take their places in the canons' stalls while Tite stood near Gonella, Riggi and the two textile experts, Testore and Vial, who see the Shroud for the first time. *[!]* Sox narrated in his book that Testore, pointing to the wound on the side asked 'What's that large brown patch?' Sox commented that Testore's knowledge of the Shroud was rather limited. Evin's presence at the sampling was debated. Tite denied he was there, but Evin claimed he was a witness, although arriving late. No decision had been made regarding the sampling site. *[!!]* Testore and Vial discussed it with Riggi and Gonella for over an hour. Riggi reported that there was 'a wide consultation of textile experts and controllers.' Vial had no advice except to suggest cutting on an edge so as not to damage the image. According to rumors that circulated in France, Vial tried to warn that the area that was chosen could well have been a repair of the fabric and not part of the main body of the Shroud."

"The criteria for the collection were subsequently outlined by Testore during the international symposium held in Paris September, 7-8, 1989. The fragment was to be part of the main body of the sheet and the risks of contamination due to repairs had to be avoided, including the lateral strip, and the partially-carbonized areas. It was not supposed to come from the image areas. The sample taken should not have been too much at the margins in the sense of height, because the edges were more manipulated and therefore had a greater risk of contamination. The dimensions had to be as small as possible, while taking into account the quantity necessary for the three different laboratories and the possibility of retaining a part that could have served to carry out further research, both radiocarbon, if the former had given conflicting results, and for other analyses, without being forced to make new cuts. From an aesthetic point of view, the sampling had to cause the least possible damage. After about four hours of analysis, the fragment was cut. It was taken from a single spot, without taking into account the fact that a single site is not necessarily representative of the whole object. It was cut adjacent to the 'Raes corner,' from which a piece had been taken in 1973, but a terrible choice, since it was a point most exposed to contamination."

"It was, in fact, at one of the two corners from which the Shroud was held up during the exhibitions. The piece of cloth was also taken a few centimeters away from one of the charred points from the drops of molten silver that fell during the fire of 1532. Furthermore, the fragment comes precisely from one of the edges stained by water used to put out the fire, where the pyrolysis products have accumulated and various materials of centuries has been deposited. Riggi detected a sort of electrophoresis in the medulla of the flax fibers of the sheet. Instead, according to the scientists of the laboratories, the chosen site was optimal: 'It is a single area of the main body of the Shroud far from any patched or charred area.' So they will write in the final report published in *Nature* on February 16, 1989. Evidently they either ignored or underestimated the contraindications for the radiocarbon investigation concentrated at that point on the sheet. Many presumed that the cut would be made near Raes' withdrawal. But, precisely because of the very high probability of contamination, STURP had recommended that the sample be taken in at least three different areas of the cloth."

"The cut was made by Riggi, assisted by the two textile experts. It took place between 10 and 11 under the supervision of the guests. Riggi proved to be skilled with scissors in detaching the left corner of the fabric. He did not wear gloves. A photo also portrays Cardinal Ballestrero with his elbows resting on the Shroud. The *Nature* report spoke of 7 x 1 centimeters cut above the point of the 1973 sampling. But in reality it was not above but at the side. And the surface of fabric actually removed was 8.1 x 1.6 cm."

"When it was detached, the piece of Shroud fabric was immediately placed with a pair of small tweezers on a precision balance. How much did the piece weigh? Riggi provided two different weights in the same report presented at the Paris symposium: 497 and 540 milligrams. In the film, you can instead read the weight on the scale: 478.1 milligrams. Three different values of the weights do not end the inconsistencies. Testore states that the piece just taken was about 8.1 x 1.6 centimeters, therefore 12.96 centimeters and that the unit weight of the Shroud fabric is 23 mg/cm^2. Since the latter value is considered correct, with a simple multiplication, it is deduced that the Shroud fragment must have weighed 298 milligrams. The real weight, however, both considering Riggi's data and that which appeared on the scales, is in any case almost double. The accounts do not match! It was true that Testore spoke of a great approximation, but here the gap was too large to be acceptable. Held with tweezers, the sample was shown to the C-14 representatives. In English, Riggi announced: 'Here is your sample!' Then he cleaned the piece, removing the external parts, a long loose thread and the jagged edge."

"According to the description that follows, the part eliminated is noteworthy. These parts were deposited in a container, being able to serve as an extra sample. At this point, Riggi reported that the cleaned piece weighed exactly 300 milligrams and measured 7 x 1 centimeters. Obviously, considering the aforementioned unit weight of the Shroud tissue (23 mg/cm^2), also in this case the real weight is almost double that expected. This inconsistency is incredible and inexplicable. An obvious question arises: but did the finished piece on the scale belong to the Shroud? Because, by weight, one cannot think of a fraudulent replacement of the sample in the time between cutting and weighing. So? The next thought can only be the following: the fragment removed could belong to an area of fabric restoration. Riggi wrote that in the eliminated part there were 'threads of other nature that even in minimal quantities could have led to variations in dating, being of late addition.' Have all foreign threads really been eliminated? In Tucson, a red silk thread and blue fibrils were found. Is it possible that Riggi didn't notice? In the Gonella/Riggi book (pg. 130), Riggi noted that the combination of seams and darning made the fabric thicker and made the intervention more difficult."

"He declared: 'It was not possible for me to ascertain the presence in that place of colored threads of any type because the very accurate observation made by me, by the textile experts and by all bystanders did not highlight the presence of threads, fragments, or anything else recognizable by different colors and inserted or placed on the removed fragment.' Could there have been an invisible mending? If Riggi and the others did not see the colored threads, a mending well-done might have escaped them! Riggi himself recognizes that the stitching threads used to join the Shroud to the Holland cloth in color and size are well masked with the Shroud fabric; so much that he regrets not having kept them to study them. Moreover, the pollutant found by STURP on the fragment of Raes originated from joint operations on the cloth over the course of centuries."

"In the final report of the laboratories, which appeared in *Nature*, we read that three samples were obtained from the strip, each of about 50 milligrams. Riggi reported that the piece taken was cut 'in parts', of which three equivalents weighing just over 50 milligrams; in another moment he specified that the parts were four, of which three were about 52 milligrams. In fact, the 7 x 1 strip was divided into two almost equal parts: one, weighing 144.8 milligrams according to Testore and 141 milligrams according to Riggi, was kept for any other tests. Seven years later, Gonella instead claimed that the preserved part was a quarter of the sample. The incredible consequence of this story is that Riggi kept the piece left over. 'We later handed it over, sealed, to Cardinal Ballestrero,' added Gonella. And Riggi specified the place of delivery: 'In the monastery of Bocca di Magra where he retired. If I remember correctly it was 1990. It was then passed on to Cardinal Saldarini's control.' For the Turin engineer, any Shroud sample should have been kept by scientists only for the time necessary to carry out the analyses."

"The other part of the strip taken on April 21, which weighed 154.9 milligrams, was divided for the three laboratories into three almost identical fragments: 52.0, 52.8, 53.7 milligrams. Riggi says in minute detail: 'By chance, each of these three parts is identical to the others because the weight of the three fragments on an electronic scale varied by about a thousandth of a gram for each piece and was equivalent to almost 53 mg on average for each.' However, the sum of these three weights, in reality, is 158.5 milligrams; therefore, the total weight is higher than that of the piece that has been divided. How is it possible? The accounts don't match this time either. Prompted to give an answer, Testore and Riggi, without getting upset, changed their version admitting that things did not happen exactly like that. For Testore, the portion chosen for the division into three parts was not the largest (154.9 milligrams), but the smallest: surprising decision, given that its weight,

144.8 milligrams, was certainly insufficient to give three parts from 50 milligrams each. The cut was markedly unbalanced: two fragments were abundant (52.0 and 52.8 milligrams), the third too thin (39.6 milligrams). They then supposedly resorted to the preserved piece, the one that was initially larger, and removed a thin strip of 14.1 milligrams which went to replenish the underweight fragment. One of the three laboratories, Tucson, therefore received two rectangles of Shroud for a total weight of 53.7 milligrams. Certainly it would have been more logical, at this point, to take an entire piece from the reserve portion to provide the third laboratory with a single fragment as well. The piece preserved at this point was reduced from 154.9 to 140.8 milligrams, and this would explain the weight of 141 milligrams attributed to him by Riggi. The latter, however, in giving the new version, contradicted Testore's variations by stating that the piece to which the three samples were obtained was the largest (154.9 milligrams). It being understood that two of the fragments obtained weighed 52 and 52.8 milligrams respectively, a small piece of 3.6 milligrams was added to the third, taken from the reserve part, thus bringing it to 53.7 milligrams. It can be deduced that the third fragment was 50.1 milligrams, a satisfactory weight for the requests of the laboratories. The need for the addition is therefore not justified. What a great confusion! Who to listen to? There is also contradiction in Riggi's reports on which side of the original strip had been divided into three, the innermost one with respect to the edge of the Shroud or the outermost one. Riggi also cut three fragments from the two samples brought by Tite, who knew where they came from and dated, for the triple-blind experiment."

"The first sample came from a tomb discovered in Nubia, and was dated to the 11th-12th century on the basis of the drawings and inscriptions on the linen from which it had been taken; the second one belonged to the mummy of Cleopatra of Thebes of the beginning of the second century AD, previously dated in the British Museum with the traditional method at 110 BC-75 AD. Even these fragments had a weight similar to that of the Shroud, 52-57 milligrams. Away from the scientists, in the adjoining chapter house, samples were placed in aluminum foil and subsequently sealed in nine numbered 'bomb proof' stainless steel containers, fire-resistant and waterproof up to 1000 meters deep. The operation, according to the official statement, was carried out by the archbishop of Turin and Tite. But the two were not actually alone. In the mixing Gonella and Riggi participated in these conditions, after it was admitted that the blind test was impossible and that the labs' representatives had seen everything. Why did the mixing scene take place separately? Bonnet-Eymard raised the serious suspicion that the staging was used by Tite to proceed with the replacement. 'His plan was to show the Shroud to the carbonists, so that with the fiction of the blind test, he could replace the Shroud sample with one similar to what the scientists had seen. They would continue to testify in good faith that it was the Shroud, while announcing that it was 600 years old. Tite has focused on replacing the sample.' Bonnet-Eymard's accusation clashes with the objective difficulty of carrying out such a project." Turin chemist Piero Savarino, who would later become the scientific advisor to Cardinal Poletto, a successor to Cardinal Ballestrero, commented, "Unfortunately, a set of facts, or rather of deficiencies and carelessness, let the suspicion survive" *[Marinelli's Valencia paper, pg. 10]*.

"At that point the fourth sample entered the scene, not foreseen by the protocol and delivered to the laboratories in an anomalous way and without those reserved mixing procedures used for the others. The contradictory information of those who wrote about the dating story is the result of the non-protocol circumstances of the acquisition of this additional sample perfectly coherent with the Shroud according to the dating made by the laboratories. Tite was photographed with nine steel containers (Shroud plus two control samples for the three laboratories), but later he will claim that there were four fragments inserted in the appropriate metal containers. Evin told of four anonymous containers and alters the official report stating that all three control samples were wrapped in aluminum foil like the Shroud. The fourth sample was delivered after Tite had mixed the others and could not be used for replacement also because it was in threads. The British Museum representative stressed this last detail to ward off suspicions, but this is enough to clarify everything. Why was the fourth, unscheduled sample brought when Tite had procured the two necessary control samples? Why on the day of the excision from the Shroud, did Vial keep the fourth sample in his pocket all morning in order to deliver it immediately? Or did Evin, who came in late, bring it? Why was it accepted even though it was in threads? And that radiocarbon age surprisingly equal to that of the Shroud? The official samples were placed by Riggi in front of the Cardinal for the gluing of the labels to the containers." Riggi would later comment "Who fantasized and was not soft in criticism and accusations, perhaps was not entirely wrong; because without documents to rely on, every fantasy was possible, every doubt was permissible and every conclusion, incorrect or unjust, when not authoritatively contradicted, could be reasonable" (Marinelli's Valencia paper, pg. 9).

"The Shroud fragments were certified by the Cardinal, the control ones by Tite. The complex ceremony continued: Riggi affixed each label by sealing the package with red sealing wax and the seal of the Cardinal archbishop of Turin. The sealing wax is melted with a small flamethrower. 'Modern technology mingles with tradition' noted the cameraman. The laboratories had asked for this too: a presence of antiquity in the modernity of the operation. Ballestrero monitored the operations from start to finish. Sealed and boxed three by three, the cylinders containing the fragments were delivered to the representatives of the three laboratories. *Nature* stated that 'the three containers holding the Shroud and two control samples were then delivered to the representatives of each of the three laboratories together with a sample of the third control, which was in the form of threads.' The word 'together' did not explain how the extra sample was delivered: inserted in the box with the three steel containers or separately? The contradictions emerging from the various accounts of what happened are not of little importance."

"Bonnet-Eymard learned that Vial had presented his fourth sample after the Cardinal had already left. Riggi had become angry about this unexpected arrival while Tite, acting surprised, intended to reject it. Among other things, no containers were prepared for this sample. Vial, thinking of Evin's effort to find the sample, insisted until it was weighed and cut. The laboratories received it in separate envelopes prepared by Tite and Gonella. According to Riggi, however, the extra sample that he does not mention in his <u>Rapporto Sindone</u>, every officially prepared wrapper has been added. Testore also confirmed the detail by declaring that the three sealed containers delivered by Ballestrero each contained four different samples. However, Tite asserted that the fourth sample was not in the boxes and that was why there were only three containers. Evin recalled that Gonella had not warned Riggi and therefore only three containers were foreseen. According to the French participants, Gonella and Tite insisted on the delivery of the fourth sample to the laboratories. The stories also diverge on the division and on the fourth sample which, among other things, should have weighed less than the official ones, given that Tite had asked Evin for one-hundred and twenty milligrams of the material; instead it weighed more."

"Riggi claims to have had this sample, made up of ivory-colored linen threads, in three envelopes already prepared for the three laboratories, each of which was not weighed since Vial had already done so. On the contrary, Testore stated that it was he who obtained, from the threads of the sacred vestment, three specimens of pure linen weighing about 70 milligrams each. Gonella claimed that all the samples were divided in Turin. Riggi specified that the three envelopes were only given to him at the time of the final closure of the containers upon request to the representatives of the laboratories if they still wanted to keep a perfectly-dated sample. At the affirmative answer, the three envelopes were inserted by Riggi in each of the transport boxes, where there were the containers with the other samples, already sealed. The whole operation, except for the mixing 'carried out by people above all suspicions' as Riggi points out, was carried out under the eyes of over thirty people, videotaped and documented by photographs. At 2 pm, the withdrawal phases were over. Ballestrero gave the representatives of the laboratories a letter to allow the passage of the customs without difficulty. No mention is made of the added envelope."

"On this occasion, the textile experts and Riggi performed other operations. Two thousand authenticated photographs were taken. The time between the delivery of the samples and the rewinding of the Holy Shroud on its velvet roller, about 3 and a half hours, is usefully employed, reported Riggi. To answer some urgent questions regarding conservation, Vial suggested consulting Flury-Lemberg. For that purpose, continued Riggi, 'some photographs were taken, microscopic observations of some parts were made and very small quantities of dust were aspirated from the margins that the textile experts had wanted that was unstitched for their analysis.' Riggi is silent, however, that that time was 'usefully employed' for something else. He for his own purposes also removed, with the help of an adhesive tape, a small sample of blood from the occipital area of the Shroud.' But from the video of those moments Riggi is seen practically face down on the Shroud, to observe and withdraw here and there. Why did he not immediately say that he had done this? It was necessary to wait more than seven years for Riggi to tell the journalist Piero Di Pasquale: 'It was during those hours we took the blood samples in order to find out the genetic characteristics of the man of the Shroud, samples that had been duly authorized, as can be seen from the filming.' Di Pasquale emphasized that Riggi 'performed, with two small pointed scalpels and without using a mask on his face or gloves, the samples on the occipital base of the figure impressed on the sheet and on three other areas of the body in the presence of about twenty other people, nuns, priests.' Gonella cautioned: 'These are not real samples. Microfragments have been collected [...]. They were free blood particles and dust. They were blood particles no longer anchored to the cloth. It is not a question of excisions, but of removal of free or marginal material.' The film, however, shows the fabric that rises under the adhesive tape pulled upwards [...]."

"On the authorization, the Turin engineer had nothing to hide: 'The Cardinal knew that we would also take some micro-material for preliminary investigations useful for designing real targeted-research protocols mainly to establish the best way of conservation. As already in 1978, the only limit obviously placed was that nothing was done to the detriment of the Shroud. Cardinal Ballestrero left around 13.30. Among us there were, among others, the engineer Monsignor Renato Dardozzi, chancellor of the Pontifical Academy of Sciences, and the parish priest of the Cathedral. I have a clear conscience.' However the fact remains that the Cardinal left while the Shroud was still out and everything that happened in the afternoon was not immediately said. And there are no minutes: 'Nobody asked us,' says Gonella. 'In addition,' explained Riggi 'we had to wait anyway, because you could not enter the Church with the Shroud in hand and pass among the people, it would have been really a surprise in a totally confidential operation, not even the public authorities and police, had been warned, so we filled the time with a specially drawn-up program and delivered it to the Cardinal much earlier, approved and then it was carried out'."

"In reality the Shroud could have been closed and sealed in the presence of the Cardinal at 2.00 pm. And anyway, there is a big difference between 'observing' and 'withdrawing'. 'We took two or three fragments of blood from the occipital area,' continued Riggi, 'according to a very old program that we had already set up in 1978, for which we had to test the blood to find out the genetic characteristics of the individual who was been contained in the cloth. This program had been filed, so there had been authorization to do them, and were improvised at that time. The materials were spilled into watertight capsules, protected from light, or initialed with the initials of the Politecnico and my personal ones, and were kept in the bank for a long time, in a bank safe deposit box, with the aim to have the possibility to analyze as soon as there was some interesting project.' A question arises spontaneously: if the samples were taken according to an 'old program because the material was not given to the scientists who wanted do those 'authorized searches,' but was it put in the bank pending a new 'interesting project'?"

"The display case and the accessories were disinfected with thymol. The live observation of the Shroud, aimed at finding the existence of parasites, strangely did not signal unwanted presence. What happened to the mites, 4-5 per square centimeter, observed in 1978? The cloth was returned to its reliquary. Since the winding cylinder is small in diameter, the protective fabrics were wound first to increase the diameter, and then the Shroud with the image on the outside. This was done last time, but the folds increased. The Shroud was relocated to the altar at 7 p.m. The scrubbing of materials was completed and the operation ended at 9 p.m."

"The next day, the Vatican press office issued the following communiqué, taken from the *Osservatore Romano*: 'April 21, 1988 in Turin samples were taken from the Holy Shroud which will be dated with the radiocarbon method. Cardinal Anastasio Ballestrero, Archbishop of Turin and Papal Custodian of the Holy Shroud, attended the operation, officially certified the origin of the samples with the collaboration of Dr. Michael Tite of the British Museum, and personally delivered them to the representatives of the analysis laboratories who came to Turin to receive them in person. The samples, with a total mass of about 150 mg, were obtained by cutting a strip of about 1 cm x 7 cm. In accordance with the blind dating procedure required by the analysis laboratories, three sealed containers were delivered to each laboratory containing the Shroud sample and two control samples, without specifying which sample was placed in each container. The identification of the samples, registered in a special confidential register, will be notified after the execution of the measurements. The control samples were provided through the British Museum and come from a 1st century AD sample and fabric of the 11th century AD; a fourth sample dated to around 1300 AD. was provided as an additional check. The sampling site was chosen in order to guarantee that the sample belonged to the main body of the Holy Shroud and that its removal caused the least possible damage to the tissue. The technical expertise for this choice and the control of the removal operations was entrusted to Prof. Franco A. Testore, of the Polytechnic of Turin, assisted by M. Gabriel Vial of the Musée Historique des Tissus of Lyon. The entire operation was videotaped and documented photographically'."

"Therefore, each laboratory was delivered four samples: the Shroud plus three control samples of which, in spite of the blind test, dates were already provided. It was not explained by where the third control arrives, i.e., the threads of the cope of San St. Louis d'Anjou. Tite had foreseen and brought two controls. Sox in *The Times* of October 15, 1988, that is, the second day after the announcement of the results, only names two controls. The truth is that there was no (and in any case its existence had not been announced) a notarial deed that meticulously documented all the phases of the operation. Tite even told us that he did not know if a report was drawn up and that, in any case, drafting it was the responsibility of Gonella and Riggi."

"This is such a failure for an operation of this importance. In 1973, only three experts were involved in authenticating the procedures just to take new photographs of the cloth. Riggi attached a report value to the video recording of the excision in his exclusive possession. However, at the request of some scholars, he denied the possibility of showing them the complete video of the operations. Testore recalls that the representatives of the laboratories, together with Ballestrero, signed a report stating that one sample had been taken and belonged to the Shroud. In addition, analysts and Tite had signed a letter of intent in which they undertook to keep professional secrecy about the results; they would proceed to publish their reports on a scientific review at the very moment in which the archbishop of Turin, first informed of the result achieved, would have given the official communication. The results were to be disclosed in less than two months. The three labs also promised not to compare their data until after submitting them to the British Museum. Sox revealed that the scientists from the excluded laboratories were saying privately that the three chosen ones would certainly communicate with each other. Hall denied it: 'There was no collaboration between the laboratories and we were impartial.' According to Bonnet-Eymard, however, the agreement was certain and complete; the laboratories did not act simultaneously, but one after the other just to compare the results."

"The waiting times for the outcome of the dating were unusually long, as Bernd Krömer, director of the physics laboratory of Heidelberg, pointed out. Usually, less than a month is enough for dating. For Gove and Harbottle a week or two, would be enough -- for Donahue a week or so. It is not to be overlooked that the delay is due to the preparation of the article so that it responded to the objections that in the meantime were being raised by other scientists and sindonologists. Tucson did the experiment first, on May 6th; Zurich twenty days later *[see comments below]* and finally Oxford. Bonnet-Eymard pointed out that they acted according to a well-ordered schedule and informed each other of everything, but without letting the slightest news filter out until they were sure of the convergence of their results. It even seems that the directors of the labs met in secrecy during the summer."

"Are there other herringbone fabrics such as the Shroud or not? It has fascinated textile experts and sindonologists in general. The answers have been different and they are not of little importance. If that type of weaving was common in the Middle Eastern area in the centuries of classicism or is documented only from the Middle Ages, one obtains, depending on the case, a pro or con regarding the authenticity of the Turin cloth. Sox described one of the two control samples as having a diagonal fabric similar to the Shroud, although with irregular edges; the other, on the other hand, was marbled, that is, with wavy veins in a different color from the background. But Tite denied that the one sample was of diagonal fabric. Vial and Evin had always denied the existence of other fabrics identical to the Shroud: according to Riggi, however, there were fabrics similar to the Shroud, so much so that the weaving with its herringbone pattern has resulted in findings that were contemporary or prior to the Roman invasion of Palestine."

"Looking for samples resembling the Shroud, Tite confessed that it was very easy for him to find a piece of linen from the time of Christ from Egypt, but much more difficult to find a piece of linen from the Middle Ages and obtain permission to procure a sample. In his search, he had turned to more than one expert. In Turin, on the day of the sampling, he would have known that Evin and Vial had succeeded in the attempt to find a contemporary find of the medieval age presumed by him for the Shroud. It was obvious that, given the weave of the Shroud, a blind test would have been impossible. So much so that no herringbone samples had been found. Not to mention that the Shroud was also very recognizable by some chemical elements present (calcium, strontium and iron). Tite himself had admitted it. So why was the unrealistic proposal made and, above all, why was it kept intact? Especially since the representatives of the laboratories had asked for and obtained permission to see the Shroud during the sampling. We continued to speak of blind dating, 'for a matter of facade before the public,' confessed Tite. Evin characterized the procedure of the blind test as 'exceptional' and while in July 1988 he considered it very necessary, in April 1989 he expressed a different opinion."

"Gonella criticized this staged scene in harsh terms: 'It was empty talk about laboratory honor, because it takes five minutes of X-ray spectrometry to identify the Shroud. The question of blind testing doesn't make any scientific sense; it was only propaganda. Scientists wanted to keep this fiction for the newspapers, knowing full well that they were all stories.' In any case, the sample should have at least been shredded. Laboratory scientists say they didn't do it because of cleaning problems; this would have required more material as there would have been greater loss of substance. According to Hall, there was a risk of losing the entire sample. The director of the Oxford laboratory also discussed the description of his and his colleagues' presence in the sacristy of the Cathedral of Turin during the sampling: 'We were not present as official witnesses (even if Riggi had spoken of 'eyewitnesses'), but as guests of the Cardinal who was sensitive to the idea that he could not be believed.' Another non-marginal point: why were the dates of the three control samples publicly revealed and announced without any need? Gonella justified the fact by claiming that the reference samples were only used to prove the absence of systematic errors, therefore to confirm and ensure the accuracy of the results. Testore also intended the function of the control samples in the same way when he reported that Tite brought two pieces of fabric -- he also forgot the additional sample -- which, 'according to the agreements', would have served as a reference to the radiocarbon tests. So, why was the support fabric (the so-called Holland cloth) not dated, also to determine the contaminants with which the Shroud has been in contact from 1534 to today? Gonella, therefore, meant the control samples only for the calibration of the instruments and not for the blind test which thus completely makes sense. The ages of the three samples were also confirmed on the accompanying letter signed by Ballestrero and Tite. The latter, 'guarantor' of the operation, on the one hand continued to support the procedure blindly, on the other invited the laboratory presenters to attend the sampling. Riggi carefully noted that 'all the bystanders' checked the cleanliness of the sample. At this point, Bonnet-Eymard wondered, what is the point of mixing in a separate room if not for the replacement designed by Tite? He claimed that it has been done in order not to change the details of the protocol. How much attention! Then, however, the transgressions of the pacts would abound [...]."

"The twenty analysts plus Tite, in their final report in *Nature* explain that for the contaminations suffered by the Shroud, explains and because of the unique character of the samples, it was decided to abandon the blind test in order to allow an effective pretreatment of the sample. However, it is not clear when this decision was made and by whom. By mutual agreement between the laboratories? But these, according to the protocol, were not to have contact with each other. Tite defended himself by specifying that the blind test was eliminated in the interest of the laboratories, thus avoiding unnecessary waste of time. 'How can this argument be considered valid,' Van Haelst wonders 'when, after all, would it have taken longer to add a fourth sample?' The impossibility of the blind test, due to the fact that whole pieces of cloth were given, was now evident. Nonetheless, even as late as October 13, 1988, Tite had the brazenness to declare: 'At the time of the measurements, the researchers at work did not know what the shroud's fabric was.' Hall noted that no explicit agreement had been made during the Turin meeting, held in 1986, in favor of the blind procedure. It seemed preferable to adopt it only to prevent criticism from skeptics who might have accused the laboratories of being biased; but the very fact that there were three excluded the need for such rigor. It was decided to abandon the procedure for taking samples and in the report in *Nature* it was written that the laboratories of Oxford and Zurich recoded the samples after gas combustion."

"According to Hall, the recoding took place only in Oxford and was carried out by an unknown someone not previously involved in the dating who supposedly didn't know it was the Shroud. In the case of the Shroud, what was the meaning of the recoding? Did not the person who performed the analysis recognize it? And what guarantee does an anonymous re-encoder guarantee? Wilson pointed out the question of the impossibility of the blind test and of the confidentiality between laboratories with these reflections: first, the Shroud is such a particular piece of linen that, in practice, anyone who worked on even a single fragment would have very little trouble recognizing its identity. This means that an element of human prejudice could theoretically creep into the dating. Although no one seriously suspects that any of the three laboratories involved, all highly respected, could deliberately falsify any results obtained, nevertheless if it happens that a laboratory gets a completely confusing date, for example fourth, ninth or sixteenth century, the temptation to consult with the rest before the primary data are delivered to the British Museum and the Metrology Institute of Turin, if not for a very human concern that they are not considered to be outside ('off limits') of acceptable tolerance of the three labs, not only because they use the same radiocarbon dating system, they used to call each other on the phone. It was therefore a very difficult clause for analysts to prohibit discussing among them the results."

"With much 'fair play,' Wilson made it clear that there was communication between the laboratories. Who guarantees that the results published are those actually obtained? The laboratories refused to collaborate with other scientists, even opposing the conduct of other analyses at the same time. All this without any scientific justification. They wanted a person to certify Tite's samples, but they did not offer guarantees. Sindonologists, Gonella, and other Vatican representatives, nor any member of STURP, nor Vial, nor Testore, nor Evin, were invited as observers during the analyses. Sox and Gove were admitted, unrelated to the analyses; they were both notoriously against the authenticity of the Shroud. No chemist was consulted to verify the appropriate cleaning of the samples from the impurities that contaminated them. Gove saw a piece of the Tucson sample after washing and it seemed very clean to him. As a physicist, he was satisfied with an eye inspection. A chemist would have checked the cleanliness with specific tests. But radiocarbon physicists refused all interdisciplinary collaboration, so that no one today can say what the exact chemical composition of the sample whose content they measured in ^{14}C."

"Gonella denounced the existence of a '^{14}C mafia.' Bonnet-Eymard joined in by saying: 'And to replace the samples you had to be among the mobsters.' Given that the degree of contamination of the Shroud was not known, the laboratories divided the samples and subjected the pieces (from three to six) to different mechanical and chemical cleaning procedures. It should be noted that in Zurich the measurements did not reveal any evidence of contamination. Wölfli himself expressed his surprise at this unexpected fact. No substantial differences emerged between the results obtained with the different cleaning procedures used. According to some scientists, it would have taken fifty percent of contamination to move the radiocarbon date one-thousand years. For other scientists, however, ten percent would have been enough. Actually, inadequate cleaning was enough to make Zurich be off on one-thousand years on test samples dated earlier!"

Source: Petrosillo, Orazio and Emanuela Marinelli. La Sindone: Storia Di Un Enigma, (Milan: Rizzoli, 3rd edition, 1998), pp. 163-184.

Comments: To continue this thread, go to Petrosillo/Marinelli entry under "1988 *(May)*."

One has to marvel at discrepancies on even seemingly non-crucial points, such as whether Evin was present at the sampling or not or whether the labs' reps were told when they arrived that the sample would not be taken that day. But it is shocking that the day of the sampling, despite having had numerous meetings and a three-day workshop in 1986, they hadn't decided from what area the sample would be taken. "Improvisation" in the most important C-14 dating ever done? That certainly would not have happened if STURP had been in control of the test. Petrosillo and Marinelli say that the sample was taken after four hours of analysis – most accounts I've read over the years usually say it was after one to two hours.

In Savarino's remark, we had someone who would eventually rank high in the Turin sindonological hierarchy, admitting the lack of rigor in the enterprise. Savarino would also later in a booklet he co-authored make a startling admission about the nature of the sample, which will be detailed in Part III. The discrepancies regarding the sizes and weights of the samples will also be elaborated on in Part III. (If they couldn't keep straight something as basic as that, it certainly should call into question the rigor of the experiment.) For those who believe the samples were switched, the fact that the control samples weighed a similar amount to the Shroud samples provided added ammunition.

Regarding the start date for Zurich's dating measurements, see the comments under the Gonella/Riggi entry for "1988 *(June/July)*."

Riggi should never have been allowed to have exclusive rights to the filming of the procedure. What good was filming it if researchers weren't allowed to watch it? I've been informed that currently the recordings are in the possession of the 3M company in Milan, but I don't know how accessible they are. There was nothing to prevent having the packaging of the samples videotaped but it was the only aspect of ten to sixteen hours of proceedings that was not recorded on tape. Tite asserted (per Marinelli's Valencia paper, pg. 11) that the filming of the sample packaging would have only been a "memorandum, not intended to be an identification proof for the samples, of which he and the Cardinal were guarantors." He also admitted that moving to a separate room for the packaging, was "quite unnecessary."

Regarding the lack of invitations from the labs, Gonella complained, "The experts of the British Museum did not trust the Cardinal and wanted to be present when the samples were taken from the Shroud, but then they did not allow a representative of the Church to watch the analysis as an observer." Piero Savarino, who would later be the scientific advisor to Turin Cardinal Poletto, remarked, "This behavior is truly incomprehensible. It is to be considered that in legal ambit any analysis performed in the absence of the other party is rejected by the courts" (Marinelli's Valencia paper, pg. 12). Although the Pontifical Academy of Sciences was supposedly excluded, I've heard from a very reliable source that they were, in fact, still involved. That fits in with all the other political intrigue found throughout the process. Flury-Lemberg was replaced not by one person, but by three: Riggi, Testore and Vial, with the latter two never previously having seen the Shroud! I've never heard an explanation why that change was made. The refusal of the labs to invite a Church representative is all the more startling considering that a TV crew was invited to Zurich by that lab.

1988 *(April 21st)*. "Prevented from coming to Lyon to fetch the controversial fourth sample, Tite asked Evin to send it to him by post. Fearing postal strikes, Evin decided to have it delivered by hand directly to Turin by Gabriel Vial on 21 April. Just when Riggi had completed his work, the Frenchman took out 'his' sample, therefore, for it in turn to be shared among the three laboratories. There was a moment of panic. The Cardinal had already left. Riggi expressed fury at this intrusion into 'his' protocol. Tite, acting surprised, wanted to refuse to consider this fourth sample which had turned up so inopportunely and for which nothing had been prepared: no little stainless steel numbered cylinders like those which already contained the 'official' control samples and the Holy Shroud samples!"

"Vial thought only of the trouble that Evin had taken to procure this sample that Tite had asked for. He was outraged and insisted that his precious fragment be weighed after all. It was then cut up to be shared among the three laboratories, each one receiving a piece in a little envelope prepared by Tite. Strange, very strange!"

Source: Bonnet-Eymard, Bruno. "The Holy Shroud Is Authentic." *The Catholic Counter-Reformation in the XXth Century*," No. 219 (Christmas 1988/Easter 1989), pg. 17.

1988 *(April 21st)*. Meacham wrote, "By March of 1988, a shroud of secrecy (so to speak) was drawn over all arrangements for dating. There were rumors, but hard information was lacking until late April when press reports confirmed that samples had been taken. The senior representatives of the three labs been summoned to Turin and were present as observers at the sample taking. They were called to the Cathedral at 4:30 am and the operation began. A lively discussion ensued between Gonella and Riggi on the one hand and the two textile experts, Vial and Testore, on the other concerning where the sample should be taken. One of the textile men is said to have asked, on noting the dark stain on the chest [blood stain from the wound in the side], 'what's this?' Gonella and Riggi finally decided to cut a single strip approximately 1 cm wide by 8 cm long, weighing 300 mg., right next to the small cut that had been made in 1973 at the corner of the cloth to provide the textile expert Raes with a sample. The reason, as Gonella told Al Adler, was that 'the Shroud was already cut there.' Adler called this the worst possible reason. The sample was adjacent to a seam that joins the main body of the Shroud with the side strip, which seems to be of the same cloth but was attached by a stitched seam at some unknown time. This seam had to be trimmed away by Riggi before dividing the sample into equal segments to give each lab."

"What is remarkable is how poor the planning and execution of this project was, despite all the brouhaha and the months of secretive preparations, in addition to the disastrous choice of sampling site and the disastrous decision to take only one sample. It is hard to imagine that, in all the months that had passed since the Turin conference, Gonella had not given due consideration to the location where material was to be removed, and that it was decided only after discussion on the very day of sampling. Riggi was brought in to do the cutting, although he had no expertise in textiles. Riggi was also given the responsibility of video-taping the proceedings, a conflict of interest one could argue. He would later treat this video as his personal property, and charge the BBC a hefty sum for use of several segments in a documentary. What is even more amazing is that, after all the exhortations by Gonella that the amount of material removed from the relic had to be minimal, Riggi cut **double** what was actually going to be given to the three labs. He then cut the 300 mg strip in half and divided one half into three segments, the other half being retained as a 'reserve piece.' Presumably, if there were any discrepancies in the results obtained by the three labs, this reserve piece was going to be used for another run. Gove's constant harping on the possibility of lab error or statistical outlier must have registered with Gonella, so he came up with this precaution. My constant harping on the need for a minimum of two sample sites obviously did not sink in, nor did the distinct possibility, as plain as the nose on your face to anyone who has done archaeological dating, that if the first run gave discordant results a second run on the same sample would very probably produce similar results."

"Unfortunately, Riggi failed to cut the half into three equal segments. The Arizona sample was only 40 mg, whereas the other two were approximately 50 mg. He then shaved about 10 mg from the reserve. Later, there were significant discrepancies between the weights of sample material made on the spot and by the labs. Even more mind-boggling is that Riggi was allowed to keep the seam trimmings, and to take sticky tape samples from another part of the Shroud with blood stain, and to run his vacuum over the Shroud in a zigzag pattern that he appears not to have planned in advance or plotted at the time. Riggi would later distribute the trimmings and the tape with blood-stained fibers to researchers in Texas, earning a stern rebuke from Ballestrero's successor. His involvement in this operation was a huge mistake on the part of Gonella."

"In addition to Riggi's shenanigans, the labs were told the age of the historical known-age control pieces, a fact that rather diminished their value as controls. Paradoxically, the pretense of 'blind testing' was maintained for the whole dating exercise, despite the fact that everyone knew that the Shroud weave was easily recognizable. Even if the samples were shredded the Shroud fiber could probably be identified by the labs, since there was so much technical data published by STURP. What happened next simply beggars belief: to maintain the pretense, Ballestrero and Tite took the samples into a private area, out of view of all the people in attendance and of the camera, and put them into vials labeled with numbers. These vials were then brought out and presented to the representatives of the three labs. This secrecy gave rise to the allegation, quite absurd on the face of it, that Tite had conducted some sleight of hand and switched the real Shroud samples with others of medieval age. There are still quite a few Europeans who believe to this day that the samples were substituted and the C-14 date that was later obtained is not from a piece of the Shroud. Loading the vials in private was a totally unnecessary and ridiculous procedure, another major error on Gonella's part."

"Standing on the sidelines through the eventful proceedings of that morning were the lab directors: Hall, Hedges, Damon, Donahue and Woelfli. Their only apparent role was that of couriers -- to await the delivery of the vials into their hands. No microscopic, physical or chemical examination was done on site, since these could of course be done back in the labs. What is surprising to learn is that, once they had brought the vials back to their respective labs, very little scrutiny of the sample was carried out. Not one lab photographed the samples they received properly, i.e., both sides and with a scale. The samples were examined under a microscope, and a few alien fibers picked out, but no lab reported anything suspicious, even though later a STURP chemist found that threads from the adjacent Raes sample had high levels of aluminum, a high occurrence of cotton fiber intermingled with the linen, some kind of coating or encrustation, a high degree of oxidation, and FTIR spectra markedly different from threads elsewhere on the Shroud. Certainly the labs were not in a position to know all the results of all previous investigations of the Shroud, but they could have consulted with STURP personnel, or they could have requested comparison fibers from other parts of the Shroud. The fact that they did neither indicates an over-confidence in their ability to date the samples through standard procedures. It seems very likely that this was a huge mistake, and as Ray Rogers, the late STURP chemist remarked: 'there will be hell to pay when the truth comes out'!"

Source: Meacham, William. The Rape of the Turin Shroud: How Christianity's most precious relic was wrongly condemned and violated, (Lulu.com), 2005, pp. 90-92.

Comments: The decision to take the C-14 samples from the one region of the Shroud that "was already cut there," as Meacham writes Gonella told Adler, is unbelievable. There seems to have been no serious examination -- before the sample removal -- to choose an area of the Shroud that was indisputably original and had not been subjected to reweaving or restoration. STURP chemist Ray Rogers' comment "There will be hell to pay when the truth comes out" will be proven correct if and when the *general public* is ever exposed to the whole truth. It is my hope that this book will facilitate that knowledge.

I received an email on April 27, 2020 from researcher Marcel Alonso of France, who said that the late Gabriel Vial said he analyzed the Shroud before the sample was taken with his own professional magnifier, and Testore looked at with his binoculars to determine if there were foreign fibers. They believed there weren't. However, as was previously mentioned, Riggi, who excised the sample, wrote that he removed fibers of other origins from the sample. Plus, chemical analysis by Adler, Rogers and others found evidence of repairs. That analysis is documented in my (and Ed Prior's) "Chronological History of the Evidence for the Anomalous Nature of the C-14 Sample Area of the Shroud of Turin" article mentioned in the Preface.

Serendipitously, I came across an email from Ray Rogers sent to Sue and me and four other researchers on October 1, 2001 in which he said, "I am just now finishing the chemical study on the archived samples from the Shroud. My results do not agree with the statements of the textile experts. I must believe that Prof. Franco Testore and Mr. Gabriel Vial did not do a 'scientific' study. I do not see how they could have observed the cloth under a microscope or done any chemical tests and made their statements. I have often observed applications of the ancient fallacy of 'argument from authority.' Authority does not adequately compete with Scientific Method. I believe that those two 'experts' are responsible for a major fiasco." Rogers also told *Inside the Vatican* (March 2005, "Was the Dating a Hoax?", pg. 25) in an interview published the month he died, that the Turin "experts are embarrassed and protecting their reputations" and that they weren't experts in the correct fields. "A textile expert knows all about spinning and weaving, thread counts, etc. They do not think in terms of chemistry. This was an occasion when the composition of the cloth was important," he said.

Rogers was also asked if he thought the Turin authorities were aware of the STURP 1978 photos ("quad mosaics" photos, which will be discussed later in the 2003 time-frame) that seemed to indicate that the C-14 corner was different from the rest of the cloth. Rogers replied (pg. 24), "[I]t doesn't matter if they ignored it or were unaware of it. Part of science is to assemble all the pertinent data. They didn't even try."

Based on Riggi's and Roger's statements, it's clear not much stock can be put in Vial's and Testore's assertion that there was definitely nothing anomalous about the C-14 sample area.

1988 (April 21st). Gove recounted Paul Damon's description of the sample taking: "Riggi was to remove the sample, but it took two hours to decide where it should be taken. Everyone knew it would be near the spot on the hem where Raes' sample had been removed and that is where it was finally cut."

Source: Gove's book: Relic, Icon or Hoax: Carbon Dating the Turin Shroud, (Bristol and Philadelphia: Institute of Physics Publishing, 1996, pp. 260-261).

Comments: It's amazing that despite the elaborate planning meeting held in Turin in 1986, authorities discussed where to take the sample from for about two hours (or four according to Petrosillo and Marinelli) at the time of sample taking. They ended up choosing an area that various people had recommended avoiding and had the required thorough chemical analyses been performed or had STURP been consulted, the conclusion that the sample was anomalous might have been made in 1988. Now, over thirty years later, based on all the accumulated new evidence, the original results are clearly in doubt.

1988 *(April 21ˢᵗ)*. In his book, Sox supplied the wording to the certificates signed by Cardinal Ballestrero given out to the labs when they were given their samples: "The containers labeled [A, O or Z]1, [A, O or Z]2, and [A, O or Z]3 to be delivered to [Arizona, Oxford or Zurich] contain one sample of cloth taken in our presence from the Shroud of Turin at 9:45 a.m., 21 April 1988, and two control samples from one or both of the following cloths supplied through the British Museum: First-century cloth; eleventh century. The identity of the samples put in the individual containers has been recorded in a special notebook that will be kept confidential until the measurements have been made."

Source: Sox, David. The Shroud Unmasked, (Basingstoke, Hampshire: The Lamp Press), 1988, pp. 136-137.

Comments: Not only were the control sample dates given to the labs on the day the samples were taken, they were announced publicly both by the Vatican newspaper *Osservatore Romano* on April 23ʳᵈ and in an article published in June 1988 in *Shroud Spectrum International* by the French C-14 scientist, Jacques Evin!

1988 *(April 21ˢᵗ)*. Ted Litherland, who was director of the IsoTrace lab in Toronto and co-inventor of the AMS method with Harry Gove, commented about the choice of the area from which the sample was taken, "All the samples came from one corner. Oh Dear! My God! That's no way to run a show."

Laverdiere wrote, "They took every care so that no accusation of possible fraud could be made. The procedures were recorded on video tape so that all the handling of the samples would be well documented. This video was later sold to BBC who produced a program on the Zurich experiment. It permitted Turin people to repay some of the important expenses incured [sic], by selling the rights to BBC." But the author then continued, "Yet the wrapping of the samples was kept more secret than the rest of the operations: the test was not fully blind anymore, but they still acted as if it was."

Another control sample was brought in at the last minute by French textile expert Gabriel Vial, who had received it from Jacques Evin, whom

Tite had previously asked to find a control sample similar to the Shroud. Laverdiere noted, "They decided there and then to include it in the controls."

Source: Laverdiere, H. "The Socio-Politic of a Relic: Carbon Dating of the Turin Shroud," (Doctoral Thesis) 1989, pg. 97. Accessible via free download at http://ethos.bl.uk/OrderDetails.do?did=1&uin=uk.bl.ethos.531916.

Comments: Laverdiere noted that the packaging of the samples was more secret than the rest of the operations but hadn't specifically stated what is known -- that the packaging wasn't actually recorded on video, which is suspicious to say the least. It is unbelievable that another control sample (from the cope of St. Louis d'Anjou, dated to about AD 1290-1310) would be added to the mix on the spot in the sample packaging.

According to the Petrosillo/Marinelli "Enigma..." book (pg. 67), "The operation, according to the official statement, was carried out by the archbishop of Turin and by Tite. Actually, these two were not alone; Gonella and Riggi also took part in the shuffling of samples." Note that there are several versions of who was involved in discussion regarding where the samples would be cut and who was present for the packaging of the samples. One would think that those facts could have been documented such that there was no confusion.

Paul Maloney told me that the late Al Adler viewed on video (14 hours according to Adler) the various procedures of the sample taking and told him it was some of the worst science he had seen in his life. The two co-inventors of the AMS method, Gove and Litherland were highly critical of the Shroud dating enterprise as well. Scientific rigor, quite simply, took a back seat to politics.

1988 *(April 21st)*. According to the documentary *La Notte de la Sindone*, when Riggi di Numana cut the sample, Cardinal Ballestrero curiously remarked, "The crime is accomplished." Testore recommended taking more than what the labs requested in case one of the three came up with a significantly different date from the other two. Italian researcher Marinelli said, "Also, regards the weights, there later were pronouncements that gave different weights. Dr. Bollone of the Turin Centro commented, "These weights do not correspond with what is seen on the scales used to weigh the material at the time of the sampling. As far as I'm aware, these inconsistencies have never been explained." The actual measurements of the 3 samples were shown as: 52 mg. for the first sample, 52.8 mg for the second sample and 53.7 mg in two parts for the third sample (which went to Arizona).

Testore commented regarding the packaging of the samples, "There was something we didn't understand: when -- how -- the three samples were put into their containers. There was only Cardinal Ballestrero and Tite, the director of the British Museum -- that was all -- in a separate room from where we were. But they came back and sealed them in public."

La Stampa journalist Marco Tosatti noted, "Once again we are faced with something that raises questions within a procedure that should have been followed absolutely rigidly to avoid any possibility of doubt or misinterpretation."

Source: *The Night of the Shroud (La Notte de la Sindone)*, documentary directed by Francesca Saracino, 2011. In 2016, it was revised and retitled "Cold Case: The Shroud of Turin," which is available at amazon.com. I have a review copy of the original version, which has an English voiceover. The revised version has English subtitles. (The material cited here can be found between approximately the twenty-three-minute and thirty-five-minute range on the original version review-DVD.)

Comments: It is not known what Ballestrero meant by his strange remark. Perhaps he was expressing his belief that such a religious relic shouldn't have been subjected to scientific testing. Testore seems certain that only Ballestrero and Tite went into the adjacent room with the samples, but other reports say that Gonella was also there (and Gonella stated he was there; also Riggi?). It seems as if there were discrepancies on all the most basic facts.

Paul Maloney wrote to me in an email of June 11, 2016, "Al [Adler] noted that [...] the door to the scales on which the samples were first weighed was OPEN and thus, right at the very foundation of the C14 science project, there were these flaws in accounting, and lack of precision, and apparently they continued right through the various labs that received those samples!!" (The doors on the scales being open would affect the air pressure and thus the measurements of the weights of the samples.)

1988 *(April 21ˢᵗ)*. Bollone wrote, "Here is the significant episode that I do not know has been noted so far. The first news to the public of the taking for 'the carbon test 14' is contained in the article 'Shroud, samples taken' in the *Stampa Sera* of April 22, 1988. The source of the anonymous columnist is certainly one of the few people admitted to the operation, seen or stated with precision: '[...] a strip of circa 1 centimeter by seven, weighing 159 milligrams, was removed yesterday', which, as we shall see, corresponds to the official version. It is clear, however, that the 'training' to the journalist went far beyond the data on the size and weight of the samples, given that exams that have yet to be started anticipate what will essentially be the final result of the investigation. In fact, he continues by highlighting the following hypothesis, preliminary to any other: 'If (the samples taken) are, for example, 700/800 years old, it must be concluded that the Shroud was not used to wrap the body of Christ'. Months later the result will be just that. It can be objected that, obviously, the articulator, a good connoisseur of the history of the Shroud, only materializes a truth that he already knows and with which he is persuaded. It is unlikely, given that no journalist, chronicler or columnist of *La Stampa* has ever expressed an opinion of this kind, never before or after, even verbally and in a strictly personal way. All these, however, are interpretations."

"It is necessary to examine the facts. The execution of the samples of April 21, 1988 is described by Giovanni Riggi di Numana who organized and excised them. Given the importance, I transcribe the report. 'At 4.30 on 21 April -- declares Riggi -- 'work started on the sheet which took a total of about 16 hours, including the opening and closing of the display case in the Chapel Altar around 10-11, after a wide consultation of the textile experts and the controllers, and under the supervision of the guests, I was given the cut of about 8 square centimeters of Shroud fabric (in the same area where Professor Raes had already taken a sample in

1973), then reduced to about 7 due to the pollution of the fabric itself with threads of another nature that could have led to variations in the dating, even in minimal quantities, being of late addition. The fragment of 7 × 1 centimeters thus obtained was subsequently divided into parts, of which three were equivalent, weighing just over 50 milligrams, to be placed in three initialed containers. With the same criterion, another three fragments were cut by me from each of two samples brought by Dr. Tite, who knew its provenance and dating to carry out the triple-blind experiment foreseen by the 1986 Meeting. The nine fragments of cloth of the three different origins were placed in nine steel containers by Cardinal Ballestrero and Dr. Tite in a secluded room. Again in the presence of all, the cylinders containing the fragments were then sealed and boxed, three by three, in special containers, suitable for transport, and delivered to the representatives of the three Radiodating laboratories (Arizona, Oxford and Zurich)'."

"'The whole operation, excluding the mixing of the fragments carried out by people above all suspicions, was carried out under the eyes of over thirty people and by the discreet objective of a camera that worked very well for every moment of the work. At 2 pm, the picking operations ended with the departure of the precious fragments towards their final destinations. Around 7 pm, with the doors of the Cathedral now closed, the Holy Shroud returned to its natural location on the Altar of the Chapel, accompanied by all of us, by the priests and by the heads of the Superintendency for the Artistic and Historical Heritage of Piedmont, managers of the Chapter Chapel, who wanted to follow the operation from start to finish. The time between the delivery of the samples and the rewinding of the Holy Shroud on its velvet roller, about three and a half hours, was usefully used to answer some urgent questions regarding conservation, which required direct observation and targeted observation of some areas of the cloth. For this purpose, some photographs were taken, microscopic observations of some parts were made and very small quantities of dust were drawn from the margins that the textile experts had wanted me to unstitch for their analyses'."

"'At the same time, the storage case and all the outbuildings have undergone a process of disinfestation from the small parasites of the fabrics that had been seen in the analyses of 1978 and which still today could be alive and present in the Shroud container. The live observation of the Holy Shroud, especially aimed at finding active parasitic presences, did not however signal unwanted presences and only future research on dust could clarify this particular aspect of the problem of conservation. At 9 pm, the clearing of materials from the Sacristy and the adjacent twill areas of the Cathedral was completed. The operation ended without clamor and without any reports from outside, according to the wishes of the Cardinal of Turin who had entrusted me with the logistics and coordination of the whole process and who had sent information on the story of the cold formalism of the Vatican Press Office in Rome'."

Bollone commented on Riggi's observations, "This procedure has aroused worries and reservations in the world of Shroud experts that must be taken into account. I will summarize them briefly."

"The sampling was carried out in a single location instead of from different points as evidently requires any scientific sampling for any kind of exam and therefore with much greater reason for a delicate examination such as that of the ^{14}C on a specimen of the size and with the past of the Shroud. In addition, a peripheral location was chosen at the long side near the seam. Centuries of depictions of ostentation show that it is one of the places where it has been grabbed hundreds of times to show it to the public. Being the site of the operation by Raes in 1973, there are signs of manipulations there. The photographs taken in 1978 seem to demonstrate a blackish (of an unknown nature) contamination in the surrounding fabric. The same Riggi declares that he had to eliminate a part of the sample because of contamination of the fabric of the same *[sample]* with yarns of other nature, and that opens the door to the possibility that the sample had coincided with an area of an invisible mending, of the kind that our grandmothers were still able to perform, which moreover one can easily expect in a place of habitual grasp. It is not enough. A quantitative analysis such as that of radiocarbon requires the utmost precision from the moment the sample is taken. There is no question that the quantity of fabric taken from the Shroud is not known exactly."

"On 7 and 8 September 1989 a symposium was held in Paris on the problems of dating. Riggi presented a memo entitled *Prélèvement sur le Linceul effectué le 21 avril 1988* in which it provides a new and more detailed report of the procedures followed which modifies its previous report. Obviously the essential data is the weight of the material actually taken. Riggi provides two different values: 497 and 540 mg. The sampling video indicates instead that the weight indicated by the scale is still different, the scale in fact indicates 478.1 mg. This is a disturbing divergence and I clarify my thoughts on this matter. It is quite clear that in such a complex undertaking inaccuracies can occur in the findings, although it is desirable that these do not affect the fundamental measures of the investigation, as has unfortunately happened, however, it is usually possible to remedy by consulting the notarial report of the transaction which, however, opened on 21 April 1988 does not appear to have been completed, so that the uncertainty remains. If it were a consultation or a judicial expert's report, an episode of this kind would represent an absolute foreclosure to the use of the data, but fortunately it deals only with the Shroud. It is for this reason that I still remain on the subject. Even Prof. Franco Testore presented his report at the Symposium in Paris, *from Le Saint Suaire - Examen et prélèvements effectués le 21 avril 1988*. It states that the piece taken is divided into two parts of the respective weight of 154.9 mg and 144.8 mg and that the piece of mg 154.9 is further cut into three parts weighing 52.8 mg, 52.0 mg and 53.7 mg. It is clear that this cannot be because the sum of the three weighings is equal to 158.5 mg, well above the 154.9 mg of the starting fragment."

"At this point the well-known Milanese sindonologist, Ernesto Brunati wrote to Prof. Testore to ask him the reason for the inconsistency. Testore did reply. He also noticed the 'inaccuracy of the explanation of the weighings' and therefore replaces the wrong text with a new one in which the three samples were not cut by dividing the 154.9 mg fragment but the smaller one of 144.8 mg. The three samples were respectively 52.0 mg and 53.7 mg while the third only 39.6 mg and so they had to be integrated with an addition of further 14.1 mg taken from the 154.9 mg sample. None of this had been said previously, not even in Riggi's previously transcribed account in which we speak of a single 'fragment of 7 × 1 centimeters subsequently divided into parts, of which three equivalent, weighing just over 50 milligrams [...]'."

"At this point Brunati, after a public debate with the two authors who evidently did not satisfy him, wrote an article on the subject. He asked why after a year and a half of conversations and conferences and after the Paris Symposium one suddenly realizes that the three weights are inconsistent with the total and that a sample consisted of two pieces. He also asked the reasons why he continued for months and months to say that the main piece of 154.9 mg and not that of 144.8 mg was cut and the reasons for having to provide three samples of the weight of 50 mg each and having just weighed two fragments of 154.9 mg and 144.8 mg it was thought to obtain them from the second. We follow Brunati in the succession of his observations. He points out that Riggi also changed explanations from the data provided previously by asserting that a box, the third, is made up of two small rectangles of linen. He points out that this second version (which is actually the third) provides data totally different from that of Prof. Testore. Brunati noted in this regard: 'The cut, he tells us (Riggi), was the largest, 154.9 mg and the small pieces weighed 52.0, 52.8 and, presumably, 50.1 mg, so the added piece was only 3.6 mg. Unnecessary, one might say, given that the sample already exceeded the fateful limit of 50 mg.' Brunati's conclusion is as follows: 'Strictly speaking, this conclusion is reached. One of the two second versions is undoubtedly fake. Both are in contradiction with too many details told to be certainly accepted without reservation. The first versions, however, previously constituted two shared testimonies, coming from two different sources seem reliable under one condition: that the three samples have been cut from a larger piece of fabric and therefore heavier than the one just cut by the Shroud. The examination of the unit weights leads to no different conclusion, a technical term to indicate the weight with respect to the surface, that is to say to each square centimeter of fabric knowing that the weight of the Shroud is about 23 mg/cm$^{2'}$.'"

"In conclusion Brunati observes that the reports made in September by the two professors at the Symposium of Paris contain such and many inconsistencies as to give rise to the suspicion that someone has replaced, in the course of the rations, the samples cut by the Shroud with others. A less pessimistic hypothesis must first be ruled out. Riggi's excision perhaps fell in whole or in part on a mending carried out in past centuries and this could be confirmed. The French Sindonologist Bonnet-Eymard obtained a photograph of the sample from the Shroud received from Tucson (Arizona) and published it. At my request, he sent me the photograph I reproduce in illustration no. 65 of this book, compared with the photograph of the right (n. 64) and the reverse (n. 66) of the cloth. The comparison shows a surface that resembles but does not seem to correspond exactly to that of the Shroud, just as it could happen in a mending case."

"Let's go back to the day of the sample-taking, that is to say, to April 21, 1988. The operations of collection, subdivision and packaging of the samples were finished. They were delivered to the representatives of each of the three laboratories together with the comparison samples. It was a linen fabric taken from the Cleopatra of Thebes mummy from the beginning of the second century AD and another linen fabric found in a tomb of Quasr Ibrim, in Nubia, datable between the eleventh and twelfth centuries AD. These dates are exactly indicated in the press release from the Vatican that announced the execution of the full visit. An observation: if it was a blind exam, why say the time? A little mystery: contrary to what was declared by Riggi in his 'report', each laboratory is given not two but three 'controls'. What happened? Just that morning from France, carried by hand by the textile expert Gabriel Vial, comes another sample of fabric. These are some threads detached from the cope of St. Louis d'Anjou, preserved in the basilica of Saint-Maximin-la-Saint Baume in Provence which has the advantage of being exactly dated to 1296-1297. It was divided and delivered as an additional sample to three laboratories. In this fact, the attempt was unsuccessful to have a sample of medieval fabric to be replaced at the right time for the Shroud. 'In the midst of so many inexplicable aspects to me, the use of a sample that, although originally not foreseen, has the merit of being exactly given -- in any case there can be no cheating since the Vatican press release declares on the same day that -- a fourth sample dated to about 1300 AD was provided as additional control'. However, it should be noted that if a sample of medieval fabric was needed, it could be found in abundance avoiding the use of loose threads. The textile experts evidently did not know that a few years before January 6, 1981, at the time of the recognition of the tomb of Antonio da Padova, who died on June 13, 1231, even four pieces of linen of the time had been found among them. With all this the hours pass quickly."

"The evening of April 21, 1988 arrives, the Shroud is put away. The samples begin the journey to the laboratories and the waiting period for the results begins. Thorny issues also begin. We know that the laboratories, contrary to the original agreements, know the dating of the three control samples. On the other hand, since the Shroud is the only 'herringbone' fabric, as their representatives have been able to ascertain, they are perfectly able to recognize the relative cloth. Furthermore, according to the agreements, they should not exchange information and instead it appears that they are proceeding with preservation, one after the other. Furthermore, always according to the commitments, they should proceed in absolute secrecy, and this does not happen. In Tucson, one of the excluded carbonists is admitted to the Gove exams and in Zurich, the Anglican pastor David Sox with a British BBC television crew. Both Gove and Sox have long been engaged in controversies against the authenticity of the Shroud. It is thanks to this undue admission that Sox will be able to release in October, at the same time as the announcement of the results, one of his books with the extremely significant title The Shroud Unmasked. The most serious aspect of this affair is that at the same time no representative of the Curia of Turin is allowed to attend the exams even if the exam supervisor will then let it be known, perhaps quite well, that this happened only because Prof. Gonella, representative of the Archbishop of Turin, had not asked for it with due firmness. We follow the thread of the events. At this point there is a series of leaks, very explainable on the basis of the conduct of the exams now indicated that anticipate the unfavorable verdict of the exams. The inertia of Turin which could reasonably have required the interruption of the procedure and the return of the material is much less explainable [...]."

Source: Bollone, Pier Luigi Baima. Sindone O No, (Torino: Societa Editrice Internazionale), 1990, pp. 280-285.

Comments: In a later book, Bollone actually disavowed the idea that a repair was the cause of the dating, but has been previously shown and will continue to be shown in the remainder of the book, there is an abundance of evidence that the area was apparently repaired. However, Italian engineer and sindonologist Giulio Fanti sent me an email on January 15, 2020 saying, "I have seen and touched (with proper tweezers) one of the samples in question. I have photographed the so-called pollution with threads of different nature. I am sure: this so-called pollution has NOTHING to do with a hypothetical reweaving. These threads really of different nature (whiter) were used to sew the Shroud's edge over itself in the following way:"

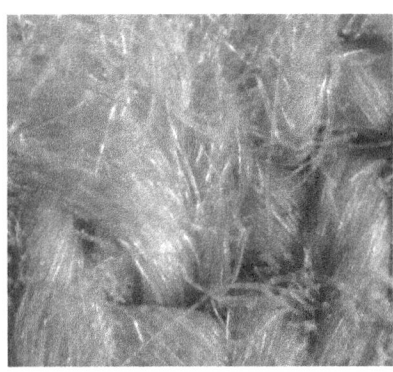 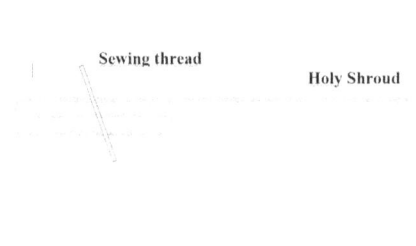

In the account of Riggi cited by Bollone, only the Cardinal and Tite are said to have gone into the sacristy to place the samples in the containers, and Gonella is not mentioned (and neither is Riggi).

There will much more on the Paris symposium later in the book. Several of Brunati's letters will also be reproduced in full.

No doubt it would have been too much of an embarrassment for all concerned if Turin stopped the process and asked for the return of the samples. If one believes the cloth wrapped Jesus, it's not too far-fetched to see him as having been betrayed again.

APPENDIX TO PART II

UNDATED (RE: SPECIFIC MONTH/YEAR, BUT ALL IN THE 1988-1989 TIME FRAME) PERTINENT INFORMATION FROM THE LAVERDIERE DISSERTATION

Pages 13-14: "An indicator of the touchiness which surrounded the whole process of dating the Shroud was the ubiquitous utterance of expressions like: 'God I don't know what you will do with that.', 'Please don't quote me on this.', 'Please keep this for yourself, don't make me any more problems, I've had already enough with all those leaks.', 'I don't think you should publish this, not before everything is well over., etc' [...]."

"Interviews had to be adapted every time to the role of the interviewee in the organization of the test, to the development of the situation, to my own evolving understanding of both the technical and '**political**' *[my bolding]* matters, and perforce, to the temperament of the individuals involved."

Page 23: "There were so many arguments about how to do it *[carbon date the Shroud]* that it created a certain uneasiness in the carbon dating community. And it could be dangerous for the reputation of the technique: It was deeply felt that the result would be heavily scrutinized by those it would displeased *[sic]*, and that whatever date was found, it was bound to displease someone [...]."

Pages 24-25: "[T]he Shroud was quite a 'glamorous' object. So much so that, for this particular test many things were reversed. It is the labs who asked for the samples, which is very unusual. 'It is the first and only time I think in the history of archeology [sic] that labs […] take the initiative of dating an object' (Gonella, interview)."

"Similarly they were <u>asked</u> for the publication of their results. 'That's the first time in my life […] you know I have published, or tried to publish many hundreds of papers. That's the first paper in my life which has been accepted before one line was written' (Wolfli, interview)."

Comments: Acceptance of a paper before it is even written is not how scientific peer-reviewed literature is supposed to work!

Pages 26-27: "[I]t is well known that the head of at least one laboratory hired an agent and found a publisher right from the beginning. He hoped to raise enough money to endow a permanent chair in his department. I was often told, also, that another expert already had his book 'How I dated the Shroud' written and was just waiting for the date to fill in the blanks."

Comments: The head who hoped to raise money to endow a chair was, as mentioned in Part I, Edward Hall of Oxford. The "How I dated the Shroud" book could possibly be referring to the one by Sox. Does anybody really think the labs would have admitted publicly to any problems that had gone on, given the stakes?

Pages 27-28: "Although they could always have argued that it did not prove anything, the biggest impact of a first Century date could have been for non-believers. One of the experts reported the conversation he had with a bishop: 'He said there is at least one person whose views on religion and Christianity might be influenced if it turned out to be 2000 years old.' And I said: 'Well who is that?' He said: 'You Gove' (Gove interview)."

Comments: Elsewhere Gove is described as a nominal Christian but this passage suggests he was either atheistic or agnostic.

Page 57: "The cloth could date to 1st Century but the image need not; both could date to the 1st Century without originating in Palestine; they could be from the 1st century Palestine but not from the grave of Jesus; and evidence for any of the above could be faked." Quote from Cole, J.R. Reply to Meacham, W. "The Authentication of the Turin Shroud: An Issue in Archaeological Epistemology", *Current Anthropology*, **24**(3):296.

Comments: This emphasizes the fact that carbon dating the Shroud was actually a no-win situation for the Vatican. Even with the most favorable results, skeptics could always come up with a rationale why the Shroud was not authentic. Performing additional multi-disciplinary tests would not have been a total antidote to this stance, but it could certainly have strengthened the case for the Shroud's authenticity.

Page 60: "[A]round 1978, Rome and Turin authorities, under the pressure of STURP, made their inquiry about the dating, they were told by specialists that the technique was not ready. And even recently, some experts feared that some laboratories could make a mess of it, which could put 'the whole technique into disrepute' (Hall, interview)."

Page 69: "'If it is an artefact we should better damn well understand it because it could have implications on the carbon dating. So that's sufficient reason for me to have both techniques *[AMS and proportional-counting]* used' (Gove, interview)."

Comments: How could the Shroud not be an artefact? We know that it dates at least from the 1350s. As the Shroud had been in a major fire and not buried in the ground as most objects carbon dated (and possibly have been boiled in oil, but that is not proven), it is a rather unique object and more tools, not fewer, should have been used.

Page 80: *[Regarding outliers, i.e., one lab having a widely different result from the other two:]* "'[I]f the 3 labs make the measurement and they all agree, you know they can be lucky and things will be O.K [...]' (Gove, interview)."

"Moreover the three chosen laboratories were all using the same technique, 'an additional reason to make the test less believable' (Gove interview) according to an expert."

Comments: But as seen throughout, because of numerous flaws in methodology, all three labs getting a similar date would not be a guarantee of the reliability of the results. Indeed, Gove cited the similarity of the cleaning technique alone as a reason to "make the test less believable."

Page 119: "Arizona and Oxford measured 14C/13C ratios by AMS and determined the 13C/12C ratios using conventional mass spectrometry. Zurich determined both 14C/12C and 13C/12C quasi-simultaneously using AMS only [...]."

"This also worried some experts: 'So they're not able to make corrections for this so-called fractionation. And Toronto *[Isotrace laboratory]* said that [...] he just feels a little unhappy that they picked labs that don't do this. When they are going to get the result there's going to be a lot of us who look at it very carefully and criticize any aspect of it. That looks a little shaky.' (Gove, interview)."

Page 122: "The spread of the measurements for sample 1 *[Shroud]* is somewhat greater than would be expected from the errors quoted" [from the official *Nature* report].

"As we have seen all along up to now the big fear concerning this test was that there could be an outlier. And that is more or less what happens here, in the sense that at least one lab has a systematic error. This was acknowledged, privately, to me."

"It could be also that two laboratories had a systematic error [...]."

Comments: Notice that it was admitted only in private and not publicly, by presumably one of the members of the three labs involved in the dating, that there had been a systematic error (and possibly two). Laverdiere unfortunately does not name the person making the aforementioned quotes.

Page 134: "Nobody questioned the use of only one technique instead of two. Similarly, the problem of the scatter of the results went unnoticed as well as the possibility that at lest [sic] one laboratory had a systematic error. That is to say that these problems went unnoticed in the public forum, as they were mentioned to me privately by various participants who also expressed concern about the way the results had been plotted, the choice of the controls, the limited blindness, the fact that the three samples came from the same place, and that two laboratory [sic] were not able to make the correction for fractionation."

Comments: This is a scathing critique by the very same people who claimed that the results could be taken with a "95% confidence level."

Pages 141-142: "Backward reading occured [sic], everything having to be reevaluated in a new light."
"An expert was saying in 1987: 'The whole image thing is a complete mystery. It appears not to be painted, it might be the one thing history has established' (Gove, interview)."
"'I just can't imagine an artist a thousand or more years ago, being able to do something in that incredible detail' (Gove, interview)'." Laverdiere commented, "After the test, however, the same expert argued that it is simply 'a clever artist' who did it."
"The expert who thought that '[I]t's [...] like something produced by heat, some kind of a scorch, and there is no scientifically known phenomena that could cause it' (Gove, interview) and that: 'God, we shouldn't reject the notion that there are miracles. Why should we? Miracles are fun.' Gove (interview) felt that Phillips [sic] argument [see comments below] was not to be taken seriously, because the latter had no explanation for how a corpse would produce radiation, and was therefore invoking a miracle."

Comments: The mention of Phillips' argument refers to Harvard University nuclear physicist Dr. Thomas Phillips' hypothesis that the Resurrection of Jesus may have caused a neutron flux that caused a fluctuation in the C-14 content in the cloth. It was proposed in a letter-to-the-editor published in the same issue of *Nature* as the official report. Phillips' unpublished letter in its entirety can be read at https://www.shroud.com/pdfs/n22part5.pdf.

Page 143: Wolfli of Zurich on the D'Arcis memorandum of the 1380s, in which the bishop of Lirey claimed that his predecessor had discovered the artist who "cunningly painted" (sometimes translated as "depicted") the Shroud: "I think this is even more convincing than carbon dating, this memorandum... For it is a much stronger argument than all carbon dating [...]. Nobody ever talked about this memorandum [...]. I read it and was quite surprised: 'why the hell I was participating in a project which was already solved!' (Wolfli, interview)."

Comments: This was an amazing statement by Wolfli. As ballyhooed as C-14 was and is, he was saying that a memorandum that was known to be unsigned, undated and not even certain that it bore the handwriting of Bishop D'Arcis, was actually a stronger argument than the radiocarbon dating test!! There was plenty of literature out there on the memorandum even in 1988 (do a search on "d'Arcis memorandum" at www.shroud.com) and it's certainly not the slam-dunk that Wolfli made it out to be. One can also find medieval statements that the world is definitively flat, but that doesn't make it true. There is also plenty of historical and artistic evidence that the Shroud pre-dates the AD 1260-1390 range the labs gave. A significant problem was the fact that most of the C-14 scientists knew little or nothing about the Shroud's history. Regarding strong artistic evidence, there is an object called the "Hungarian Pray Manuscript," dated to AD 1192-1195 that seems to depict the Shroud. Again, there is an abundance of material accessible on it at www.shroud.com.

APPENDIX TO PART II

Page 147: "[T]he big winners still are the three laboratories selected in the final protocol. Oxford might not have a lot of experience with their machine as it took them years to set it up [...]."

Comments: That fact made Oxford a questionable choice, especially considering there were plenty of other good labs with more experience.

Page 148: "As for the Turin people they remained quite angry against the laboratories involved, feeling that 'they succeeded *[sic]* in transforming the scientific research into a farce' (Gonella, interview)."

Pages 156-157: "I am really sick with this thing that everybody who has clearly stated that the Shroud must be a fake and is a foreigner from Italy is automatically objective, and everybody in Italy is suspect (Gonella, interview)."

Pages 161-162: Discussing the radical act of stopping the experiment due to the scandalous behavior of some: "We would have done that if we were a lay museum, but the Cardinal was a religious institution so every time that we [...] said something, there was immediately, it was not explicit, but [there was the implication] *[brackets in original]* that the Church was afraid (Gonella, interview)."

Comments: Clearly, Church politics allowed a test to proceed that should have been halted due to numerous problematical issues.

Page 165: Regarding Harry Gove being invited to Arizona, against the provisions of a signed agreement, to view their dating: "[In Arizona, they said:] *[brackets in original]* 'after all it is a colleague, and we had him signing a pledge that would not telling anything to anybody [...] we asked him because he was so critical of the new protocol, that we thought that it was best to invite him in order to keep peace in the family [...].' '[I answered]' *[brackets in original and referring to Gonella]* 'we [...] know quite well this mentality of keeping the peace in the family, we call it Mafioso behavior. That is our definition of Mafia [...] you are a mafia. Because the origin of mafia is just that. It is a family, it is a group, and in the group you must agree, and this comes before any considerations of morality with the rest of the world. Thus the killings are just a logical consequence [...] the main thing is to keep agreement with each other, with the carbon community'."

Comments: Although Gove signed Arizona's agreement not to divulge any information, it was revealed in Sox's book and Gove's own book that he did reveal information to his colleague Shirley Brignall. Also, as previously mentioned, Zurich allowed a BBC crew to come to their facility. Apparently, signed agreements weren't worth the paper they were printed on.

Pages 167-168: "Interestingly the exclusion of four laboratories, which led to the invitation of Gove in Arizona and the accusation of a *mafiosi* behavior, also led to complaints, within the group of laboratories, that some people were not sufficiently concerned with the good of the carbon dating community, but were rather devoted to an even smaller group. As we have seen, when the time came for the three chosen laboratories to accept, or refuse to continue they were faced with the choice between loyalty to their institutions, their laboratory, and allegiance to the larger carbon dating community. But it was felt that one lab would go ahead along, showing its allegiance to a smaller group than what was expected. And others reproved of 'his constant desire to push his lab [...] his constant behaving in improper way [...]' (Gove, interview)."

"This lab was perceived as too self-centerd *[sic]*, to an extent that made some wish to see it replaced by another. 'I would like to see X replace Z. Because [the head of Z] *[brackets in original]* has made so snearing *[sic]* comments about the public, snearing *[sic]* comments about the small counter technique, God that would be just marvelous'."

"It was also felt that all those arguments would affect the fabric of the community. 'I guess the thing I resent most [...] is the fact that he [Gonella] *[brackets in original]* has caused this kind of unfriendliness to develop between leaders of the three labs, of the seven laboratories. We were colleagues and in some cases close friends. We're now looking at each other in different ways [...] more suspiciously, and that's very [...] I mean the kind of human interaction that we had which is terribly important has been soured [...]' (Gove, interview)."

Comments: It appears there was plenty of blame to go around.

Page 229: "Alleged failure from the point of view of ethics also prompted someone to say, for example that -- not for himself but for other people he has met -- 'it takes an intellectual effort to beleive [sic] the results of somebody who behaved like that' (Gonella, interview)."

Comments: This is an extremely strong statement. He was saying that it's justified to doubt the results of the labs based just on their behavior alone. Unfortunately, he didn't emphasize it enough.

Page 253, endnote 4: "Although even among the laboratories there might not have been such a consensus about whom to include in the test. One told me that he would have liked 'to chuck' out three of those laboratories, Zurich because they had a wrong result in the pre-test, Rochester because they had the 'wrong machine' and Harwell because 'so far their results have been untrustworthy'. (They were still having an argument about the outcome of a carbon dating test they both had done on the same object.)"

Page 256, endnote 3: Regarding a British television program on the Shroud called "QED": "What hope then of a final answer for the Shroud of Turin which carbon dating has shown does not come from Jerusalem 2000 years ago [...]." But they conclude on a more pessimist view of the ability of science to solve that problem: [...]. 'This is a mystery science is never likely to explain'."

Comments: Some of the mystery can be attributed to the fact that, despite some skeptics' claims, it is not known how the image got on the cloth, so there is still the possibility that the image-formation process could have impacted the C-14 dating in some unknown way.

Page 259, endnote 13: "In the mist *[sic]* of all the arguments about the test one of the expert's previous admiration for a second one turned so sour that the latter was even said by him to have 'as much fair play as a barracuda'."

Page 264, endnote 16: "[D]uring my interviews, […] I was often told that 'any reasonable person' would accept the result of the carbon dating test."

Comments: More than thirty years later, there are numerous "unreasonable" people out there.

Part III
Post-April 21, 1988

1988 *(April)*. Archaeologist Paul Maloney, General Projects Director for the Shroud group known as "ASSIST" had met with Dr. Luigi Gonella on November 21, 1987 to discuss the upcoming sample-taking. During the very time period the sample was taken, Maloney wrote in a document titled "A Brief Evaluation of Current Plans to Carbon Date the Shroud of Turin," "If scientific rigor is followed to the letter, it would have been preferable to have had at least two separate institutions which have experience in carbon dating, overseeing the statistical analysis. The British Museum is eminently qualified in this regard and Dr. Michael Tite is a highly respected member of that organization. However, and with all due respect to the Institute of Metrology in Turin, 'G. Colonetti', by their own admission they do not have experience in the special field of radiocarbon dating statistics. This fact was relayed to me by Prof. Gonella himself. Dr. E. M. Scott, of Glasgow University would have been a wonderful choice. Her experience and trained eye would have enhanced immeasurably the quality of the statistical analysis. I respect all labs involved and I appreciate their honest intent to maintain as objective a stance as possible. But if there is to be no blind dating and knowledge of which samples are actually from the Shroud is readily available to members of each laboratory I think it is really too much to expect that there will not be a leak of information and an informal exchange of notes. This is only human. There are many factors which can affect objectivity. It is of no small concern, for example, that it is rumored that one of the labs has agreed to sell the story for an undisclosed sum of money. And in this regard reporters, persisting to be the first to obtain a story, might well be the link to transmit the information from lab to lab.

So long as it is publicly known that there is no attempt to follow scientific rigor and adhere strictly to the rules of blind testing or to submit to peer review, the results will be suspect no matter what the outcome is. To our knowledge guidelines for reporting those results have not even been laid out so that we can be certain apples will be compared with apples when all the work is done."

Source: ASSIST document "A Brief Evaluation of Current Plans to Carbon Date the Shroud of Turin," 3rd revised version, April 23, 1988. The author has a copy of this document that was sent by Maloney.

Comments: Sadly, just as Meacham's advice to Cardinal Ballestrero in 1985 was ignored, Maloney's advice in 1987 to Gonella was ignored. Maloney was accurate in his predictions: there were leaks; Hall from Oxford received one hundred thousand pounds from ITV, the BBC's rival; and the testing results ended up being suspect. A French physician, Oliver Pourrat, wrote a paper titled "A True Blind Radiocarbon Dating of the Turin Shroud is Feasible," in "La Datazione Della Sindone -- Atti del V Congresso Nazionale di Sindonologia -- a cura di Tarquino Ladu -- Cagliari 29-30 April 1990, Palazzo dei Congressi" (pp. 96-98). He laid out a fourteen-step scenario in which it could be accomplished. Italian author Tessiore also asserted a blind test could have been done in his article "Un Esame All Cieca E' Possibile" in the September/October 1990 issue of *Collegamento Pro Sindone*, pp. 35-36.

1988 *(April)*. Harbottle recounted an intercomparison test done by the British Museum in May 1983, in which they mailed out two samples of unknown origin ("blindfold") to six different labs as a preliminary test for the Shroud dating. He wrote, "The intercomparison was organized with direct reference to the Turin Shroud dating because 'it is generally agreed that any such measurement ought not to be undertaken by a single laboratory, or even by the use of a single technique.' The sponsors also recognized that only through a blindfold intercomparison could the variation among laboratories be honestly determined. The results were to cause many a C14 scientist to utter a fervent 'Amen' to these cautions." The results were released at the Trondheim conference in 1985.

Harbottle also recounted the 1986 Turin planning meeting. It "called for seven samples of the Shroud and seven laboratories (five TAMS, two counter), but necessarily one-to-one. A total of 50 mg (of carbon) would be taken. Dummy samples would also be distributed, with total blindfold encoding by the British Museum. Most importantly 'the taking of the samples will be done so that representatives from the seven laboratories will have complete knowledge of the process'. This step, to ensure the integrity of the 'chain of evidence' and the others were aimed, simply 'to ensure a result that is scientifically rigorous' and 'to maximize the credibility of the enterprise to the public'. Seven laboratories was [sic] not considered excessive -- indeed the number was nearly minimal, given the results of the intercomparison test and the possible need to identify outliers."

"Through the winter of 1987 the laboratories awaited their call to Turin. But on April 27 a cloud appeared on the horizon: Prof. Gonella announced in an interview with 'La Stampa' of Turin that 'only two or three laboratories would participate' in the dating. This news triggered a spate of letters and telegrams pointing out the risk of an outlier occurring, which could probably not be identified as such if there were only two other measurements. Since Prof. Chagas, President of the Pontifical Academy, had helped draw up the 7-lab Protocol, it was obvious that the Vatican had during the winter for reasons unknown placed a wall between the Academy and the Archbishop. This makes one wonder why the Vatican has an Academy of Sciences, if not to provide peer-review on an experiment involving its own property: the holiest relic in Christendom."

"Prof. Gonella's plan received the imprimatur of the Archbishop in a letter on October 10, 1987 to all participants. Three Shroud samples and three laboratories (Tucson, Oxford and Zurich) were specified, but the representatives of the laboratories were explicitly not to witness the sampling procedure. The criteria for choosing the three was 'internationality and experience' [...] here the Archbishop was misinformed since Harwell, left out, had more experience in C14 than the three chosen labs put together. Also the Archbishop opted exclusively for new technology (TAMS) over old, as against the Protocol, that would have used both TAMS and mini-counters."

"The Archbishop's new plan not only disappointed those laboratories that were left out, but even aroused concern among those chosen: the lead scientists from the Tucson, Oxford and Zurich laboratories, Profs. Donahue, Hall and Wolfli, penned a joint reply to the Archbishop's letter. In it they expressed concern at the reduction from seven labs to three, and said that it would 'seriously reduce the credibility of the enterprise'. The mentioned the 'many critics in the world who will scrutinize these measurements'. Abandoning the original Protocol 'will certainly enhance the skepticism […].' They then again warned of 'unrecognized, non-statistical errors'. 'If only three laboratories participate, and one of them obtains a divergent, non-understandable result, the entire project could be jeopardized'. They closed by leaving the door open: 'We would very much like to take part in the program…' (to date the Shroud) 'but are hesitant to proceed'."

"Prof. Gonella, however, prevailed. On January 22, 1988 he met Profs. Damon and Donahue (Tucson), Hall and Hedges (Oxford) and Wolfli (Zurich) in M. Tite's office at the BM. The concern over non-statistical errors, loss of public credibility, safeguards and the blindfold in the procedure, Tite's own earlier advocacy of more than one technique, difficulties in spotting outliers ('only very high deviations can be rejected') with few data -- all were swept aside. Oddly it was agreed that the 'spirit of the original Protocol -- will be retained'. Dr. M. Tite and the Archbishop will certify the samples, each laboratory will have a control sample, and a 'qualified textile expert' will wield the scissors."

"We have then, in my opinion, a shaky experiment, badly designed, innocent of peer-review, and having a reasonable chance of failing to produce a result convincing everyone, for all time, of 'the truth'. The experiment is to be performed by the same scientists who not three months earlier clearly and effectively objected to its terms. But Oxford is heavily supported by the British S.E.R.C. *[Science and Education Research Council]* and Tucson equally so by N.S.F. *[National Science Foundation]*; one will be very surprised if Shroud dating is not prominently mentioned in their next round of grant applications."

"No matter what date is obtained, there will be many, skeptics or devout, who will be disappointed. They already have at hand more than sufficient ammunition to attack any date they do not like. When it is all over one may be forced to ask, like Pilate, 'What is truth'?"

Source: "Carbon Dating the Shroud of Turin" by Garman Harbottle, Senior Chemist, Brookhaven National Laboratory, Upton, NY 11973. Unpublished article originally written for "Perspectives" Section of *SCIENCE* Magazine."

Comments: This article was sent on April 22, 1988 by Harbottle to the late Paul Maloney, who sent me a copy. For some reason, the article was not published, but given Harbottle's credentials and his involvement in the Shroud C-14 dating, I felt it was worth reproducing. Harbottle handwrote a note to Maloney on the title page, saying: "Tite's plan is much worse than I had envisioned!" I have reproduced significant excerpts as shown above. As early as February 1989, Oxford indeed mentioned the Shroud dating in their application.

So, the 1983 comparison test was six labs and two C-14 methods. The 1988 Shroud dating was three labs and one C-14 method (which was only about eleven-years-old at the time). Although Harbottle's fear of one lab out of the ultimate three that were chosen obtaining a divergent result did not officially happen, the results were questionable at best. See entry under "2019 *(March)*" regarding the analysis of the labs' raw data. The comment about the Tucson, Oxford and Zurich hesitating to proceed was laughable. There is no way they would have passed up the publicity they would get by being involved.

Although the January 1988 meeting intended to keep the "spirit of the original Protocol," when the actual sample was taken, not only was the original protocol ignored, but so were some dictates of the January meeting. Although Swiss textile expert Mechthild Flury-Lemberg was originally chosen to excise the sample, Italian scientist Giovanni Riggi di Numana (not a textile expert) ended up doing it. No reason was ever given for the switch. Harbottle's comment, "experiment is to be performed by the same scientists who not three months earlier clearly and effectively objected to its terms," is interesting in light of the fact that he and a co-author had a peer-reviewed article ("Carbon Dating the Shroud of Turin" in *Archaeology Chemistry* IV), which was written before the announcement of the Shroud C-14 results, but had a concluding paragraph appended that read, "As of October 21, 1988, the ^{14}C measurement have been made, and the results sent into the British Museum Research Laboratory. It has been announced that the date obtained falls within the 14th century. The agreement of dates from the three laboratories is an impressive demonstration of the new ^{14}C dating technology." But this was written only a short time after he had said, "We have then, in my opinion, a shaky experiment, badly designed, innocent of peer-review, and having a reasonable chance of failing to produce a result convincing everyone, for all time, of 'the truth'." Was he really convinced by the data alone that the results were accurate? One could be excused for believing something went on behind the scenes that caused him to do an about-face.

1988 (April). Gove wrote, "On 25 April at 11 am, Harbottle called. He had learned from Otlet that the shroud samples had been removed on 21 April 1988. Hall had flown into London on 25 April with the samples in hand and he received a lot of publicity. The archbishop had been, according to Harbottle, furious about Hall's trying to commercially capitalize on the venture. Harbottle also said that the BBC were going to film the measurements at Zurich. He said that, according to Otlet, there was no possibility this time of any outliers because the three labs would consult together so the answers would come out the same. I must say I thought that Otlet was being either paranoid or surprisingly cynical."

Gove also talked to Donahue, who said that Oxford might have problems because they had made some recent adjustments, but the system had not yet been fully tested. Gove was sent on the 26th a clipping of an English newspaper, *The Independent*, which featured an interview of Hall. The article said that Hall had his three samples, the Shroud and the two control samples. The author wrote "He has no way of telling which is which, they are simply numbered 1, 2, and 3." But Gove commented, "Since the samples were not unraveled it would be instantly apparent to Hall which one came from the shroud -- as he well knew. Hall continued to play the 'blind measurement' game." He also told *The Independent* that he hoped an English Sunday newspaper would pay him a large sum of money for the rights to the dating story. Hall told the paper that he would have been 'hopping mad' if his lab had not been chosen, and he agreed with the decision to go with three labs instead of seven because 'You only need one lab to get it badly wrong to confuse everybody and the chances of that are higher with 7 than with 3.' Gove commented, "Hall had conveniently forgotten that in the British Museum interlaboratory tests, it was only because six laboratories had been involved that it had been possible to identify the one outlier measurement. I found the article quite amusing. It was just the kind of publicity that Teddy revelled in."

The BBC was planning to do a documentary on the dating. Neil Cameron of the BBC called Gove on the 27th and said he wanted to film on Saturday May 14. Gove recounted, "He said the BBC would provide $500 plus anything else he could scrape up. He claimed that Woelfli was concerned that STURP might have switched the samples."

Source: Gove's book: Relic, Icon or Hoax: Carbon Dating the Turin Shroud, (Bristol and Philadelphia: Institute of Physics Publishing, 1996), pp. 252-256.

Comments: There have been many questions and suspicions regarding the raw data released by the labs. See entry under "2019 *(March)*."

It's amazing that someone like Wolfli of Zurich, who was so prominently involved in the dating, didn't even realize that the uninvolved STURP group could not have possibly switched samples. It's sad to think he thought STURP would consider doing something of that nature.

I called the late Paul Damon on October 23, 1988 and asked him if the labs consulted with each other about their results, which had been prohibited by the protocol. His answer was a cryptic, "No, not between the three labs at all." One can get the impression from his wording that *some* entities did discuss the results early on. But others claimed that the labs themselves consulted with each other.

Regarding the identity of the samples, Hall could have clarified to the author of the article that he, in fact, would know which one was the Shroud sample, but didn't. And at the press conference held in London on October 13, 1988 (see entry below for that date), Hall had the audacity to talk about the "scientifically trustworthy." Hall was quick to try and capitalize on Oxford's involvement with the Shroud dating.

Gove noted that Hall liked publicity but Gove, by his own admission, as mentioned in several entries in Part I, did as well. On page 268 of his book, Gove wrote regarding the planned BBC documentary, "I felt that since I had been involved with this thing for so long that I had better be in this programme as well." Gove seemed to have no problem criticizing others for behaviors that he himself engaged in.

In a paper not published until 1989 but written before the results had been released, Gove concluded in *Radiocarbon* **31**(3): 969, "The radiocarbon dating of the Turin shroud which the author had envisaged as a convincing test of the power and efficacy of AMS for carbon dating small samples of precious artifacts turned out to be a complex and, in some respects, a rather divisive enterprise. It may be that, although there are many questions that science can answer, there are some that it need not and, indeed, probably should not tackle. Be that as it may, whatever age the shroud turns out to be, the result will be contentious in some quarters, in part because of the inadequacies of the procedures being followed [...]." I would disagree with Gove that it was "rather divisive." It was TOTALLY divisive. Both Church officials and representatives from the C-14 labs exhibited varying amounts of egos, agendas and backstabbings. Note that Gove admitted that the procedures were inadequate.

1988 (April). In notes dated "27 April 1988," Tite had the following entry for 10.30 (a.m.) on the 21st: "The sample was removed from the shroud by Riggi using scissors from above the edge where the Raes sample had been taken. Any material surviving from the side strip was removed. Sample was weighed (400 mg). Material which was possibly contaminated by later stitching was removed from the left hand side and bottom edge of this strip. This strip was then re-weighed (approx. 300 mg). It was decided to give 150 mg to the 3 laboratories and retain the remainder of the sample for possible subsequent measurements in Turin."

Source: "TURIN SHROUD – Thursday 21 April 1988." Official notes of Dr. Michael Tite regarding sample-taking of 21 April. From release of British Museum documents in 2017 due to Freedom of Information Act request.

Comments: The 400 mg figure was not accurate and the 300 mg was stated as an approximation. One would have liked to have seen more precise figures. Indeed, there ended up being huge discrepancies regarding the sizes and weights of the samples, which, needless to say, are rather basic scientific measurements.

1988 *(April)*. Wilson wrote to Gonella, "Dear Luigi, I am sending this letter via Father Rinaldi, and with his encouragement, as a memorandum of some personal thoughts on the timing of any testing of the Shroud ancillary to the carbon dating. As you are already aware, although I continue to be unhappy with the non-inclusion of a proportional counter laboratory (specifically Harwell), otherwise I have broadly supported your decisions on the carbon dating. There has been little encouragement from me for those who have sought to return to the seven lab protocol. I have also warmly approved your new openness relating to press statements on the Shroud and was amused by your catching everyone by surprise on the 21st!"

"I do however feel very strongly that a great mistake could be made if you leave ancillary testing until after public announcement of the results of the carbon dating. Let us consider what would happen if, say, all three laboratories produced a fourteenth century date. Regardless of whether that date happened to be true, public confidence in the Shroud would immediately be shattered. And even if, say, ancillary tests were allowed at some later stage which happened to contradict the carbon dating, it would be too late. The ancillary tests would simply be interpreted as 'rigged' to restore the Shroud's credibility. What happened with regard to the Vinland Map is a case in point. It has become so firmly fixed in the public mind as a fake that whether or not this happens to be true, it would need a minor miracle to change it."

"My suggestion, therefore, is that when the carbon dating personnel reassemble in Turin, you should invite at one and the same time just a few key personnel to carry out a strictly limited program of ancillary tests from the proposals as originally outlined by STURP, ASSIST (?) and ourselves. Importantly, these personnel would not be told the results of the carbon dating until their arrival in Turin, thus safeguarding the results' confidentiality. The invited specialists would then apply their expertise to cross-checking the carbon dating. For instance, in the case of a fourteenth century dating, our mediaeval paintings specialist, Anna Hulbert, would have a key role in trying to identify unequivocal evidence of the hand of an artist from that period. Our two archaeological textile specialists would attempt to satisfy themselves that typologically the Shroud's weave characteristics could not be from anything of the order of a first century date. And by the taking of a few microscopic image samples our Dr. Allen, in consultation with specialists back here in Britain, should be able to arbitrate on the McCrone vs. Heller & Adler controversy. In fact, I would be quite happy to limit our British representation to just these four, on the understanding that among all groups the overriding purpose of these tests would be almost exclusively to check once-and-for-all for the hand of an artist, in the light of whatever date the radiocarbon work has indicated. This is assuming that all three chosen labs have furnished a consistent and meaningful date; if there is any discrepancy in this latter [situation] I would hope that Harwell might be added to those invited to Turin."

"I cannot emphasize enough that however misguided they may be, most of the general public in countries such as my own and the United States expect the carbon dating to 'prove' the Shroud authentic or otherwise. And it is precisely because of this false expectation that some low-key ancillary appraisal, of the kind outlined above, really must precede public announcement of the carbon dating results, and be published in unison with those results. Together, and assuming they are not contradictory, these two independent approaches should then carry a genuine authority impossible from carbon dating alone. I for one would not accept a fourteenth century date without having demonstrated to me some unequivocal independent evidence of the hand of an artist, and I would similarly not expect anyone who believes the Shroud to be a forgery to change his mind solely on the evidence of a first century date. While I have not supported Paul Maloney's recent appeal to the Holy See, his arguments do endorse one point of paramount importance: that carbon dating should be used merely as a guide, and most certainly not as an arbitrator in itself."

"If you do decide on such ancillary testing, obviously the timing must be your decision, but the best plan would seem to be for this to be held in June or early July, synchronous with the re-assemblage of the carbon dating laboratories for discussion of their findings, with the release of all results perhaps in September or early October. I would happily come to Turin to discuss with you, at the shortest notice, a final choice of invited specialists, if you feel all this may be viable. Obviously our specialists would appreciate as early notice as possible; it has been difficult enough retaining their confidence after a two year history of false starts."

"I hope you will convey these thoughts to His Eminence the Cardinal. The carbon dating laboratory scientists will, I am sure, want just to get the results 'off their chests' without the distraction of any ancillary work, and if carbon dating were totally reliable I would be the first to agree with them. But it is not. And whatever date the laboratories arrive at, I believe the Cardinal owes it to Christians and non-Christians alike to allow some independent cross-check before release of what could be one of the most momentous announcements in all history. Assuring you of very best wishes from all of us here in England, yours as ever."

[signed]
Ian Wilson

Source: Letter on April 30 from Ian Wilson to Prof. Gonella, *British Society for the Turin Shroud Newsletter*, https://www.shroud.com/pdfs/gonellaltr.pdf.

Comments: This was more great advice that was unfortunately ignored by Turin.

Gonella had seriously considered using Anna Hulbert, as I found in Gonella's archive an undated memo (but probably written not long after the above Wilson letter) telling a group in Turin, "As scientific adviser to His Eminence the Cardinal Archbishop of Turin, I am responsible for coordinating those who internationally are to be involved in a new scientific examination of the Shroud of Turin. One of the British experts to be involved is Miss Anna Hulbert, B.A., conservator of papal paintings, polychromed sculpture and mediaeval wall paintings, and a member of British Picture Restorers. Miss Hulbert's task will be to examine the Shroud for any recognizable signs that it is from the hand of a mediaeval painter. There are certain mediaeval works that Miss Hulbert would like to examine microscopically prior to her work on the Shroud, and it would be greatly appreciated if you could make such facilities as she needs available to her. Yours faithfully, Prof. Luigi Gonella." One can only wonder what impact she might have had if she had been allowed to examine the Shroud and stated that she had <u>not</u> found indication of an artist's hand.

1988 (May). Senator Moynihan wrote to the U.S. ambassador to the Vatican asking him to try to get Rochester and Brookhaven involved in the dating of the Shroud. But it was too late. Arizona made their first measurement on May 6th. The Vatican and Turin seemed satisfied with letting the three chosen labs complete the process.

Source: Gove, Harry. <u>Relic, Icon or Hoax?: Carbon Dating the Turin Shroud,</u> (Bristol and Philadelphia: Institute of Physics Publishing), 1996, pg. 324.

1988 (May). Gove arrived in Tucson on the 5th in order to observe Arizona's first run on their samples. He was interviewed by newspaper reporter Bill McClellan, who was the son-in-law of Doug Donahue of the Arizona lab. McClellan was aware that Gove had told another reporter that "Gonella was a second rate scientist" and that he "was a Professor of Metrology -- whatever that is -- at the Turin Polytechnic" and asked Gove about it. Gove admitted that he had made the remarks but told McClellan he would appreciate it if he didn't repeat them because "they were somewhat injudicious and impolitic." *[But those remarks were already in the public domain!]*

Gove had bet that the date would be about AD one-thousand and Brignall had bet it would be from the time of Jesus. The loser would buy the other a pair of cowboy boots. Although Gove was somewhat off from the official date, he was closer than Brignall, and she bought him the boots. Gove wrote "The reader, by now, will have guessed that despite the agreement I had signed, I told Shirley the result that had been obtained that day. She and I had been associated with this shroud adventure now for almost exactly eleven years -- there was no way I could not tell her [...]. She has told me that even now, her heart still tells her it is Christ's shroud."

Source: Gove, Harry. Relic, Icon or Hoax?: Carbon Dating the Turin Shroud (Bristol and Philadelphia: Institute of Physics Publishing), 1996, pp. 262, 265.

Comments: Gove once again showed here his penchant for being overly critical of non-C-14 scientists. (On pg. 265 of his book he wrote, "In my opinion both Meacham and Dinegar had been out of their league concerning the Turin Shroud since 1978. Their attempts to stay in the running would be sad if they were not so annoying.) Since one seldom sees the word "impolitic," I'll define it per Webster New World Dictionary: "not politic; unwise; injudicious, inexpedient." Two definitions of "expedient" are: "based on or offering what is of use or advantage rather than what is right or just" and "guided by self-interest." The view of Gove here, not surprisingly, was unflattering. Although he usually spoke whatever was on his mind, the episode with McClellan makes one wonder if there were other instances of inconvenient truths that Gove never disclosed. Ironically, even though Gove was dismissive of Gonella being "a Professor of Metrology -- whatever that is;" it was the same thing as Gove: a Professor of Physics!

Gove admitted that he broke his confidentiality agreement as "there was no way I could not tell her." There actually was a way for him not to tell her -- he could have honored his agreement and simply not told her. In light of all the evidence, Brignall's heart was probably more accurate than either Gove's guess on the date or the official date.

1988 *(May).* According to Petrosillo and Marinelli, "Gove and assistant Shirley Brignall arrived in Tucson on May 5 at 4 p.m. and met with Damon, three days before Sox was to arrive in Zurich and meet Damon. And so the 'strange couple' Gove-Sox 'controlled' two laboratories. Damon and Donahue, who had dated one-thousand and twenty samples in 1987, immediately discovered which of the three pieces of cloth came from the Shroud. A red silk thread and some blue fibrils contribute to further identification. 'The red thread comes from the material that covers the Shroud in the reliquary, and the blue fibrils from the support fabric,' Sox informed. In reality, it is not the support fabric, but a blue satin border. Riggi had assured that the very precise observation of the sample had not revealed the presence of threads of different colors. In Arizona, Damon and Donahue divide the Shroud into four pieces, each of 0.5 square centimeters. The sample received by them must therefore have been two square centimeters. How is this possible, if the initial strip of 7 x 1 centimeters had been divided in half and one of the two fragments then divided into three parts for the three laboratories? In this way, each of these should have had only one square centimeter of Shroud available. It is further proof that the accounts did not correspond. Two of the four pieces of Tucson were treated as in Zurich: diluted hydrochloric acid, diluted sodium-hydroxide and further-diluted hydrochloric acid with intermediate rinse. Damon used, on the other two pieces, special commercial detergents, hydrochloric acid to remove calcium carbonate, ethyl alcohol to solubilize fats and, furthermore, distilled water. Twenty-five to thirty percent of the weight of the original fabric was lost, but the linen remained yellowish in color. In Oxford, however, the piece of Shroud became 'whiter than white.' There the samples, after a first cleaning, were divided into three parts and all treated with hydrochloric acid, sodium hydroxide and hydrochloric acid at 80° C with intermediate rinse; subsequently, two of the three parts are bleached in sodium hypochlorite."

"On 6 May at 9.50 a.m. in Tucson the first dating had been made. Dinegar believed the Shroud was authentic, as Donahue and his wife believed it was from the time of Jesus. According to Damon, however, it dated back to the fifth century. Conflicting views also between Gove and his secretary. Brignall said she was convinced of the authenticity of the Turin cloth while Gove was so sure of the medieval age he engaged in a bet with her. The stakes were not particularly noble: a pair of boots. And Gove won. Immediately after the conclusion of the exam in the laboratory in Arizona, rumors began to circulate that Tucson obtained some incompatible results. In *Nature*, this laboratory will give four measures for the Shroud and five for the samples. Was the fifth measure an outlier?"

"On the 8th the BBC arrived with Sox in Zurich for the 'Timewatch' program, a clear violation of the obligation of confidentiality. Thanks to the information gathered in Zurich, Sox can write his book in August with the results of the exams that will be made known officially only on October 13th in Turin. The BBC filmed the opening of the black plastic box containing the samples, sealed with red tape and marked 'Z' for Zurich. Inside, there are three stainless steel cylinders about 8 centimeters long with the Cardinal's seals."

"The camera filmed Diego Jaggi, Wölfli's assistant, cutting the red scotch tape. Once the box is opened, the cylinders come out, a brown sachet and the certificate signed by Ballestrero which says: 'The containers labeled Z1, Z2 and Z3 to be delivered to ETH representatives contain a sample of tissue taken in our presence by the Shroud of Turin at 9.45 in the morning on 04/21/1988 and two control samples from one or both of the following fabrics supplied by the British Museum: first century and eleventh century. The identity of the samples placed in individual containers has been recorded in a special notebook which will be kept secret until the measurements have been made.' The certificate had been drawn up by Gonella and signed by Ballestrero and Tite. It did not mention the fourth sample. The pieces of cloth are extracted from the three containers and photographed. The sachet with the threads of the cope of St. Louis d'Anjou, supplied by Evin, was also opened. Wölfli complained that the threads have been placed in such a container, that is, in a sachet. The samples are weighed with a high precision scale, a Mettler AE 163. The director noted that the sample Z1 is smaller in size than the measurements given in Turin, 'Moisture lost in flight?' someone asks. The other samples instead correspond to the weights detected on April 21st in Turin. Questioned in July 1989 Wölfli would declare that he did not know the weights of the samples because he did not ask for them. An answer that cannot fail to appear surprising. In November, however, he will let people know that his piece of Shroud was 52.8 mg. The question of the weight of the three samples was also addressed to the managers of the Oxford laboratory. They replied that they had yet weighed them. The piece of the Shroud, after removing a portion for SEM (scanning electron microscope) observations and a preliminary treatment to remove coarse dust, weighed 48.5 milligrams. Its initial weight probably was around 49-51 milligrams. The Shroud fragment examined in Tucson weighed 52.3 mg.; however, that figure did not correspond to any of the three officially-communicated weights. The samples in Zurich were photographed and micro-photographed. Anyone who was familiar with the Shroud would have no doubt which of the three came from it. Wölfli admitted, "All you need is to look at the photographs of the Shroud fabric published in *National Geographic'*."

"After cleaning with ultrasound, the director decided to divide each of his samples in half in order to be able to use a part in case of need for further tests. He had accepted to be involved only on condition that he had enough material to work on. If things had gone wrong, as in 1982, he wanted to be able to have another chance. He was also worried about the results obtained with his mother-in-law's linen tablecloth. The fabric, no older than fifty years old, turned out to be three-hundred and fifty-years-old per ^{14}C. He had justified himself by saying the detergents used on the fabric could have thrown off the date. But Zurich's credentials did not look impeccable. Sox asked Wölfli if the other labs also decided to split the samples the way he did. He said he didn't know and didn't want to know on the grounds that labs were not supposed to be in communication with another until after the results had been sent to Tite and the Colonetti Institute. A conference on ^{14}C was take place in Dubrovnik in June, and Wölfli asserted that if the measures are not completed at that time, he will not go there so as not to be pressured and forced to say something prematurely. The Zurich laboratory had made one-thousand and two-hundred datings in 1987, including some of the relics of saints that proved too old."

"Susan Trumbore, specialist in sample preparation, began cleaning in front of the cameras. Each sample medium is further divided into three parts. A third is treated with fairly strong chemical solutions: 5% hydrochloric acid, 2.5% sodium hydroxide and 5% hydrochloric acid again, with temperatures of 80° C and intermediate rinse. Another third with mild solutions: 0.5 percent hydrochloric acid, then 0.25 percent sodium hydroxide and 0.5 percent hydrochloric acid again with room temperature and intermediate rinse. The last third does not receive further treatments after the initial one with ultrasound like the other samples. Sometimes the solutions were strong enough to dissolve fragile fabrics such as mummies. This must not be done for the Shroud, even if as much contamination as possible must be removed. The first 'line of defense' was cleaning with ultrasound. Contamination of textile samples can be the result of stains, chips, detergents or carbonates due to washing in extremely hard waters. As already noted, the Shroud in Zurich was not particularly dirty. This outcome of the decontamination appears completely strange and in contrast with the observations made previously."

"Sox had written that a strange assortment of debris had been found on the Shroud fabric, from fungi to nylon pieces, 'the surface of the individual fibers has a 'dirty' appearance with abundant deposit of polluting but intimately foreign material connected with the individual fibers of the fabric,' declared E. Morano in 1978. The quantity of foreign material is truly remarkable and such as to represent, at a first empirical evaluation, more than ten percent of the mass of the thread. Tite stated instead that the piece of Shroud had the normal contamination of any sample. Ultrasound detaches fungi, pollen, earth and ash which can postdate the sample. This is placed in distilled water and cleaned for an hour. With heavily-soiled samples, the debris goes away in a small cloud of dust. This was not expected with the Shroud. Hydrolysis with mild acid follows to dissolve any calcium carbonate due to hard water and to easily remove leached organic substances. What was needed for the phial test is cellulose, therefore the fibers must not be destroyed in the process. This occurs when the prepared sample is converted into carbon dioxide. The super-clean specimen -- often a piece of linen becomes snow-white -- is sealed in a quartz tube with copper oxide and a silver thread. The carbon dioxide obtained is then reduced to graphite, which is compressed into balls of one millimeter in diameter. At this point, it becomes target for the radius of cesium atoms."

"Sox described in a suggestive way the last moments of the preparations: 'Neil Cameron and I were around ETH at midnight on our first night in Zurich looking at the surroundings for possible filming sites. There wasn't a soul around, but a few lights were on in various labs and an odd low howling noise came from somewhere. Our imagination saw it as unwilling sample ready for the blast of the accelerator -- or more likely it was the last wail of die-hard sindonologists unwilling to face a wrong date.' This story is not so much praise to the hospitality of Wölfli as an accusation against the Swiss laboratory's incredible lack of confidentiality and respect for the agreements. In this, however, he is not alone. Tucson had invited Gove, opponent of Gonella and Turin's Cardinal, as well as particularly eager for revenge after the exclusion of his laboratory, from the dating. Ballestrero disapproved. Gonella also protested with Damon, who replied that it was a gesture of friendship towards his colleague to 'keep peace in the family.' Gonella replied angrily: 'We in Italy know very well what is behind this phrase *keep peace in the family*.' We call it 'mafia behavior.' Irritated, the consultant of Ballestrero continued: 'It was a mafia. When there are people who consider good relationships with their friends and family more important than the ethical duty towards others, then this is a mafia.' 'On that occasion,' he added, 'we asked that the British Museum express at least its disapproval of the invitation addressed to an outsider.' Dr. Tite replied that it was not his job to be a policeman for the labs."

Source: Petrosillo, Orazio and Emanuela Marinelli. La Sindone: Storia Di Un Enigma, (Milan: Rizzoli, 3rd edition, 1998), pp. 184-190.

Comments: To continue this thread, go to the Petrosillo/Marinelli entry under "1988 *(June)*."

Petrosillo and Marinelli said that Gove and Sox controlled two of the three labs. It's worth pointing out again that a big issue with Oxford was that Hall was on the board of directors of The British Museum, overseer Tite's institution. Tite's lack of concern over Arizona openly going against the confidentiality agreement was troubling.

The Vatican newspaper *L'Osservatore Romano* published an article on May 2nd, in which the dates of the control samples were listed. On page 270 of his book, Gove stated, "[T]he labs might have been wise to consider the possibility that the ages of the controls quoted in the article were wrong and had been supplied to *L'Osservatore Romano* as a ruse to entrap them." That statement probably does not require a further comment from me.

1988 *(May)*. Newspaper reporter Bill McClellan, the son-in-law of Arizona C-14 scientist Doug Donahue, travelled to Arizona during the period that Arizona was performing its testing on the Shroud. He interviewed several of the principle scientists involved, including Harry Gove, who had been invited to be present at the Arizona testing. Gove continued his mantra that STURP were comprised of religious zealots, but then also made a startling admission immediately after that: "'Almost without exception, they were people who honestly believe it is Christ's shroud,' he said. 'It's a well-known fact that scientists can produce whatever result they want. If you believe that passionately in something, you can steer the results. My God, we've all been guilty of that'."

Source: McClellan, Bill. "Secrets of the Shroud." *St. Louis Post-Dispatch*, May 15, 1988, pp. 1, 13-14 on pg. 13.

Comments: Gove never explained how he knew that almost everyone on STURP believed the Shroud to be authentic. Barrie Schwortz, STURP's documenting photographer, told me that team members weren't even asked to disclose their religious affiliations, because it was a scientifically-based project. An independent journalist, Robert Wilcox, actually interviewed many of the STURP members about the authenticity question. In "The Shroud [...] How scientists see it now" published in *Catholic Twin Circle*, April 4, 1982, pg. 3, Wilcox revealed, "According to recent interviews with 26 of the team's core of approximately 32 scientists, half, or 13 believe, or 'lean toward believing,' that the Shroud was in fact the burial cloth of Jesus." Thirteen of twenty-six does not qualify as "almost without exception," especially since Wilcox didn't even interview an additional six of the core thirty-two. Wilcox also mentioned (pg. 3) that, to a person, all thirteen who either believed the Shroud to be authentic or leaned that way conceded "that the Shroud may eventually be found to be something other than Jesus' real burial Shroud [...]." Gove was guilty of overgeneralization, and it's a safe bet that he hadn't seen Wilcox's article or had any other hard data to back his contention.

Gove's overgeneralization was not surprising, but his apparent inclusion of himself in the acknowledgement that scientists can steer results based on their passion is. Since Gove was passionate about his assertion that STURP was comprised of religious zealots, is it unreasonable to conclude that Gove, at least sub-consciously, steered his actions to insure that the results would not produce a first-century date? Given that an enormous amount of publicity, grants, other financial considerations (e.g., Oxford eventually being given a one-million-pound donation to establish a chair) and a perceived battle between science and religion were involved, the Shroud dating was a prime candidate for passion ruling an experiment. Consider this incredible statement found in an article by Cullen Murphy titled "Shreds of Evidence" in *Harpers*, November 1981 (v.**263**), pg. 55: "On one occasion, an Italian shroud-researcher took some tape samples from a colleague at gunpoint." *[!!!!!]*

1988 (May). A letter-to-the-editor by Gove was published in *Nature* in the May 12th issue. Gove wrote, "Sir-Dr M.S. Tite (*Nature* **332**, 482; 1988) has revealed the new procedures decreed by Cardinal Ballestrero, Archbishop of Turin, for radiocarbondating the Shroud of Turin. These differ so remarkably from those of the original protocol agreed by all parties at the Turin Workshop held in the fall of 1986 and chaired by Professor Carlos Chagas, president of the Pontifical Academy of Sciences [H.E. Gove, *Nucl. Instr. Meth. Phys. Res.* **B29**, 193 (1987)], that a brief comparison of the two seems in order. 1. The involvement of seven laboratories has been reduced to three. This eliminates the possibility of detecting a mistake made in the measurement by one or more of the three laboratories. As Tite knows, such mistakes are not unusual. 2. The use of both decay counting and accelerator mass spectrometry (AMS) has been changed to AMS only. The two methods are distinct and independent. 3. The amount of cloth each AMS laboratory receives has been increased by almost a factor of two. With this much material, several more laboratories could have been included. 4. Representatives of the three laboratories will not be permitted to observe the sample removal from the shroud. Tite will be the only independent scientist present at this operation. *[Note: Gove was in error on that point.]* 5. The shroud and control samples will not be unraveled and thus, despite Tite's comments to the contrary, the shroud sample will be much more easily identifiable. 6. The scientific body connected with the Roman Catholic Church which has a high reputation in the world of science, the Pontifical Academy of Sciences, has unaccountably been excluded from official participation in any aspect of this important and controversial radiocarbon measurement. 7. The acknowledged textile expert selected at the Turin Workshop to remove the shroud sample has been replaced by some unnamed person. All these unnecessary and unexplained changes unilaterally dictated by the Archbishop of Turin will produce an age for the Turin Shroud which will be vastly less credible than that which could have been obtained if the original Turin Workshop protocol had been followed. Perhaps that is just what the Turin authorities intend."

H.E. GOVE, Nuclear Structure Research Laboratory, University of Rochester, Rochester, New York 14627, USA"

Source: "Radiocarbon-dating the shroud." [Letter], *Nature* **333**, 110 (12 May 1988).

Comments: Here Gove laid out the many changes that were made after the 1986 Turin planning meeting. It was written before the actual dating, but not published until after it. He suggested at the very end that the Church did not want the results of the test to be credible. The politics evident in every phase of the testing, and coming from all parties involved, is what was incredible.

1988 *(May).* Wilson wrote Tite about seven weeks after the sample-taking, "I don't know to what extent you may be party to the decision on the exact location from which the dating sample should be removed from the Shroud, but although I note you say in *Nature* 'all stages will be fully documented by video film and photography', I hope particular attention can be given to properly professional close-up photographs being taken of the sample site (I am concerned that the video work, for instance, seems to be being done on a friend of Luigi Gonella's home-movie outfit). As I tried to convey to Luigi, the site for removal inevitably has to be at the Shroud's edges, yet the edges are the very feature of the cloth that have so far gone unexamined and undocumented because of hitherto invariably having been covered by a more recent fabric surround. They may contain some quite crucial information relating to (a) the Shroud's original weaving method; and (b) how it has been displayed in earlier centuries. In particular, historically, I want to know if there are any remains of a fringe. It would be tragic if the very taking of the carbon dating samples destroyed some crucial part of this evidence."

"Second, while there is obviously no way in which should be involved in the actual carbon dating work itself, as you note in your *Nature* letter you inevitably have to have a meeting to discuss the results, once the laboratories have fed you their raw data. At that point, whatever date or dates are arrived at, it seems to me that all you physicists will need some form of input from historians and related specialists. For instance, if the date comes out as fourteenth century, it will carry so much more credibility if you can say that specialists in mediaeval textiles now recognize features indicative of a particular mediaeval provenance; and that the handwork of a mediaeval artist has now been conclusively identified. Conversely if the date comes out at first century, it will again be so much better if specialists in ancient near eastern textiles can corroborate this, accompanied by some new and respected assessment of how the image may have been formed, to resolve the prevailing McCrone v[s]. Heller & Adler conflict."

"In this connection you may recall that the afternoon we met Luigi was promising that something of this kind would be allowed before public release of the results, as a result of which I dropped my pressure on him to include Harwell. When I last spoke to him on the telephone a few days ago he told me, to my quiet chagrin, that any such secondary testing would no longer be possible. Yet more recently I have heard that he could be again about to change his mind, but all I would urge most strongly against is any throwing of the carbon dating results -- whatever dates are arrived at -- totally cold into the public arena. As you are well aware, public expectations of carbon dating's accuracy and infallibility far exceed the actuality, In view of the considerable time-scale between the taking of the samples and the 'end of 1988' promised release of results, there seems every opportunity for some form of independent check to the results, and there is no shortage of willing and suitable specialists for this. In the textile field for instance, those whom we hoped to send from Britain, Dr. John P. Wild and Gillian Eastwood from Manchester University's archaeological department, are genuinely independent and uncommitted on the question of the Shroud's authenticity, I am sure everyone would be prepared to respect confidentiality prior to the results being formally released."

"My third, and more minor plea, is simply the personal one of hoping you may keep some form of diary of your experiences in Turin (and indeed after) so that at some later stage I might perhaps be able to include something of these in a third (and final!) book on the Shroud?

Obviously you may want to publish something of this kind yourself, in which case there is no way I would want to pre-empt you, but my main concern is that someone puts some clear account onto paper before memories become hazy, as is too often the case [...]."

Source: Letter dated May 30, 1988 from Ian Wilson to Dr. Michael Tite. British Museum archives, released in 2019 via Freedom of Information Act request.

Comments: Wilson alluded to keeping precise notes on paper, something obviously basic, but yet something not done very well by multiple people involved in the dating.

1988 *(Early June)*. Maloney contacted Gove to see if he had any information on the progress of the testing. Gove told Maloney that as far as he knew Arizona had received about forty mg. According to Bracaglia, there was an assumption that if each lab received about forty mg., there could be some material for the Harwell proportional-counter lab in England. But both Maloney and Gove were unaware that Arizona had received an additional twelve mg. to equal the fifty-two that the other two labs had received. Gove asserted that the sample Arizona received was cut into four pieces and distributed to various locations on campus for safekeeping. He said one sample was tested ten times to be sure of the results.

Maloney was shocked to learn from Gove that he had been invited to witness Arizona's dating, which actually was prohibited by Turin. Gove further told Maloney that Arizona would do a second test in a few weeks, possibly after a C-14 conference that was to be held in Yugoslavia in late June.

Source: Bracaglia, Giorgio. Uncovering the Paradox within the Archives of the Holy Shroud Guild. (Honeoye, N.Y.: Holy Shroud Guild), 2019, pg. 239.

Comments: More details about the Arizona sample per Bracaglia can be found under one of the entries under "1989 *(June)*."

A recent article (July 2020) on the Arizona samples can be found at: https://www.academia.edu/43533848/SHROUD_OF_TURIN_C-14_ARIZONA_2_SAMPLE_REVISED_ZONE_OF_COLLECTION.

1988 (June). Per Petrosillo and Marinelli, "Tucson's results were sent on June 10th to the British Museum. Sox's story stops there. In the final part of the book, the 'anti-Shroud reverend' railed against the old triumphalism, mocking the history of the cloth 'invented' by sindonologists and the 'ridiculous' previous discoveries. He denigrated the members of STURP and trumpeted that it was Gove who snatched the ^{14}C from their hands. He reaffirmed his faith in McCrone's theory, adding that he was happy that the Shroud is false 'because God does not work in this way.' Evidently he is sure he knows how God acts [...]."

"The reconstruction of the story written by Bonnet-Eymard, however, continues with the account of what happened at the conference on radiocarbon in Dubrovnik in June 1988. Gove felt he was there as the main spokesperson and coordinator for the dating of the Shroud. The ^{14}C battle, in a way, was won by him and the hostilities that marked it revealed the existence of a clan of carbonists. When the Dubrovnik conference takes place, Tucson has already submitted their results to the British Museum. Gove did not reveal them, but their outcome can be guessed by the serene tone of his statements. The analyst recognizes that the three laboratories chosen are as qualified as the four excluded and the same dating will come out and if there is no agreement between them — 'and it is certain that there will not be', Gove pointed out -- then 'the result will be credible.' He had reservations, just a month earlier, as expressed in *Nature* but he admitted that perhaps the general public would not find the outcome credible and already declared himself already extraneous to the disputes he foresees will be addressed to Ballestrero and his advisers because they produced a different protocol from that of Turin. It also minimizes, while criticizing it, the exclusion of the Pontifical Academy and the textile expert Flury-Lemberg. A far more serious defect, for the reliability of the test, consisted in the origin of the three samples from the same point of the cloth, an area which was not recommended by STURP."

"Having the three laboratories adopted the same cleaning method lead to the conclusion that uncontaminated contamination will alter the three measurements in the same way. But Gove minimized these serious scientific objections. So much now he knows that if the laboratories converge on a date, it would be that 'expected' by him and his clan, however he criticized the fact that he did not stick to the blind test, so much so that the ages of the three control samples were disclosed at the time of delivery. The ages assigned by analyses could, in his opinion, be accepted with caution. The only real reproach that Gove still addressed was that of having limited the laboratories to three, and with the only AMS method. In this way you could not discover an aberrant result. He knew the result of a laboratory, that of Tucson, and is satisfied with it; he repeated that the current state of the measurements provides a credible result. But if there is a difference with the other two, it will not be possible to know which of the three is wrong and the final result will be invalidated. Gove also criticized the implication of Colonetti, another member of the 'home' group. However, he put all his trust in Tite. He did not miss the opportunity to throw arrows at STURP scientists whom he defines as 'self-appointed religious fanatics'."

Source: Petrosillo, Orazio and Emanuela Marinelli. La Sindone: Storia Di Un Enigma, (Milan: Rizzoli, 3rd edition, 1998), pp. 190-191.

Comments: To continue this thread, go to the Petrosillo/Marinelli entry under "1988 *(July)*."

1988 *(June and July)*. According to Gonella, Arizona had been the first lab to date their sample, on May 6th, followed by Zurich in late June and Oxford at the end of July. Rumors began in May that Arizona had gotten a medieval date; Arizona asserted they had not leaked any data. None of the labs invited a representative from Turin to be present for the datings. Gonella, however, discovered that Arizona had invited Gove and his assistant for their dating. While they did have Gove sign an agreement that he would not disclose anything before publication, Arizona broke their signed confidentiality agreement with Turin by letting Gove attend. Gonella called the Arizona lab to confirm if it was true. On the 15th, Damon and Donahue wrote to Gonella, justifying their invitation to Gove with the desire to "keep the peace in the family" and apologizing if it was considered inappropriate. According to them, Gove initiated the idea. Damon and Donahue also sent on the 15th to the Cardinal their testing results. The Cardinal thanked them but expressed his disappointment to them that they violated the confidentiality agreement. Gonella asked Tite if he was going to express his displeasure at Arizona's actions, but Tite said although he found their behavior reprehensible, he thought about it for several days and decided that it was not his job to "act as a police officer."

Wolfli communicated that he would not proceed with his dating until after the Dubrovnik Radiocarbon Conference June 20-25 so as not to be badgered on the occasion by questions regarding the Shroud dating. When the BBC broadcast a Shroud documentary that summer before the official results were announced, it showed the Zurich lab and gave the impression that it was filming the dating of the actual Shroud sample. Sox stated in public that he was in Zurich when the measurements were made and in his book he describes in detail the opening of the sample containers in Zurich, giving the impression that he actually witnessed this operation, and wrote that Zurich took the measures twenty days *[see comments below]* after Arizona, that is, a month before Wolfli said he had done them. Wolfli told Gonella that the BBC television team with Sox as a scientific consultant *[!]* had visited him in May to take generic shots of the laboratory; Gonella asked him the favor of writing to Tite on which days he had taken the measurements and in which the BBC and Sox had been in his laboratory, to dispel shadows on Zurich about the rumors that were running on the results of the measurements, but Wolfli did not oblige. In fact, at the end of May, Sox told Fr. Rinaldi that he had known that a medieval date had been measured, without specifying whether the information came from Gove or from Zurich.

Gove actually presented a Shroud paper about the dating at the conference, and although he complained about the number of labs having been reduced from seven to three, he declared that the result would be acceptable. As he had been claiming through May that the results would be suspect, Gonella was suspicious of this apparent turnaround by Gove. Gonella said it was unheard of for a scientist who was not officially involved in a project to give a paper on it at a conference. Gonella called

Tite on July 7th to discuss the possible date of the final meeting to discuss all the labs' results and noted that the prior knowledge of the date in anti-Catholic circles would have allowed them to organize a campaign for the propaganda exploitation of results; Tite told Gonella that he thought it inevitable that the result would be exploited in an ideological manner and then warned him that the rumors were not false; the date measured by Arizona was around 1300 and was probably correct because the dates they got on the control samples matched to the known dates. Gonella quietly informed Ballestrero. Rumors continued throughout the world about a medieval date through the end of July when Oxford did their dating.

Source: Gonella, Luigi and Riggi di Numana, Giovanni. Sindone: il mistero continua, (Milan: Fondazione 3M), 2005, pp. 76-77.

Comments: Gonella stated that Zurich's dating was in late June and also claimed that Sox wrote in his book that he and the BBC film crew observed the opening and cleaning of the samples, which was twenty days after Tucson's first dating of May 6th. I was unable to find in Sox's book a reference to the twenty days. Sox stated that he and the BBC crew arrived on May 8th. According to a log that Wolfli (per a document released in 2017 from a Freedom of Information Act request), the first cleaning was on May 13th. So was Sox and BBC there for 6 days? It seems like a long time to have tied up a BBC crew. Wolfli seemingly was trying to give the impression that the BBC arrived long before the Shroud sample was actually opened. Was he perhaps embarrassed that the lab had broken the confidentiality agreement?

As we have seen, Petrosillo and Marinelli recounted in their "La Sindone ..." book that the BBC was there when the sample was opened. Wolfli also had announced he wouldn't do any dating until after the Dubrovnik conference he was attending on June 25-28. But according to his log, the first dating (there were multiple dates for cleaning and measuring samples) was on May 25th (actually nineteen days after the first Tucson dating), which means: 1) he wasn't being 100% truthful in saying a dating wouldn't take place until after the conference and 2) he wouldn't even have been at Zurich for the first round of datings. I suppose one could question whether, as director, he should have been.

As one can see, discrepancies abound everywhere one looks in the Shroud dating.

It doesn't seem like there should be a question about whether violating a signed, written confidentiality agreement was inappropriate.

I was at a gathering the weekend of the July 4th Independence Day holiday in Indiana with other sindonologists when we received a call from Fr. Rinaldi, who disclosed he had already heard the rumor that the date would be medieval.

1988 *(July)*. Leaks begin in papers in England that the dating gave a medieval date, even though Oxford hadn't even performed their testing yet. The labs claimed they did not know who was responsible for the leaks. Accusations against Oxford began to surface. Robert Hedges of the lab refuted the accusation in a letter to *The Times*. Hedges said that Oxford was still installing new equipment.

Source: Bracaglia, Giorgio. Uncovering the Paradox within the Archives of the Holy Shroud Guild, (Honeoye, N.Y.: Holy Shroud Guild), 2019, pg. 239.

1988 *(July)*. French C-14 expert Jacques Evin told Michel Leclercq of *Paris Match* that the "blind test" aspect was important because of "public opinion."

Source: Bonnet-Eymard, Bruno. "The Victory of the Holy Shroud Won By Science." *Catholic Counter-Reformation in the XXth Century*, September-October 1989, No. 223, pp. 28-29.

Comments: Since when is a scientific experiment designed with "public opinion" in mind? According to Bro. Bruno, Evin "recognized that they had to 'cheat' ('*truander*' to use his own French expression) to achieve the 1260-1390 dates" (*CRC*, February-March 1996, no. 238, pg. 9).

Leclerc also asked, "Were you present at the scene?" (i.e., for the removal of the samples). Evin replied, "No, I arrived a little too late." But in February 1989, he told Colombani in a broadcast, "I was present at the removal." (Bonnet-Eymard, Bruno. "The Shroud Daters." *Catholic*

Counter-Reformation in the XXth Century, June 1989, No. 220, pg. 27, fn. 5.) So was he late or was he on time?? He also told Colombani, "I myself brought a piece of 14th century cloth coming from the cope of St. Louis d'Anjou."

French mathematician Arnaud-Aaron Upinsky had plenty to say regarding the labs' statistics. "The significance level of the dating of the Shroud was only 5%! Barely, while the significance of the control samples were 30%, 50% and 90%. This meant that the significance level of the Shroud dating was, respectively, six times, ten times and eighteen times lower than that of the three levels of three other samples! An explanation is essential for the layman. What does the level of this meaning mean in statistics? Three laboratories had measured the same object. For their three results to be valid, they had to be not too far from each other. However, they were very far away. So much so that statistically there was only a 5% chance that these laboratories have measured the same thing. This is what this 5% 'level of significance' meant. There was only a 5% chance that the measure would make sense [...]. The 'level of significance' -- the test at Ki2 is the control criterion that measures this risk and makes it possible to accept or refuse an experiment. In the case of measuring the shroud sample, his answer was: 5% chance! The measure should therefore have been refused. But, in these conditions, how could the custodian have spoken of a 'certainty rate' of 95%? This 95% applied, in fact, to the 'confidence interval that gives the space of time in which it is estimated that the true date is located.' This statistically means that this 95% was that if the measurement was valid, then there was a 95% chance for the true date of the Shroud to be between 1260 and 1390. If the measurement was not valid, if the three laboratories had not measured the same thing, so this 95% no longer made any sense [...]. Such a disturbing finding that this 5% level of significance had not even caught the attention of the signatories to the report. Which allowed measuring its seriousness. On the one hand, we had a Cardinal who accepted a date interval with a certainty rate of 95%; on the other, a report which indicated a significance level of the dating of the Shroud of 5%. With this paltry significance level of 5%, there was a great contrast with the peremptory claim of 1260-1390."

The late French journalist Daniel Raffard de Briene added, "This article *[in Nature],* which we repeat, has remained unique, does not respect any rule of scientific expertise. We should have been able to read the reports of the three laboratories with the details of the methods used, the equipment used and the circumstances in which the tests were carried out, with the raw results obtained with each pass, in short everything that would have allowed us to control the seriousness of the work carried out. None of this was found. Failure to comply with these rules and in the absence of a complementary publication, this article has no scientific authority and, therefore, the results it leads to, whatever they are, cannot be accepted." (**Source** for Upinksy and de Briene comments: Ojeda-Mari, Victor, Linceul De Turin -- L'Imposture Du C14, (Les Edition le Gant et la Plume), 2018, Kindle locations 2573-2648.

1988 *(July)*. Ballestrero wrote to Damon and Donahue, "Dear Sirs, While thanking you for the information that you sent your results to the British Museum, I have to complain of the intervention in the measurement of people not belonging to your Laboratory. This fact, that we had to learn from the press, was at our eyes a breach of the confidentiality that had been promised. The breach is the more obnoxious in view of the hostile attitude taken by Professor Gove in my regards, openly expressed in recent letters to scientific journals where the procedures for this test were heavily misrepresented."

Yours truly,

[signed]
Anastasio Card. Ballestrero

Source: Letter dated July 26, 1988 from Cardinal Ballestrero to Professors Paul Damon and Doug Donahue of the Arizona lab, with copy to Dr. Michael Tite, from British Museum archives, released in 2019 via Freedom of Information Act request.

Comments: I have a copy of the confidentiality agreement that members of the Arizona team (as well as Gove) signed on May 5th and 6th. Given that Ballestrero stated in his letter that only learned about Gove via the press, it appears that he was not informed about Gove having signed it back in

May. It could also be that Gonella did not know Gove signed it because the corresponding typed name was not underneath the signature. See also entry under "1988 *(August)*" regarding a comment Damon made to Gonella about Gove's presence at Arizona.

1988 *(July).* Petrosillo and Marinelli wrote, "In Oxford, sample number one was photographed with magnifications of twenty and sixty times. The fabric appears clean. Gonella suspected that all pollution with the cleaning systems used may not have been removed. The greater dispersion of the data obtained by this laboratory on the Shroud sample shows that the Turin sheet was considerably more polluted than the other samples. Hall said it was brown, but not because of the contamination but due to the effect of age, atmosphere and light. With the 'cleaning' treatment, twenty percent of the fiber surface was dissolved. After dating the dissolved part, after having removed the solid body from it, a date similar to that of the main Shroud sample was obtained. Probably, always according to the director of the Oxford laboratory, only one percent of the surface was contaminated. Hall was contradicted by Evin and Gonella: the fractions eliminated by cleaning have not been dated. The material is small and difficult to extract from the solvent. In data, the identical answer. This only proves that there is no contamination, it was the same sample, before and after the cleaning, that had to be eliminated with the cleaning system used. Even the twenty-one signatories of *Nature* initially thought that Shroud was contaminated, because it has been exposed for centuries to a wide range of potential sources of pollution. In their report, however, they declare that they found the sample clean. This is in singular contrast to the observations made by other researchers."

"At least two laboratories, Zurich and Tucson, failed to commit to confidentiality by inviting two outsiders. But it was after the completion of the Oxford exams, carried out in July, that the first confidential information was proposed. Bonnet-Eymard was very explicit in his accusations: Oxford, knowing that their results coincide with those of the other two, gave rise to the revelations, which were proof of the understanding between the laboratories. If there had been the slightest doubt of a divergence, a premature disclosure would have risked blowing everything up. This is why Tucson and Zurich, in agreement with Oxford, were careful not to let people know about the result obtained. Except, of course, for the special guests such as Gove and Sox. The following statement by Tite is astonishing: 'The results from each laboratory were communicated to the others with a proposal for an agreed date of the Shroud obtained from the samples, but I have not heard from anyone yet.' It was therefore the guarantor of the respect of the procedure that candidly declares that the carbonists had contravened the obligation of confidentiality. As with McCrone's 'machination' of 1980, Bonnet-Eymard recalled, the blow started from London with a similar orchestration. In the *Sunday Telegraph* of July 3rd, the historian Kennet Rose claims to have 'picked up signs' that the Shroud linen was medieval. Riggi was not surprised. He confessed that he expected such rumors 'because much envy have arisen around this scientific event.' There is the attitude of certain Protestants against Catholics and the academic rivalry of the various universities, schools and scientific foundations. The suspects to have launched the rumors fall on the Oxford laboratory. Hall denied it with a letter published in *The Times* on July 9th: he asserted the testing had not yet begun in Oxford. The delay was due to the search for a new ion source and other work in progress."

"On July 27th the BBC aired the broadcast Sox had made, thinking that in those days the results, already known by him, would be made known. At the last moment, the title of the Sox program is changed from 'Verdict on the Shroud' to 'Fragments of Evidence'. The interviewees included Gonella and Ballestrero. The latter seemed to have a psychological detachment from the story by declaring that 'whether or not it is a fabric of the first century does not have much importance. The Church has nothing to fear from the truth. Provided it is the proven truth.' The transmission, as it was easy to guess, was all oriented on the medieval date -- therefore the journalistic indiscretions resumed circulating. Gonella at first refused to believe in their merits and said 'I know the directors of the three centers and their seriousness is not in question, so it seems unlikely that they have failed in the bond of secrecy.' He would soon change his mind."

Source: Petrosillo, Orazio and Emanuela Marinelli. La Sindone: Storia Di Un Enigma, (Milan: Rizzoli, 3rd edition, 1998), pp. 191-193.

Comments: With all of Gonella's previous statements about how the C-14 representatives were "mafia," it's shocking that he was reluctant to believe that they had broken their bond of secrecy.

In a letter sent by Tite shortly after the BBC broadcast (released as part of the Freedom of Information Act release in 2019), Tite wrote to the producer of the program, "I thought that the reference to the Oxford results being 'delayed' was a very appropriate response to the somewhat excessive publicity that they sought when at last actually undertaking their measurement." On a similar note, Sox wrote in a letter to Tite on April 6 (also released as part of the Freedom of Information Act release in 2019), "I was very disturbed that Harry [Gove] sent that letter (January 27, 1988) to Sir David Wilson, and as one of his oldest friends told him so. It was totally uncalled for. But equally disturbing is what a friend in the BBC told me about Teddy Hall's 'offer' to them. I am certain Turin would be horrified." The labs and individuals clearly wanted to capitalize on the publicity they were getting for having been involved in dating the Shroud.

1988 *(August)*. Petrosillo and Marinelli continued, "Oxford ended its testing on August 8th. At the end of the month, returning from holidays in Greece, Tite found all three results on his desk. After a few days, he returned the statistical analysis and calendar dates to the laboratories. Checkmate to the Church. Before Tite was able to officially view the report of the three laboratories, a new, and this time circumstantial indiscretion put the mass media and international public opinion in turmoil. On August 26th, the *London Evening Standard* announced to the on the front page, that the Shroud is a fake executed in 1350. The scoop bears the signature of Richard Luckett, a professor from Magdalene College in Cambridge. The news spreads and was picked up by all Italian newspapers and by many foreigners. Hall was surprised and declared: 'I don't know who the hell Richard Luckett is -- he has nothing to do with us.' Gonella in vain tried to deny the indiscretions. 'I am really annoyed by this dripping of rumors and inferences. I have not been officially informed of anything and consider those from London to be non-news.' If indeed the Oxford scientists had leaked something about the results of the analyses, there would be a clear violation of the procedures and pacts they have claimed and signed. This 'Mr. Luckett,' added the irritated Gonella the day after, 'has no titles to speak of and makes judgments. In twelve years that I have worked on the Shroud, I have never heard of him. For me, after all these inexplicable allegations, there is a group in the world that pursues antireligious purposes in bringing out such uncontrollable news. The intent is to put the ecclesiastical authorities in difficulty and the scarce consideration for them is evident. While all over the world it is said that the Shroud turned out medieval and was therefore a fake, the legitimate owner and the pontifical custodian have not yet been informed. And the consequent lapse of time between the indiscretion and the publication of the results has been characterized as delaying maneuvers of the Church, reluctant to accept the 'scientific verdict' on the Shroud'."

"Hall said, 'The person who set the date tells me that Professor Luckett is completely wrong. I cannot say what the real date the Shroud dates from, but it is not the one indicated by him.' For Hall, none of the researchers violated the secret and the Cambridge docent worked only by fantasy. In the meantime, Wölfli denied Sox's claims that he attended the exams in Zurich and from that drew further reason to believe that the Shroud was not authentic."

Source: Petrosillo, Orazio and Emanuela Marinelli. La Sindone: Storia Di Un Enigma, (Milan: Rizzoli, 3rd edition, 1998), pp. 193-194.

Comments: To continue this thread, go to the Petrosillo/Marinelli entry under "1988 *(September)*."

Hall claimed that Luckett's 1350 date was "completely wrong," yet when the dates were announced in October the date range was 1260-1390. How was 1350 "completely wrong?" Luckett himself was quoted (in Marinelli's Valencia paper, pg. 12): "Laboratories are rather leaky institutions." Since Hall claimed that the date was wrong and thus those associated with the test didn't disclose the secret, does it not follow that the exoneration was incorrect? In fact, Hall was quoted (per Marinelli's Valencia paper, pg. 12) as saying, "Frankly, I think it was a hopeless prospect to keep the result secret. You couldn't. With the best will in the world." Secrets can be kept and confidentiality agreements can be honored when integrity is deemed to be important, but neither one happened in this case.

Ian Wilson, in the Laverdiere dissertation (pg. 164) said, "They are a bit of a club anyway … They all know each other, whether they're proportional counter ones or AMS ones. [So] it is quite a big danger … that one of the laboratories is going to ring up the others and say: 'What's your reading?' And you get them sort of going back and lining up their result. I am not saying it's the way it will happen but it is a danger." Gove admitted in his 1999 book From Hiroshima to the Iceman (pg. 163), "There is no way of completely ruling out the possibility that the three laboratories discussed the results of their measurements amongst themselves and made appropriate adjustments to the data before releasing it to the British Museum and, in particular, that they conspired to produce a fourteenth century date for the shroud […]."

Regarding Wolfli's denial about Sox and Zurich, we saw earlier that Sox brought a BBC crew to Zurich in May and Wolfli had previously acknowledged that they had at least taken some generic scenes. But Wolfli was denying at least a perceived claim that the BBC filmed the Zurich dating. We also saw previously that Wolfli had stated that Zurich wouldn't perform the test until after the C-14 conference of June 25-28 yet his log showed the first dating on the 25th. So we have seen outright lying and/or less-than-forthcoming behavior from all of the three directors of the three labs. One certainly has the right to say that since the directors were guilty of the aforementioned behaviors, "I'm mistrustful of anything you tell, even the final results of your datings." Riggi said (Marinelli's Valencia paper, pg. 12), "The laboratories committed themselves on their honor to provide that nothing would have leaked. Instead, they have exploited the research, they use the rumors to promote themselves. For sure they don't come out clean."

1988 *(August)*. On the 5th, Gove viewed a BBC documentary on the dating, which included footage of the sample-taking. Gove noted, "The casualness of this operation was emphasized by the fact that he was not wearing gloves! So much for the sterile environment and procedures in handling the shroud Riggi had emphasized at the Turin workshop."

Source: Gove, Harry. <u>Relic, Icon or Hoax?: Carbon Dating the Turin Shroud,</u> (Bristol and Philadelphia: Institute of Physics Publishing), 1996, pg. 276.

Comments: I have also seen a video clip and a still photo of Cardinal Ballestrero leaning on the Shroud with his elbows firmly on it. The casualness of the situation was extremely shocking.

1988 *(August)*. The BBC broadcasted a program called "Shreds of Evidence" in their "Timewatch" series. It indicated that the Shroud would likely date to the medieval period. Because the program featured only one Shroud expert, the Rev. David Sox, some members of the BSTS believed he was the leaker. They believed he had informed Luckett and actually gave him material for his article in the *Evening Standard*. How had Sox received his information? Gonella suggested to Damon that Sox had received his information from Gove, who had present for the first testing in Arizona. Rinaldi confronted Sox about the accusation.

Source: Bracaglia, Giorgio. <u>Uncovering the Paradox within the Archives of the Holy Shroud Guild</u>, (Honeoye, N.Y.: Holy Shroud Guild), 2019, pp. 240-242.

Comments: Sox wrote Rinaldi in early October denying that he was the leaker. Bracaglia commented that "there never was a definitive answer regarding who was responsible for the leak." See Petrosillo/Marinelli entry for "1988 *(September)*" for a comment by Sox regarding the accusation.

1988 *(August)*. Edward Hall, head of the Oxford lab, noted odd fibers in the C-14 sample. Hall enlisted the opinion of Peter South of Derbyshire lab, who concluded, -- the rogue fibers were fine dark yellow strand *cotton* [...] and *may have been used for repairs in the past [...]*" [my italics].

Source: "Rogue Fibers found in the Shroud." *Textile Horizons*, December 1988, pg. 13.

Comments: This is another indication that even experts were acknowledging that the area was apparently not homogeneous. The finding of the cotton was even mentioned in the famous *Nature* report.

1988 (August). Damon sent a letter to Gonella, "I received a Fax copy of the article by Charles Langley entitled 'Shroud of Turin Really is a Fake'. I am dismayed as you must be concerning all of these rumors and rumors of rumors with the sources unattributed or incorrectly attributed."

"Everyone here, who was privy to our results, signed the enclosed pledge of secrecy, including Harry Gove. I am quite confident that no one of us was source of any of these rumors. We are disturbed that you and Cardinal Ballestrero were offended by our including Harry Gove as an observer to the initial analytical work. At least, he now knows the care that we took on these important samples, and I have good reason to believe that he is not guilty of a breach of confidence. The enclosed letters *[not reproduced here]* are one of the factors that lead to my confidence. I can also assure you that he was very discrete in Dubrovnik. I tried to telephone you without success but Jacques Evin assured me that he had reached you and kept you informed."

"We hold you in great respect and in no way want to be a cause of distress to you or to Cardinal Ballestrero."

Sincerely,

[signed]
Paul E. Damon, Professor.

Source: Letter of August 26, 1988 from Paul Damon to Luigi Gonella. Luigi Gonella archives, of which the author has a copy.

Comments: I find it troubling that Damon had agreed in principle to sign a confidentiality agreement, but proceeded to allow Gove to observe. I don't believe it reflected well on Damon's integrity.

1988 *(September)*. Petrosillo and Marinelli wrote: "In early September, Gonella replied to every indiscretion by pointing the finger at the laboratories, including the failure of the blind test that had long since been shelved as a pious illusion and a way to avoid suspicions of illegal agreements. 'If some researcher has spoken it means that he took the trouble to check which of the samples delivered to each of the three labs came from the Shroud. We had trusted each other, now we are disappointed'. Towards the end of the month, the revelations become more concrete: this time it was Dinegar, a STURP member, who confirmed that all three tests results were medieval. Gonella exploded in bursts: 'We still have not communicated anything. It is unwise behavior.' And again: 'They had given us their word, now they have betrayed it'; 'They treat us like an underdeveloped people'; 'Here we do not go into the merits of the results, which will be for science to confirm in their validity' [...]."

"It had become increasingly clear that a delay imposed on the official publication of the results would appear to be the Church's refusal to admit the medieval origin of the fabric. The confirmation of this supposition came from the newspapers [...]." (Gonella continued), 'In September, the tone of all the questions that journalists from all over the world asked me, especially from abroad, was:' 'Does the Church decide not to finally announce these results? Is it afraid of the truth? Is it putting pressure on the laboratories?' 'I'm really starting to think that there was someone who wanted to put us in an embarrassing situation, to get us nervous. Certainly, if during this very tense year, someone in the hierarchy of the Catholic Church had said: *No, we refuse this result,* the others would have taken a deep breath of relief:' 'Here, we have proof that when the Church finds herself in front of a scientific result that she does not like, then she does not accept it'. 'And in this trap a person of Cardinal Ballestrero's diplomatic experience didn't fall for sure. They did everything possible to arm it and make us fall into it. This operation was conducted in a scientifically unspeakable and condemnable way from many points of view. The story has greatly diminished the esteem for the scientific world in general'."

"A few months later Hall gave a different version of the facts. Hall attributed the leaks indirectly to a Gove press conference without explicitly accusing him. The American said: 'I could even bet that it is medieval.' As for the indiscretions, he managed to claim that it was difficult to deny the rumors, given that they were true. Finally, we learn that behind the rumors there was the Sox saga. Sox stated, 'May I be damned if I let the blame fall on me.' One thing is certain: the book of Sox was already printed at the end in September, that is, more than two weeks before the official publication of the results. The reverend has shown that he is very well informed on many confidential aspects of the affair thanks also to information passed to him by Gove. Having fought the rumors for weeks, Gonella finally surrendered: the accuracy of the leaked news has been confirmed by Tite himself. However, he was irritated and outraged by the behavior of the laboratories. His speech was halfway between the outburst and invective: 'The keepers of the Turin cathedral have acted more seriously who kept silent about the removal of seven centimeters of the cloth than a group of scientists, who allowed themselves to violate the secret is to announce in scandalistic newspapers that the Shroud is a medieval fake. They treated Turin as a Third World country. Cardinal Ballestrero was treated as they would not have dared to do with the director of a provincial museum. They treated him like a country parish priest'."

"'These gentlemen scientists did not trust the cut of the Shroud *[to the Church alone]*, did not trust the Cardinal and came *en masse* to Turin, but then they did not allow a representative of the Church to participate in the research and to assist in the analysis as an observer. I have already said it and repeated, for me there is an anti-Catholic conspiracy of certain well-defined environments, and they cannot accuse the Church of having evaded the examination of ^{14}C'."

"After much discussion, when no one doubted the outcome of the analyses, Tite finally communicated to Turin the results of the operations on September 28th [...]. The secretary of Cardinal Ballestrero, Fr. Giuseppe Caviglia, was instructed to personally communicate the results of the exams to Cardinal Casaroli, Secretary of State. 'It was now time,' said Gonella, 'to make the public announcement. The proposal of the laboratories that it should be the ecclesiastical authority to communicate the results to the public was well accepted, because, whatever the outcome, it was appropriate to frame the religious implications well, so poorly understood by the public (both Catholic and anti-religious)'."

Source: Petrosillo, Orazio and Emanuela Marinelli. La Sindone: Storia Di Un Enigma, (Milan: Rizzoli, 3rd edition, 1998), pp. 194-197

Comments: According to Marinelli's Valencia paper (pg. 12), the original plan for informing the Cardinal was that the labs were supposed to have sent the data from their testings to the Colonetti Institute in Turin for statistical analysis, they did not do so. The representatives of the labs had also been expected to meet in Turin to prepare a scientific report that was to be given to Cardinal Ballestrero, but Tite sent the letter on September 28th instead. She also said that there were rumors that the representatives from the labs had secretly met during the summer in Switzerland.

Regarding Sox's role, we will soon see that several individuals implicated him as the leaker.

1988 *(September)*. The Sunday *London Times* had a headline on the 18th: "It is official -- the Shroud of Turin is a fake." Gonella protested vigorously with the English press and laboratories, especially Oxford. Hall phoned Gonella from Australia telling him that he himself did not know the date because he had left everything to his assistant Hedges and that he thought the source of the rumors was Gove. Tite admitted that rumors of such consistency could not be invented by journalists, but said that it was inevitable that someone would speak. The leaders of Oxford continued to deny having given any information, saying that it was journalists who spoke of the 14th century, this being the "most probable" date. In September, the strong journalistic interest in the Shroud took a very nasty turn because the media by now was already convinced that the date was medieval, and they went on to insistently ask why the Catholic Church did not decide to announce the result and to hypothesize "pressures" of the Vatican for the laboratories to change the data, the accusation of wanting to hide or filter the news was spelled out in various places, for example by Prof. Firpo in *La Stampa* of Turin on September 25th.

The delay in the official communication of the results to Ballestrero was thus openly exploited in an anti-Catholic sense and Turin could do nothing but protest against the failure of the laboratories to the established procedures and to repeat that he had not yet received any communication rectors did not contact the Colonetti Institute and did not send their results in parallel with the British Museum. In April there was a maneuver aimed at excluding the participation of the Colonetti Institute: following journalistic comments (taken from the Italian press by the English one) that attributed to Colonetti and the British Museum the role of "guarantors of the operation," a group of Turin researchers wrote to the Director of the Institute, Prof. Bray, asking him not to involve the Institute in such an ambiguous position, and Bray wrote Gonella expressing his perplexities in this regard and asking for clarifications on the role that Colonetti should play; Gonella reassured him that he was not asking for "guarantees" but only the analysis of the results, and Bray told him what data he needed to be able to perform a correct analysis. Bray was surprised he hadn't heard from the labs, considering they had requested another institution besides the British Museum to assist with the data analysis. With the prevalent rumors about the medieval dating, Colonetti again expressed the desire to stay out of an event that took on the aspect of a propaganda operation rather than a scientific research project; Bray agreed to remain in the project only as a personal favor to the Archbishop of Turin, as the withdrawal of an institute based in Turin at a time when the mass media denounced the Shroud as fake would certainly have been misinterpreted -- he consulted with Prof. R. Levi, a well-known metrologist at the Turin Polytechnic.

When Bray had not heard from the labs in June, Gonella had written on the 30th to their representatives (and copied to Tite) listing the information requested by Bray and specifying the limits of its possible performance. The laboratories sent only part of the requested information, and they did it through the British Museum, without direct contact with the Colonetti. The first data this Institute received was a summary of the results mailed by the British Museum on August 22nd along with additional information from Arizona and the complete results with additional information from Oxford. On September 7th additional information and measurement data corrected with respect to the first version (with British Museum being copied) was received by fax from Zurich. On September 14th, the revised version of the Zurich's results was received by fax from the British Museum. Finally on September 20th, the complete Arizona data by fax from the British Museum was received. Colonetti Institute immediately sent their comments to the British Museum, and then on September 29th they forwarded them independently to Ballestrero, having learned that the British Museum had communicated the results to him. Bray put forth his comments in the strictest metrological language, making it clear that he could only speak on the basis of the data communicated to him (of which he listed the source and date of transmission), and therefore could not provide any evaluation of the measurement method; this meant that he was unable to analyze any errors.

The institute noted that the dispersion of the results of the three laboratories for sample No. 1 (the Shroud) was greater than for the other samples, indicating that this sample was more sensitive to the differences in procedure between the various laboratories, and commented that this could have avoided with a better standardization of the measurement method (this anomaly was not detected by the British Museum statisticians). The comment allowed a professional metrologist to understand that Colonetti's rigorous criteria would have given the result a wider range of uncertainty. There was no need to explain this criticism outside the more strictly technical environments, Bray observed, discussing the systematic with Gonella, but he had to limit himself to evaluating the statistical coherence of the results. This appeared to be a comment, because in any case it was not such to considerably shift the date found, and therefore its emphasis on the general public would have been misunderstood in the media climate of the moment. Once the situation created in September with regard to the press, which worsened day by day, Turin repeatedly asked Tite to proceed as soon as possible to convene the final meeting to officially define the results. It could not be excluded that the anticipated indiscretions on the results and the delay in the official communication were part of a conscious maneuver to put the Catholic Church in a bad light, causing some unofficial authority or spokesman to issue statements on the authenticity of the Shroud that would later be denied or anti-scientific statements that could be interpreted as *a priori* rejection of undesirable results. Prudence required that Ballestrero scrupulously observe the agreed procedures, especially when they were dealt with by the others.

Tite tried to blame the Colonetti for the delay. Bray, in the meantime, was tormented by telephone calls from journalists who had received his name and telephone number from London (not from Turin) and asked for information on the progress of the work; to them he replied with only one sentence: "In collaborative works at this level there are precise deontological rules that our Institute is used to respecting." Tite concluded that it was useless to call a meeting to define the result because there were no discussions to be made on the measurement data, which were very consistent between them, and that he could simply communicate them officially to Ballestrero with a letter. Turin did not oppose this solution because a further delay to hold a meeting that scientists considered to be of a purely formal nature would certainly have been interpreted in the journalistic climate of the moment as a delaying maneuver on the part of the Church. Tite then communicated the results with a letter dated September 23rd, which Ballestrero received on the 28th. The letter contained the date ranges assigned to all four samples, a thank-you on behalf of himself and the laboratories for having been allowed to participate "in what was a very exciting project," and the regret of problems caused by the badly-informed comments of the press in recent weeks. Even before receiving this letter Ballestrero had dealt with Casaroli on the question of how to communicate the results to the public, which had to be taken care of in every detail to make clear the Church's position in the confusion created by the journalistic clamor on the affair. A first draft of a press release was submitted to the Vatican Authorities on September 26th. (The final text was finalized by October 10th.) Also on the 26th, Tite sent to Gonella a draft of the official report to be published in a scientific journal; Tite asked Gonella if he wanted to add any comments but Gonella declined.

Ballestrero asked if it was not convenient to give the announcement directly from the Vatican, and Casaroli decided that the communiqué would be issued by Ballestrero in Turin, to emphasize the Turin competence on the Shroud, at a press conference presided by the Director of the Vatican Press Office, the late Dr. Joaquín Navarro, who emphasized how he spoke on behalf of the Holy See.

Source: Gonella, Luigi and Riggi di Numana, Giovanni. Sindone: il mistero continua, (Milan: Fondazione 3M), 2005, pp. 78-82.

1988 *(September)*. Ian Wilson wrote to BSTS members, "Dear Members, In view of many still unresolved questions concerning the Shroud carbon dating results, this is an interim letter, for members' guidance and information, prior to a full Newsletter, which will follow as soon as official results are released, and the situation is clearer. As members can scarcely fail to have been aware, ever since early July there have been a spate of press rumors that the Shroud has been carbon-dated to sometime in the mediaeval period. The rumors have chiefly come from this country, and began with a 'gossip' piece in the *Sunday Telegraph* 'Albany at large' column of July 3, intimating 'In spite of the intense secrecy surrounding the investigation I hear signs that the linen cloth has been proved to be mediaeval.' Other media around the world picked the story up and assumed the source must have been the Oxford laboratory, obliging Professor Hall and Dr. Robert Hedges to write a letter to *The Times* protesting that as at that date [July 9] they had not even begun the processing of the Shroud samples, due to the installation of new laboratory equipment."

"Towards the end of July the rumors were rekindled as a result of pre-publicity surrounding the BBC Timewatch television program 'Shreds of Evidence'. The program itself (transmitted 27 July) leaned heavily in favor of mediaeval date, despite the fact that Oxford had still not yet completed its work on the Shroud. The program had just one 'expert' consultant, the Revd. David Sox."

"Hardly had this wave of publicity died down before on 26 August the *London Evening Standard* ran as its front-page lead story "Shroud of Turin Really is a Fake". Accompanying this was a seemingly authoritative article by librarian Dr. Richard Luckett of Magdalene College, Cambridge, cryptically remarking that 'laboratories are rather leaky institutions' and 'a probable date of about 1350 looks likely'. This again generated media stories all round the world, yet both the Oxford laboratory and Dr. Michael Tite of the British Museum insisted that they knew nothing of how Dr. Luckett had come by his information, and had had no dealings with him. When in a telephone enquiry to Dr. Luckett I asked whether the Revd. David Sox had been his source, he hastily changed the subject."

"On Wednesday 14 September the *Sunday Times* contacted me requesting me to supply three panels of information on the Shroud, its known history, and the background to the carbon dating, to accompany an article in which the Science Correspondent would set the scene for the formal announcement of the dating results at the end of this month. The material was faxed to them, then just prior to publication I was told that the plan had been radically changed because of new information that the Shroud had been dated to between 1000 and 1500 AD. The Science Correspondent refused to divulge his source for this, and since my own contribution had become reduced to one panel that bore little resemblance to the original, I withdrew my name from this. On 18 September the *Sunday Times* carried the front page headline 'Official: Turin Shroud is a Fake', accompanied inside by the Science Correspondent's full page feature 'Unravelled: The Riddle of the Shroud'."

"This included some of the background material supplied by me, plus the new 'leaked' information on the dating, which although described as 'official' was backed up by no directly quoted source. Since checks with Professor Hall of Oxford and Dr. Tite of the British Museum again established that neither had been responsible, I complained to the *Sunday Times* Editor with particular regard to the 'official' headline. This prompted a conciliatory phone call from the Science Correspondent who when challenged directly, admitted that his source had been the Revd. David Sox. He said he had in front of him the Revd. Sox's already complete book about the Shroud's mediaeval date, awaiting publication the moment this news becomes formally released."

"Sadly, as evident from a *Daily Mail* article of September 19, Professor Gonella and Cardinal Ballestrero in Turin have recognized the succession of apparent 'leaks' emanating from England to malicious breaches of confidentiality on the part of the Oxford laboratory scientists and Dr. Tite. It seems clear that they have been mistaken, and that the true source of possibly all the leaks is the single non-English clerical gentleman whose identity will now be self-evident. This individual's means of obtaining his 'inside' information (which can only have come from Arizona or Zurich), and his motives for flouting the confidentiality that all others have respected, can only be guessed at. His only explanation to me was that he 'thinks' he knows the result by a 'fluke'. Not being party to the same source (s), I can neither confirm nor deny the information's truth, only deplore the insidious and underhand means by which it has been disseminated."

"If indeed all three laboratories date the Shroud to the Middle Ages, then of course this news would be serious, unsettling and deeply disappointing to those of us who have pursued seemingly valid evidence for the Shroud's authenticity. But it would be wrong for any of us simply to follow unquestioningly the inevitable slick 'fake' judgments of the media. As was made clear in *Newsletter* no. 14, carbon datings are not infallible, and certainly not 'proof' in themselves of anything. Only if and when someone demonstrates beyond all question (perhaps by replicating Dr. McCrone's findings) how an artist produced an image as extraordinary as that of the Shroud, or how, a crucified body did so, will the Shroud enigma genuinely be near to a solution. Until then, subject of course to your own support, your Society will continue its own quiet existence, its research efforts now redoubled to persuade Turin to allow the ancillary image analysis work which should have accompanied the carbon dating, but which (as explained in the last Newsletter) Professor Gonella ultimately refused."

"As previously remarked, *Newsletter* no. 20 will follow in the wake of the release of the true official results, due from Rome and the British Museum either at the end of this month or early next. A BBC QED film, with new information on the Liverpool image, will be transmitted at around this same time."

Yours sincerely,

[signed]
Ian Wilson, B.S.T.S.

Source: "On the recent 'Leaks'..." Memo from Ian Wilson to BSTS members on September 23rd. https://www.shroud.com/pdfs/bstsleaks.pdf.

Comments: The "Liverpool image" referred to in last line of the Wilson letter was a pancreatic cancer patient whose body image was imprinted on his mattress after his death. See: https://www.shroud.com/pdfs/imprint.pdf.

1988 *(September)*. Tite sent an official letter on British Museum stationary to Cardinal Ballestrero. "Your Eminence: I am writing on behalf of the laboratories at the University of Arizona, Tucson, the University of Oxford and the Federal Institute of Technology Zurich (ETHZ), to advise you of the results of the radiocarbon dating of the samples that you provided them with in April 1988. The dates for the four samples given below represent the calendar age ranges with at least 95 percent confidence (i.e., the true ages have at least a 95 percent probability of lying within these ranges). These calendar age ranges were determined from the mean of the radiocarbon dates for the three laboratories using the high-precision calibration curve published by Stuiver and Pearson (*Radiocarbon*, Vol.**28**, pp. 805-838, 1986)."

1	Shroud of Turin	1260-1390 AD
2	Nubian linen (dated historically to 11th/12th centuries AD)	1020-1160 AD
3	Egyptian linen (dated by the conventional radiocarbon method in the British Museum to 1st century BC/1st century AD)	10 BC-80 AD
4	St Louis d'Anjou cope (dated historically to late 13th/early 14th centuries AD)	1260-1290 AD

"I am further able to confirm that the 2-sigma error ranges (i.e., representing 95 percent confidence) on the radiocarbon dates obtained by the three laboratories overlap for each of the four samples."

"I would also advise you that a paper giving the full details of the radiocarbon dating results and their statistical assessment is currently in preparation and will be submitted to a scientific journal, possibly *Nature*, as soon as possible. I will of course keep you informed of developments in this respect."

"Finally, I should like to take this opportunity, on behalf of the three laboratories and myself, of thanking you for allowing us to participate in what has been a most exciting project. At the same time I also wish to say how very much I regret the problems caused by the uninformed press comments during the past few weeks."

I remain my Lord Cardinal,
Yours faithfully,

[signed]
Michael Tite
MS Tite

Source: Letter dated September 23rd from Dr. Michael Tite to Cardinal Ballestrero, from British Museum archives, released in 2019 via Freedom of Information Act request.

1988 *(October)*. Meacham wrote to Dreisbach, "Dear Fr. Kim, I am most impressed with your commitment and willingness to defend the Shroud in the midst of a 'Shroud typhoon' which will probably have hit the shore by the time you receive this. The preponderance of evidence angle is without doubt the way to go, but along with that we must focus on why the C-14 date cannot be taken at face value."

"Fr. Otterbein wrote me that 'people will forget all the evidence in favor of authenticity if the Labs announce a medieval date.' I think we must persist in pointing out that no single test or measurement is infallible, especially not C-14, and the bulk of the scientific evidence still indicates that the Shroud image is the body imprint of a crucifixion victim. The C-14 sampling has been so badly done that no one should conclude anything until at least two more samples have been taken from different sites on the cloth, pretreated with some of the most sophisticated methods available, separated into chemical and perhaps thermal fractions, and dated. If after all that the result comes out the same time then we must consider that the Shroud does not date from the time of Christ, recalling however that this will still not have <u>proven</u> but would be an explanation consistent with the Shroud's radiocarbon age."

"I hope that STURP and ASSIST will not abandon their interest and efforts to do further testing, since there is of course still much to be learned, and certain types of tests may well shed light on the nature of the C-14 sample and possible anomalies in that area."

"Now what occurs to me is that all those who have studied the Shroud need to come together in pressing for the ancillary testing and for another round of C-14 measurement, with scientific attention focused on the sites where samples for C-14 are lifted. This is of course exactly what should have been done the first time round, but now it is crucial that it be done and without the usual delays in getting things moving."

"What I have in mind specifically is some sort of petition addressed to Cardinal Ballestrero from all those who retain an interest in and respect for the Turin Shroud. This would also be good for media consumption, as it would show that scientific interest in the Shroud has not evaporated, that those who have studied the relic closely have not closed the book on its possible authenticity, and it would create a new sense of drama that something else is happening and will shortly produce new results that may or may not corroborate what has just been announced."

"We might say something like this:"

TO CARDINAL BALLESTRERO, ARCHBISHOP OF TURIN

We the undersigned have followed with great interest the scientific studies of the Turin Shroud. As scholars, researchers and enthusiasts of this unique object, we are dismayed by the impact of this carbon-14 dating and the near-universal tendency now to dismiss the Shroud as a medieval fake. We believe that there are many questions which deserve further research, and urge Your Eminence and other responsible authorities to permit such investigations to proceed as soon as they can be organized, hopefully by the Spring of 1989.

In particular, we are most concerned that the true radiocarbon age of the Shroud <u>may not</u> have been established by the recent tests, owing to the exclusion of the small counter method and the sampling of only one point on the cloth. We urge that further sampling be permitted of at least two other areas of the Shroud, and that this second round of dating be under the direct supervision of archaeologists experienced in the use and field applications of C-14 dating. A wide array of sophisticated analyses should also be conducted at these sampling sites to insure that they are not in any way anomalous. Only with such comprehensive examinations can we obtain evidence on which to base conclusions regarding the possible authenticity of the Shroud of Turin -- which remains one of the most fascinating objects in existence.

We pray that Your Eminence will give urgent consideration to this petition so that the physical reality of the Shroud may be clarified without undue delay.

"Let me know what you think of this statement and whether or not there is any support for this kind of petition from the other research groups."

Sincerely,

[signed]
William Meacham

Source: Letter of October 5, 1988 from William Meacham to Rev. Kim Dreisbach. Luigi Gonella archives, of which the author has a copy.

1988 *(October)*. Casaroli wrote several days before the official announcement a draft of the press release that would announce the dating results, and sent it to Ballestrero saying that the Pope knew about it and authorized to publish it. The text published by the Vatican newspaper *Osservatore Romano* of the press release unfortunately contained a printing error, for which the technical term "uncertainty" was written "certainty," a term that has no technical meaning and caused misunderstanding.

The press conference was scheduled for October 13th; the Pope had been away for the Vatican and they wanted to wait until he returned there. The press conference was to be held at 11 a.m. in the complex of Mary Help of Christians. While the methods of this communication were being decided, Tite called Gonella repeatedly asking when the announcement would be made, and he also told him that Hall had warned him that if Ballestrero failed to communicate the results as soon as possible, he himself would hold a press conference; upon learning that the date had been set, Tite called again to ask what time the press conference of Ballestrero would be held, so as not to do theirs first. Tite and Hall held a press conference in London to communicate the results in the afternoon of the same October 13th.

Source: Gonella, Luigi and Riggi di Numana, Giovanni. Sindone: il mistero continua, (Milan: Fondazione 3M), 2005, pp. 81-82.

1988 *(October)*. I received a confidential letter from one of the major participants in the C-14 dating process, who had revealing comments on the 1986 planning meeting in Turin as well as the January 1987 letter from Cardinal Ballestrero to Pope John Paul II. This participant wrote me:

"At the Turin conference it appeared that Prof. Chagas, a most distinguished biologist and President of the Pontifical Academy of Sciences, was in charge. At this meeting, which was to draw up the plan or protocol for C-14 dating by consensus, only Gonella raised the issue of 'two or three laboratories, to conserve the precious cloth'. All other scientists and Prof. Chagas forcefully voted for the seven laboratory protocol, with blind distribution of coded samples. But Gonella had another agenda, and had no intention of abiding by the protocol. His reason was not conservation of the precious cloth (he eventually cut off 50% more than the amount specified by the protocol) but a deep personal antagonism to one of the scientists who was eventually to be eliminated. (That scientist was not me!). He worked through the Archbishop of Turin, who blindly followed his advice. Around November-December of 1986 the Archbishop sent a letter to the Holy Father. The Pope replied in January 1987, essentially telling the Archbishop to proceed as he thought best. In this the Pope clearly acted out of a desire not to contradict the wishes of the Archbishop, but in so doing he obviously went against the advice of his own Academy of Sciences. When he found out what had happened, Prof. Chagas is reported to have said he felt like 'he had been kicked in the belly'. It is also reported that at one point (earlier) the Archbishop had tried to have Prof. Chagas replaced as President of the

Academy. The higher authorities of the Mother Church obviously know how to play hardball."

"The January 1987 letter of the Pope to the Archbishop was the basis of instructed by the Holy Father to limit the latter's claim that he had been the testing to three laboratories. I am sorry, but that was not the case. The thrust of the request went other way (Turin to Rome) and was merely permitted or endorsed by the Pope. In this letter the Pope also agreed to any further participation in the dating exercise."

"All the above is for your information only, Brother Joseph *[I was a monk at the time]*, and is not to be attributed to me. The original cause of Gonella's antagonism is that the scientist in question referred to Gonella as a 'charlatan' and this was duly reported to Gonella by one of the STURP scientists, who obviously did not know what he would accomplish by it."

"After the actual data is released, we will know a little more. But Otlet pointed out to me that, with no blindfold, the three labs were free to check with each other and harmonize their dates if they initially disagreed for any reason. The volume of leaks that have occurred tells you that there were ample opportunities for intercomparison and that the vows of secrecy the three labs took were just so much rubbish."

"If I were, as you may be, a believer, then my main objection would be the lack of blindfold coding of the samples. That in itself renders the test unscientific, and forever suspect. One very distinguished C-14 scientist has said that the three-lab plan would never have passed peer review. It is a pity that Gonella, who once admitted that he understood nothing about C-14 dating, was in charge. What you now will have is a result that will please some, disappoint many, and <u>prove</u> nothing."

Source: Personal letter to author, dated Oct. 4, 1988.

Comments: Even though this scientist is now deceased, I am honoring his request not to identify him. Although this scientist explicitly told me the results will prove nothing, publicly he took the stand that the C-14 results were trustworthy. One can certainly wonder if this scientist was compensated in some way, shape, or form for his public stance.

1988 (October). A few days before the dating results were announced, Meacham composed a letter to the British Museum, which he copied to STURP, Gonella and several news agencies, in which he wrote, "In sum, the British Museum has much to answer for in its involvement:"

1. Why did it acquiesce in the reduction of samples to be taken from seven to three, against the recommendation of the Turin Commission?

2. Why did it agree to the elimination of the small counter laboratories, which employ a more reliable counting system?

3. Why did it agree to only one sampling site, thereby raising the possibility of an anomalous zone being dated?

4. Why did it agree to the sampling of a scorched area of the cloth, again in conflict with the recommendation of the Turin Commission?

5. Did it approve the choice of a textile "expert"? And is it satisfied that his visual inspection of the sampled area is sufficient to rule out any possibility of a restoration/re-weaving of that area?

6. Why did it not follow its own guidelines in the inter-comparison experiment and insist that samples be taken well away from selvedges? Or is 2-3 cm. considered to be "well away"?

"Clearly the full weight of the Museum's expertise was not brought to bear on the project and its involvement does not add any credibility whatever to the results."

Source: Meacham, William. The Rape of the Turin Shroud: How Christianity's most precious relic was wrongly condemned and violated, (Lulu.com, 2005), pp. 95-96.

Comments: Meacham had consulted both with Barrie Schwortz and Vern Miller, the two main photographers from STURP, regarding the sample area. Miller, noting the highly stained nature of the corner, told Meacham (pg. 96 of his book), "I don't see how they could have chosen a worse location." Meacham also remarked (pp. 96-97), "It should never have sampled for dating, and under no circumstances should it have been the only sample taken. It took me a few hours of investigation to reach that basic conclusion, yet people like Gonella and Gove, who had spent countless thousands of hours involved in this carbon dating fiasco never did this very basic piece of homework."

Meacham received a letter on October 12th from Lukasik of STURP, who among other things, said, "I knew you would not think much of the radiocarbon dating. It always surprises me when otherwise [sic] competent people fail to observe the most obvious scientific procedures [...]. Clearly further dating will have to be done in much closer scientific collaboration with the dating laboratory than was the case with the current three laboratories. Everyone wanted an arms-length relationship for credibility but it seems to have turned into a minds-length one, unfortunately [...]. The **political** [my bolding] turmoil surrounding the Shroud is intense" (http://freepages.rootsweb.com/~wmeacham/religions/wmslcorr.pdf, pg. 6).

1988 *(October)*. Petrosillo and Marinelli wrote that according to Gonella, "The Cardinal had dutifully proposed that the result should be announced by Rome, but they decided that he was the one to announce it in Turin, asking him to make a draft statement, and to make clear the perfect communion with the Holy Father in this matter, he sent to Turin Dr. Navarro-Valls to preside over the press conference: the room of Mary Help of Christians where the results were announced on October 13 had at that time, in all respects, the role of the Vatican press room. The draft press release circulated a few times between Rome and Turin for formal adjustments, and the final text, approved by the Holy Father, was brought to Turin by hand by Dr Navarro-Valls. I would like to underline these details because some have later accused the Cardinal of having acted in contrast with the Pope. Ballestrero was right in saying that he did not act in contrast with the Holy See in, but it was incontrovertible that his 'mild way' of playing down the result seemed to contrast with the attention and veneration of John Paul II for the Shroud."

Petrosillo and Marinelli commented, "The rumors were, of course, well founded. Per radiocarbon analysis, the Shroud was medieval, woven in the space of years between 1260 and 1390 AD. This is the full text of the statement made by the Cardinal to journalists in the hall of the Salesian mother house in Valdocco in Turin: 'With dispatch received by the Pontifical Custodian of the Holy Shroud on 28 September 1988, the laboratories of the University of Arizona, University of Oxford and Polytechnic University of Zurich who carried out radiocarbon dating measurements of the fabric of the Holy Shroud, through Dr. Tite of the British Museum, coordinator of the project, have finally communicated the results of the their operations. This document specifies that the calibrated date range assigned to the Shroud fabric with a 95 percent confidence level and between 1260 and 1390 AD. The most precise and detailed information on this result will be published by the laboratories and by Dr. Tite in a scientific journal with a text being developed. For his part, Prof. Bray of the Istituto di Metrologia G. Colonetti of Turin, in charge of the review of the summary report presented by Dr. Tite, confirmed the compatibility of the results obtained by the three laboratories, whose certainty is within the limits provided by the method used. After informing the Holy See, owner of the Holy Shroud, I give news of what has been communicated to me'."

"'In remitting the evaluation of these results to science, the Church reaffirms its respect and reverence for this true icon of Christ, which remains the object of the worship of the faithful in the attitude always expressed towards of the Holy Shroud, in which the value of the image is pre-eminent with respect to any historical value -- an attitude that causes the free theological inferences advanced in the context of a research that had been proposed as solely and rigorous -- consistent with scientific mind. At the same time, the problems of the origin of the image and its preservation still remain largely unsolved and will require further research and further studies, towards which the Church will manifest the same openness, inspired by love for the truth will have shown allowing radiocarbon dating as soon as a reasonable operational program was submitted to you. The unfortunate fact that many news related to this scientific research has been anticipated in the press, especially in the English language, is the reason for my personal regret because it also fostered the insinuation, certainly not serene, that the Church was afraid of science by trying to hide its results, an accusation in clear contradiction with the attitudes that the Church, even in this circumstance, has pursued with all firmness'."

Petrosillo and Marinelli then commented, "The following day the text appears in *Osservatore Romano*, without any comment. The Holy See had kept the maximum confidentiality on the matter. The Vatican newspaper, apart from this press release, had not published a line on the Shroud since 1987, and will even come to 'censor' the Pope, as we shall see. This reticence of the newspaper appeared excessive to many observers. When the Cardinal read the statement, the comments and responses to the journalists on hand follow. Ballestrero began with two observations. First of all, he declared that 'the Shroud is a cult object, a sacred icon of the face of Christ it is and remains so.' Then, to those who ask him if the sheet is authentic or a fake, he replies that 'the question is not entirely pertinent, it is not entirely objective, as the Shroud has its authenticity in the dimension of the icon that is.' The Cardinal criticized those who took advantage of scientific research 'to create a theology because they went astray'."

"Gonella intervened to express some evaluations on the analyses: 'Scientifically I would feel much more relaxed and at ease if this dating job had been done in the context of a complete and extensive and in-depth chemical/physical investigation of the cloth as expected originally. The ^{14}C labs preferred to work independently and did not want to collaborate with other scientists, which, from the point of view of scientific methodology, left me very surprised and certainly not satisfied. The radiocarbon test was proposed with an objective, realistic program, for the first time in 1984. I take this opportunity to repeat it because I have repeatedly said that the Church had always opposed it, had always refused this test until recently. The ecclesiastical authority, as far as I have seen in these years that I acted as a consultant, has never refused any proposal to date to ^{14}C because no proposal was received until 1984. Before, no proposals had been made, they had been written articles in various newspapers saying that the Shroud should be dated; but there is a big difference between writing an article in a newspaper saying 'it would be appropriate to date the Shroud' and to present a real, documented proposal with the person in charge, saying 'we can do so and so'. This proposal was presented in 1984 as part of a large multidisciplinary research program, presented by the group that had already worked in 1978 and by other people. These researches were mainly intended to better understand the physical/chemical structure of the sheet and image precisely with basic purpose of understanding the formation mechanism and also the possible means of conservation. Of course, it would have given a background, implications on the ^{14}C. The laboratories originally had a much wider background to evaluate all the posses this research, they had almost a commitment to work, then they told us, in all possible ways, that they opted operatively from every other test, and so the research separated. Say nothing of what you can expect or not. It is a question of general principles of scientific methodology whereby, in any congress, in any scientific discourse, it is said that science must be multidisciplinary, it must be interdisciplinary. So I found with amazement, with regret as a scientist, that once there was the real possibility of doing a big multidisciplinary job, a sector of scientific research preferred to act complete isolated. I do not know why. And so regret for a lost opportunity'."

"Ballestrero was asked who organized the research, how much it cost and whether there would be another exhibition. Regarding the latter he does not answer. About the organization of the test, he says: 'The research has been carried out by several scientific entities, both European and North American. We know that Shroud groups have multiplied and this has led to an increase in interest around the world. The project that was accepted, because it is the only one that managed to revolve around concrete proposals, was precisely that relating to the analysis of the ^{14}C. The rest remained there because of opposition between groups pursuing different goals. Some essentially of piety and devotion, others historical in nature, disinterested in science; other groups with no apologetic interests. Attention to the Shroud has manifested itself in different ways, even with anti-Catholic positions such as: the Shroud is a fake, the Catholic Church advocates forgeries, therefore, the Catholic Church is not credible. These simplified syllogisms are particularly incisive in a certain world. The attempts that the Church of Turin has made to carry out research on an interdisciplinary level have not been followed up. I do not recriminate anything or anyone because it is not for me, but I must make the observation. I wonder, with some perplexity, what will happen now.' 'Because someone may say: *You have conducted research in a way that was not the appropriate one, now you have collected what you have collected*. I hope that there will be no controversy against science because I said *ad nauseum*, faith has nothing to do with the Shroud. Superimposing the Shroud on the evangelical, biblical, exegetical argument does not make sense, therefore, I do not see from this point of view how big problems can arise. The mutual respect of the various sciences is a fact that I continue to hope possible and still think possible. Time will tell. The past years to fine-tune this research and to bring it to a conclusion have taught, at least to me, many things. I was perhaps naive but I believed that, at certain levels, purely personal and particularistic considerations affected less. I had to change my mind. But I still have time to convert. So I don't worry about it that much'."

"Gonella answered the question on the economic aspects of the operation. 'The normal laboratory fee for a ^{14}C test is in the order of one million lire. In any case, for this test, since they had asked for it, the laboratories waived the fee. Because, I remember, this ^{14}C exam was not asked by us to do it, the laboratories asked us: *We would like to date this object for reasons of due scientific knowledge*. It was, among other things, I think, the first time in history that an analytical laboratory went to the owner of an object to say *I want to date this object*. Usually, the opposite occurs. This was one of the very strange facts. The sampling was organized in Turin and it was self-financed. On the other hand, all the scientific research on the Shroud, even that carried out in 1978, was self-financed.' Ballestrero added: 'I can confirm what Prof. Gonella said. This whole laborious procedure this time didn't cost anything. It will be a miracle of the Shroud. Someone will tell me: *does the Church still believe that the Shroud works miracles?* Well, I don't want to cause a stir, but in any case I can assure you that the Shroud has worked miracles and still does. It is a scandal what I am saying but I say it, because *amicus Cicero, amicus lato sed magis amica veritas*.' 'As far as I am concerned,' continued the Cardinal, 'I think it is not the case that the Church questions these results. It leaves the analysis of the same to science when it has in its hands the most technical and most minute of the procedures implemented. And since for the Church this authorization for analysis in ^{14}C has a very relative importance, I do not think that we will have to take the trouble to review the position of very respectable scientists who so far deserve only respect'."

Petrosillo and Marinelli continued, "The 'respectable' scientists, however, did not behave exactly flawlessly, and Gonella's judgments that appeared in all the newspapers will have a different tone later on. Here the Cardinal's consultant merely says, 'As far as I am concerned, we must carefully distinguish between the evaluation of the scientific result and any judgments on the behavior of the labs with respect to the agreements made. I have expressed a scientific wonder that laboratories did not want to work in an interdisciplinary field: but it is not a question of respect for the agreements; they demanded this, they wanted it and we accepted this fact. The agreements not respected by the laboratories were those of the methodologies for communicating the results, and also because they wanted a blind test they only wanted the results to be communicated to the public at the end of the year. They were the ones who assured that they would not present any report except through a work published on scientific review. They therefore asked that the archbishop of Turin, custodian of the Shroud, communicate the results to the public [...]. As the Cardinal rightly said, we do not care who or how; we must note that, according to what has been read in the press for four months, these procedures have not been followed. But this circumstance, which can displease us and has brought us a lot of weight, has nothing to do with scientific evaluation. They are two completely separate factors'."

"Gonella's position appears contradictory. How can we trust with our eyes closed the result provided by laboratories accused of behavior deemed 'mafia' or 'simply unheard of and unspeakable'?" A journalist put Ballestrero on at least slippery ground: *after such a result can the Shroud still be called a relic?* How will the Church consider it from now on? It is clear that, to be a relic of Christ, the Shroud must have covered his body, otherwise it is only a cloth with an image of the crucifix. 'Without going into really complicated matters,' replied Ballestrero 'because even the concept of relic is now a very widespread pluralistic concept, as far as the image is concerned, it seems to me that it can be called an icon. It is an image, that is, a telltale sign of a face, a face that has a religious and spiritual meaning, and I that the right term is precisely to call it an image or icon of the face of Christ, of the person of Christ. With this term, we enter a little into the logic of icons as an integral fact, the cult of the Church. I don't see any difficulty. I remember very well that in 1978, when there was the last exhibition, having had to make the homily most of the evenings at mass, I never once used the term *relic*. And to those who asked me why, I replied: *for me the icon is the true value; to say if it is a relic I would need to know for certain things I don't know*. Today, however, I should think a little differently, because even the term *relic* has undergone a bit of extensive evolution by liturgists, theologians, and historians. But forget it. For me it is an icon'."

"The position of John Paul II appeared different. On April 28, 1989, answering the question of Orazio Petrosillo during the flight by plane to Madagascar, he said in a bold way: 'Relic it certainly is.' To the question: 'Do you think it is authentic?, the Pope replied: 'If it is a relic, I think it is. If many think so, their convictions in seeing in it the imprint of the body of Christ are not without foundation.' These significant sentences of the Pontiff were censored by *Osservatore Romano*. On May 3, reporting with large quotation marks all the topics of John Paul II's interview with journalists, the Holy See newspaper cut only this topic. Incredible but true. In the aftermath of the press conference of 13 October 1988, the journalists report the sensational news of the medieval age of the Shroud. The imperturbability of Ballestrero, expressed by phrases like this, is passed over in silence: 'I never considered the Shroud a relic'. 'And given the results of the research, perhaps I was a prophet; Someone asked me if I am dismayed by the results of the exams. The consternations of the Church are not these. On the contrary, they are completely serene; I don't think there is a universal pastoral problem. True, 93 percent of the in the world knows that the Shroud exists: a recent investigation has revealed it. Ten years ago it was 5 percent. The interest around the area has revived a lot, but I think it is the result of the influence of the means of communication rather than an increase in worship and devotion. One must be careful. In essence: if I made the Shroud the object of a circular letter to the bishops on the concerns over the pastoral consequences, they would say that I have too much time on my hands'."

"Following his apparently good-natured disposition, Ballestrero succeeded on more than one occasion in defusing situations that would not have facilitated peaceful judgments. In this case, however, in the eyes of numerous Catholics and non-Catholics, he exaggerated. The

Cardinal was influenced by his previous belief that the Shroud is only an icon of Christ and showed the attitude of those who accept too lightly an anomalous result, in contrast with previous verdicts worthy of equal scientific respect. The Shroud's custodian thus gave the impression of underestimating the enormous unfavorable impact that such a response would have caused in millions and millions of faithful. Otherwise, the jokes pronounced with defusing intent but which have aroused misunderstandings about the Cardinal's real interest in the much revered sheet cannot be explained."

"John Paul II's statement to the authors of this volume was completely different. When, on February 9, 1994, we handed him a copy of the Polish edition, he said that 'it is very important to deal with the Shroud because the Lord leaves it along with the sacraments.' Ballestrero's statements raised a certain indignation that he then tried to mitigate, claiming to have been misunderstood. The interview granted to the *Voice of the People* on November 6, 1988 was a sign of this. When asked 'Why did you trust science?', the Cardinal replied: 'Because science has asked for trust. And it is easy to realize that the accusation of science towards the Church has always been that the Church is afraid of science because the truth of science is superior to the truth of the Church. So having given an audience to science seems to me to be a gesture of Christian consistency. Living according to the principle that *not trusting is better* is not Christian. I would like to underline, however, that the Church has not accepted the results with closed eyes. The Church believed -- also to free herself from an accusation of fear and disloyalty -- to give audience to science. Science has spoken, now science will judge on the results. Nobody made me say that I accept these results. I didn't say it and I don't say it because it's not up to me, I'm not the judge of science. That this having given an audience to science did not cost the Church is not true; however the Church is serene, reiterated and reaffirms that the cult of the Holy Shroud continues and that the veneration for this sacred linen remains one of the treasures of our Church. And I underline again what I have said many times: if the Shroud entered the liturgy of a Church, this is significant in its importance and validity. The discourse of science goes its way: and it is very clear it is far from exhaustive compared to this bewildering Shroud which evokes the face of Christ, and not only the face, which evokes the mystery of the Lord's passion and death, and perhaps even the resurrection. And this is the reason for my seriousness even if, evidently, the interpretations given to the publication of the results have at times been read as *consensus of the Church* which, in reality, the Church did not give and should not give. In this regard, I must also say that recent news about future research projects do not concern me. The current problem is to give the Shroud an adequate conservation system, which guarantees the maximum possible safety for the cloth'."

"In saying that future research did not concern him, Ballestrero thought of his imminent resignation from the Turin metropolitan office. But, as was said in the introduction, with an exceptional and unexpected decision by the Pope, communicated to him by Cardinal Agostino Casaroli, secretary of state, on May 15, 1989 he was made custodian of the Shroud despite the new archbishop in the Turin seat, Monsignor Giovanni Saldarini. For a few weeks after his entry into the diocese, the latter acted as the custodian of the ancient linen, celebrating the liturgical feast and favoring other research. The extraordinary papal decision, suspending the practice, was never made public. Evidently the Holy See did not mean that Ballestrero's reputation should remain tied to the medieval dating of the Shroud and to the suspicion of a not-particularly-careful and authoritative influence of the story."

"The anomaly was canceled a year and a half later, with the wise assignment of papal custodian 'for the conservation and worship of the Holy Shroud' to Saldarini's responsibility. A question spontaneously arises: 'why did Ballestrero announce the result of the research on ^{14}C while all those of the previous studies had been communicated only by scientists?' According to Gonella's clarifications, it was the laboratories that asked that the results be made known to the public by the archbishop of Turin, contrary to the custom followed up to then in the Shroud research, always placed under the complete responsibility of the scholars. 'It would have been fine for us,' said the Turin teacher, 'that the labs themselves had communicated the results. It was they who asked for this procedure because they told us: *We are pure scientists, we only want to speak in scientific press journals; therefore we pray that you will then keep the reports with the press*. All right. As everyone knows from the journalistic experiences of last year, within a week of when the first analyses were made in Arizona, journalistic revelations began about the medieval date'."

"So with the Cardinal pushed in front of the spotlight of the international press to communicate the results of research carried out by others, the game was played by those who were ready to give the archbishop the title of 'the enlightened one with the purple *[vestment]*.' Obviously it had been deduced that there was a consensus from the Church. On the other hand, Ballestrero had always considered the Shroud to be only an icon: 'For me the Shroud is simply an icon of Christ. It is an image. But it is an unguided icon.' 'Behind an oriental icon,' wrote Vittorio Messori in the aftermath of the radiocarbon verdict, 'I know there is a monk who painted it. But what is behind this *icon*, that I should continue to accept and venerate as if nothing had happened? There is a Shroud scam, with cynical oriental manufacturers of relics that, starting from the corpse of a beautiful young man, first they get the plaster cast, then they melt it in bronze, then put the brown semblance on the cloth and retouching it with human blood? Will there not be chilling suspicion -- the testimony of a crime: a poor Christ (never a term would be more appropriate) martyred on purpose as the Gospels say, and then treating his body, in order to obtain a lucrative fake? Speaking of *venerable icon of Christ* when it is announced that a certainly-bloody sheet is medieval is at least inappropriate'."

"After the announcement in Turin, a press conference is organized at the British Museum. Tite was between Hall and Hedges. Behind them a blackboard with the inscription: '1260-1390!'. The exclamation mark is full of derision. According to Hall, no scientist worth of the name can believe the Shroud's authenticity any more. It was a fake, period and that's it, especially if this hasty attitude of analysts is linked to their decision not to have the other tests carried out, refusing to collaborate with other scientists. The radiocarbon test, then, did not come out of any date, but precisely the age desired by the deniers of the authenticity of the finding: the date, that is, of the appearance of the Shroud in France. There was, in short, the prejudice of the sheet was medieval. Evin strongly denied Bonnet-Eymard's suspicions. 'Scientists are honest people,' he said, 'without a shadow of a doubt.' But Bonnet-Eymard countered, 'But they didn't respect each other either!' A doubt about Ballestrero and Tite's intentions was legitimate. There were surprising, dark, and unacceptable events. To make the 14C result accepted as absolute and definitive, beyond any control, it took a *coup de force*: Ballestrero and Tite took on the responsibility of persuading the public that the Shroud was a fake. Someone else, albeit in more moderation, reproached the Cardinal for accepting the anomalous response perhaps too light-heartedly and with an excessive respect for scientists. Ian Wilson explained with an example: 'We would not expect a pilot of a Jumbo jet who, crossing the Atlantic, realizing that the fuel lights indicate zero, to immediately dive the plane into the sea without carrying out the minimum verification'."

"Of course, the concordance of the results of three laboratories appeared as the decisive proof of the medieval date. But Wilson pointed out that the three labs dated the same piece of fabric and used exactly the same new AMS method, which had been introduced in recent years, not yet fully accepted by the scientific world and only sporadically used on textiles. The AMS labs were engaged in a strenuous struggle with competitors, such as Harwell and Brookhaven, who used the more conventional and more experienced proportional counting technique. Everyone was eager to participate in the Shroud dating project to garner the opportunities offered by their competitive techniques and to advertise. The laboratories, then, not only knew which sample was from the Shroud, because they had observed the fabric in Turin, but they were informed, without any need, of the exact dates of the other two samples from the accompanying letter compiled by Gonella and signed by Tite and Ballestrero. Among other things, it was very serious that Sox came into possession of a copy of this letter. Whatever the case, Wilson concluded with an act of trust towards Tite and the laboratories, attributing the medieval date to contamination or isotopic modification of the fabric."

"Days after the official announcement of the results, Sox launched on the market, with suspected timeliness, his book, full of information. The title spoke for itself: The Shroud unmasked - Uncovering the greatest forgery of all time. The introduction dated back to the previous August.

Source: Petrosillo, Orazio and Emanuela Marinelli. La Sindone: Storia Di Un Enigma, (Milan: Rizzoli, 3rd edition, 1998), pp. 197-212.

Comments: To continue this thread, go to the Petrosillo/Marinelli entry under "1988 *(November)*."

I find it particularly striking that Pope John Paul II asserted that Jesus left the Shroud along with the sacraments, which would have only been his personal opinion, since the Church would never decree that one has to believe in the authenticity of the Shroud or any other relic.

Marinelli noted in her Valencia paper (pp. 13-14) that Tite gave an interview to "Radio Courtoisie" in 1989 saying he didn't remember who wrote the "1260-1390!" on the board. Hall triumphantly pronounced that nobody scientifically trustworthy could now deny that the Shroud was a fake. The Cardinal's statement appeared the next day in the official Vatican newspaper, *Osservatore Romano*, ostensibly accepting the results, but the Vatican would question the results in another pronouncement made in August 1990 *(see entry for that period below)*. Riggi stated, "We believe that a single test, unconnected with the other 25 proposed, cannot give a reliable answer."

According to Meacham's book (pg. 97), "It was presented to the public as a 95% probability that the flax used to make the linen was harvested within the quoted time frame. This was of course only a statistical probability of measurement scatter, and had no bearing at all whether there was contamination, isotope exchange, re-weaving, or any of the various other possibilities that might put the date in question."

Due to many problems -- including that the three labs would not officially release their raw data, multiple contradictions related to the reported sizes and weights of the sample, and an eventual discovery of both cotton and dye in the sample area -- doubts were raised by many regarding the validity of the test. Regarding Tite's supposed not remembering who wrote the "1260-1390!", see the Kersten/Gruber entry under "1989 *(November)*."

Hall's statement about the "scientifically trustworthy" was scientific arrogance at its worst. Meacham (pg. 98) remarked, "Hall showed his ugly, ignorant side with this remark: 'There was a multi-million pound business in making forgeries during the fourteenth century. Someone just got a bit of linen, faked it up and flogged it.' Yes, this incomparable object, arguably the most intriguing object in existence, was merely 'faked up' by someone and then sold off. Brilliant Oxford scholarship!" And to think people like Hall were given so much input while STURP, whose many members had put in countless hours of research, blood, sweat, tears, and much of their own money when they had studied the

Shroud, were not allowed to take part. 'Fiasco' doesn't even begin to describe it."

1988 *(October)*. Bollone related, "This leads to October 13, 1988. Cardinal Anastasio Ballestrero reported the outcome of the analyses in a very crowded press conference. They confirm what is already widely known, namely that the dating places the Turin cloth over a period of time between 1260 and 1390. This is the text read by the prelate: 'With dispatch received by the Papal Custodian of the Holy Shroud on September 28, 1988, the laboratories of the University of Arizona, of the University of Oxford and of the Polytechnic of Zurich who carried out radiocarbon dating measurements of the fabric of the Holy Shroud, through Dr. Tite of the British Museum, coordinator of the project, have finally sent off the results of their operations.'

"'This document specifies that the calibrated date range assigned to the Shroud fabric with the 95% confidence level is between 1260 and 1390 AD. Results will be published by the laboratories and by Dr. Tite in a scientific journal with a text being developed. For his part, Prof. Bray of the 'G. Colonetti' Institute of Metrology of Turin, in charge of revising the summary report presented by Dr. Tite, confirmed the compatibility of the results obtained by the three laboratories, whose certainty falls within the limits set by the method used. After informing the Holy See, owner of the Holy Shroud, I give news of what has been communicated to me. In remitting the evaluation of these results to science, the Church reaffirms its respect and reverence for this venerable icon of Christ, which remains the object of the worship of the faithful in keeping with the attitude always expressed towards the Shroud, in which the value of the image is pre-eminent with respect to any historical-find value -- an attitude that makes the free theological allegations advanced in the context of a research that had been proposed as uniquely and rigorously scientific. At the same time the problems of the origin of the image and its preservation still remain largely unsolved and will require further research and further studies, towards which the Church will manifest the same openness, inspired by the love of truth, which she showed allowing radiocarbon dating as soon as a reasonable operational program was submitted to her. The unfortunate fact that many news related to this scientific research has been anticipated by the press, especially in the English language, is a reason for my personal regret because it has also favored the certainly not serene insinuation that the Church was afraid of science by trying to hide the results, an accusation in clear contradiction with the attitudes that the Church has pursued with firmness even in this circumstance. We all remember that the newspapers of the following day spread the result with extreme emphasis, even if there are some who dissent openly'."

"Meanwhile, a press conference took place in London at the British Museum. On the podium are Hall, Tite and Hedges and on a blackboard behind them stands out in large letters the writing: '1260-1390!'. The Shroud was declared a fake without appeal. It's a cue. In fact with this attitude the coordinator of the exams and two lab representatives highlight an animosity that does not help the serenity needed to critically consider the whole story. On the evening of February 15, 1989, Professor Hall gave a lecture at the British Museum Society Hall, decidedly provocative: 'The Shroud of Turin: a self-conviction lesson'. He exposes himself in witty, indeed decidedly caricatured terms, presents a whole series of inadmissible historical errors, accuses sindonologists of fideism and reconfirms the full validity of the now-shaky radiodating. At the end of April 1990 he was invited to the 5th Congress of Sindonology held in Cagliari on the theme of the dating of the Shroud. He adheres and sends a 5-line summary in which he acknowledges that after the publication of the results of the dating of the Turin Shroud a number of doubts were expressed regarding the negative nature of the attribution made with the ^{14}C. Regarding the validity of the doubts he affirms that 'apart from the historical problems, there have been others expressed mainly by non-scientists'. The day of the Congress he does not appear. A behavior is defined that defines itself and which is confirmed by a subsequent episode."

"When the criticisms of radiodating no longer allow defenses, the life-size photograph of the Shroud is presented in an exhibition entitled 'Fake? The art of deception', organized by the British Museum itself. The catalog states that the forger of the Shroud and those of the man from Piltdown, a famous scientific hoax probably organized by Conan Doyle, the father of Sherlock Holmes, acted with the same motivations! A long time after the announcement of 14 October 1988 (added 5 months later), the 16 February 1989 edition of the well-known English journal *Nature* publishes the scientific report of the exams with an article bearing the signature of 21 authors and confirms the dating between '1260 and 1390.' A new surprise. Tite is no longer coordinator of the research but one of the authors of the work. The list is missing Riggi, Prof. Gonella and Prof. Bray, and this is undoubtedly another index of seriousness, the second in the whole affair. A letter of criticism sent to the magazine by the historian Ian Wilson, the greatest English sindonologist, is not published. The news of the radiodating in the Middle Ages and the *Nature* article come immediately and motivated reactions. Let's see the main ones."

"None of the 21 authors is an expert on the Shroud and in the bibliography quote a single sindonological text. There are the highly criticized results of the research and studies carried out by the Commission of Experts appointed by the Archbishop of Turin, Cardinal Michele Pellegrino, published in 1976. The article presents the Shroud in erroneous terms. It claims that the Shroud was 'first shown in Lirey, France in 1350.' No. As we have seen in chap. 9, at that time the owner of the sheet Geoffrey de Charny is a prisoner of the British and has not started the construction of the Church of Lirey which welcomes him between 1353 and 1356. The article adds that 'the image can be considered the negative of a photograph' that is indisputably an extremely reductive and substantially improper concept to describe an image of an unclear nature which behaves like a negative photograph but which contains positive impressions. There is enough to be able to say that the authors of the article have not very precise ideas on the subject of their research. We continue reading. We learn that 'to confirm the feasibility of dating the sheet with these methods, the British Museum in 1983 coordinated a cross-check [...]'. That's right. A sample of Egyptian linen from 3000 BC and a Peruvian from 1200 AD are then radiodated, which is much more recent than it is. The Zurich laboratory is a thousand years off due to not having cleaned the samples well. The silence on results of so little accuracy is certainly disturbing."

"It turns out that the radiocarbon laboratories, which fiercely fought for their dating not to be carried out in conjunction with other types of research, did not hesitate to have the fabric examined by a textile expert and to examine it with the scanning electron microscope. Requested in this regard, they omit to report on the results. Radiodating with the ^{14}C does not take into account the fact that the Shroud, exposed for centuries to the environment, has collected large quantities of foreign materials, which contains fungi and small mites, which has been exposed to incense fumes and candles on numerous occasions, generation after generation, and to two different fires, one of which (that of 1532) in a confined environment with subsequent sprinkling of buckets of water which has transported large quantities of debris that are still evident today. We know that in order to rejuvenate a bi-millennium fabric in the Middle Ages, pollution of 26-33% of recent carbon is needed, but one wonders if we are not really close to values of this kind. In any case, the parallel dating of an ancient cloth [...] brought to the melting temperature and then suddenly cooled can ascertain if this causes recent carbon intake from the environment. On the other hand, the history of the Shroud includes a large number of episodes perfectly capable of polluting it with recent carbon and as such rich in ^{14}C. I will give you some examples."

"Antoine di Lalaing, lord of Montigny, reports that Holy Friday April 14, 1503 the Shroud was solemnly displayed in Bourg-en Bresse and adds: 'And to prove its authenticity it was boiled in oil, it was thrown several times in the fire, but it was not possible to delete or remove the imprinted image'. St. Francis de Sales remembers having poured on the Shroud 'many drops of sweat that fell from (his) face' during the exhibition of May 4, 1613, a particularly hot day. During the repair work of 1694, Blessed Sebastiano Valfrè shed great tears on it. The use of imposing copies, images, crowns and rosaries on the Shroud is known at every ostension. Fossati recalls the copy of 1822, preserved in the sacristy of the Castello di Agliè, to which a parchment with the following inscriptions is linked: 'From Turin, on January four, one thousand eight hundred and twenty-two. It was stretched by us, without a canvas, over the Holy Shroud, so that they perfectly matched in all its parts'."

"Still Fossati proposes a survey previously formulated by the Italian sindonologist Gaetano Intrigillo. During the whole period of the exhibition between 27 August and 8 October 1978, that is to say for more than a thousand hours, 99.995% nitrogen is circulated in the display case of the Shroud. A total of 300 cubic meters of nitrogen flow in the display case for a total of 5 liters of impurities. Even if we want to forget that a sample has not been examined, it does not appear or in any case polluted the fact that the calculations performed by the carbonists lend their side to harsh criticisms that empty them of any vindication. Bonnet-Eymard points out that the statistical analysis of the published results shows a discordance between the dating between the coherent ones of Tucson and Zurich (XIV-XV century) and the irreconcilable one with them of Oxford (XIII century). Tite has calculated a mathematical average, but it is not correct. On the other hand, if you carry out Pearson's banal statistical test, you realize that these are irreconcilable data. At the Paris Symposium, Bourcier de Carbon, professor of statistics, had strong reservations about the processing of data performed by the three laboratories. It's not over.

Remi Van Haelst from Belgium notes that Wilson-Ward was used. They are two Australian authors. They developed a little-known procedure that is not part of the classical statistical analysis methods. Van Haelst telephoned the two authors and obtained the necessary information. Repeated the calculations, obtained different results. At this point it would be necessary to have the raw results available, but the three laboratories refuse to let them know. It is clear that the elements available on the topic of radiodating the Shroud with the ^{14}C outline a totally negative balance. In fact, the caretaker's lack of confidence in the authenticity of the report, the modest contractual weight shown by your representative, the errors in the samples, the unclear management of the analyses, the poor knowledge of the material under examination by the laboratories and, let's face it, the propensity of the *carbonists* to say the last word in such a complex sector constitutes a series of unbridgeable liabilities. In essence, it clearly appears that no confidence can be given to the validity of the result that places the *origin* of the Shroud between 1260 and 1390. What is most surprising is the stubborn will manifested by the three laboratories and their coordinator M. Tite to present the result obtained, which does not harmonize, on the contrary, contrasts sharply with the results of the other research sectors, as definitive and irrefutable."

"A few years ago, Michael Winter, an expert in radiodating, stated: 'In theory dating through ^{14}C should be rather precise since the half-life is relatively short. There is talk of an accuracy of more or less 150 years. In practice, the results are so scattered that many of them have never been made public. If a date resulting from the ^{14}C confirms our theories, we make it appear in the research text. If it is only partially contradictory, we relegate it to the note. And if it deviates completely from the expressed thesis, we put it aside.' Marie Claire Van Oosterwyck-Gastuche, a medieval specialist, says: 'A dating isolated from the archaeological context is not sufficient to define the age of a finding'. To further devalue radiodating in the Middle Ages, another figure is added. In the weeks following the statement by Card. Ballestrero, the American Protestant archaeologist William Meacham, professor of the University of Hong Kong, with 16 years of experience of radiodating with the ^{14}C revealed that in 1982, a thread of fabric taken from the sample studied by Gilbert Raes in 1973 was subjected to radiodating with the ^{14C} method using an accelerator of the University of California. The results were contradictory and in any case very different from those obtained in 1988 by the three laboratories of Tucson, Oxford and Zurich. In fact, the two ends of the same thread gave totally different answers: 200 AD on the one hand and 1000 AD on the other. The result was never made public because the examination with the ^{14}C at the time had not been authorized by Card. Ballestrero. Moreover, Prof. Gonella was informed of this in 1986. All this induces to believe that in the situation that lead to a scientifically unacceptable carbon dating have contributed, on the one hand, the quiescence of those who had the legal ownership *[of the Shroud = the Holy See]* and, on the other hand, the pressure of the wishful thinking of some parties added to the determination of others."

"All of this is long gone. A sample survey carried out by the sindonologist Don Gaetano Intrigillo on 1284 visitors to the sanctuary of the Madonna del Rosario in Trani allowed to obtain precise opinions. 2% believe that the investigation should be repeated. 3% believe that the result has been manipulated by the Church with the aim of aligning more with Protestants. 5% think that the ^{14}C test can't go wrong. 11% replied: 'I didn't know about the Shroud [...] I wanted to know more about the topic and now I'm perplexed. Can there be so many falsified concordances?' 14% believe that the researchers, all Protestants, wanted to make a mistake. Some reprove that the Church trusted without reservations. 26% were convinced that everyone has acted with frivolity: the Church, which wanted an unnecessary dating and the scientists, who made the dating without knowing the history of the Shroud. 39% believed that radiodating was simply an accident on the way."

Source: Bollone, Pier Luigi Baima. Sindone O No, (Torino: Societa Editrice Internazionale), 1990, pp. 286-291.

Comments: All of the post-October 13, 1988 events alluded to by Bollone will, of course, be elaborated on later in the book.

Bollone quoted both "Michael Winter" and the late "Marie Claire Van Oosterwyck-Gastuche." I've read in several sources that "Michael Winter" was actually a pseudonym for Van Oosterwyck-Gastuche. Bollone attributed the quote that includes "If a date resulting from the ^{14}C confirms our theories…." to Michael Winter, but as shown in the "1961 *(December)*" entry, it apparently originated with a "Prof. Brew." When asked her opinion of the 1988 results, Van Oosterwyck-Gastuche bluntly replied, "It was made by the enemies of the Church. The AMS *[labs]* have not measured a medieval date, it was manufactured in a very clever way. This was the purpose of the extremely complex statistical calculation -- and incomprehensible for most. Only statisticians were able to follow the mysteries. They all declared the medieval age devoid of scientific foundation." (**Source**: Ojeda-Mari, Victor, Linceul De Turin – L'Imposture Du C14, (Les Edition le Gant et la Plume), 2018, Kindle location 2344.

1988 *(October)*. Archaeologist Paul Maloney, General Projects Director for the Shroud group known as "ASSIST," sent out a press release on October 14, which stated, "[T]he carbon date obtained by three labs for a sample taken from the Shroud of Turin may not prove the medieval date of this cloth. This is the position of the Association of Scientists and Scholars International for the Shroud of Turin, LTD. (ASSIST). Given the unanswered questions about contamination on and in the cloth, the nature of the sample itself, the lack of controls from the shroud, and the accumulation of evidence which has supported a case for its antiquity, the ASSIST group believes a date for the shroud in the medieval time frame is not yet scientifically tenable. ASSIST believes further carbon tests are necessary." Appendix F *[not reproduced here]* specified four major procedural issues: "[W]e would like to make the following observations about the carbon dating test just completed: 1) No detailed and extensive chemistry has ever been conducted to determine the kinds of contaminants present on the Shroud, methods of such detection, and methods of their removal complete with scientific controls for same; 2) There was no peer review by the radiocarbon community of the 3-lab plan prior to the radiocarbon tests; 3) There was no random selection of sampling sites; at least two other sites ought to have been carbon dated, and such sample as was tested came from the single most contaminated place on the Shroud and may represent an anomaly; sites beneath the patches, effectively protected since April 17, 1534, were passed for testing; There was no blind testing; each test sample was delivered to each laboratory completely intact and each control cloth identified as to the century within which its date should fall."

Source: Press Release sent to UPI by ASSIST: "The Carbon Date for the Shroud of Turin," October 14, 1988. Author was sent copy by Maloney.

1988 *(October)*. Meacham indicated that Fr. Rinaldi wrote him that "The Cardinal has been crucified in Italy for his stand, for swallowing hook, line, and sinker, and almost gleefully proclaiming: 'We now know the truth! The Shroud is not what we thought it was, but at the very least it remains a beautiful icon.' Of course Gonella shares the blame and I must tell you he is very unhappy [...] he had since been in the USA where he met some of the STURP people who took him apart and blamed him for everything that has happened." Meacham commented, "And rightly so, as he failed in the most crucial aspect of any scientific project -- to involve people who have experience and expertise." Meacham then sent another press release to news agencies and newspapers titled "SHROUD C-14 DATE UNRELIABLE ARCHAEOLOGIST CLAIMS."

Some of the highlights of the release were Meacham saying "that C-14 dates should not be regarded as infallible; *[C-14 scientist]* Prof. P. Betancourt and his colleagues remarked on the fact that 'so many dates have proven to be useless because of contamination and other causes;' and Wolfli, director of the Zurich lab, had stated in a recent paper, 'no method is immune from giving grossly incorrect datings when there are non-apparent problems with the samples [...] this situation occurs frequently'."

Source: Meacham, William. The Rape of the Turin Shroud: How Christianity's most precious relic was wrongly condemned and violated, (Lulu.com, 2005), pp. 99-101.

Comments: Even though C-14 scientists know there are many anomalous C-14 readings, the three labs that dated the Shroud were at the outset trying to convince the public the Shroud results were airtight.

1988 *(October)*. Meacham decided to send a follow up press release which, among other things, stated, "Meacham said the recent C-14 tests proved nothing at all about the Shroud as a whole, since all three samples dated by Arizona, Oxford and Zurich had been taken from the same spot on the cloth -- a corner that had been scorched in the Church fire of 1532."

"It is also possible that this area was re-woven by a medieval restorer, since it is just next to a selvedge edge and side panel that were added to the Shroud at some time after its original manufacture."

"The Shroud may not be one homogeneous cloth as far as its chemistry is concerned. We already know of significant variation from one point to another, and the radiocarbon content likewise may vary significantly."

"The recent testing was very poorly planned. It is astonishing that samples from at least two or three different points on the cloth were not taken for dating. Archaeologists who make frequent use of C-14 results are accustomed to samples occasionally giving aberrant results, and would normally not attach much importance to a single date, or this case, three dates on a single spot."

Meacham said he had repeatedly urged Gonella not to rely on one single site for dating the Shroud [...]. Criticism of Gonella surfaced earlier this year when 4 of the 7 labs originally planned to do the testing were dropped from the program. Meacham cited a letter he had just obtained that was written by one of the labs' directors to the British Museum in January of this year, in which the current C-14 project was described as "a rather shoddy enterprise [...] which the British Museum may live to regret."

Meacham then commented about this second release, "After two weeks the subject of the Shroud had run its course in the press. I realized that it was going to be difficult to move the mass media or the mass mentality until there were new developments. It seemed to me very likely that Ballestrero and Gonella would favor a new round of testing, including all the tests that STURP was planning, plus another C-14 run on samples from different sites on the cloth. And why not? There was everything to gain and nothing to lose. This was eminently logical, but as often happens in Church or any other **politics** *[my bolding]*, logic does not always prevail."

Meacham wrote a third piece, for which he hoped Fr. Rinaldi could find a contact in Rome to get it published. Fr. Rinaldi wrote to him, "I called a friend on the editor's staff and asked what the chances might be that it would publish. "Not one in a million," he told Meacham. "You must have noticed that, except for the official communiqué on the results, the *Osservatore* has been silent on the whole issue, while other newspapers both Catholic and secular have registered all sorts of protests and criticism of the way Cardinal Ballestrero handled the October 13 press conference. For the time being the *Osservatore* (i.e., the Vatican) wants to stay out of it."

Source: Meacham, William. The Rape of the Turin Shroud: How Christianity's most precious relic was wrongly condemned and violated, (Lulu.com), 2005, pp. 102-103, 106-107.

Comments: Logic did not prevail. Meacham had commented about the Shroud having run its course in the press. Certain topics never seem to run their course in the press, like the celebrity-worship, e.g., the Kardashians/Jenners in the U.S.A or anything to do with the British Royal Family. But something like the Shroud, which has the potential to change millions of lives, is deemed not important enough for continuous reporting.

Meacham's statement: "The Shroud may not be one homogeneous cloth as far as its chemistry is concerned. We already know of significant variation from one point to another, and the radiocarbon content likewise may vary significantly" would be borne out later by various pieces of evidence.

A remark made to Meacham by an unnamed correspondent (pg. 117 of his book) is worth mentioning. That person stated [...] "that the three laboratories came up with what has been announced as practically identical results, and that these results coincide with the accusation of Pierre d'Arcis -- whose ghost we thought we had laid to rest -- smells strongly of some procedural deviance prior to the actual tests." In other words, this person suspects that the labs manipulated the data so that they were close to each other as well as to the date when the Shroud is clearly documented in the historical record. In fact, questions about the statistical aspects will be touched on further below.

1988 *(October)*. Gonella made a visit to the U.S. and was interviewed by a paper in Albany, New York. Gonella made various critical remarks about the C-14 scientists. "The constant leaks of information from scientists who, with Church permission, examined the Shroud and their attitude of mistrust and suspicion, 'gave us the sad impression that we were taken for a ride,' Dr. Gonella told *The Evangelist*."

"We got the feeling that they were only interested in good advertising", Dr. Gonella said of the behavior of scientists at the three labs in Great Britain, the United States and Switzerland who examined the Shroud."

"He said that, from the beginning of the project, the Church was made to appear as if it were an obstacle to scientific investigation. Scientists, questioning the objectivity of the Church, made sure to personally take samples from the Shroud for fear that Church officials might substitute samples from an older cloth."

"For Dr. Gonella, that suspicion was an insult which resulted in the 'Bishop of Turin being treated in a way that they would not have treated a provincial museum director'."

"He said the Church was forced to accept such conditions for fear that Cardinal Ballestrero would be charged with putting obstacles into the path of science."

"'They put the Church of Turin into a very awkward position,' he said. 'We were charged with trying to hide the truth. I had too many times to apologize to the bishop for the behavior of my fellow scientists'."

"While the results were supposed to be gathered and then formally presented by Cardinal Ballestrero, news leaks, especially in the British press routinely occurred throughout the investigation. 'We didn't know anything. (The scientists) kept the press Informed without informing us,' said Dr. Gonella, who said that the scientists were interested in advertising their labs' involvement in the project. Some, he said, also were intent on making theological statements debunking the value of the Shroud."

"The scientists also insisted on having their own textile experts study the Shroud. According to Dr. Gonella, such measures were an insult to Italians, who have long been in the textile Industry."

"'Italy was being treated like an underdeveloped country. Turin is not a backwater town.' Dr. Gonella insisted."

"The conflicts about the Shroud, concluded Dr. Gonella had little to do with scientific method or the search for objective truth. Rather, he said, the conflicts 'had no relevance to science and faith. It had a lot to do with the relationship between the scientific and the clerical world'."

Source: Feuerherd, Peter. "Shroud expert from Turin hits scientists' methods." *The [Albany] Evangelist,* October 20, 1988, pg. 8A.

1988 *(October).* Art historian Anna Hulbert, who was trained in the 1960s at the world famous Courtault Institute (a college of London University), said "If the image on the Shroud is purely the work of a medieval artist, it raises more problems for me as an art historian than if it is genuinely the Shroud of Jesus of Nazareth." She was asked her evaluation of C-14 dating. She commented, "Carbon dating, like X-rays or any other analytical technique, should be regarded as one tool among many. It is chiefly useful in the dating of undisturbed archaeological material. In the case of the Shroud, one should calculate carefully whether any of its known wanderings or adventures, such as the 1532 fire, could give a distorted reading to whatever date the radiocarbon laboratories come up with."

"It is science, and not the Catholic Church, that's trying to prove the authenticity, or otherwise, of the Shroud of Turin, and it would be quite ridiculous to dismiss the Shroud as a medieval artifact on the basis of a non-too-reliable carbon test."

Source: Jennings, Peter. "Art historian not convinced the Shroud is a fake." *Our Sunday Visitor*, October 23, 1988, pg. 24.

Comments: The Shroud was anything but "undisturbed archaeological material." Hulbert also confirmed the perception of many, not held by the general public, that the C-14 test is not always reliable.

1988 (October). Many objections began immediately after the results were announced. One of them involved Hall of the Oxford lab, who was being accused of commercialization for selling film rights to a commercial station in England for the right to film the testing there. Some believed that Hall preferred three labs to seven because more attention would be focused on Oxford, which would enable them to receive grants and public funding, which in turn would enable them to expand their research center. Other objections would surface throughout the years.

Source: Bracaglia, Giorgio. Uncovering the Paradox within the Archives of the Holy Shroud Guild. (Honeoye, N.Y.: Holy Shroud Guild), 2019, pg. 243

Comments: I addressed some of the major objections to the C-14 dating results made by various individuals over the years in a paper I presented at a Shroud conference in Ancaster, CANADA in 2019 and is accessible at https://www.academia.edu/40272184/The_Invisible_Reweave_and_ Other_Challenges_to_the_Turin_Shrouds_C-14_Medieval_Dating_A_ Review.
    On October 14th, the *New York Times* came out with an article titled "How Carbon 14 Was Used to Fix Date of Shroud." The author used the word "fix" in one of its main meanings to mean "set," but "fix" also has a slang denotation denoting deceitfully changing the outcome of a situation. Given all of the evidence presented in this book, it's understandable if many, in their evaluation of the 1988 dating, would lean toward the latter rather than the former.

1988 *(October and November)*. Bro. Bruno called Gonella several times in preparation for a planned meeting on November 27th. Bruno said that Gonella would always reply to questions, "You must keep quiet now. Say nothing and write nothing and let the scientists get on with their work, otherwise you will make it look as though the Church is against science. Already articles are appearing in Italy headed, 'Down with Science!' It's catastrophic. It is the very worst reaction. So, you just keep quiet and leave us to get on."

Source: Bonnet-Eymard, Bruno. "The Victory of the Holy Shroud Won By Science." *Catholic Counter-Reformation in the XXth Century*, September-October 1989, No. 223, pg. 27.

Comments: Bro. Bruno sent a telegram to Cardinal Ballestrero after the November 27th meeting asking him to address various problems with the dating. Bro. Bruno received no reply.

1988 *(November)*. C-14 scientist Dr. Haas wrote to Meacham on the 11th, "Dear Dr. Meacham, Thank you for your interesting thoughts on the Shroud dates contained in two letters. Through lucky coincidence I met this week Professor Donahue and Dr. Linick from the Arizona AMS Lab, furthermore I had correspondence via Bitnet *[a predecessor to the Internet]* with Professor Wolfli. Through these connections I have learned that there is solid evidence for the material tested being indeed much younger than the expected age close to AD zero."

"The level of contamination during the 16th century by fire, boiling in oil, restoration attempts etc. would have to be added or exchanged. Such a severe chemical change should have left more visible marks than the ones shown on photographs."

"The Zurich lab divided the sample in different fractions, one was run without pretreatment, others with varying degrees of treatments. The ages all fell close together. This is unlikely for a heavily contaminated sample. The match of the results of the three labs is good enough to exclude even a remote chance that the true age may the 'hoped-for' one."

"For the present I would like to recommend further DNA studies on the stains. Ever since the new result became public, the making of the Shroud has been represented as a painting process. I would like to entertain the possibility that a dead person or perhaps a life martyr with the appropriate cuts or wounds could be wrapped in the cloth, to produce the observed patterns. The religious habits of that time do not exclude such an interpretation."

"Chemical tests for combustion products or for evidence of conservation effort should be carried out as well. All such non-destructive tests should include tarnished and untarnished portions of the Shroud. The hypotheses proposed in your draft protocol about isotope exchange might be tested by using medieval or earlier textile samples and C14-spiked reagents or gases. Many experiments can be set up in this way without having to make another request for Shroud samples so soon after the recent sampling. If such experiments clearly demonstrate the possibility of large scale isotopic exchange then a new request for mere Shroud samples might be in order. At the present time I would be surprised by such an outcome."

"The large number of radiocarbon dates on organic materials, charred and uncharred, with acceptable results make the sudden discovery that this was just pure luck rather than unlikely. I agree with you that it would be a dramatic setback for the radiocarbon dating community. But I cannot see it as a likely event."

"You may want to contact biomedical laboratories where C14 tracer studies are routinely performed. Radiocarbon dating laboratories do not want to work in this field because the expected high C14 activities pose the danger of equipment contamination. Furthermore I have the impression that the three labs are not keen on starting another dating project unless solid evidence for sampling errors can be demonstrated."

"These comments are being prepared rather hastily since I am leaving on Sunday for field studies during 3 weeks. Please keep me posted on the further developments of the study."

Sincerely,

[signed]
Herbert Haas, Ph.D., Director

HH/mnm
copy: Professor Willy Wolfli, ETH Lab, Zurich

Source: Letter of November 11, 1986 from Dr. Herbert Haas of Southern Methodist University C-14 lab to William Meacham. Luigi Gonella archives, of which the author has a copy.

Comments: Meacham replied to Haas on November 29th. See further below for full text.

1988 *(November)*. Tite continued to send to Gonella drafts of the official report that would be published in a scientific journal. On November 14th, Gonella wrote to him thanking the information and providing the specific information requested, but declaring that he did not intend to make any comment on the text, so as not to say that Turin had tried to influence the scientific presentation of the data. Tite had asked him if he expected to list his name as an author, and he replied in the negative because he wasn't used to signing scientific papers he hadn't made a direct contribution to, and he hadn't signed any other work regarding research on the Shroud. Tite also asked Bray about comments/signing the official report. Bray sent Tite a summary of a few lines of his commentary for inclusion.

Source: Gonella, Luigi and Riggi di Numana, Giovanni. Sindone: il mistero continua (Milan: Fondazione 3M), 2005, pp. 82-83.

1988 *(November)*. "In November, the manuscript of their final report was not yet submitted to the editors of *Nature*. Tite tried to justify by saying, 'It took time to put together the work of 21 people.' Gonella expressed his disappointment at the delay: 'I find inconceivable that today, less than a month after the official announcements on radiocarbon research, we still have to wait for the publication of the entire extract concerning the works and the overall evaluations of the laboratories. The agreements had included that the results would be shown in draft form to the Cardinal. In fact, the press conference and the publication of the article in a specialized magazine should have taken place simultaneously. Well, that article has not yet been seen. A simple letter was received by the Cardinal while the indictments were already raging that today, one month later, discretions about the fake'."

Source: Petrosillo, Orazio and Emanuela Marinelli. La Sindone: Storia Di Un Enigma, (Milan: Rizzoli, 3rd edition, 1998), pg. 212.

Comments: To continue this thread, go to the Petrosillo/Marinelli "La Sindone …" book entry under "1989 *(February)*."

According to Marinelli's Valencia paper (pg. 14), Gonella said when interviewed by the Italian newspaper *Il Sabato*, "It was blackmail. They put us against the wall just with blackmail. Either we accepted the test of C-14 on the terms imposed by the laboratories, or it would break out a campaign of accusations saying the Church fears the truth and is an enemy of Science."

1988 *(November)*. Fr Rinaldi sent out a newsletter to the Holy Shroud Guild members stating that Cardinal Ballestrero had already regretted having said that there was no reason for the Church to doubt the results. Gove wrote, "Rinaldi claimed that people in both religious and scientific circles were up in arms because of mistakes in test procedures and 'the unprofessional behavior of the scientists involved in the tests'." Fr. Rinaldi noted that *L'Osservatore Romano* had not published anything besides the announcement of the results, which he interpreted to mean that the Pope wasn't pleased with the way the whole C-14 process had been handled. Gove commented, "I wondered exactly what it was about the test procedures that might have made the Pope unhappy. If they were the same one that had initially made me and others unhappy, the Pope had only himself to blame. It was he who cut Professor Chagas and the Pontifical Academy of Sciences out of the action and handed control to Luigi Gonella. Only the fact that the scientists in the three laboratories had conducted their measurements in such a professional manner and had independently reached agreement on the dates of the four pieces of cloth they had been given, saved the situation from disaster. Rinaldi's charge that the scientists' behavior had been 'unprofessional' was quite preposterous. All along it was the behavior of Gonella and his STURP cronies that had been unprofessional. I again thanked providence for ensuring that the carbon-14 results had not been tainted by any involvement by STURP. If the Pope, by continuing to exercise benign neglect, permitted the shroud to suffer further assaults by STURP, so be it. At least they would not be desecrating a relic."

Source: Gove, Harry. Relic, Icon or Hoax?: Carbon Dating the Turin Shroud, (Bristol and Philadelphia: Institute of Physics Publishing), 1996, pp. 294-295.

Comments: In one short paragraph, Gove managed to criticize the Pope, Fr. Rinaldi, Gonella, and STURP, while using inflammatory language about STURP by referring to them as Gonella's "cronies" and stating that any testing by them would be "assaults" AND lauding the behavior of the three labs, despite the many questionable behaviors by them documented in this book. The actions of the three labs did NOT save the situation from disaster. He "thanked providence" for keeping STURP away to boot.

It seems that a lot of the principal players were very much at ease blaming various other players for many of the problems.

There was an interesting comment by Sox, whose book ends, "A vicar who knew of my interest in the Shroud once approached me after reading all the material which proclaimed its authenticity, and said he still found it very difficult to believe the Shroud was real. 'God doesn't operate this way, does He?' he asked. He was right. He doesn't." Personally, I'm always leery of someone who claims to know exactly how God operates.

1988 *(November)*. German author Kersten paid a visit to Belgian textile expert Dr. Gilbert Raes, who had been given a sample of the Shroud to study in 1973, and had possession of it for several years after that. Kersten inquired regarding the whereabouts of the sample. Raes revealed some interesting information regarding it. Sometime in 1974, he had received a letter from Rev. David Sox, who was secretary of the BSTS, and who was pro-Shroud at the time. He asked Raes to receive a Dr. Walter McCrone from Chicago, who Sox described as a "radiocarbon specialist." McCrone visited Raes in September 1974 telling Raes he could date minute samples and asked Raes to give him the sample. Raes was leery about the request and wanted to consult another specialist. Raes consulted with Belgian C-14 expert Prof. Daniel Apers, who agreed to meet with McCrone. Apers advised Raes not to give McCrone the sample, since McCrone's proposed testing would have a plus/minus factor of 700 years, which was too large for a possible two-thousand-year-old date.

Kersten asked Raes if he knew McCrone strongly believed the Shroud was not authentic. Raes expressed the opinion that McCrone's anti-authenticity stance was because he wasn't able to get the sample from Raes and that Sox, who eventually changed from pro-Shroud to anti-Shroud, might have changed camps for the same reason.

Raes had these things to say about the C-14 testing, "I cannot understand why representatives of the dating laboratories were present during the sampling in April. As I heard, everything was to be kept in strict secrecy, to avoid influencing the researchers. But the weave of the Turin Shroud is so characteristic that it can be recognized immediately. I think they should have taken the specimens apart to leave only the individual threads; then they really would have been unrecognizable. But as they were anyone could recognize the Shroud specimen at once. That is not a blind test! And then they probably talked among themselves too. If there were differences of 600-700 years, they had to harmonize the results so that the public was not suspicious. I am fairly sure they compared notes. Finally, there is still the question why the four other laboratories from the seven originally selected were suddenly excluded! I find more and more reasons to make me doubt the correctness of this dating procedure. What makes me most suspicious is that the laboratories were in contact with each other."

Source: Kersten, Holger and Elmar R. Gruber. Jesus Conspiracy: The Turin Shroud & The Truth About The Resurrection, (Rockport, MA: Element, 1994), pp. 48-50.

Comments: McCrone was not a C-14 specialist -- he was a microscopist. McCrone always claimed the Shroud was a forgery because of artists' pigments he found on sticky-tapes. It's important to state that it is known that many artists who painted the Shroud over time were allowed to touch their copies to the Shroud to "sanctify" them, so it's not surprising that artists' pigments were found, but STURP, which directly examined the cloth (McCrone only studied sticky-tape samples) said the trace pigments had nothing to do with the image. I've read over the years that Sox changed from pro to anti based on McCrone's findings. But Raes seems to have picked up some very strong impressions regarding McCrone and Sox, and seemed to believe that both McCrone and Sox were both anti-Shroud due to personal animosity due to not having been given the samples. Given Sox's role in the whole C-14 affair, the repercussions (if Raes' perception was correct) of McCrone/Sox not being given the sample were enormous. It's interesting that Sox had close connections with both McCrone and Gove, who didn't actually get along. It's not unlike the situation with Garman Harbottle, who was close both with Dinegar of STURP and Gove, who despised everyone and everything STURP. Some strange bedfellows, indeed. Raes was also clearly suspicious of the behavior of the labs.

1988 *(November)*. Archaeologist Eugenia Nitowski wrote, "In any form of inquiry or scientific discipline, it is the weight of the evidence which must be considered conclusive. In archaeology, if there are ten lines of evidence, carbon dating being one of them, and it conflicts with the other nine, there is little hesitation to throw out the carbon date as inaccurate due to unforeseen contamination. The Shroud should not be given less than standard procedure. Clearly in this instance, the carbon date is conflicting with the weight of the evidence [...]."

Source: [Nitowski, Eugenia.] "The Shroud of Turin and Carbon Dating." *The Wanderer,* November 24, 1988, in "the Forum" section.

Comments: There are many more than ten lines of evidence when it comes to the Shroud, which is generally acknowledged as the most intensely-studied artifact in human history in terms of the number of person-hours devoted to its study. One would think that if the Shroud were a medieval forgery, that there would be other solid scientific evidence pointing to that, and while many debunkers believe that there is, many scientists and researchers continue to believe that the weight of the evidence indicates that the Shroud is authentic. I believe the emphasis on the reliability of the

C-14 in the case of the Shroud is part of the fabric (no pun intended) of the politics that plagued the whole enterprise.

1988 *(November)*. Meacham replied on the 29th to Herbert Haas' letter of November 11th, "Dear Dr. Haas, Thank you very much for your letter of Nov. 11. I would certainly agree that massive contamination or exchange would <u>normally</u> leave visible evidence, as would of course any restoration work in the area. But as Wolfli *et al.* pointed out in their 1985 paper:"

"'The existence of significant indeterminant errors can **never** be excluded from any age determination. No method is immune from giving **grossly** incorrect datings when there are non-apparent problems with the samples originating in the field. The results illustrated [in this paper] show that this situation occurs **frequently'** *[Wolfli originally italicized the words that Meacham bolded/underlined to Haas].*"

Meacham continued, "It is the non-apparent that often presents the problems in 'rogue samples', such as the recent series from Mycenae which are unquestionably 3300-3600 years old, but according to Stuart Fleming the C-14 dates clustered around 300 B.C. Willi *[Wolfli]* also discussed with me in Turin the discrepancies that turn up regularly in dating Egyptian material of historically known age."

"My view is simply this -- if we had a single sample from Mycenae or Egypt or Santorini that gave a date in conflict with the other evidence what would the reaction of archaeologists and radiocarbon scientists invariably be? Take more samples and run more sophisticated tests! The Shroud presents another problem almost unique in radiocarbon dating, and that is its exposure to fire long after the date of the material. The only example I can find in the literature is the Akrotiri case mentioned in the protocol I drafted. You will recall that Harbottle also asked for information on this from about 40 lab directors, and none could cite specific instances. Approaching this problem in an honest and scientific manner would not raise the spectre of a 'setback' for the radiocarbon dating community. I mentioned the 'impact' that a possible earlier date would have with the thought that it would be this strange relic that opened some new vistas in the radiocarbon field."

"Your suggestion of setting up experiments with ancient linen samples being subjected to all sorts of conditions that might induce large scale contamination or exchange is worth pursuing. I have searched the literature for similar experiments to investigate how contamination of samples might occur, but found little. More effort seems to have been put into analyzing a few cases of obvious or suspected contamination, with very enlightening results, e.g., the Meadowcroft rock shelter samples which Haynee found to be contaminated by older material. The attraction of this approach is that one is working with the actual ancient sample rather than attempting to recreate all the possible scenarios that might have affected it. Al Adler reeled off to me half a dozen possible chemical reactions which could have introduced younger carbon into the Shroud linen. Nonetheless I intend to propose to STURP and ASSIST that some experiments along the lines you suggested be conducted."

"Whether or not another round of C-14 sampling probably depends on **political** [my bolding] factors in the Vatican rather than scientific considerations anyway, and I can well understand the reluctance of the labs to get involved in further efforts to secure samples. But on the long shot that we do in fact receive permission for another round I hope you and Willi would be willing to be involved in the pretreatment and measurement, respectively."

"I enclose two papers [not reproduced in this book] : 1) a further scenario [the merit of which I have no idea] for a possible false C-14 age for the Shroud, and 2) a recent paper I did on the dating of Neolithic pottery, incorporating the results obtained by Evin's lab. I also discuss the singular lack of success we have had in getting consistent dates on shells and d^{13} [sic] readings on animal and flesh bones."

"You did not mention any progress on our pottery samples now with you, but I trust that these are in the pipeline for treatment in the next few months. The Macau date is one of the earliest for painted pottery in China, so we are naturally concerned to know if clay carbon contamination is a factor!"

Sincerely,

William Meacham

Cc: W. Wolfli; A. Adler

Source: Letter of November 29, 1986 from William Meacham to Dr. Herbert Haas of Southern Methodist University C-14 lab. Luigi Gonella archives, of which the author has a copy.

1988 (December). *Nature* received on the 5th of December the paper from the twenty-one scientists "even though the scientific text still had to be reviewed by peers and published in a specialized periodical before it was communicated to the public."

Even though the Colonetti Institute in Turin had originally been chosen to be one of three institutions to analyze the results, the laboratories had asked that the work be entrusted to Tite because of his supposed independence. Gonella noted, "Any Italian would have been looked upon with suspicion and anyone from Turin would be doubly suspect. We accepted because we were in the situation that, if anyone had objected, his objection would have been interpreted as an obvious proof of the desire to cheat."

Source: Petrosillo, Orazio and Emanuela Marinelli. The Enigma of the Shroud: A Challenge to Science, (San Gwann, Malta: Publishers Enterprises Group), 1996, pp. 110-111.

Comments: Why was the paper sent to *Nature* even before it was in its final stages? It gave the impression that a conclusion was made without regard to the details. Recall that Wolfli had claimed that *Nature* accepted their paper even before a single line was written! Petrosillo and Marinelli noted that none of the signatories had done any previous research on the Shroud; the group that was knowledgeable about the Shroud, STURP, had not been allowed to participate. Had STURP been allowed to, they surely would have been able to discover sooner the labs' questionable practices that took a long time to surface.

1988 (December). A letter from Dr. Raes was sent to *Shroud News* editor Rex Morgan. Raes told Morgan, "[M]any questions arise concerning these tests. Why was the number of designated laboratories reduced to three? What is the exact place from which the samples were taken? From the photograph on page 9 of *Shroud News* (No. 49) it seems it was about at the same place that my sample was taken. Also, at this place a piece of about 7 cm width of cloth was added, probably to centralize the image. There is no evidence that this piece of cloth is of the same age as the remaining part of the Shroud, and may have been added centuries afterwards [...]."

"To me it seems evident that before communicating their results to Turin the laboratories contacted each other in order to avoid too much difference between their results. Indeed, big differences would contribute to doubts about the credibility of the laboratories and it is logical that the three laboratories tried to avoid such a possibility [...]."

"Many other questions may be put, on which we will probably never receive a satisfactory answer."

Source: "From Emeritus Professor Gilbert Raes." *Shroud News*, No. 50 (December 1988), pg. 20.

1988 *(December)*. Meacham wrote a letter to the Turin prelate:

Dear Cardinal Ballestrero,

As one of the participants in the Turin Workshop on Carbon Dating the Shroud, I was very distressed to observe how the testing unfolded. Most disturbing were two aspects: 1. That only one spot in a very suspicious corner of the cloth was sampled and 2. That the results were taken almost universally as conclusive proof that the Shroud is a medieval fake.

I am writing to you with the fervent hope that you will allow another round of carbon dating, with different parts of the cloth to be sampled and special attention paid to the possible contamination problems that might affect the Shroud.

There are very good scientific reasons for further carbon dating of the cloth. I am sending full details to Prof. Gonella, along with an outline proposal for further C-14 testing.

Speaking as an archaeologist with 16 years' experience in the direct application of C-14 to ancient objects and sites, I would maintain that nothing has been definitively established about the Shroud's age by the recent test results. The massive public notion that the Shroud has been 'proven' by science to be medieval or a fake is most unfortunate.

I hope and pray that you will authorize further scientific investigation on this question, so that the radiocarbon age of the Shroud may be more accurately measured.

Yours in Christ,

William Meacham, Centre of Asian Studies, University of Hong Kong

Source: Letter of December 18, 1988 from William Meacham to Cardinal Ballestrero. Luigi Gonella archives, of which the author has a copy.

1988 *(December)*. Meacham wrote to Gonella, "Dear Luigi, Enclosed is a batch of papers relating to the carbon dating exercise, including some correspondence which may interest you. *[Material referred to here is not reproduced in this book.]* Gove considered the whole affair a 'shoddy enterprise' and I have been quoting this to the press. How can he and his comrades now be saying that the Shroud has been definitely dated?"

"The fact is that the first round of C-14 has not definitively proven <u>anything</u> about the cloth. Because the site of the one sample is not representative, we cannot speak with certainty of the radiocarbon age of the cloth. (It would be a great extrapolation to speak of that from any single site.) We cannot even speak of the undisputed radiocarbon age of the 22A area, in view of the serious and unresolved discrepancy in the measurements on Lindow Man. What we can say is that the 'AMS radiocarbon age' of the 22A site indicates a medieval date. But to have the Shroud universally branded a fake on the basis of such debatable scientific evidence is a great tragedy for the Shroud, and for those of us who have diligently researched the question over the years."

"This happened as we know because of the high publicity profile that the C-14 test attracted, and the **political** *[my bolding]* in-fighting that surrounded the effort. As you remarked several times during our phone conversation, had it not been for the intervention of the labs and Chagas, the C-14 testing would have been done properly as an integral part of the other scientific studies, and we would have complete data on the site from which samples were taken. The only thing I would say is that, although the damage has been done and it is severe, let's now proceed and do it right. Let us take samples from 3 or 4 points on the cloth; let each of these sites be thoroughly examined; let the samples themselves be thoroughly screened; let's have intensive pretreatment and date the resultant fractions -- And IF, at the end of it all, we still have a medieval age, then I think we will have to acknowledge the high probability that it is really a medieval cloth."

"But if the results differ at all from what the three labs found, the impact will be enormous, especially on the British Museum, Hall, Gove and others who have confidently, even arrogantly declared the shroud to be a medieval fake. If we find any older carbon in any of the residual fractions or by the small counter measurement then they will look extremely foolish in having proclaimed that the Shroud was proven, by a single sample measurement, to be of medieval origin. Haas was quick to note that this might appear to the public as a setback for the radiocarbon profession. Actually it would only be a confirmation of what those of us who use C-14 regularly know -- that one can never trust single dates no matter how reliable the sample appears. Without corroboration from a suite of other measurements and other data the date from one sample means very little. Even simple archaeological features such as a burial, house, hearth, or kiln would normally be dated by repeated measurements on several samples, even if there were little or no controversy about the proposed dating, and even if the first date fell right in line with the suspected age. The enclosed descriptions are just a small sampling out of many cases where the anomalies have turned up through continued dating of samples from ordinary non-controversial features."

"Meditate for a moment on the data concerning Lindow Man. If the samples had been counted by Harwell, who would ever have suspected that the radiocarbon age would be other than Oxford's reading – which itself now seems to be off by 400 years. Harwell's is off by 800 years. The other evidence taken together is quite compelling for a date not later than 300 B.C. The problem is clearly in the sample itself, and as Wolfli *et al.* wrote, this kind of gross inaccuracy can and does occur frequently. The only way to know if the Shroud has non-apparent problems of this nature is to do more measurements on different parts of the cloth and employ both counting systems."

"Or consider the Alaskan site -- what reason would there have been to doubt the dates on the wooden house-posts at 600-1000 B.C.? But in following the usual practice of seeking further measurements on other samples from the same context, the charcoal from the hearths was found to give different ages, up to 1000 years older! Or consider Akrotiri -- if only a few samples, say P-2563 and P-1885, had been run it would appear to the casual observer that radiocarbon dating had confirmed the destruction of the town at the estimated time of 1500 B.C., without a hint of the enormous problems that the site now poses for C-14."

"These two sites have something also to note in relation to the Shroud: fire may have played a role in skewing the C-14 dates in some manner as yet unexplained. At Akrotiri the long-lived samples were 200 or more years old at the time of the destruction (by fire) of the town, and yet, three of the samples (1619, 2562, 2566) gave dates well <u>after</u> the fire. Their radiocarbon ages are thus at least 400-600 years were too young. Conversely for the Alaskan site where the charcoal from various hearths consistently gave dates 800-1000 years older than the unfired wood in the posts. If the Shroud's radiocarbon content has been affected at all by the 1532 fire, it is unlikely that it would be uniformly affected over the whole cloth. The only way to establish that this is or is not the case is to do more measurements. The earlier run that was done on the Raes thread at California *[the secret 1982 dating]* certainly indicates that variations may be found. Rest assured however that Gove Hall, Tite and company will oppose any further dating, and jeer at the effort, but privately they will certainly be worried at the prospect that such variations may turn up."

"The series of dates on a single plank of wood from Malaysia is just one of a number of examples in which serious discrepancies have been found within a supposedly homogenous sample. The collector of these samples (Brian Peacock) told me that the wood was of a species that never exceeded 100 years of age, that not more than 50 years were represented in the plank, and that the samples were taken at 1m intervals from one edge. Of course the labs did not confer on pretreatment beforehand, and some of the discrepancy might be due to differing lab procedures. But the magnitude of the discrepancies suggests that, like in the Alaskan house and hearth samples, the problem lay in the samples themselves. You said on the phone that the Church did not want to be seen to initiate new C-14 tests, that this might suggest to some that it did not accept the scientific results of the last round. The above example should demonstrate very clearly that extensive and repeated testing is the rule in archaeology, even when there is no ostensible reason to doubt the first set of results. This is all the more true when a major question is at stake and hinges on the C-14 results. So further C-14 testing would be in order even if there was no particular reason to query the recent results. There is however every reason to doubt these results."

"Contamination looms as the major consideration, especially with small samples. Take Tyrer's comments, for example. His contamination scenario is different from mine, and would depend on how much moisture was present in the reliquary at the time. But the questions he raises are pertinent, and we can penetrate them with the technology at hand IF we can get adequate samples. To my mind, this is the *sine qua non* of future testing. Without another C-14 run any 'ancillary testing' would seem to be accepting that the main question of the Shroud's dating has already been settled, and that research is continuing only to settle the question of how the image was faked. Until we have more C-14 measurements the Shroud's reputation will be under an unproved aspersion; we should deal with the question of contamination or isotope exchange as completely as we can, THEN talk about possible forgery or other medieval origin."

"Another possibility is that there is medieval intrusion and/or restoration in the corner where the sample came from. The more I think about this question the more ridiculous it seems to me that only one sample was lifted, from that very curious area near the sewn border where the side strip is missing and which is discoloured as if scorched. It is not enough to say that there is no visible disturbance to the weave, just as it is now not sufficient to qualify a female athlete by simply verifying that s/he has female genitalia! We know from human biology (and from archaeology) that there is more to those questions than sometimes meets the eye! Stuart Fleming told me he was quite certain that a frayed edge could be repaired by a good medieval restorer without leaving traces of the work. The point is simply this: another measurement on another sample would completely and easily dispel this notion, or give it real credence. Just imagine the impact if we really did find that another sample gave a result of 200 AD, as the California lab supposedly found."

"It is not the Church that would be initiating another C-14 round, but scientists and scholars who have the studied the Shroud and who do not accept the recent C-14 results as having conclusively settled the issue. In addition to Tyrer and myself, STURP, ASSIST and BSTS have called for further confirmatory C-14 testing. Surely this must count for something with the Cardinal."

"I have spent several hours writing up a draft protocol that would determine the radiocarbon age of the Shroud as accurately as modern technology permits. Otlet told me he would be willing to be involved if invited, but he does not want to be seen to be pressing for another round. He said he resented Hall's implications of sour grapes, and being misquoted in the press. He can certainly be trusted to do a careful, totally scientific analysis without playing **politics** *[my bolding]* or publicity. I believe Evin and Duplessy would be similarly reliable and low-keyed participants. With them and STURP plus my IsMeo colleagues and myself we would have an excellent team that could work together with you and the Cardinal to get to the truth of the Shroud's date -- WITHOUT the byzantine intrigues and jockeying for publicity that characterized the last round. So the question is simply: can we move forward or is the Cardinal against any further sampling?"

"Finally, I have given some thought to the idea of submitting a Shroud sample to a lab without informing the lab of its identity. I feel on reflection that this would not be entirely ethical, and would open up a possible criticism of the result that the sample might have been substituted. To me the best way is simply to organize a good team and work together to see the best scientific measurement gets done. One has to trust the labs ultimately, and it is better that they address themselves to the possible chemical and thermal contamination problems that might pertain to the Shroud rather than wasting a sample to get another standard reading. Surely we could trust Wolfli or Duplessy not to cook the results in any way, especially if they knew that Harwell was also measuring the same samples. Evin and Haas have close links with Duplessy and Wolfli, respectively, and their labs are the world's best at thermal separations."

"I hope you will stress to the Cardinal the scientific merit of further C-14 measurement, and help us to organize a truly state of the art investigation of the Shroud's radiocarbon age. For the minimal intrusion outlined in my draft protocol, we could definitively establish the C-14 content as uniform or varied, and confirm or overthrow the results of the first round. If the Shroud is confirmed as medieval, we will have to move on to deal with it as an extraordinary contradiction of everything we know about medieval art. I recall the conclusion of a famous Canadian metallurgist, Ursula Franklin, who spent the better part of her career researching the invention of bronze. She finally concluded that, weighing all the improbabilities and coincidences that would have had to overlap to allow for the invention of bronze, it was easier to believe that it never really happened! She was not far from von Daniken!!"

"Of course there will always be some who continue to believe in the Shroud's authenticity no matter what, and it will not ever be possible to absolutely rule out a massive isotope exchange in the fire. But science works with probabilities, and we can obtain a high probability by sampling other parts of the cloth, assuming the worst possible contamination, and employing the small counters as well as the AMS. Now that it is well-nigh universally deemed to be a fake, we should move quickly to establish the truth or falsehood of this. To me and I am certain to most people with deep respect for the Shroud, it is infinitely more important to address this question than to worry about the intrusiveness of another two small sample removals [...]."

Sincerely,

[signed]
Bill

Source: Letter of December 18, 1988 from William Meacham to Luigi Gonella. Luigi Gonella archives, of which the author has a copy.

1988 *(December)*. German author Kersten made a trip to Zurich, to interview Prof. Wolfli. Kersten mentioned that Tite couldn't find any control sample that would match the Shroud and thus the blind procedure was not needed. Wolfli replied, "Yes, but, the coding was needed at least because [...] shall we say for the journalists for one thing." Kersten commented, "I reflected, that experienced scientists apparently allowed journalists to dictate their methods, and so jeopardized the credibility of their work. This was something new to me, quite irrational; it did not sound like the customary self-confidence of science at all."

Kersten asked Wolfli could give him an advanced copy of the upcoming report due to be published by *Nature*. Wolfli replied, "The paper is still just a draft, not the final version, and so I do not wish to give it out. Not because it contains something secret, but because it is possibly not the final version. To be more specific: it may be that a serious error still remains in it, which we have overlooked and which would still have to be eradicated [...]. [T]here is the problem that if something still has to be changed in the text, and you published the earlier false version then one would have to withdraw it, because there would be some mistake in it which only the reader had noticed and none of us, then [...]"-- according to Kersten, "Again the sentence was left hanging in the air."

Kersten commented, "This set me thinking. So it was errors which had to be eradicated. I had a very simple idea of scientific work. One does a test, and at the end one obtains a result, which can be expressed as a numerical value. If everything was done correctly, there can be no mistake. Surely he could not have meant typing errors? For a practised team which had already composed many papers, writing down numbers correctly should not present any problem. What could he have meant by 'false version'?"

Kersten went on at length about the "blind" testing: "The whole procedure of secretly distributing the specimens in the small, screw-capped containers was a farce. This play-acting was no use even for the benefit of the press and public. The BBC film team was present in Wolfli's laboratory when he broke the seals of his three containers and laid out the cloth pieces before him, and anyone could see which one belonged to the Turin Shroud. It was rather poor play-acting, and unnecessary. It is astonishing that these crucial events were not better planned, if only to fool the public. Or had some crude blunder occurred during the planning? This secret distribution of specimens would only have made any sense if, as Prof. Raes remarked, the fabric pieces had been unthreaded before being placed in the containers. Then the scientists in their laboratories could not have distinguished the experimental specimen from the control specimens."

"I asked Wolfli why they did not do just this. His matter-of-fact reply was: 'We discussed this very question during our preparatory meetings in London in January this year. But finally we decided to leave the specimens intact. Even two years ago at the September meeting in Turin, attended by all seven of the selected laboratories, we discussed the sampling procedure. It was found that while cleaning the unraveled material too much waste was incurred. You can see this loss of material on our lab picture with the test sample separated into small pieces: if you add up the weight of the small pieces, a considerable quantity is missing. Besides, it was nice to be able to keep track of the specimen right up to the time of vaporization in CO_2, so that no swapping could take place. Yes, that was itself a good way to check that it was really the right specimen.' I find this quite baffling. It seems that the decision to keep the Shroud specimen intact until its experimental destruction had been made long before, years before in fact. Nonetheless, control specimens were procured at great expense, and the shoddy farce of a secret distribution was acted out. All the participants knew it was totally unnecessary. If they wanted to date other textile pieces from different periods, they could just have been handed out to the researchers in transparent containers. Containers of different colours could have been sealed before the running cameras, and in the presence of a notary, and we would have been spared a lot of mystery. Instead Shroud and control pieces disappeared behind locked doors, until a nervously smiling Dr Tite reappeared with some tin boxes on a tray. Who was trying to fool whom? The researchers behaved as if they had only realized that the test was not quite 'double blind' afterwards, and then said their action was justified to avoid the danger of a switch. These same scientists spend their whole time analysing blind specimens, never asking whether they could have got their specimens mixed up. When it came to the Turin Shroud, they had agreed for obscure reasons not to perform a blind test, and still wanted to let the public believe it was one. Why had such a major undertaking, after years of planning, ended up with this contradictory test programme? Was it just sloppiness? That was unthinkable, when one considers the precision with which scientific tests are normally carried out, without the benefit of lengthy preparations. Such thoughts left me with a very uneasy feeling about the affair."

"Regarding the third control sample (threads from the cope of St. Louis d'Anjou, dated at c. 1290-1310) that had been added at the last minute during the sample taking, Wolfli said, 'About this the British Museum told us': *Here's something extra, if you like you can practise on it, the age is precisely known.* Kersten commented, "It was just one more strange fact to add to the string of other confusing things: these highly specialized laboratories were offered a chance 'to practise', although they were in the habit of dating hundreds of specimens month by month and were hardly in need of any practice."

Kersten asked Wolfli how long Cardinal Ballestrero and Tite were in the sacristy putting the specimens in the nine containers. Wolfli told him about 30 minutes. Kersten commented, "Half an hour, to put nine pieces of cloth the size of postage stamps into small tubes. Incredible!"

Wolfli admitted that his lab dated only half of the sample they had received. When Kersten asked him where the other half was, "The reply came with a secretive smile: only he and his wife knew that!"

Source: Kersten, Holger and Elmar R. Gruber. Jesus Conspiracy: The Turin Shroud & The Truth About The Resurrection, (Rockport, MA: Element, 1994), pp. 54-59.

Comments: Kersten came away from his interview of Wolfli "with a very uneasy feeling about the affair."

Wolfli said that only Cardinal Ballestrero and Tite put the specimens into the containers but Gonella maintained that he was there also. Kersten speculated that the reason Ballestrero and Tite were in the sacristy so long, was that an exchange of the samples was being orchestrated.

Although Wolfli told his wife where the extra sample was being kept, it doesn't appear he told others from the lab!

Australian blogger Stephen Jones has made some interesting observations regarding the spread of the measurements. See http://theshroudofturin.blogspot.com.au/2015/11/the-1260-1390-radiocarbon-date-of-turin.html.

1989 *(January)*. German author Kersten sent photographs of the Zurich specimen "Z1," one of the samples dated in 1988, to Belgian textile expert Prof. Gilbert Raes to compare with photographs of the sample that Raes had received in 1973. Kersten quoted Raes, "I have compared the specimen which I received in 1973 with Prof. Wolfli's photos. I must state that the general appearance is quite different. What could be the reason for this difference? In each case the main difference lies in the differing number of threads per centimeter in the directions of warp and weft. It is not easy to count the number on a photograph, but I did not find the same number as on the piece I received in 1973. I may conclude from it that the two specimens cannot come from the same item. That is my impression when looking at the specimens."

On the 18th, Kersten flew to London and talked with Susan Black, Secretary of the BSTS. *[Note: Kersten mistakenly gave her last name as "Brown."]* He asked her about her opinions of the various British investigators involved. "She described Hall as an arrogant careerist, not at all interested in the cloth itself. He was only interested in gaining publicity for himself and his institution, and he thought the members of the BSTS and all those interested in the cloth were mad. His powers of imagination seemed to stop at his laboratory doors. Now that the Turin Shroud has been dated to the Middle Ages, Hall claimed that he had known it all along. But really he knew absolutely nothing about the cloth, he did not have a clue, Susan told me and added: 'This is a *volte-face* for him. I suppose if the Shroud had been dated to *[AD]* 100 he would have gone the other way -- he would have said how wonderful it was and how he always believed it'."

Kersten asked Black how it was possible for Sox to have his book ready weeks before the official announcement was even made. "He was very unpleasant when we last spoke to him because we had just told Reuters or someone that we thought he was implicated in the rumours. So he is not very pleased with us [...]. He is a very emotional guy [...]. He tends to get very excited about things and then there is a big depression [...]. He knew the results then pretty well, he obviously knew what they were, but the book indicates that the only way he knew was from Harry Gove, who was with Paul Damon in Arizona. Harry Gove's laboratory did not get the sample that they wanted. They were the people who first asked Turin to do the sample, it was all their idea, and their original paper and everything else, and they were very annoyed when they didn't get a sample. It was because all the three laboratories chosen were using the same method, which again is very strange, it doesn't seem sensible at all. Paul Damon from Arizona is a terribly nice, gentle, very scientific man, who doesn't really understand all this kind of religious fervor, and Gove managed to persuade him to allow him to be in attendance when the result came through, that's how he knew. Harry Gove is very sociable, and Sox found out about the bet with Harry Gove's assistant, and that's what he based his whole premise on."

Regarding Tite: "He is almost a businessman; he is a scientist but he is obviously capable of avoiding the truth, because he certainly avoided telling the fact that the dates of the other samples were known." Kersten asked Black if Tite was "bribable." She replied, "Oh, I don't know about that. You just don't know, every man has his price, there may be something that I don't know." Black also commented, "Everything that could have been done wrong has been done wrong, and none of the things that were suggested by the Harwell laboratory, which was an objective outside laboratory, were done at all. Something strange there!"

Source: Kersten, Holger and Elmar R. Gruber. Jesus Conspiracy: The Turin Shroud & The Truth About The Resurrection, (Rockport, MA: Element, 1994), pp. 61-66.

Comments: None of the British investigators Black described received glowing recommendations.

1989 *(January)*. Kersten listened to a tape of a talk that Dr. Michael Tite of the British Museum gave to the BSTS in November 1988.

Kersten summarized the main part of the talk: "If one looks more closely at his statements, first he stated that the piece cut off was divided into three parts. As we were later to learn, it was in fact cut into four parts, that is, it was first halved, and then one half was divided into three parts. Then he said that only he and the Cardinal took part in the secret packing of the specimens. He failed to mention that Prof. Gonella was also present. He can surely not have forgotten whether there were two or three of them in the room? And if he had forgotten, so much the worse for his credibility! Thirdly he said that the strip removed was 1 x

7 cm in size. In fact it must have been almost twice as wide and over 8 cm long! This discrepancy must have been known to the 'co-ordinator' [sic] of the dating test, or at least he must have noticed it at some stage."

Kersten continued, "Finally the audience was allowed to ask questions. Besides various questions about a possible contamination of the cloth specimens, one of those present said that he was surprised to read in David Sox's book that Tite and the Cardinal had signed a document which expressly declared that the specimen of cloth really did come from the Turin Shroud. He asked: 'Does such a document really exist?' Tite's reply, in his exact words was: 'I don't know, I'd have to go back to the video.' The questioner said again: 'But according to him it was signed by yourself!' Tite replied: 'Well I was going to say I'd have to go back to the video, this is why we had a video taken, I mean I have a feeling that I did sign something, yes, which is why I had a video taken!' Laughter in the hall. One might consider it strange that the person guaranteeing the experiment could no longer remember anything about this extremely important detail. Moreover it would be quite pointless to refer to the video on this point, since there were (according to Tite's statements) no witnesses present at the distribution of the specimens in the containers, and so that procedure was not filmed. Surely Dr Tite must at least have remembered that. Tite was obviously disturbed and somewhat ruffled. Chairing the discussion that evening was Ian Wilson, and he was polite enough to pass quickly over the embarrassment and ask if anyone had any further questions."

"A member of the audience then raised the question whether the laboratories had been in contact with each other during the test phase. After categorically denying it at first, Tite admitted that there had probably been leaks contrary to the agreement, and in the ensuing unrest in the hall he conceded that the so-called blind test too was really no blind test! Surely he must have known this already before the sampling, when he was supposedly unable to organize the procurement of identical fabrics. Why then stage the whole show with the secret packing of the samples in the containers away from the public eye? What purpose could such play-acting have served? There is no reasonable answer to this question. The responsibility for the exchange of information among the laboratories, which Tite admits to, also rests on his shoulders. He was the guarantor, the referee so to speak, who was supposed to see that the agreed experimental procedure was exactly adhered to."

Source: Kersten, Holger and Elmar R. Gruber. Jesus Conspiracy: The Turin Shroud & The Truth About The Resurrection, (Rockport, MA: Element, 1994), pp. 68-70.

Comments: Kersten added, "In the event it was as if no agreement were followed at all." I also have a copy of Tite's talk, which was given on November 7, 1989. Tite also had seemed certain they hadn't been told the specific dates of the control samples, even though it's well established that they had been (as shown in Part II). That's not the only time Tite had problems with his memory. In 1989, he was asked in an interview who had written the "1260-1390!" on the blackboard at the press conference when the dates were officially announced. In the interview, according to Marinelli in her Valencia paper (pg. 13), he said he couldn't remember who did. Yet, in a BBC 2016 interview (http://www.bbc.co.uk/programmes/p03lqvkb), he stated that he had written it. How does he forget one year after the event who wrote it, and twenty-seven years later remember that he himself wrote it??

A questioner says that photos of the C-14 sample area suggest there could be a dye in that corner. Tite asked a rhetorical question: "What is the source of the dye?" A second questioner then noted that chemist Ray Rogers in 1978 had said that the Shroud has a mussy appearance; the questioner wondered about the carbon content of the apparent dye. Curiously, Ian Wilson, who was moderating, intervened, wanting to go on to another question, and Tite was not given an opportunity to answer the question about the dye.

1989 *(January)*. On the 14th, the English periodical *The Tablet* published an interview of Edward Hall by journalist John Cornwell. Hall was asked why there hadn't been blind measurements as had been recommended in the 1986 Turin workshop. Hall said that the workshop had also recommended that the samples not be unraveled and since the Shroud weave was so distinctive, it could not be concealed, so there was a problem. But he went on to tell Cornwell that their tests <u>had</u> been blind. Gove commented in his book on this aspect, "After the samples were burned to carbon dioxide gas they were recoded separate from the carbon dating team, somebody who was sworn to secrecy. Thus neither Hall, nor Hedges, who actually made the measurement, knew which of the four gas samples was the shroud. Hall said the other two labs did not take this precaution (he was wrong about that because in the paper that was published in *Nature* on 16 February 1989 it was stated that Zurich followed the same procedure as did Oxford). He made it clear that he thought this was a very wise and clever thing they had done at Oxford."

Gove continued, "My personal view was that it had been a very silly and unwise thing to do. It meant that the two senior scientists at the Oxford AMS facility did not know whether they were actually measuring the samples Hall had so dramatically announced he had brought back to England from Turin. The same situation apparently also applied at Zurich. Whatever it was they measured at Oxford and Zurich fortunately bore a one to one relationship to the samples measured at Arizona, where the fate of the samples was followed by Donahue from the cradle to the grave so to speak. It is not too outrageous to argue that the shroud sample was only measured at one laboratory -- Arizona. The other two, as far as the senior scientists knew, measured a sample of carbon dioxide gas that just happened to give a carbon date close to the shroud's known historic date. I obviously do not believe this, but it was risky to have carried out the measurements in such a way. I am sure that Hall, Hedges, and Woelfli would have argued that you had to trust someone. That was certainly true, but the people I would be most inclined to trust were the senior scientists at the three labs because they had the most to lose by improper behaviour. In the case of two of the labs, it could be argued that the senior scientists did not exercise a close enough control to merit such trust."

Source: Gove, Harry. Relic, Icon or Hoax?: Carbon Dating the Turin Shroud, (Bristol and Philadelphia: Institute of Physics Publishing), 1996, pp. 296-297.

Comments: The narrative of the "blind sampling" is a good example that the labs did not, as Gove previously proclaimed, conduct their measurements in a professional manner. Note that Gove said, "It is not too outrageous to argue that the shroud sample was only measured at one laboratory -- Arizona." If one controversial sample dated by three labs isn't dubious enough, then one controversial sample ostensibly dated by only one lab-- this being the opinion of the co-inventor of the method -- is all the more dubious.

1989 *(January)*. When interviewed, Hall was reminded that many C-14 datings are found to be in error. He said, "It is beyond the scope of my conversation to explain why some of these things occurred -- most of the errors are due to the validity of the sample-taking procedures. All right. I can't deny that mistakes have been made [...]. But Hall and others refused to admit that any mistakes were made in the case of the Shroud, despite all the damning evidence. Pertaining to the blood on the Shroud, he said, "But whether it's human or pig's blood -- who knows." Hall was pressed on the point that no one still knew how the image was formed. He replied, "[T]his assumes that I'm interested in solving these remaining mysteries in the first place and I'm *not*, to be quite honest. I haven't given it much thought, and I certainly don't intend to now that I know it's a fake. I actually find it totally uninteresting now."

Source: "John Cornwell Interviews Edward Hall" in *The Tablet,* 14 January 1989, pp. 36 and 38.

Comments: The question has to be asked: what sort of scientist is content with only part of a solution to a mystery??? Hall also questioned whether the blood might be human or just "pig's blood." If Hall had read some Shroud literature, he would have known that Adler and Heller had proven, in a peer-reviewed journal, that the blood was definitely primate (Heller, J.H. and A.D. Adler. "Blood on the Shroud of Turin." *Applied Optics,* Vol. **19**, no.16, August 14, 1980, pp., 2742-2744). Despite having granted an interview to *The Tablet*, when Remi Van Haelst wrote to Hall with some questions about the dating, he "replied by writing that could not spend his precious time, to answer the questions, posed by scientists of little standard, blinded by faith [...]. He also used just one word to describe Van Haelst's critique of the *Nature* report: "nonsense" (http://www.sindone.info/VHAELST6.PDF, pp. 2 and 21). In Hall's obituary (https://www.independent.co.uk/news/obituaries/professor-edward-hall-9260740.html), it ran his infamous quote, "There was a multi-million-pound business in making forgeries during the 14th century," he bluntly told a British Museum press conference. "Someone just got a bit of linen, faked it up and flogged it." And again, "Some people may continue to fight for the authenticity of the shroud, like the Flat Earth Society, but this settles it all as far as we are concerned."

1989 *(January)*. At the start of his trip to London, Kersten had written to Hall at Oxford, hoping to set up an interview. Kersten called his office and was informed that Hall was abroad and wouldn't be back for ten days. Kersten called again after those ten days and got the exact same story. Kersten commented, "Keen to find out whether Hall had really left or had got people to lie about his whereabouts, I looked his number up in the local phone book. I called the number and asked to speak to the professor. A young, friendly man at the other end of the line told me that Hall must be at work, but would certainly be back home that evening. Now it was quite clear I had been systematically lied to. Why was the Oxford professor so evasive? Or did he even have something to hide?"

Source: Kersten, Holger and Elmar R. Gruber. Jesus Conspiracy: The Turin Shroud & The Truth About The Resurrection, (Rockport, MA: Element, 1994), pp. 71-73.

Comments: Hall simply could have refused to meet with Kersten. The fact that the latter was lied to was cause for suspicion. But Hall did later send Kersten a letter with some information about the weight of the samples (although not specific) as well as several pictures (pg. 82). However, Hall did not come across as generally truthful.

1989 *(February)*. Hall wrote Ballestrero on the 22nd, thanking him and Gonella "for arranging the whole program in a most competent and dignified fashion," a statement he made at a press conference at the British Museum the previous week and asserting that neither he nor colleagues at Oxford had made any comment on the results.

Source: Gonella, Luigi and Riggi di Numana, Giovanni. Sindone: il mistero continua, (Milan: Fondazione 3M), 2005, pg. 84.

Comments: Check the timing and remarks from this entry and the previous one. On February 15th, Hall told the British Museum crowd that the Shroud was a fake. On February 22nd, he told Ballestrero that neither he nor his colleagues had made any comments on the results. The latter was a lie, pure and simple.

1989 *(February)*. In mid-February, Gove called Donahue to say he heard the official report would be in the February 16th issue of *Nature*. Donahue confirmed this and also said that issue would have a letter from Thomas J. Phillips of Harvard and Fermilab hypothesizing that the resurrection of Jesus could have produced neutrons and an excess of C-14. The Tucson press contacted Donahue about the letter and he told Gove "it was the first time he had used expletives." Gove added, "He was indignant that *Nature* would publish such a letter."

Source: Gove, Harry. Relic, Icon or Hoax?: Carbon Dating the Turin Shroud, (Bristol and Philadelphia: Institute of Physics Publishing), 1996, pp. 299-300.

Comments: Donahue was a Catholic, so one wonders why he would have a problem with the hypothesis of Phillips, who was affiliated with two of the most prestigious institutions in the U.S.A., which means the hypothesis cannot easily be dismissed. And why indignation at a piece putting forth a scientific hypothesis? If Donahue's findings were solid, he should have had nothing to fear.

Nature printed a response by Dr. Hedges to Dr. Phillips letter. Phillips sent another letter to respond back to Hedges, describing various tests that could be performed to test his hypothesis, but *Nature* refused to print it. Phillips concluded, "Until these further tests are made, we cannot conclusively state the Shroud of Turin is medieval."

1989 (February). The *Nature* issue dated February 16th had the official report and was titled "Radiocarbon Dating of the Shroud of Turin," which had twenty-one signatories, with Paul Damon of Arizona being listed first. According to the report, "These results therefore provide conclusive evidence that the linen of the Shroud of Turin is mediaeval."

The report gave the impression that each lab used up all of its samples but Damon had said "We have preserved a piece of the sample, if there was a dispute, to show it to the Church authorities." Wolfli also admitted that he had preserved a portion of the sample. See comments on this entry for explanation by Ramsey from Oxford.

Source: Marinelli, Emanuela. "The Setting for the Radiocarbon Dating of the Shroud." Presented at 1st International Congress on the Holy Shroud in Spain -- Valencia -- Centro Español de Sindonologia (CES), April 28-30, 2012, pg. 12. www.shroud.com/pdfs/marinelliv.pdf.

Comments: The *Nature* report is accessible at https://www.shroud.com/Nature.htm. Even before all the facts that have been brought up in this book, the labs never should never have been so confident to say that the results were "conclusive evidence" that the Shroud was a fake.

In a posting on www.shroudstory.com from May 7, 2013, Christopher Ramsey from Oxford said, "As far as I am aware the whole material was used for the dating here -- that is what the weighed components suggest -- and we don't have any remaining sample in our archives with these sample numbers. I think the position taken here was that we only had permission for dating and not other research -- and at that time, the measurements needed as much material as possible. It is true that it is normal practice to retain some material for further checks in routine dating and we normally do this -- so I can see that other labs may have made different decisions." Since Ramsey was involved in the 1988 dating, one would think he could have come up with something stronger than "As far as I am aware [...]." It was another example of the horrible record-keeping by the labs.

The late Al Adler told author Mark Antonacci in the 1990s in one of their many phone calls that all three labs retained pieces after the dating, but did not specify the source of his information.

Gove commented in his book (pg. 301), "The article was rather opaquely written -- difficult to comprehend in complete detail even by experts in the field [...]." Was the complexity the result of input by twenty-one authors -- or was something else going on? It's striking that Gove himself asked "was something else going on?"

Bro. Bruno ("The Victory of the Holy Shroud Won By Science." *Catholic Counter-Reformation in the XXth Century*, September-October 1989, No. 223, pg. 30) pointed out various problems with the report. In his words:

1. The article reveals an obvious bias on the part of the authors when they begin by giving a mediaeval date for the "shroud."

2. The article reveals a lack of protocol for the experiments so that the way is open for improvisation, especially concerning the place from which the sample is taken.

3. The article reveals that the "double blind" procedure was abandoned.

4. The staging of the *mescolamento** was kept, however, in order to make it seem that the "double blind" procedure had been respected.

5. The laboratories were forewarned of the date of the two "official" control samples.

6. The article mentions a third "unofficial" control, which is dated and recognisable by the fact that it is unraveled and packaged in an envelope, unlike the others. The date of this third control is 13th-14th century, which is the date expected for the Holy Shroud by opponents of authenticity.

7. Tite therefore lied to the Ansa agency on 30 March 1989 when he declared that "in the presence of Cardinal Ballestrero, Archbishop of Turin, he had placed four fragments in the metal containers provided."

*Mescolamento refers to the placing of the samples into the containers.

"A statistician working with Bro. Bruno (*Catholic Counter-Reformation in the XXth Century*, April 1991, No. 238, pg. 13) wrote: 'We stress the absence of a precise, complete official report. The imprecision in describing the samples seems especially disconcerting. In my profession, I have to call on laboratories for their expertise in mechanical products or pieces, and I can assure you that the first part of the report always consists of a description of the sample: its visual appearance and then as seen through the microscope, its weight, dimensions and colours, with photographs of the most significant details --photographs that will be authenticated by date and laboratory stamp. I did not fail to express my surprise at this to Dr Tite when we met him ten days ago, and his answer left me speechless: 'You know, I'm just a plain man; I didn't want any directive document; no, I didn't note down exact values; it wasn't important for us then.' I doubt whether my boss or the clients for whom I work would be satisfied with such answers!"

1989 *(February).* According to Petrosillo and Marinelli, "None of the 21 scientists who signed the report as authors were known for research on the Shroud. No sindonologist, no STURP member was permitted to participate in the work. Tite also signed the scientific report as guarantor, in a few months he became coordinator. He claimed it was only fair that he sign because of the demanding job of collating everything that 20 people of 3 labs had done." The Colonetti Institute, chosen to be one of the three institutions that were to analyze the data, was not allowed to handle the results independently was only invited to confirm the calculations made by the British Museum. Tite remained the sole controller of the whole affair. The laboratories wanted the task to be entrusted to him as an independent. 'Any Italian was considered suspect' recalled Gonella, 'and any Turin person was doubly suspicious. We accepted because we were in the situation where, if one refuses, his refusal is interpreted as obvious proof one wanted to cheat.' Tite, head of the British laboratory since 1975, had previously worked with Hall and on 1 October 1989 he also succeeded him on the Oxford chair. These reports, before and after the radiocarbon test, certainly do not support the neutrality of the 'guarantor' towards one of the three laboratories. There were violations of the established agreements. The planned final meeting in Turin between representatives to discuss the results did not take place. They were simply sent to the British Museum for statistical analysis."

"Bonnet-Eymard advanced this suspicion: the concertation of the three laboratories had been so close that it was not thought to maintain the fiction of the 'last discussion'. The meeting was no longer needed also because the samples had already been identified. In the brief historical description in *Nature*, the authors do not mention the important episode of the fire suffered by the Shroud in 1532. They allude to the fact that research has been carried out in previous years, but they don't say with what results. The account of the procedure followed on April 21 in packing the samples does not refer to any established protocol, neither to that of Riggi nor to the previous ones. The publication of the article of only four pages in *Nature* did not meet the expectations of those who expected a much broader and more detailed report or, even more, the reports of the individual laboratories with the primary data obtained from the measurements and also the photographs of the samples. The 'rough' measurements of the analyses remained well hidden in the laboratory files. No one could therefore check the reliability of the final result. Tite objected that the deal was for a single article and that the magazine could not accept a longer one. *Nature* was chosen because it is more widespread and is published weekly, while *Radiocarbon*, a specialist magazine on the subject, is published every four months. Results for samples 2 (Nubia), 3 (Thebes) and 4 (France) were expected: 1026-1160 AD. for 2; 9 BC - 78 AD for 3; 1263-1283 A.D. for 4. For sample 1, i.e., the Shroud, two possible intervals are provided, 1262-1312 A.D. and 1353-1384 AD, then summarized in the famous 1260-1390. Bonnet-Eymard stressed, however, that it was arbitrary to embrace, in a single interval of over 130 years, uneven responses. 'The concordance between the results of the three laboratories' we read in *Nature* is exceptionally good. The expected errors from sample 1 (Shroud) was a little larger than one would have predicted. The difference between the measurements obtained in the three labs on the Shroud sample was uneven and an indication of an abnormal amount of ^{14}C."

"Van Haelst recalled that in the 1983 pre-examination, conducted by the six laboratories which had applied for the Shroud dating, the Peruvian sample, eliminated because it had provided a more recent date than expected, revealed an equally problematic behavior which was described as follows: 'The variation between the samples is higher than expected on the basis of the expected measurement errors.' Also in that case, therefore, the sample did not have a uniform distribution of radiocarbon inside it. The other coincidence between the Peruvian and the Shroud samples is therefore interesting: both provided a more recent date of the expected one. Laboratory experts did not provide detailed information on their measurements. Anthos Bray himself, director of Colonetti, made his review only on the data already 'amalgamated.' In Oxford they found cotton fibers in the Shroud. Hall claimed that these were colored cotton fibers. 'This cotton,' explained Peter H. South, director of the textile analysis laboratory in Ambergate (Great Britain), 'is a thin, dark yellow thread, probably of Egyptian origin and rather ancient. Unfortunately it is impossible to say how the fibers ended up in the Shroud, which is basically made of linen. They may have been used for restorations in the past or simply remained woven into linen threads when the artifact was woven.' 'Perhaps,' noted Van Haelst, 'the presence of a type of non-European cotton was of less importance for scientists engaged in radiocarbon measurements'."

"The Belgian chemist regrets, however, that no further information is given in *Nature*, but only the news that cotton has been found. Bonnet-Eymard argued that in Oxford in the Shroud fabric traces of a black thread were observed in some places, passing through the white linen as if it had been introduced or basted there. The passing of time failed to quell the bitterness for abuses suffered. On more than one occasion, Gonella continued to deplore that the laboratories gradually demanded that the collaboration of any other scientist be denied and that the samples were used exclusively for the ^{14}C exam. In addition to moving away scientists from other disciplines, the three laboratories claimed that they had no controls of any kind by representatives of the ecclesiastical authority. Focusing on the fact that Ballestrero had granted them the maximum freedom of research and skillfully exploiting the suspicion that the Vatican's word could not be trusted, Tite and the laboratories obtained to conduct their exams in complete freedom, without any control. They justified themselves by claiming that dating is an operation completely separate from other research and that the presence of other scientists would have diminished the confidentiality of the examination. Even if they themselves have not kept it! Ballestrero's consultant therefore defended himself: 'Respect for freedom of research has thus forced us to separate the dating from other research by postponing their execution. We have accepted checks on our work which has not been offered any counterpart'."

"For his part, Frank C. Tribbe, an American prosecutor known to have written a book on the Shroud, criticized Gonella 'because he did not allow any sindonologist or archaeologist to be present for a consultation or to observe the sampling of the samples.' Ballestrero's consultant himself was not invited to the laboratories. Tite argued, with apparent candor, that the representatives of the laboratories insisted on being present in Turin, while Gonella did not do the same to attend the analyses. 'We have not dreamed' confessed Riggi, 'to ask to check their work.' 'Besides, added Gonella, 'anyone who knows a complex laboratory knows that being on exams doesn't mean anything, because if I want to manipulate an analysis, I can do it under the eyes of anyone.' It is a matter of trust. Riggi expressed some other reservations about the exam: 'We believe that it alone, separated from the other 25 exams proposed, cannot give a reliable answer. After what happened Gonella bitterly concluded 'I think that nobody has the courage to talk about the good faith yet. Among the many anomalies of the story, more or less influential on its correct development, we should also note the unusual initiative of the carbon daters. It had never happened that it was the analysis laboratories that asked to date a finding, nor that there were so many, seven, to want to do it in parallel. Normally it is a scholar, as a submitter who presents the finding to the labs and asks for a response.' 'But this time,' noted the Turin engineer, 'the laboratories specialized in ^{14}C wanted to act as 'submitters of themselves.' The fact is that, from the beginning, the story of the dating of the Shroud was marred by the publicity aspects, to which the conclusion of the scientific consultant of Ballestrero who respects complete confidentiality is lapidary: 'We were not satisfied at all with the ^{14}C scientists and they proved to be all too sensitive.' The conclusion of Gonella, which reflected the conviction of the Church authorities: 'No, we were not satisfied at all with the ways in which the labs conducted the study on the Shroud'."

"Some attitudes by analysts also burdened the judgment, such as: the demand to assist with the sample-taking due to lack of trust towards the representatives of the Church, and, on the contrary, not having wanted trusted technicians from the Curia to in turn attend the labs' analyses. 'When did you ever see it,' asked Gonella, 'that a dating lab wanted to be present at an excavation because he didn't trust the archaeologist who took the samples? When did the laboratories ever refuse to collaborate?' The management of the research was completely entrusted to non-Catholic experts, without any real guarantee of confidentiality and respect for the agreement not to identify during the examination which of the fragments delivered was of the Shroud. A criticism of this kind was addressed to Ballestrero and Gonella: 'There is no doubt,' declared Bruno Barberis, president of the Turin International Center of Sindonology, 'that the whole affair was handled superficially and not in keeping with the importance and peculiarity of the object to be examined. Even from Vatican sources explicit criticisms have come about in this regard. The fact that it was not considered appropriate to involve an official body, such as the Pontifical Academy of Sciences, in the management and control of the entire operation is certainly one of the main causes of the great fuss raised in recent months. The indiscretions, the voices, the interventions of people totally extraneous to the exam could be foreseen and avoided through a more careful management.' The question arose: if the malevolent intention of the analysts was clear from the outset, why did you leave them free hand in conducting the exams? 'The Church has found itself,' Gonella defended, 'facing a challenge launched by some people who, with their requests, have done everything possible to be told no, to be able to say that the Church was afraid of science. Then, faced with this danger, it was decided to carry out the scientific examination at any cost, even with protests'."

"The Turin engineer did not skimp the strong words: 'It was a blackmail. They put us back up against the wall with blackmail. Either we accepted the ^{14}C test under the conditions imposed by the laboratories or a campaign would have been unleashed with accusations against the Church of fear of the truth, of being an enemy of science'. In the October 13 release, however, Ballestrero had spoken of a 'reasonable operational program'. One suspect seems legitimate: is it not possible that the 'ideological passion" of the scientists had an effect on the results? We asked the accuser: 'I don't know, it's very difficult to say; what is certain is that when people behave so differently than usual, everything is possible to think. But it is not scientifically and morally possible to make such a reckless judgment.' On the behavior of the laboratories, Gonella was explicit: 'They behaved like dogs. I protest for their absolute non-professionalism in the field of ethics. I protest about the infamous way they did it. I told them to their faces that they are *mafiosi*.' Gonella was particularly angry with Tite and Hall: 'It was a certain environment, let's call it 'scientific', which revolves over England, which has promoted this mystification. The gentlemen of Oxford and London have behaved very badly; in their attitude, there is an attack on other scientists without even having read their articles. I had a great esteem for the University of Oxford that I no longer have. Scientists came out of this test very disqualified'."

"Gonella was exacerbated by the mocking tone released by Hall in a conference of 15 February 1989 to the British Museum Society when he began showing a cartoon of the *Observer* in which a scientist in a white coat confessed accusing himself of 'having committed ^{14}C twice.'* After a series of gross errors concerning the history of the Shroud, the director of the Oxford laboratory, in that conference, went on criticizing STURP, which he said was particularly guilty for giving the Shroud 'a false scientific credibility.' Regarding the reduction of 25 from seven to three laboratories, Hall defined the decision 'completely wise,' adding that Harwell's laboratory could not be included 'because we are both English' and because the material needed was five times greater than what Oxford needed. He also argued that that morning, during the excision in the sacristy of the Turin cathedral, the carbonists were 'kept well away.' On the problem of dating Lindow man, which resulted in dates hundreds of years different between Oxford and Harwell, Hall defended the measurement obtained from his lab. According to him, many errors in dating are caused by the incorrect sample collection procedure. Hall then narrated a significant episode of his attitude towards the results of other sciences. In 1983 the police brought him a skull for dating. A pathologist was absolutely careful, based on the photographic comparison, that it was that of a woman murdered a few years earlier. The dating placed the skull in 410 AD. Obviously in the judgment of the scientist, it was necessary to believe in ^{14}C and not the pathologist. 'Oxford,' Hall boasted, 'performs a thousand dates a year, all going well; we seem to have been right without exception.' But what happened in 1983 during the test exam of the six laboratories candidates to date the Shroud was not mentioned: Oxford also gave a wrong date on the Peruvian sample."

"The British C-14 expert expressed no doubts about the effectiveness of the cleaning performed on the Shroud sample: 'The microphotographs showed many crystals of sodium chloride and calcium carbonate on the surface of the Shroud, carefully removed with a wash in solvents. It would have taken 60 percent of modern contamination to arrive at a false medieval date from a truly first-century fabric.' A level of contamination he called 'ridiculous.' Hall also said that he would be surprised even if he had found contamination greater than 10 percent on the Shroud and if there had been 1 percent of contamination left after cleaning. Regarding the possible alteration of the ^{14}C content caused by an explosion of radiation, he pointed out that such an eventuality would be conceivable only with the involvement of a really large amount of energy. But the odds –- in his opinion -- would have been one in a thousand million because such a process could have provided the right dose to obtain precisely the date of the fourteenth century, the most likely one for the creation of his Shroud, if it is a fake'."

"The conclusion of the conference was mocking as its beginning. Hall reported that he and Tite had received surprising letters, including one condemning them to hellfire 'many times over.' He pretended to console himself by noting that 'there seems to be no Christian ayatollahs around.' Hall's security was contradicted by Gonella, for whom the scientific procedure adopted by the three laboratories was not flawless. In particular, according to Riggi, the pretreatment operations of the three samples would be questionable, i.e., impurities-reduction techniques. However, the two Turinese argued that the scientific result of the analyses is not seriously contested. There would be no objections capable of anticipating the date of the finding by one thousand years: 'At the most, it will be possible to go back a hundred years'. From other statements by Gonella, it is deduced that there are several reasons not to be calm about these analyses and not only for the lack of for the non-interdisciplinary physical-chemical examination of the samples in order to eliminate the natural and artificial contaminations that have encrusted the Shroud in so many centuries of history and vicissitudes. 'Misconduct,' insisted the professor of the Polytechnic, 'there have been plenty of them. The colleagues of the ^{14}C group behaved in a disgusting way. Those scientists have hatched a real conspiracy to discredit the Shroud.' Several times Gonella reconstructed the phases preceding the sampling with observations highly critical of the laboratories accused of having suffered a 'successful drunkenness:' 'At the beginning, when they asked us to examine a Shroud sample, they guaranteed the utmost seriousness and completeness of the analyses, together with and collaboration with the custodian of the Shroud, i.e., Archbishop Ballestrero, and with his scientific consultant, i.e., myself. Caught by celebrity fever, those scientists began to renege on their commitments: no more interdisciplinary examinations, only ^{14}C. They also stormed Rome with pressure for Turin to give their conditions. They made use of the then president of the Pontifical Academy of Sciences, Professor Chagas, to get the undersigned out of their way and go their own way.' It was natural to ask Gonella: why then did the Holy See and Ballestrero agree? 'Because Chagas acted alone, bypassing the other academics. And the Holy See was continually threatened by the laboratories themselves, which they repeated: if you do not leave it to us, only to us, the results will not be acceptable. So, in the end, Ballestrero had to give in, despite suffering a lot. Also because those gentlemen were doing everything to support the thesis that the Church was putting the clamps on science'."

"As for Chagas, he left the presidency of the Pontifical Academy two weeks after the announcement of the results. A rumor immediately circulated that it was 'torpedoed' for the Shroud affair. Undoubtedly his attitude in the affair was disconcerting due to his disagreements with the engineer Gonella and with Cardinal Ballestrero (who complained to the Vatican), the double game and the agreements under the table with Gove, declared opponent of the authenticity and also interested for reasons of publicity related to the dating. The public tribute paid to the resigning Chagas by the Pope for the 16 years as president certainly does not exclude the reservations of the Secretary of State on his behavior in the Shroud affair. So does the international scientific world reject that verdict? To this question, Gonella replied: 'For now, no, the basis for reaching such a conclusion is currently lacking. But certainly the vast majority of colleagues are not persuaded, neither by the procedures adopted, nor by the conclusions. Those gentlemen, moreover, are complaining to the four winds that the last word is now said on the matter. Theirs, obviously.' It is not so? 'Quite the opposite,' specified Gonella [...]. 'It is not enough to say that the sheet is a medieval fake. The problem is to understand how it was created. It is not the last word.' It is interesting now to see how Cardinal Ballestrero, a decade after these events, reconstructed the whole story in an interview curated by his secretary, Father Giuseppe Caviglia."

"'The interest was rising,' recalled the Cardinal 'and at the same pace the requests and pressures for the examination were granted. A serious slander against the Church was also on purpose: enemy of science because fearful of the truth, worried not to lose the relics that make money [...] and it gave the impression that in a climate like ours with ecumenical ramifications, it was not going well. At that point I felt it my duty to inform the Holy Father of the matter. And the Pope, the first time I spoke to him, said to me': 'But what about? It's a relic! Can a relic be subjected to such a technical, so material analysis?' 'Holiness, the choice is yours!' 'Months and months pass. Finally I received a letter from the Secretariat of State that informed me that the Holy Father, after lengthy reflection or consultation, had decided in a positive sense, believing that there could be no risk for the faith from the examinations, nor a risk of disrespect for the relic and therefore that I proceed as well.' 'Proceed! I didn't know what to do! I then asked them to allow me to avail themselves of the advice of the Pontifical Academy of Sciences. There, in fact, I thought there were well-versed scientists in that matter and therefore I would have felt protected at my sides. Having consented, I spoke with the President of the Academy to report specialists belonging to the Pontifical Academy. But the President, to my amazement, said to me, *No, no, I'll handle it!* My shoulders dropped because he was a biologist. And so began the first series of troubles. Speeches, reminders [...]. In the end I decided: to gather in Turin the directors of the seven existing laboratories in the world to develop the rigorously scientific procedure to follow. Having come to Turin, after three days of meetings and discussions -- great difficulties arose because all seven wanted to take part. I had to make a great effort and made various enemies, deciding that in order not to ruin the Shroud too much only three labs would be chosen. The Holy Father confirmed the decision and so we proceeded. The laboratories with more experience on the subject were chosen for the number of exams already performed and for internationality: one

Swiss, one English, one American. It was then a matter of developing the phase of the sampling and the problem was that of guaranteeing an analysis that could be controlled just as a procedure. Then these scientists thought of extending the analysis to three samples. That is, each laboratory would receive three samples, one of which from the Shroud and two from fabrics with dates previously established. By subjecting the three samples to a unitary analysis, knowing the certain date of two, if the analysis of the three had respected the known dates of the two, it had to be concluded that the third one was also valid. So it was done'."

"'A French technician of the National Textile Authority, which is one of the most famous in the world, and another in Turin, were supportive to identify with the greatest possible precision where to cut the Shroud, without causing serious damage. And off we went. So we chatted! The directors of the excluded laboratories surrounded the three laboratories and, unfortunately, rumors and indiscretions began to circulate. And in the meantime I, who should have received all the informative material, received nothing! Finally, after speaking several times with the director of the British Museum in London, who was the chief coordinator of the whole operation (the British Museum in London is the most competent body in the matter), the report finally arrived, which I published. In the meantime, how much chatter: the Church does not publish it, the Archbishop of Turin does not keep his word ... And when I published it' ...!"

Source: Petrosillo, Orazio and Emanuela Marinelli. La Sindone: Storia Di Un Enigma, (Milan: Rizzoli, 3rd edition, 1998), pp. 212-222.

Comments: *For those puzzled by the remark "having twice committed Carbon-14," in the Catholic sacrament of "Reconciliation", the person confessing to a priest is supposed to say how many times they committed a certain offense. While the new investigations (apart from the controversial "restoration" in 2002) have still not materialized, it does indicate that the Church authorities acknowledged that the C-14 scientists were limited in their scope.

1989 *(February)*. I received a confidential personal letter from another prominent researcher who was involved in some aspects of the C-14 dating process, who commented on the effect of Prof. Gonella's involvement:

"You really gladdened my heart when you said that you learned from Dorothy that Luigi was OUT! As someone said it last year, May he rest in Sindonological purgatory forever.' (Ha!) Actually, I mean him no malice but it is so very sad that as I learn each little detail about the C-14 testing the terrible damage this one man has done to Shroud research. I learned another tidbit last night that just made me utterly disgusted -- Luigi Gonella had been told on three previous occasions not to use Thymol!!!! Can you imagine!!? He completely disregarded the advice and went ahead and allowed it!!! Astounding! Shocking! Disgusting! Revolting! Why??? There was no peer review (as you have well said by quoting numerous observers in your paper *[I had an article on the Shroud C-14 dating published that month in a Catholic periodical called "Fidelity"]* -- Harry Gove included) at all whatsoever! If history proves that the image on the Shroud has been damaged by the Thymol he will go down as one of the most irresponsible men ever to have had anything to do with the Shroud and, ironically, all the things he criticized others for proposing to do with the Shroud, Luigi bested them all with his own headstrong plunge into the abyss of selfcentered *[sic]* illbegotten *[sic]* planning in his nearsighted but misguided conviction that he was 'saving the Shroud'!"

Source: Personal letter dated Feb. 17, 1989 to author.

Comments: As with the author of the confidential letter of Oct. 4, 1988 from a major C-14 participant, this researcher is also deceased, but I again have honored his request not to be identified. "Dorothy" refers to the late historian and sindonologist Dorothy Crispino. The reference to Gonella being "out" refers to the fact that he was no longer the scientific advisor to the Cardinal of Turin. This same researcher also sent me an email on September 9, 2003 mentioning information he had given to Gonella after the former had "interviews with Garman Harbottle, Harry Gove, Henry Polach, Minze Stuiver, Stuart Fleming, Bob Otlet, Marian Scott (the latter being the statistician with the International Radiocarbon Calibration Programme based in Glasgow, Scotland)." The researcher continued, "I had cautioned him that if he didn't handle things correctly he would have a **political** *[my bolding]* and scientific disaster on his hands. Gonella was so stubbornly committed to "saving" (meaning "conservation") the Shroud by sampling only the Rae's *[sic]* Corner that he lost the whole thing and we are where we are today by his ineptitude." The aforementioned individuals were some of the top C-14 experts at the time of the Shroud dating.

Recall that in 1984, STURP's proposal for the location of the sample-taking would not have visually affected the appearance of the cloth. Had STURP's advice been taken and the area next to the Raes sample avoided, much of the controversy might have been avoided.

According to a press release put out by Meacham after the C-14 dating, Fleming stated that a medieval restorer "could certainly have re-woven a damaged edge to a standard not visible to the naked eye."

1989 *(February and March).* Gonella wrote that the *Nature* report was substantially correct, except for the fact that a staggered account is given of how the operation was arrived at, it gave the wrong dimensions for the sample cut, and the brief report by the Colonetti Institute is heavily censored and paraphrased, suppressing the observation on the greater dispersion of data in the Shroud sample. Tite sent a copy of the report to Ballestrero, thanking him for the opportunity to have participated but complained "that the attention of the media caused so many problems for all of us."

Ballestrero was hoping to have research on the Shroud continue, but he was turning seventy-five, the mandatory retirement age for bishops. On March 8th, he sent a letter to Casaroli recommending the continuation of the work when he no longer was archbishop of Turin.

Source: Gonella, Luigi and Riggi di Numana, Giovanni. Sindone: il mistero continua, (Milan: Fondazione 3M), 2005, pg. 83.

Comments: Tite later wrote to Gonella in April 1990 (see that time frame below for complete text) asking if they should publish a short adjustment note for the sample size erroneously given in the article. Gonella replied it should be done only if other inaccuracies were rectified together; for some reason, the matter was never followed up. Why was the Colonetti report censored and paraphrased if everything was straightforward? That wasn't completely answered until a Freedom of Information Act request in 2017 forced the release of the raw data -- see entry under "2019 *(March)*." The greater dispersion of data in the Shroud sample refers to the wider range of dates for it compared to the control samples. This suggests that something was anomalous regarding the Shroud sample. It might also be said the actions of Tite and the labs caused untold problems for those interested in the truth about the Shroud.

1989 *(March)*. Hall received one-hundred-thousand pounds from ITV, the BBC's rival, and also one-million pounds from forty-five businessmen and "rich friends" to establish a chair at Oxford upon Hall's retirement. Who took over Hall's position? -- it was Dr. Michael Tite from the British Museum, who had been the supposedly-independent overseer of the Shroud dating. Gonella commented, "Since the beginning, this story of dating the Shroud has been vitiated by publicistic aspects, to which the C-14 laboratories showed to be even too much sensitive." Gonella further criticized the C-14 scientists, "who took the liberty of violating the secret and of announcing to scandal-seeking tabloids that the Shroud is a medieval fake. In my opinion, there is an anti-Catholic conspiracy of specific milieus." Gonella didn't specify which milieus, but in a later interview, Cardinal Ballestrero was asked "In this whole affair could the Freemasonry have had a hand? And external pressures?" The Cardinal replied, "I think it's indisputable!"

Source: Marinelli, Emanuela. "The Setting for the Radiocarbon Dating of the Shroud." Presented at 1st International Congress on the Holy Shroud in Spain -- Valencia -- Centro Español de Sindonologia (CES), April 28-30, 2012, pg. 13. www.shroud.com/pdfs/marinelliv.pdf.

Comments: The religious persuasions of the businessmen and rich friends that donated the million pounds are not known but one can wonder if the donation would have been made had Oxford come up with a first-century date, especially given the fact that Cardinal Ballestrero believed that the Freemasonry, which is generally opposed to Christianity, was involved. The facts that Hall was on the Board of Trustees of the British Museum, and that Tite replaced Hall certainly qualify as conflicts-of-interest.

Regarding the Freemasonry aspect, two Protestant authors wrote in a 2016 book (The Final Roman Emperor, the Islamic Antichrist, and the Vatican's Last Crusade, Kindle Edition, by Thomas Horn and Chris Putnam, Loc 2737-2741), "There is at present in Catholic circles a constant, subtle and determined campaign in favor of Freemasonry. It is directed by the progressive element which is currently enjoying a great influence in French and American Church circles and beginning to show its hand in England too [...] This element consists of a number of priests, including a Jesuit, Editors of Catholic newspapers and several writers of note." This gives added weight to Cardinal Ballestrero's speculation about the Freemasons.

According to the "Night of the Shroud" documentary, Cardinal Ballestrero wrote a letter to Vatican Secretary of State, Cardinal Casaroli, outlining his beliefs about the Freemasons and accused some in the Turin Centro, which handled day to day Shroud matters, of "mismanagement." The documentary also states that Cardinal Ballestrero wrote a letter to his secretary in 1997 saying he believed there had been an "anti-Catholic" plot.

1989 (April). Responding to the suggestion that the cope sample could have been confused with the Shroud sample, Tite told *Il Messaggero* Vatican correspondent, the late Orazio Petrosillo, that it was absurd to think that the samples could have been exchanged by error or by malice. Petrosillo responded, "To eliminate suspicion, it is not enough to say that the two samples could not have been confused. Instead, it is necessary to explain why this sample was procured outside the protocol rules, why the piece had to be perfectly like the Shroud, why the dates obtained for the cloth of the Shroud and for this sample should perfectly co-incide *[sic]* and, what is more, correspond with an absolute precision with the period determined in advance by the opponents of the Shroud of Turin's authenticity. There are too many mathematical coincidences for our suspicions not to be aroused."

Source: Bonnet-Eymard, Bruno. "The Shroud Daters." *Catholic Counter-Reformation in the XXth Century*, June 1989, No. 220, pg. 28.

Comments: Petrosillo, as Vatican correspondent, interviewed Pope John Paul II many times.

1989 *(April)*. In the *Nature* report the twenty one authors admitted that the "blind test procedures were abandoned in the interests of effective sample pretreatment." Evin, who had said the previous July that the blind test was important, wrote to various colleagues that there were drawbacks: "I would rather it did not take place, otherwise it could always be said that the threads were easily exchangeable."

Source: Bonnet-Eymard, Bruno. "The Victory of the Holy Shroud Won By Science." *Catholic Counter-Reformation in the XXth Century*, September-October 1989, No. 223, pg. 29.

1989 *(Spring)*. Gonella continued to complain that the C-14 labs had insisted they had to work alone. "In addition to keeping the scientists of other disciplines away, the three laboratories also required not to be controlled in any way by the delegates of the ecclesiastical authorities. Counting on the fact that Ballestrero had guaranteed them the greatest freedom of investigation, and ably exploiting the suspicion that the word of the Vatican could not be trusted. Tite and the laboratories obtained permission to conduct their investigations in complete liberty and without any controls."

Gonella asserted, "They justified themselves by maintain that the dating was an operation that was completely separate from other investigations and that the presence of other scientists would have risked the confidentiality of the examination. Even if they themselves did not respect it!" Gonella went on, "Through respect for their freedom of investigation we were forced to separate the dating from the other investigations and to postpone this implementation. We accepted controls on our own actions, but they did not offer the same behaviour."

Riggi asserted, "We are of the opinion that the test by itself, isolated from the other 25 proposed tests, cannot provide a reliable answer."

Gonella continued, "After what has happened, I think that nobody has the courage any more to speak of the good faith of the laboratories."

"In the history of C-14 dating, never had laboratories themselves requested to date a specimen and never had seven labs at once wanted to date the same object. Normally a scholar submits a proposal to the laboratories and asks for their responses."

According to Gonella, "But this time, the laboratories that specialized in Carbon 14 dating wanted to act as their own submitters. The fact is that, right from the beginning the Shroud dating affair was vitiated by its publicity aspects to which the Carbon 14 laboratories showed themselves to be excessively sensitive [...]."

"We are not at all satisfied with the way the laboratories conducted their study of the Shroud. It is true to say that the attitude of the analysts influenced this judgement; for instance they insisted on being present, at all costs, during the cutting operation because they did not trust the officials of the Church; but, on the other hand, they did not invite the experts who enjoyed the trust of the Church to be present during their examinations. Since when has a dating laboratory wanted to be present during an excavation because it did not trust the archaeologist who was excavating the specimens? Since when have laboratories refused to collaborate? The management of the investigation was left completely in the hands of non-Catholic experts, without any effective guarantee of secrecy or that they would respect the agreement not to determine, during the tests, which was the Shroud specimen out of those submitted to them. Such criticism was directed towards Ballestrero."

The president of the "Centro Internazionale di Sindonologia in Turin," Dr. Bruno Barberis said, "There is no doubt that the whole affair was managed in a way all too superficial and not suited to the importance and the uniqueness of the object being examined. Explicit criticism on this point were *[sic]* received even from Vatican sources. The fact that it was not deemed opportune to involve an official organization, such as the Pontifical Academy of Sciences for example, in the management and the checking of the entire operation is surely one of the principal causes of the great dust storm that has been raised these last few months. The indiscretions, the rumours, the interventions by persons having nothing to do with the test could have been foreseen and avoided with a more careful management."

If the bad faith of the C-14 scientists was evident from the beginning, why did the Church allow them proceed? Gonella answered, "The Church found itself faced with a challenge issued by a number of persons who, by their demands, were doing all they could to be told 'no' so that they could say that the Church was afraid of science. Therefore, faced with this danger, it was decided to proceed with the scientific examination at all cost, even at the risk of protests."

Gonella continued, "It was blackmail. They put us with our backs to the wall with their blackmail. Either we accepted the Carbon 14 test with the conditions imposed by the laboratories or they would unleash campaign of accusations against the Church saying it was afraid of truth and that it was the enemy of science." In the communication of October 13th, however, Cardinal Ballestrero had spoken of a "reasonable operational programme [...]."

Gonella was asked if the "ideological passion" of the C-14 scientists could have negatively affected the results. Gonella replied, "I do not know, it is very difficult to say; what is certain is that, when people behave in such an unusual manner, it is possible to think anything. But, it would not scientifically or morally be proper to deliver such a rash judgment"

According to Gonella, "They behaved like dogs. I protest against their complete lack of professionalism in the field of deontology. I protest against the infamous method they followed. I told them to their faces that they were *mafiosi*."

Source: Petrosillo, Orazio and Emanuela Marinelli. The Enigma of the Shroud: A Challenge to Science, (San Gwann, Malta: Publishers Enterprises Group), 1996, pp. 113-117.

1989 *(Spring)*. A prominent Shroud researcher, who did not want to be identified, told only a few other Shroud researchers, including myself, about a curious phone call he had received one day at about 1:30 in the morning. His recollection was that it was not long after the C-14 dating results were announced in October 1988 and sometime in the following spring. I will call the researcher "Harry." Harry indicated the (male) person, who did not apologize for calling so late, sounded distraught. The person told Harry he had been involved in falsifying the results of the 1988 dating. Harry thought the accent might have been German, and thought the person was in his forties, but wasn't sure because of the accent and emotional nature of the call. The person would not reveal his name (the person claimed it wasn't important) or from where he was calling. He kept asking Harry if he would forgive him for having done a disservice to humanity. The person even mentioned the word "espionage" in relation to the event. The only detail he gave about the procedure was saying that the real Shroud sample was thrown in the trash. Harry tried repeatedly to get the man to identify himself and when he tried to get more details, the man said he couldn't say more as he could get in some real trouble. Harry said the person said he also planned to call other Shroud researchers, but as far as it is known, he never did. Harry has wondered over the years whether the call itself could have been a fraud, but he is firm that the person sounded distraught to the point that said he wouldn't have been surprised if the guy would have said "I've got a gun and I'm going to shoot myself." Even now, Harry just isn't sure what to think.

Source: Several personal communications, including May 13, 2016.

Comments: "Harry" told me he didn't want to be identified because he can't prove anything. Harry is a person of the highest integrity, and I have absolutely no doubt the call happened. I mention it because of the explosive nature of the content and also because of its possible relevance to a theory of Australian blogger Stephen Jones (see entry further below for "2014").

I believe the caller's story about the real Shroud sample having been thrown in the trash does not fit well into the context of three separate labs having performed the testing.

1989 (May). Following up on Ballestrero's request to Casaroli in March to have research into the Shroud continue, Casaroli informed Ballestrero that the Pope was maintaining him as Papal Custodian of the Shroud despite his retirement, and wanted research to focus on the image and conservation of the cloth. This action was taken as many Catholic groups were attacking Ballestrero, and it was intended to show the Pope supported him.

Source: Gonella, Luigi and Riggi di Numana, Giovanni. Sindone: il mistero continua, (Milan: Fondazione 3M), 2005, pp. 83-84.

1989 *(May)*. Gonella said in an interview, "The gentlemen in Oxford and London misbehaved; in their attitude there is an attack to other scientists without even reading their articles. I had great respect for the University of Oxford that I no longer have. The scientists came out of this test very discredited." He went on, "The vast majority of my colleagues are not satisfied, either by the adopted procedures, or by the conclusions. These gentlemen, moreover, shout from the rooftops that now the last word was pronounced *[sic]* on the question. Theirs, of course." He also emphasized the procedures lacked a preliminary chemical/physical examination and the pretreatments used to remove impurities were questionable.

Gonella was not done with the criticisms. He accused the labs of "intoxication by success" and added "Misconducts -- there were tons. The colleagues of the C-14 behaved in a disgusting manner. Those scientists have hatched a true plot to discredit the Shroud. At first, when they did ask us to examine a sample of the Shroud, assured us of the utmost seriousness and completeness of the analyses, along with the collaboration with the Custodian of the Shroud, that is the Bishop of Turin, and his scientific advisor, i.e., the undersigned. Driven by celebrity fever, those scientists began to turn their backs on their own commitments: no more interdisciplinary examinations, only C-14. They flooded even Rome with pressures so that Turin had to accept their conditions. They used the-then president of the Pontifical Academy of Sciences, professor Chagas, to get the undersigned out of the way and go their own way."

Gonella was asked why the Vatican and Cardinal Ballestrero accept the labs' demands. "Because Chagas acted alone, bypassing other academics. The Vatican was continually threatened by the laboratories themselves, who went on repeating: if you don't leave it to us, only to us, the results will not be acceptable. So, in the end, Ballestrero had to surrender, though suffering badly. And I *[had]* to submit. Also because these gentlemen did everything to support the argument that the Church was throwing a spanner in the works of science."

Source: Marinelli, Emanuela. "The Setting for the Radiocarbon Dating of the Shroud." Presented at 1st International Congress on the Holy Shroud in Spain -- Valencia -- Centro Español de Sindonologia (CES), April 28-30, 2012, pp. 14-15. www.shroud.com/pdfs/marinelliv.pdf.

Comments: Ironically, although the labs apparently maintained that the results would not be acceptable unless they were allowed to act independently, there is extreme skepticism to this day about the results they provided.

1989 *(May)*. Gonella noted that the new archbishop of Turin, Monsignor Giovanni Saldarini, while celebrating a Mass on May 4th, indicated that new 'investigations would continue and that this time they would be entrusted to people who are more open intellectually.' At a conference held on the 10th at the Rosetum in Milan (there would be another there on May 15th of 1990), Gonella critiqued in some detail the official report. "In the report in *Nature*, I have noted two things that do not please me at all. The first is that unusual statement: 'We have proved conclusively that the Shroud is of medieval origin.' Since when does a physics laboratory deliver judgements of an archaeological nature? I have never seen a scientific report in which anybody said 'What I have said is the last word.' Usually, when a researcher has something really conclusive to say he leaves it to be understood from the context, because it should be left to others who should say it. My second point concerns their analysis of the error. I am not much convinced. Perhaps it is not worth talking about it, particularly in this climate of polemics, because it is absolutely ridiculous to make so much fuss if the date is 1260 plus or minus 150 years or plus or minus 250 years. The result would not change in the public's mind; it would only give an impression. But they were very preoccupied by the coherence of the statistics, that which, in classical terms, is known as 'the analysis of the accidental error', and they did not bother at all about the analysis of the systematic error. It was the laboratories that said: we would like the dates to be statistically analysed by an independent body, namely the British Museum. How much, then, is the British Museum independent?"

Gonella continued,"We said 'no:' two institutes should analyse the results. And as for the second one I proposed the name of the Colonetti Institute of Metrology. Prof. Chagas immediately objected: 'Well, an institute in Turin would be suspect'. I replied: 'I am not suggesting a secondary-school laboratory. I am talking about one of the five principal metrological institutes in the world.' The analysts enthusiastically accepted. However, while it was written in the terms of agreement that the results should be sent in parallel to the British Museum and to the Colonetti, what actually happened was that they were sent to the British Museum, while the Colonetti only received results that had already been worked out."

He went on, "The laboratories asked me whether I intended to sign the final report. I said 'no.' I have never signed any work in the analysis of which I have not participated directly, and I will certainly not sign this. The Colonetti has answered in the same manner and has not signed. I must add that they considered it opportune to leave out a part of the Metrological Institute's report. Colonetti pointed out that it could not pass judgement on the method of measurement employed, that is of both the operations and the apparatus used. Significantly, the Colonetti has declared: 'On the basis of the data presented to us, we have nothing to object regarding the statistics of the results.' When an actual metrological analysis is made, the analysing institute is involved right from the start and has to analyse thoroughly the entire method used. Therefore we can only judge that everything is all right as far as the statistical analysis is concerned starting from the results of the readings, but we can say absolutely nothing about the other aspect, that is about the uncertainty which accounts for the systematic error. And they added a note: the spread of data between one laboratory and another is very much wide for the Shroud specimen than for any of the others. This discordance could probably have been reduced if a more precise procedure had been followed in the treatment of the specimens. This means, in simple terms, that the Shroud specimen was notably more contaminated than the others."

Gonella concluded, "As a metrologist, I do not accept the affirmation in the article that, since the same date was obtained both with the uncleaned and the cleaned samples, the absence of contamination had therefore been proved. Instead it proves only that there is no contamination of the type that can be removed by the system of cleaning employed. Their quite accurate systems of cleaning have been calibrated on archaeological specimens that are usually contaminated with soil. In principle, using a certain system of cleaning, certain measurements should be carried out on the cleaned specimen and others on the uncleaned one. If the same date is obtained, it does not mean that the contamination was not relevant, but that the type of contamination that could be eliminated by that system of cleaning was not relevant. But if there is a type of contamination that the system of cleaning does not naturally remove, it will not be noticed. Actually, these are but details; they could only offset the uncertainty. For example, instead of 1300 plus or minus 150 years, it culd [sic] be 1300 plus or minus 400 years. Not that it will change matter. But as a metrologist, neither I nor the Colonetti Institute are very impressed by either the validity or by the precision of the measurements."

Source: Petrosillo, Orazio and Emanuela Marinelli. The Enigma of the Shroud: A Challenge to Science, (San Gwann, Malta: Publishers Enterprises Group), 1996, pp. 120-123.

Comments: According to an article in *30 Days in the Church and in the World* ("Not by Carbon Alone," June 1989, pg. 28), Cardinal Saldarini said in a homily, "However the Shroud was formed, we need to explain this unique scientific and historical object, now more surprising and mysterious than before, by an interdisciplinary, free and united research program. For the Catholic Church, then, the Shroud remains an open question." It's a shame that the Church had to opportunity to have that "interdisciplinary, free and united research program" in 1988, but allowed Chagas, Gove and the three labs to go in the opposite direction. I strongly believe the Church has a grave responsibility to try and rectify the situation, since it was ultimately responsible for what happened.

1989 (May). The Secretary of State wrote Ballestrero, "I refer to the esteemed letter of 8 March in which we were informed about the reactions, including critical ones, that took place among Shroud scholars, following the publication, in the 16 February 1989 issue of *Nature*, of the scientific report on the investigations for the dating of the Shroud, conducted by three labs with overseer Dr. M.S. Tite of the British Museum; and at the same time he asked for instructions on the answer that should be given to those who ask that new and more detailed investigations be carried on the Shroud."

"By venerable disposition, I take care, first of all, to tell you that the Holy Father has decided to confirm Your Most Reverend Eminence, *'donec aliter provideatur'*, in the role of Pontifical Custodian for the Conservation and Cult of the Holy Shroud."

"With regard to future researches to be carried out on the Holy Shroud, which must always be previously authorized by the Holy See, it seems appropriate that they be conducted -- under the direct responsibility of Your Eminence in the capacity of Custodian -- by a small group of highly-qualified experts, which are of unassailable international prestige and above partisanship."

"These experts must first examine the report published in *Nature*, regarding the analyses mentioned above, discuss their contents and record any difficulties that may arise regarding what is stated in it, and they will be able to 'situate' in the most appropriate way the same analyses within the complex of the complex of exams already carried out and which are intended to be performed on the Shroud."

"At the same time, an attempt will be made, as far as possible, to obtain a collegial and coordinated interdisciplinary evaluation of the various experiments, which should however be aimed both at obtaining the desired 'illumination' on the formation of the Shroud image, and at identifying useful indications for a better conservation of the same Holy Shroud."

"[...] I ask Your Most Reverend Eminence to keep this Secretary of State informed of further developments in this matter [...]."

Of Your Most Reverend Eminence,

[signed]
A.*[gostino]* Cardinal Casaroli, Secretary of State"

Source: Letter of May 19, 1989 from Cardinal Casaroli to Cardinal Ballestrero. Luigi Gonella archives, of which the author has a copy.

Comments: The critique of the *Nature* report was made by qualified scientists early on. It's amazing that Cardinal Casaroli could be talking about his wish-list for optimum testing for multi-disciplinary testing of the Shroud only thirteen months after they squandered to the chance to do that. STURP's twenty-six-test proposal, which would have also included conservation aspects, would have given them <u>everything</u> he detailed.

1989 (May). A Shroud conference was held in Bologna, Italy. For some reason, both Gonella and Riggi were not invited. Kersten had hoped to meet with Riggi there but instead made arrangements to meet with him in his office in Turin. Riggi first told Kersten that he would be willing to show him (and his co-author Gruber) the ten to twelve hours of video footage from the sampling. (Bro. Bruno of the "Catholic Counter Reformation in the XXth century" normally cited in his publications a figure of sixteen hours, a figure apparently he got from Riggi.) But according to Riggi, Gonella objected, so Kersten and Gruber were only able to see an edited version. Kersten commented, "One has to ask, why an apparently uninterrupted video recording (excluding the episode in the sacristy) was made if people were not going to be allowed to view it as it stood. Would it not have been better to drop the documentation entirely, just as the verbal protocol had been simply dropped? What use was the assurance that everything proceeded correctly and was 'perfectly' documented, if no one was allowed to check? They had evidently also decided against having a notary confirm the events."

Source: Kersten, Holger and Elmar R. Gruber. <u>Jesus Conspiracy: The Turin Shroud & The Truth About The Resurrection,</u> (Rockport, MA: Element, 1994), pp. 84-86.

Comments: It's interesting that Riggi had told Kersten and Gruber that there were ten to twelve hours of video but told Bonnet-Eymard at one point there were sixteen. That's quite a difference.

The fact that the actual putting of the samples into the containers was not filmed was strange. In one of his interviews, Tite basically maintained that this aspect was so secret, that it couldn't even be documented on video!!! The question is why? The labs wouldn't be seeing the video before they worked on the samples. Because it wasn't recorded, there is no way to document if some mix-up occurred. *[See separate entry for a letter from Tite to *Nature* under "1990 (July)".]*

The meeting between Kersten/Gruber and Riggi produced some significant information regarding sample measurements. Kersten asked Riggi if the strip was actually wider than the reported 1 cm. Riggi replied "About 1.2 cm, but not straight, uneven." The total weight of the initial sample removed was 478.1. After the sample was cut in half, one of the halves was cut into three pieces and the scales showed the weights as 0.0520, 0.0528 and 0.0537 grams. Riggi kept the other half "for the future."

1989 *(May)*. At the Bologna conference, Fr. Werner Bulst, S.J., stated, "What is still lacking is adequate documentation about the operation of removing the samples. A correct documentation on this has never been done. No notarized acts as in 1989, and then there were only nine photographs! The gravest suspicions still weigh on the conduct of Doctor Tite."

Source: Bonnet-Eymard, Bruno. "The Shroud Daters." *Catholic Counter-Reformation in the XXth Century*, June 1989, No. 220, pg. 34.

1989 *(June)*. On the 29th, the Turin Polytechnic University held a press conference to discuss the sample-taking from April 1988. It was chaired by Gonella, Riggi, and Giorgio Tessiore. Tessiore wrote an article for the Italian Shroud periodical *Collegamento Pro Sindone*, in which he highlighted some of the events. He wrote, "[...] 1 square cm of the new sample had to be discarded because of the presence of contamination and different color threads." Riggi kept some of the edge trimming that was part of the discarded material because it was part of the side strip that was fastened by stitches to the main part of the cloth. Tessiore went on, "From the rest are fashioned three small rectangles of nearly 52 mg, each sufficient for two dating*[s]*; small pieces of similar dimensions were taken from linen certified by the British Museum, one dating from the first century and the other from the twelfth." Tessiore did not seem to be aware that the first cutting for Arizona was only about forty mg., so an additional twelve mg. was taken from the reserve sample to give them fifty-two mg. total. It is during this time that Bracaglia believes that Gonella extracted six additional threads for his personal use

The measurements laid out by Tessiore do not agree with a diagram put together by Riggi. Riggi's first cut was approximately 8.4 cm. x 2.2 cm., weighing four-hundred and seventy-nine mg., which exceeded the supposed two-hundred mg. that the labs had asked for. After the edge trimming was removed, the dimensions measured 7.9 cm. x 1.3 mg., with a weight of three-hundred mg. However, regarding the latter, Riggi's diagram is not consistent. Riggi cited in his diagram that the Oxford portion was underweight, but in fact, it was the Arizona fragment that was located furthest from the reserve that was under-trimmed. Thus, an additional fourteen mg. was taken from the reserve. Each lab received approximately fifty-two mg. for testing, and the rest of the material kept in reserve weighed in at 140.6 mg.

Source: Bracaglia, Giorgio. Uncovering the Paradox within the Archives of the Holy Shroud Guild, (Honeoye, N.Y.: Holy Shroud Guild), 2019, pp. 237-238.

Comments: As far as I am aware, Tessiore was not actually present at the sample-taking. Notice that Tessiore mentioned both contamination and foreign threads having to be removed. He said one of the control samples was from the twelfth century, but from the certificate given to the labs by Ballestrero, it was the eleventh century.

Another discrepancy concerns how many milligrams were taken from the reserve piece to give to Arizona such that their sample weight matched the other two labs. This entry mentions ten (Tessiore), fourteen (Riggi), and there was also 14.1 (Testore in another account previously mentioned).

1989 *(June)*. Evin sent a letter to various researchers regarding the fourth sample (the cope). He claimed that in February 1988, Gonella had asked him to procure a control sample from France. After obtaining some threads from the cope, he said that he and Vial arrived in Turin in the morning. He said that Gonella had not informed Riggi of the fourth sample, so the latter only prepared nine steel containers. So, Tite and Gonella instructed that the cope threads be put in a separate envelope.

Source: Bonnet-Eymard, Bruno. "The Shroud Daters." *Catholic Counter-Reformation in the XXth Century*, June 1989, No. 220, pg. 28.

Comments: Evin claimed here he had arrived with Vial but he told *Paris Match* in July 1988 that he was late. Vial claimed that Evin was unable to come (CRC No. 223, pg. 28). As opposed to Evin saying Gonella had requested a sample. Tite would later say at the Paris symposium (September 1989) that he had asked Evin to get a sample (CRC No. 223, pg. 28). In fact, we do have a copy of the letter from February 1988 that Tite wrote to Evin. Although Evin said Tite, along with Gonella, had instructed the cope samples be put in an envelope, Tite had told the ANSA agency on March 30, 1989 that he had put all four samples into metal containers! (CRC No. 220, pg. 29). Were those gentlemen's memories really that bad? And if you're confused by all this, you should be. And the next entry makes it even more confusing.

1989 *(July)*. In the *Paris Match* July 1988 interview (mentioned in the "Comments" in the above entry), Evin said, "There were cut 3 small samples from a strip of 1 x 7 cm, weighing about 150 mg. Dr. Tite gave to each representative of the 3 laboratories not one but **four** *[Source author's bolding]* containers, all identical and anonymous, bearing only a different number."

Source: Van Haelst, Remi. "Radiocarbon Dating and the Shroud of Turin, pg. 17." This is a typewritten early draft from 1989 from which Van Haelst used most of the material in other later published articles. The draft was circulated among various sindonologists, and the author has a copy.

Comments: Van Haelst quotes Evin as telling him in a personal communication, "Because I and Mr. G. Vial (Textile Museum -- Lyon) are involved in the matter, I like to give some justifications. About February 1988, Prof. Gonella asked me for a piece of linen, historically dated XIVth century and originating from France. Abbe Boyer (CNRS-Draguignan-France) was of great help, because it was not easy to obtain the authorization to cut a piece from such a rare historical object. Finally, the authorization was given by Mr. Prevot Marcilacy (Chief Conservator Monuments of France) to remove a small piece of the cape of Louis d'Anjou, guarded in St. Maximin. Mr. Vial and myself brought this sample on 21 April to Turin. Prof. Riggi was unaware of this 4th sample and therefore did not have sufficient containers. Therefore Dr. Tite and Prof. Gonella wrapped this 4th sample, in the form of threads, in 3 envelopes and gave them, together with the 3 numbered containers to the representatives of the 3 laboratories. I like to underline that switching between samples was impossible, because the 4th sample was in the form of threads, while the other samples were pieces of linen."

So, looking at just this entry and the previous one (never mind other entries in the book also dealing with the number of samples and the containers), we can ask: 1.) Did Gonella ask Evin to get a control sample from France (Evin letter to researchers in February 1988 and Evin in personal correspondence to Van Haelst) or did Tite ask Evin to get the sample (Tite per interview at Paris symposium in September 1989, for which a letter exists)? 2.) Did Evin and Vial arrive together in Turin on April 21st (Evin in interview with Bro. Bruno pre-June 1989) or did Evin arrive late (Evin in personal correspondence to Van Haelst)? 3.) Were there only nine steel containers (Evin in interview with Bro. Bruno pre-June 1989) or twelve steel containers (Evin in *Paris Match* interview July 26, 1989)? At least Evin was wise enough to say in the *Paris Match* interview, "It is not sound to claim more certitude than is measurable with the methods used today" (pg. 2 of the Van Haelst draft document), which the labs did with their "95% confidence level" pronouncement.

Prof. Anthos Bray of the Colonetti Institute was a reviewer of the *Nature* report and suggested leaving out the sentence, "The results of radiocarbon measurements at Arizona, Oxford and Zurich yield a calibrated calendar age range with at least 95% confidence for the linen of the Shroud of Turin of AD 1260 - 1390 (rounded down/up to nearest 10 yr).", but this suggestion was ignored. This fact was learned via the 2017 Freedom of Information request of the raw data of the British Museum.

1989 *(September)*. To gauge the general reliability of C-14 dating, an intercomparison trial among thirty-eight labs took place in Scotland. The organizers concluded that the margin of error was two to three times greater than previously claimed. Of the thirty-eight labs, only seven produced satisfactory results.

Source: Coghlan, Andy. "Unexpected errors affect dating techniques." *New Scientist*, September 30, 1989, pg. 26.

Comments: This was another blow to the labs' facade of the infallibility of the 1988 C-14 testing. Oxford declined to take part in this intercomparison. One has to wonder why they didn't take part after their supposed success with the Shroud.

1989 *(September)*. Despite Hall's proclaimed disinterest in the Shroud in his interview in January 1989 with John Cornwell, he actually promised to attend the International Shroud symposium held in Paris in September 1989 to answer doubts about the validity of the experiments -- but canceled at the last minute claiming he had to attend an important meeting, and sent no communication to the Congress.

Source: Bonnet-Eymard, Bruno. "The Holy Shroud -- Silent Witness." *Catholic Counter-Reformation in the XXth Century*, April 1997, no.295, pg. 25.

Comments: This source (pg. 20) also revealed that it was only learned at the Paris symposium that one of the labs had received their Shroud sample in two pieces and was not reported in Riggi's report of April 26, 1988 or by Tite in the *Nature* report.

1989 *(September-pre-Paris-conference)*. Crispino wrote, "It is not of international TV cameras nor press reporters with pencils poised that I want to tell about; but a Parisian radio station called Radio Courtoisie: an FM broadcast that every Thursday evening brings cultural programs to a constant audience of 500,000. On the Thursday before the

Paris symposium, different aspects of Shroud studies were discussed by the Rev. A.-M. Dubarle, Dominique Tassot, Phillipe Bourcier de Carbon and myself. Every third Thursday, Radio Courtoisie presents a three-hour program by Chantal Dupont, with a repeat on Friday morning. For her program to be broadcast on September 21 and 22, Mme. Dupont interviewed a number of symposium speakers, among whom were John Jackson, Gilbert Lavoie, the Rev. Adam Otterbein, Jacques Evin, Mike Tite.... The following excerpts are from Mme. Dupont's interview with Dr. Tite. Her opening question requested his conclusions concerning the results of the carbon dating of the Shroud. Dr. Tite replied: 'C14 has certainly shown the Shroud to be medieval. I am convinced that the Shroud is medieval'."

Why did you give the results as a journalistic scoop? There were unfortunate leaks, but the scoop was due to the press, not us. (Dr. Tite speaks crossly.)

Who put the exclamation mark after the date on the blackboard? I can't remember who did that, the press, Hall, or me ... it reflected the mood of the moment.

Were you interested in the Shroud before your participation in the tests? I was not interested in the Shroud before but I am now and will continue to follow events.

Why was the blind test not carried out, according to the protocol? There were two aspects. We decided in January that we would rather not have a blind test because that could be done only by unraveling the samples, and that makes it much more difficult to clean. The decision to abandon the blind test was made in Turin at the moment of the cutting. As for Ballestrero and I going into a separate room, that was quite unnecessary.

Why were the laboratory representatives present? The laboratory representatives wanted to be there just for personal reasons, just to be part of the event....

You accepted a fourth sample in an envelope. Why did you not stop the meeting? A medieval sample is very difficult to find. There was none in the British Museum that we were permitted to destroy. I asked Vial, Evin, Wilson and others. When I arrived in Turin I learned that Evin had found some medieval threads. We had to make a decision quickly -- it would have been impolite to refuse it -- people would have been furious. It was done for scientific reasons and good will. (He laughs amiably.)

Why were the labs informed about the dates of the control samples? It would have been better if the labs did not know the dates. If I did it again, in hindsight, I would do it differently. But that it invalidates the result is absolute nonsense.

The Shroud is consistent with what we know about Jewish burial in ancient times. But C14 is the only method that gives a scientific date. Inferences from other information are not hard facts.

Do you believe the Gospels? No, I cannot believe the Gospels. Any historical document is full of errors ... even today....

According to Matthew something happened. Would the Resurrection have any influence on C14? If one accepts that there was some supernatural event ... I suppose that if there was a flux of neutrons it could have influenced the date. As yet, no one has tested neutrons ... to see if that changes C14. If there was a supernatural event and a change of neutrons it would be a major coincidence if it came to this precise medieval date.

In iconography as early as the sixth century we see icons of Christ with characteristics similar to the Shroud Face. How do you explain that? I don't know, you tell me! (Ha ha.) I personally believe we should leave relics as relics. Let people believe....

But in the case of the Shroud there is so much other evidence for a first century date. The only specific date is by C14. The cross, the nail, the crown.... It is best to leave well enough alone!

What did you feel when you saw the Shroud? (long pause) -- Ah! -- (pause) I mean ... it is a remarkable image. I never looked at it the way I look at paintings of Christ; that is just an esthetic experience rather than religious. I was there to do a specific job. It is different when you are in an art gallery or in Churches. I did not see it as an esthetic experience. It's tied in with my work.

Would you be interested in further tests to solve some of the Shroud problems? I do not feel the need. If it were a supernatural event ... but I don't believe in that. If neutrons did change the carbon content, that could perhaps give an apparently aberrant date on the Shroud.

Source: [Crispino, Dorothy]. "Radio Courtoise." *Shroud Spectrum International*, Issue #32/33, September/December 1989, pp. 36-37. https://www.shroud.com/pdfs/ssi3233part6.pdf.

Comments: Tite mentioned several times that C14 is the only test that gives a specific date. But one can't simply dismiss all of the other scientific and historical information related to the Shroud, which is the most intensely-studied artifact in human history. If the Shroud were a medieval forgery, one has to explain why there is so much evidence that points in the other direction -- results do not exist in a vacuum.

Regarding the question of the Resurrection possibly having affected the Shroud C-14 content, it is interesting to note that in 2016, when a team of scientists was allowed to open the tomb where Jesus was believed by most scholars to have been buried, one scientist was quoted as saying that "the tomb had a strong, unexplainable electromagnetic field that messed up their equipment"
(https://churchpop.com/2016/12/05/astounding-mysterious-magnetic-readings-at-recently-opened-tomb-of-christ/).

1989 (September). At the International Scientific Symposium held in Paris on September 8-9, the Scientific Committee made a declaration that there were reservations on the statistical analysis of the results, especially on the 6.4 value of the "chi-square" test, which indicate that the samples were not homogeneous. The Committee requested the release of all the raw data from the three labs as well as a commentary written by Prof. Bray of the "Colonetti."

Source: Marinelli, Emanuela. "The Setting for the Radiocarbon Dating of the Shroud." Presented at 1st International Congress on the Holy Shroud in Spain -- Valencia -- Centro Español de Sindonologia (CES), April 28-30, 2012, pg. 16. www.shroud.com/pdfs/marinelliv.pdf.

Comments: The facts that the raw data were not released immediately nor Prof. Bray's commentary were suspicious in the extreme. It is still difficult to get data from the labs. Researcher Pam Moon from England recently tried to get Oxford's pictures of their Shroud C-14 samples. After being stonewalled for some time, she finally got them by filing a "Freedom of Information" request.

Van Haelst had put some questions to Prof. Bray not long after the dating results and wrote, "Prof. Bray did not answer my SPECIFIC questions about: *The combination of 8 into 4 Arizona dates. *The arbitrary enlarging of the Arizona error from 17 to 31 years. *The 'switch over' from the Ward & Wilson method to the Classic method. *The silent rejection of a number of clearly NEGATIVE Chi^2 tests. *The very bad Arizona Chi^2 test results for samples 3 & 4. *The strange absence of any contamination, reported by earlier examinations. He refused to open his files, without the written permission of the British Museum. In vain, I asked the written permission of the British Museum" ('Radiocarbon dating the Shroud of Turin,' http://www.sindone.info/VHAELST6.PDF, pg. 45.)." As we have seen, researchers were finally able to secure in the past few years both the raw data from the labs and Oxford's photos of their sample.

In the aforementioned article (pg. 46), Van Haelst also wrote, "In the Dutch technical paper 'Natuur en Techiek,' I found an article about the Shroud of Turin, authored by Dr. Bottema (University of Groningen. Holland). He noted that Oxford dated the Shroud about 1150 AD. Because an unknown photo of the Oxford samples was shown, it is clear, that Dr. Bottema received 'inside' information from a former member of the Oxford AMS team. This strange result, about 100 years older than the Oxford mean date 750, raises questions! In vain I tried to *[access]* the

Oxford files, in order to obtain more information. Dr. Hedges and Dr. Bottema *[became]* suddenly silent."

1989 (September). Petrosillo and Marinelli wrote, "An international conference of scholars and scientists of many disciplines was held in Paris about a year after the fateful verdict. The validity of the radiocarbon dating results was supported by the scientists who conducted the exam, but harshly criticized by others and with valid arguments. Most of all, the claim that dating is a definitive proof was rejected. On the contrary, it was reiterated that a dating isolated from the archaeological context of an object is not at all sufficient to define its age. The medieval date declared for the Shroud conflicts with the results of all the other studies, which provided a very different scenario. Some speakers criticized the laboratories and the 'guarantor' Tite for not disclosing the primary data obtained from the analyses, despite the repeated requests of the other scientists. Most of the scholars agreed in admitting that the quantity of ^{14}C present in the samples is that actually found by the laboratories; but the result must be interpreted. In fact, the greater uncertainty of the AMS method does not lie so much with the measurement of radiocarbon, as with the stage of interpretation of the present ^{14}C."

"The hypothesis that the percentage of ^{14}C is not the same in all points of the cloth is rejected for theoretical reasons by almost all of the radiocarbon specialists. Unless the object examined has been heavily polluted. But this is precisely the case of the Shroud. The result of radiocarbon, due to its isolated nature, does not represent a sufficiently-established scientific truth, and, moreover, the existence of contradictions cannot be accepted in the field of experimental sciences. Tite, on the other hand, sees nothing else: he believes that radiocarbon is the only certainty. It is the act of faith of an agnostic. Gonella repeatedly expressed his strong criticisms of the behavior of the laboratories, calling them 'unscientific' for various reasons, but has accepted the results, like all those of previous research which have no less value. Ballestrero's consultant denied the contradictory nature of his attitude with the following example: 'If we were faced with a murder or theft, it is not by identifying the thief that we can be sure I also know who the murderer is.' The example is irrelevant, or, on the contrary, if there is a link between theft and murder, one is authorized to strongly suspect that both are the work of the same person!"

"Among the numerous objections raised to the laboratories, the Turin engineer also addressed that of having claimed to interpret the results, declaring that the Shroud is a forgery. With this they say that someone intentionally wanted to deceive. But it should be explained according to which mechanism would have been made for this purpose. Tite returns to the subject one year after the announcement of the results with a letter addressed to Gonella himself: 'I am writing to make clear the fact that I personally do not believe that the result of the radiocarbon dating of the Shroud of Turin shows that the Shroud is a fake. As you have appropriately pointed out, describing the Shroud as a fake implies a deliberate intention to defraud and radiocarbon dating clearly does not provide any evidence to support this hypothesis. As far as I am concerned, I have always tried carefully to avoid using the word *fake* in discussing the radiocarbon dating of the Shroud, but I fear that the description of the Shroud as a fake has still crept into some newspaper articles based on interviews given by me. So I can only apologize once more for any problem that these news reports caused to you and to others in Turin.' One can only take note of an evident backward movement."

"There is the surprise for the passionate interest revealed by the analysts, here. The result of radiocarbon was used in a particular way by Hall in showing that the Shroud is not a relic. 'In reality the laboratories were more interested in advertising in the press than in scientific truth', concluded Gonella bitterly. A British newspaper revealed, for example, that Hall hoped to get a large sum for his esteemed lab by selling rights to the history of ideological dating to a Sunday newspaper that has nothing to do with science. One hundred thousand pounds and perhaps more seems to have received by them, again for the laboratory, from independent British television which wanted the anticipation on the 'definitive' verdict pending publication of the results. Forty-five businessmen and wealthy friends donated him a million pounds, to ensure that the laboratory continued to operate after his retirement. Hall said he did not believe that the advertising derived from his research on the Shroud explained his success in collecting money. Others, like Ian Wilson, instead believe that Hall used the Shroud dating to make an advantageous advertisement for his laboratory. In the context of the Paris symposium, the mining engineer Dominique Tassot declared himself not at all convinced of the real impossibility to carry out a blind test in the case of the Shroud. He therefore defined the fact that it was not carried out as 'abnormal.' 'There was a very simple way,' he said, to make it happen: 'to entrust a fourth laboratory with the care of preparing, cleaning and reducing the samples in fibers before giving them to the three laboratories in charge of the measurement precisely called'."

"Radiocarbon dating is indispensable. To perform a ^{14}C dating a lot of archaeological and historical information is needed. In chemical tests there is no need for exogenous information; this is the weak point of the ^{14}C method: its specialists do not agree to proceed unless the probable result is given to them. It is an anomaly in the sciences, all the more significant as the ^{14}C claims to have an aura of infallibility among other chemical analyses. The chemical measurement that provides the percentage of ^{14}C in relation to total carbon, must be subsequently interpreted. The amount of radiocarbon present, in fact, comes both from the natural radioactive decomposition in every aging, and from any phenomena that occurred in the history of the analysed object that may have altered its content in ^{14}C. This second aspect was not discarded by radiocarbon experts. It is a hypothesis that no one today can rule out, indeed many scientists take it into account to explain the fact that the results of the three laboratories for the Shroud sample were not as close as they should have been. This further element helps to reiterate that there is no scientific solution to the problem of the Shroud that is unidisciplinary, as it is a unique object of its kind. It reserves surprises for experts of all disciplines; why shouldn't scientists from ^{14}C have surprises?"

"It is no wonder that people who know the gaps in their dating method speak of complete certainty of their results. There is no absolute dating method. One may think that, feeling attacked, they put themselves on the defensive. There means that are beginning to be used, such as that of thermoluminescence. Evin and other ^{14}C scientists reject specimens on which errors are frequent, such as shells and stalactites. The comparison between the two methods, AMS and small counters, was not entirely convincing, so there was no right to eliminate one. But there was a war going on between the laboratories that used the two different methods. Another circumstance underlined by Tassot is that the laboratories were not satisfied with giving the numbers, but they interpreted the results stating that the Shroud was a fake without taking into account other sciences, which can also provide a date. Chemists or physicists are not a qualified 'moral' authority to judge whether the Shroud is a fake or not. Fields of critics did not raise only isolated voices. Concerns about the results of the radiocarbon dating of the Shroud expressed by the scientists of the scientific committee that organized the Paris symposium, including Evin, with an official declaration signed unanimously. It reminds us that, already before symposium, advanced professional statisticians such as Philippe Bourcier de Carbon, had strong reservations about the method by which the statistical analysis of the results obtained by the three labs was conducted."

"If you want to examine the distribution of a certain variable on a whole, using measurements carried out on samples, the homogeneity of these samples with respect to the considered variable is an essential condition of their representativeness of the studied set. The verification of this is done with the Pearson test on the variable y. From the data published in *Nature*, and in particular from the value 6 provided by the x^2 test for the three pieces of the Shroud, it appears that these samples were not homogeneous at all. The maximum limit of ransom is, in fact, of 5.99. There are therefore more than 957 out of 1000 probabilities that they are not representative of the radiocarbon date of the whole sheet. At this point, any further statistical estimate appears to be worthless and therefore it would have been necessary to suspend the judgment. But there is more. From the value of x^2 we go back to the 'significance level,' the level of significance, a percentage that indicates 'the probability of randomly obtaining a dispersion between the three dates such as that observed, assuming that the errors cited are due to all the sources of random variability.' Engineer Ernesto Brunati pointed out: 'A 90%" significance level', as for sample 2, means that it was estimated that there was a 90% probability that the dispersion was random and only 10 percent that systematic factors took over. So, a significance level of 5 percent, as per sample 1, means that only 5 percent is estimated that the leakage is due to chance, compared to 95 percent that a defect has occurred system. So it's all right if the level of significance is high, while

5% is the minimum limit of acceptability; and for the Shroud it is reported in *Nature* just this value. Checking the calculations starting from x^2 6.4, however, Brunati made a sensational discovery: in reality the level of significance that results for the Shroud is 4.17 percent, so this figure was falsified. The engineer then wanted to go further and surprises were not lacking. Recalculating the average ages on the basis of the measurement data provided in Table 1 of *Nature*, the Arizona sample shows 646 +/- 17 yr BP instead of 646 +/- 31, which is the value published in *Nature's* table 2. The difference is not of little consequence: if the correct value is used, the x increases to 9.13 and the significance level decreases to 1.04 percent! 'How can we believe,' asks Brunati, 'everything they tell us, how we can accept as valid the final result proclaimed by the Cardinal, if we discover that what they told us, in the few points that we have been able to verify, do not correspond to truths'?"

"Therefore, the engineer wanted to get in touch with Anthos Bray, who, for a more detailed statistical analysis, addressed it to the British Museum, also specifying (further surprise) that the opinion on the results he was asked personally; therefore the l'Istituto Metrologia G. Colonetti had not been involved in the research at all. The British Museum, for its part, replied that it should turn to Tite and in the meantime, incredibly, reiterated that in any case 'a much larger error would have been necessary to change the results significantly.' Brunati was therefore convinced that the samples analyzed were not the original ones; this is confirmed by the widely-documented differences in weights and measures. Nobody has a monopoly on the measures. The Parisian scientific committee believed that the anomaly highlighted in *Nature* alone was enough to justify the request to repeat the dating. However, only after having clarified the serious problem posed by the divergence of the results of Oxford from those of Tucson and Zurich. In the case of a new radiocarbon experiment, the committee asked that the samples to be examined be previously subjected to non-destructive tests, in particular to the chemical study. It was also asked that a precise procedure of the operations be established, fully controllable, with the competition from the British Museum, the Pontifical Academy of Sciences and STURP. Delegates from these three institutions, and possibly from others to be established, should be instructed to verify at all times that the established procedure is being followed exactly. Finally, no monopoly on the interpretation of the measures taken should be granted to anyone; therefore, all the raw data obtained should be published immediately and without corrections. What happened with the last exams must not be repeated: the publication of the primary data obtained from the ^{14}C analyses did not take place despite the repeated requests received from many parties. Such disclosure is essential given the high value of x^2. There were also reasonable doubts about the level of confidence declared by the laboratories (95 percent) which, according to accurate statistical assessments by Van Haelst, did not appear justified."

"Bonnet-Eymard comments on the whole story explaining the medieval result with two hypotheses: the replacement of the sample or the lie on the conclusions of a poorly conducted experiment. His suspicions weigh on the conditions for taking the samples, their transmission and the calculations that transformed the ^{14}C measurements into real age. On the latter point, the French sindonologist advanced the following hypothesis: the laboratories would have found high rates of ^{14}C, so much so that their interpretation would have been oriented for too recent dates (for example, 15th century), with an obvious anachronism. The statistical commission charged with harmonizing the three results, taking a look at the known history of the Shroud, would therefore have aged them according to a method of correction which had become habitual, if not official, until they were possible! The hypothesis of replacing the sample is fed by the inconceivable inconsistencies found in the weights and measures of the Shroud samples: they were twice the weight of the unit weight measured for the Shroud. There was a clever way of easily swapping the samples, noted Georges de Nantes: Tite could have declared as coming from the Cleopatra mummy between a medieval fabric delivered to him by Victoria and Albert Museum and he would have exchanged it for the Shroud. In this way the real sample of the Shroud would become sample 3, which provided the date of the epoch of Christ attributed to the mummy of Cleopatra, while sample 1, officially the Shroud, and resurrected medieval because in reality it would have been dated that of the Victoria and Albert Museum. Gonella's harsh and detailed criticisms of the composition of the laboratories, the numerous and arbitrary variations of the examination protocol, the substantive objections on the statistical analysis of the results in addition to the multiple questions about the reliability of such an exam on a fabric with the history and characteristics of the Shroud, however, form such a long chain of evidence and clues that they are completely dissatisfied with the outcome of this date. Anyone familiar with radiocarbon literature knows that the 'infallibility' of ^{14}C is uncertain."

"Among the numerous objections made to the laboratories, there is also that of having claimed to interpret the results, declaring that the Shroud is a fake. With this they claimed that someone intentionally wanted to deceive. But it should be explained with what mechanics. Arnaud-Aaron Upinsky, mathematician and epistemologist, President of the Institut Euclide of Paris, said that the radiocarbon scientists, not responding to the objections, placed themselves outside the scientific debate."

Source: Petrosillo, Orazio and Emanuela Marinelli. La Sindone: Storia Di Un Enigma, (Milan: Rizzoli, 3rd edition, 1998), pp. 224-232.

1989 *(September)*. The late Dr. Alan Whanger, M.D., a long-time Shroud researcher and amateur photographer, contended at the symposium that photographic and video evidence indicated that the C-14 sample area was anomalous and was given a few minutes by the organizers to show his findings. Jacques Evin, a French C-14 expert, who was involved in the Shroud sample-taking, was then asked about the possibility of a reweave and remarked, "I quite agree that the labs did not take the weaving techniques into account and they did not date the threads *per se* [...]. Thus, if the weave was rewoven with threads from modern restoration, this would be reflected in more modern results."

Source: C.I.E.L.T Paris symposium videotape belonging to the author

Comments: This question would not have arisen if the labs had done a proper chemical characterization of the samples. There is now, in fact, strong evidence that there were rewoven threads in the C-14 sample area. Testore also made an interesting statement at the symposium: "Another interesting thing that we discovered was something pertaining to the splicing technique used to change the warp during the weaving process. There is splicing through superimposition technique, which is used in a very frequent manner nowadays, but which was obviously not as frequent at that time when the cost of manpower was considerably less than that of new materials. This splicing and superimposing of restoration threads onto a portion of the C-14 sample may account for the difference observed in warp pattern, thread appearance, and C-14 content." Although the statement seems a bit ambiguous, he did seem to acknowledge that a reweaving possibly affected the dating results.

1989 *(September)*. Kersten attended the Paris conference and asked Prof. Vial, involved with the mysterious fourth sample *[= third control sample]* taken from the cope of St. Louis in Saint-Maximin how many threads he had removed. Vial could not remember specifically but said it must have been about 200 mgs and were divided into four parts of 50 mg each. He kept one part and the other three were given to the laboratories. But Wolfli had written to Kersten on August 17, 1989 that the weight of his specimen was 68.8 mg.

Source: Kersten, Holger and Elmar R. Gruber. Jesus Conspiracy: The Turin Shroud & The Truth About The Resurrection, (Rockport, MA: Element, 1994), pg. 89.

Comments: This discrepancy in weights is another example of the lack of rigor in measurements. Interestingly, Tite had requested a sample that was similar in weave and color to the Shroud and preferred that it be from the fourteenth-century.

1989 *(September)*. According to Bonnet-Eymard, after reports by Riggi and Testore at the Paris conference, it was possible to line up four different versions proffered by them regarding the weights of the samples:

"1. Thursday, 7 September 1989, Paris: Franco Testore relates how he weighed the sample cut by Riggi. The strip removed from the Holy Shroud measured 8.1 X 1.6 cm and weighed 300 mg, twice the amount requested by the laboratories. Riggi therefore cut the piece into two fragments, which Testore weighed: 154.9 and 144.8

mg. From the first fragment, weighing 154.9 mg, Riggi prepared three samples weighing 52; 52.8 and 53.7 mg respectively."

"2. We then pointed out to Testore that the total weight of the three samples (158.5 mg) exceeded the weight of the piece from which they came (154.9 mg). Testore took note of this and immediately altered his report accordingly, even though it was in the middle of being printed. His answer was that the samples had been extracted not from the big fragment but from the small one (144.8 mg) to which a piece from the larger fragment had been added (14.1 mg)."

"3. For his part, Riggi answered that he had indeed cut the samples from the big piece (154.9 mg) but had completed it by taking 3.6 mg from the small piece."

"4. At the same time, Riggi sent the printers his fourth version, which made no mention of any piece added on. Everything was just right: a strip of 7 X 1 cm had first been removed in conformity with the report published by Riggi himself on the day after sample removal, and also agreeing with the report by the twenty-one co-authors published in the review *Nature*. Contrary to the evidence, Riggi stated that the piece weighed 300 mg, which he had divided into two fragments of 150 mg, one of which was then cut into three. As 'chance' would have it, each of these three fragments was 'identical and each equivalently weighed almost 0.053 g [...].' How can three samples of 'almost 0.053 g' weight 0.150 g? We are in the thick of fiction here."

Source: Bonnet-Eymard, Bruno and Georges de Nantes. "Holy Shroud -- The Turin Tricksters in Disarray." *Catholic Counter-Reformation in the XXth Century*, November-December 1989, No. 224, pp. 7-8.

1989 *(September)*. Prof. Raes confirmed in a presentation at the Paris symposium that Sox and McCrone had worked together, starting in 1976, to get the samples held by Raes, for a C-14 test, as told by author

Rex Morgan: "Professor Raes said that since the November 1973 sample had been taken strange rumours had been circulating about it. Raes then proceeded to tell, for the first time publicly, I believe, the astonishing story of the Sox/McCrone conspiracy of that period, information which Raes had given to me many years ago but which I have never published. In 1976 Raes received a letter from Sox asking him to talk to McCrone. According to Sox he was in a position to date the cloth accurately. Obviously, said Raes, this letter was to prepare him to allow McCrone to use the Raes sample. He was skeptical and contacted Prof. Apers, the Belgian C14 expert. On September 18, 1976 he received Sox and McCrone in his home and suggested they meet Apers. This took place at the end of September 1976 and Apers subsequently stated that McCrone had not convinced him of the accuracy of his protocol. Raes then contacted Turin to make his fears known because he expected Sox to insist on making the sample available. Raes was immediately requested to return the sample to Turin which he did in October. On 12th October he received another letter from Sox saying that McCrone had answered all the objections to his method and would Raes now please release the sample. Raes told him he did not have it and to contact Turin direct. Raes has never heard from Sox since nor does he know what happened to the sample after it was returned to Turin."

Source: [Morgan, Rex.] *Shroud News*, No. 55, October 1989, pg. 23

Comments: See the entries for "1977 *(June)*," "1977 *(July)*", and "1978 *(June)*" pertaining to this in Part I.

1989 *(September)*. Bro. Bruno interviewed Gonella at the Paris symposium. Gonella wouldn't answer specific questions but cryptically said, "Between 1984 and 1988, a whole lot of things happened which I cannot enlarge on. Some of them, however, were inexplicable and unexplained."

Source: Bonnet-Eymard, Bruno. "The Victory of the Holy Shroud Won By Science." *Catholic Counter-Reformation in the XXth Century*, September-October 1989, No. 223, pg. 26.

Comments: According to Bracaglia (pg. 247), Gonella also made the shocking remark in an interview after the conference that the next step before any future Shroud testing was "to kill all the Sindonologists." *[!!]* Bracaglia believes that Gonella's own arrogance caused some problems. The former believes the debate is still open about whether Gonella was really in charge and whether he really had any say regarding where the sample was taken from. Bracaglia believes based on his investigations and sifting through the Guild archives that Gonella's role was limited, and that he was just following his superiors' orders.

1989 *(September)*. Bro. Bruno also interviewed Italian textile expert Franco Testore regarding the weights of the samples. Although Riggi, in his book Rapporto Sindone (1988) and Tite in the official report in *Nature*, **337**, pg. 612, both said the sample taken from the Shroud weighed 150 mg. Testore claimed it was actually 300 mg.

Bro. Bruno asked Tite who put the samples in the containers. Tite said that both he and the Cardinal did and that Gonella was also there, which Gonella confirmed. Tite had told the Ansa agency on March 30, 1989 that he had put four fragments in containers in the presence of Cardinal Ballestrero (no mention of Gonella). In fact, only three samples were put into containers and the fourth sample was put in an envelope. At one point during the Paris congress, Tite corrected himself and acknowledged the fourth sample had been treated differently. Riggi *[pg. 40 of article cited below]* said that Cardinal Ballestrero and Tite *[again no mention of Gonella]* had put the samples into the containers. Kersten (pg. 58 of his book with Gruber) had quoted Wolfli as saying that only the Cardinal and Tite were in the room but then later told Bro. Bruno that Riggi was also in the room -- see entry under "1989 *(December)*" that starts with what Evin had revealed to Bro. Bruno.

At another point, Tite mentioned that "The control samples were tested simultaneously by the laboratories in order to consolidate the comparative tests made in 1983." Although Hall had maintained that there had been no collaboration (*The Tablet*, January 14, 1989, pg. 30), Dinegar, speaking at the Paris congress on September 7th, plainly spoke of "intercommunication between the laboratories." Gonella confirmed this in an interview with *Il Giornale*. See also entry under "1989 *(December)*" regarding what Evin told Bro. Bruno.

Source: Bonnet-Eymard, Bruno. "The Victory of the Holy Shroud Won By Science." *Catholic Counter-Reformation in the XXth Century*, September-October 1989, No. 223, pp. 27 and 30.

Comments: Note here we have Testore saying the sample weighed **twice as much** as the figure given by Tite and Riggi!

Who had been in the sacristy is another piece of the data that should be clear, but has had numerous discrepancies. This was one of the most important scientific tests of all times, and people had trouble remembering the sizes and weights of the samples, who put which samples into how many containers, and who was in the room when that happened! Tite's admission that the labs tested the control samples at the same time contradicted another statement by Tite made at the Paris congress that "the three laboratories had undertaken not to compare results until after they had been transmitted to the British Museum."

The episode about how many samples went into how many containers is especially perplexing and will be addressed further below.

1989 *(September)*. There is a major discrepancy about who brought the controversial third control sample. "The Catholic Counter-Reformation in the XXth Century" specified the three points below. According to Bonnet-Eymard:

1. At Colombani, in a broadcast on channel 6, Evin declared, "I myself brought a piece of 14th century cloth from the cope of St. Louis d'Anjou." But in Paris, on 7 September, Gabriel Vial declared, "I brought the fourth sample. Jacques Evin was unable to come."

2. Evin falsified the report in *Nature* by adding to the number of control pieces. The original English version reads: "Samples weighing 50 mg from two of the three controls were similarly packaged." (*Nature*, 16 February 1989, v. **337,** pg. 612). Where the English has "two", Evin shamelessly puts "three". The reason

for this "mistranslation" is that Evin is not an eye-witness since he was not present when the samples were removed on 21 April. He therefore corrects the text of the *Nature* report in order to remain consistent with. Tite's lie to the Ansa agency on 30 March 1989 when he declared that "in the company of the Archbishop of Turin, he had placed four fragments in the metal containers provided." But to confound Tite and Evin there is the photograph published by Riggi in his Rapporto Sindone, where Tite can be seen presenting on a tray nine small tubes destined for the three laboratories: three tubes for each laboratory, one "sample no. one" and two "control," not three. The third will emerge in the meanwhile from Gabriel Vial's pocket and will be 'packaged' differently from the others -- in an envelope.

3. The object of Jacques Evin's ambivalence is the "blind" test, which he must of necessity explain in line with Tite's words and deeds. In July 1988, he explained to Michel Leclercq of *Paris Match* how important it was to use this procedure on account of "public opinion". And he went on to explain, "when the results are published, non-specialists can in fact be told:" 'You see, there were three samples from different periods, from places unknown to the scientists, and yet the laboratories succeeded in determining their date. That proves how reliable the carbon 14 method has become. Therefore, the date proposed by common accord for the shroud by the laboratories is surely the right one!' (*CRC* 220 Eng. ed., pg.30).

Bonnet-Eymard continued, "In the report published in February 1989 by *Nature*, the 21 authors admit that the 'blind-test procedures' were abandoned 'in the interests of effective sample pretreatment.' In April, therefore, Evin wrote to his correspondents that there would have been disadvantages in this procedure: 'I would rather it did not take place, otherwise it could always be said that the threads were easily exchangeable [...]'."

"He even stated in Paris on 7 September that that was what he had always thought: 'In my opinion, the double blind test procedure was totally pointless. It seemed absolutely unnecessary to me to have a double-blind test. Anyway, in the end it did not take place. Each laboratory had its proper responsibilities and was chosen because it had a certain reputation and inspired confidence. Myself, I would have preferred that when the samples were taken by the Cardinal they had been put straight into the containers, in the presence of the representatives from the laboratories, and given to them directly. For it has to be known that the important thing when addressing an analysis laboratory is whether or not I have confidence in it. If I have confidence, I don't really see why I should believe that it will rig its results. The laboratory will deliver the results as it itself has them. This business of the double-blind test was required under rather strange conditions, but in the end I am glad to say that it didn't take place, because it would have been superfluous'."

"Evin stated at the Paris symposium, 'In my opinion, the double-blind test procedure was totally pointless. It seemed absolutely unnecessary to me to have a double-blind test. Anyway, in the end it did not take place.' An astute attendee, noticing his change of stance from previously, submitted a written question to Evin, who said 'It is being said that my speech does not agree with what I related in *Paris Match*. Well, I don't know so well [...]. I don't see why [...]. I'll ask you again in private [...]. I don't remember what it's about [...]. There is so much [...]. I don't understand'."

Source: Bonnet-Eymard, Bruno. "The Victory of the Holy Shroud Won By Science." *Catholic Counter-Reformation in the XXth Century*, September-October 1989, No. 223, pp. 28-29.

Comments: The blind test refers to the fact that the labs weren't supposed to know which sample was the Shroud. The double-blind test refers to the labs not knowing the identification of the other samples as well. That, of course, did not take place because the labs were inexplicably given the dates of the control samples!

1989 *(September)*. Bro Bruno had been handing out at the Paris symposium one of his articles in which he recounted all the various discrepancies and inconsistencies given out by those who had been involved in the excision of the sample. Riggi was supposed to have presented a paper titled "Sample taking from the Shroud, 21 April 1988," in which he was supposed to provide commentary on a twenty-minute video. Right before he was to speak, Riggi, who held a written report in his hand, told the person next to him, "I shall not be speaking." The video was shown without the expected commentary by Riggi.

Tite made some comments, for which Bro. Bruno made some observations: "The control samples were tested <u>simultaneously</u> by the laboratories in order to consolidate the comparative tests made in 1983" (cf. "Summary" CRC 220 Eng. ed.). I have underlined the word 'simultaneously' because it contradicts what was insistently repeated elsewhere in the same talk, namely, that 'the three laboratories had undertaken not to compare results until after they had been transmitted to the British Museum' (cf. *Nature*, **337**, p.612 col. 1)."

"It is true that Tite does not say whether that undertaking had been kept. I have written saying that it had not (cf. CRC 219 Eng. ed. p.18-19). Although Hall maintained that there was no collaboration between the laboratories (*The Tablet*, 14 January 1989) Dinegar clearly spoke in Paris on Thursday 7 September of 'intercommunication between the laboratories.' And Gonella confirmed this to be so before a representative from *Il Giornale*. If Hall had been present, we would have asked him this question. He would have been obliged to explain himself in front of Dinegar and Gonella, which is one of the reasons why he abstained from speaking."

Source: Bonnet-Eymard, Bruno. "The Victory of the Holy Shroud Won By Science." *Catholic Counter-Reformation in the XXth Century*, September-October 1989, No. 223, pg. 29.

Comments: Bro. Bruno wrote, "[F]or the moment it is plain that Riggi is afraid to explain himself in front of an audience alerted by our 'summary' of the several anomalies in the proceedings of 21 April 1988." Clearly, both Riggi and Hall did not want to face questions about the data.

Italian Engineer Ernesto Brunati wrote in a paper titled "Considerazione Sui Vari Rapporti Pubblicati in Merito Allle Operazione Di Datazione Della Sindone" ("Considerations on Shroud Dating Reports") in "<u>La Datazione Della Sindone</u> -- Atti del V Congresso Nazionale di Sindonologia -- a cura di Tarquino Ladu -- Cagliari 29-30 April 1990, Palazzo dei Congressi" (pg.

118): "[T]he lack of precision exceeds all limits and the inconsistencies, especially the conceptual ones, are such as to strongly question the vaunted scientific nature of the whole operation; so it's result must be carefully evaluated."

Van Haelst wrote, "At the last minute I was allowed to distribute my lecture among the participants of the Symposium. During the Paris Symposium I had the opportunity to debate with Dr. Tite, Prof. Evin and Prof. Gonella. I presented them all my statistical analysis. Prof. Gonella became furious. He blamed Dr. Tite about some violations of the protocol of the testing and some promises never kept. I will not repeat the words used …." ("Radiocarbon dating the Shroud of Turin," http://www.sindone.info/VHAELST6.PDF, pg. 3).

1989 *(September)*. The late Prof. Jerome LeJeune, a distinguished French scientist of the Pontifical Academy of Sciences, said, "I found it exceptionally strange that Tite would sign a protocol requiring a double-blind test and then abandon that commitment in the middle of taking the Shroud samples." "It is, if I may say so, experimental scientific psychology," Lejeune observed sardonically at the Paris Symposium on the Holy Shroud in September 1989. "It means that these scientists are faced with something that impresses them so much that they behave in a way that is technologically futile." "Unfortunately," continued Lejeune, "it goes much further than this. This phenomenon of changing mind over the protocol in the middle of the sample removal -- well, what can that have been for? I don't see […]. But it pushed them into a major methodological error which, in my opinion, makes the experiment as it was done quite pointless. The fact is that having inopportunely renounced the double-blind procedure, they calmly told the laboratories the ages of the witness samples!" Prof. LeJeune, speaking on Radio Courtoisie on September 11th, said, "I put this objection to Dr. Tite, I said to him, 'but why did you reveal the age of your control samples, which meant that you no longer had any controls?' He gave me this staggering answer: 'You are the first person to have made this criticism, and I confess I have no answer. Your criticism is well founded'."

Source: Jasper, William. "Science and Faith," *The New American*, **6** (26), December 17, 1990, pg. 27.

Comments: Prof. LeJeune was not, in fact, the first person to level that criticism. Ian Wilson, editor of the *British Society of the Turin Shroud Newsletter*, did so in their October 1988 issue.

1989 *(September)*. Petrosillo and Marinelli asked Tite about the so-called "blind test." Tite had stated that there were no communications between the labs. Marinelli then said that the *Nature* report had said that labs decided to abandon the blind-test procedure; she wanted to know how and when the decision was made if they hadn't communicated with each other. Tite responded, "In practice, the decision to blindly abandon the tests was made the moment samples were taken. They realized that the sample could be immediately identified for the unusual weaving system and therefore the decision that the non-blind proof was blindly made in Turin. But this did not cause us to change the details of the protocol. And therefore, although the proof was not blind, the Cardinal, me and also Prof. Gonella retreated to the separate chamber to put the samples into the sealed cylinders. For us, the protocol was a path to follow, even if in some details it had been disregarded [...]." **Brunati's comment** *[see source below for this entry]*: "When Tite traces the decision to abandon blind dating to the workshops in Turin, it seems as if he is convinced by doing so. At first it seems that he himself remained extraneous to the decision, but then he says that despite everything they did not want to change the protocol; which, in other words, means that he was aware of the change. What he doesn't say is if the others (Cardinal, Gonella & C.) knew it. But if they knew, why did the Turinese continue to test blindly? Then remains the other question: the blind test had been requested by the laboratories themselves in the meetings preceding the operation. Didn't they know then what kind of fabric it was? When did they realize that the fabric was so clearly identifiable? Logic would say: after seeing it, after, that is, cutting the samples. So there was a meeting before leaving Turin. But if the decision to abandon the blind test was made after the cut, then it doesn't make sense to have said that 'despite' the non-blindness, it was decided not to abandon the protocol, retiring to a separate room. It seems to me, therefore, that Tite's answer is not only not convincing, but leaves room for further perplexity."

Petrosillo: - Prof. Gonella said and repeated that the laboratories have always denied the permission of Vatican representatives to attend the dating tests, although the presence of the representatives of the laboratories had been allowed to collect the samples in Turin. **Tite**: - I don't know exactly how these negotiations took place, but I have the impression that if Gonella had insisted, the permission would have been granted But, I repeat, this is my impression, I don't know exactly how things went. Ask Prof. Gonella. Moreover, when the representatives of the laboratories asked to be present in Turin, they were specific and decisive in their request. Also, because it was a historic event. Being present at that event meant physically seeing the Shroud and performing the withdrawal almost personally.
Marinelli: - Wasn't it a problem of lack of trust? **Tite**: - It is not so much for that, but for the fact that being so many witnesses to us, no one will ever be able to say that they are samples of different origins from the Shroud. So the laboratories insisted on being present, while, in all probability, Prof. Gonella was not nearly as persistent.

Petrosillo: - I would like some clarifications on the fourth sample. **Tite**: - I was asked to provide two samples: one from the time of Christ and one from the Middle Ages. The samples from the time of Christ were very easy to find: there are many pieces of fabric from that time available.
Petrosillo: - Was one of these from Cleopatra's tomb? **Tite**: - I'm not sure, but there are so many pieces available and I got that sample very easily from the British Museum; there are none in the British Museum. The second piece was from the Middle Ages and it was much more difficult to find a piece of fabric reasonably well dated historically. There are none in the British Museum, none at Victoria and Albert Museum in London, there are none in the Cluny museum in Paris. I was really in crisis. I wrote to Prof. Gonella, and he spoke to Prof. Evin in Lyon; Evin wrote to me and I told him that any help would be appreciated. I contacted other people in England, such as Elisabeth Cooper who then procured the samples I brought to Turin. I extended my question to several sources and had the opportunity to examine several samples: but I wanted one and in the

end I got a sample from Nubia, from some Christian tombs in Islamic territories: it was a little more recent than we would have liked and it wasn't dated so precisely and we weren't even able to get a sample large enough to allow us to date with the conventional system: so I had my doubts. And so I came to Turin with my two samples. I knew that Mr. Evin was trying to have his sample, but I practically knew that Evin and Vial had managed to get them when I was in Turin. And this was therefore the fourth available sample. It was in the form of threads but we agreed to give all four samples to the laboratories; although the rest were pieces of fabric, they were all treated the same way. They were delivered in an envelope together with the other samples; the Cardinal was present. **Petrosillo**: - But Riggi doesn't talk about it... **Tite**: - But Prof. Testore.

Marinelli: - Was the envelope not in the box with the three cylinders? **Tite**: No, because the box was made to hold only the three cylinders. **Marinelli**: - There was no mention of this fourth sample in the Cardinal's letter. **Tite**: - No, because when the letter was made or he knew it would be there too. **Marinelli**: - Was Evin present? **Tite**: - No, there was Vial but Evin was not there, although I think he was in Turin. It also seems to me that he was present in the afternoon. I willingly gave this fourth sample, because it would improve the possibilities of control and also because it was much more certainly dated than what I had brought. **Petrosillo** - But what was embarrassing is that this fourth sample gave exactly the same result as the Shroud. **Tite**: - We were looking for a sample that was, as an origin, the closest to one of the dates of <u>presumed origin of the Shroud. That, therefore, was exactly the type of ideal control we were looking for. EXACTLY THE SAME DATE!!!</u> *[The underlining and capitalization are in the original article.]* The argument is reversible. If they had found what the Shroud had looked like in the first century, there would have been someone who would have said that his sample would have been exchanged for the Cleopatra sample. **Marinelli**: - Sox says that one of the two control samples was herringbone. **Tite**: No, it wasn't. I have already said that we had a lot of difficulty finding a sample of the medieval age without worrying about the type of weave. Professor Vial has already talked about the difficulty of finding herringbone linen fabrics before the 16th century. **Marinelli**: - I think you have read the book of Sox. **Tite**: - No. **Marinelli**: - Sox is very specific in this. **Tite**: - Perhaps he was referring to the Cope of St. Louis, which, however, only gave the impression of being herringbone for the gold quilting. But above all, the sample of this cope was given in the form of threads. **Marinelli**: - But Sox does not refer to the fourth sample, but to one of the other two control ones. **Tite**: - Then you're wrong.

Petrosillo: - Why didn't the various laboratories publish separate reports with more complete weight data, photographs, etc.? **Tite**: - *Nature* wasn't prepared to publish a longer article than they published. It was a struggle already to have the press of what they put. **Marinelli**: - Given the importance of the topic, we expected either a longer article or three different articles. **Tite**: - But *Nature's* article already exceeds their usual standard and the agreement included only one article. There are however much more details than is normally published about other radiocarbon dating. **Marinelli**: - Why did you choose *Nature* and not, for example, *Radiocarbon* or other publications? Tite: - Because *Nature* is faster, it's weekly, while *Radiocarbon* comes out once or twice a year, now I don't remember. **Marinelli**: - But for us, who had to wait until February, it didn't seem so quick. **Tite**: - Yes, but you try to write an article with twenty people, divided into three groups, plus the British Museum. I signed because I put the reports of the three laboratories together with the measurements; I added our part, because the British Museum developed the statistical part, of the results. Then I had to send the draft to the various laboratories, which sent it back. Finally, *Nature* took at least a month to submit the final text to the scientific editorial board, which formulated its notes which I had to take into account. In total, *Nature* had the article on December 5 and, if you consider that dating was made in September, it is not bad. Then, after their examination, they sent me back the article which was redone and was finally accepted by *Nature* on January 18.

Marinelli: - In his book (you must read it) Sox states that when Zurich checked the weights, he realized that that of the Shroud was slightly lower than the weight of Turin. **Tite**: - I don't think the labs knew the exact weights of the various samples. All they knew was that they weighed about 50 mg. **Marinelli**: - If they had known the precise weights, it would have been a system to identify each sample. **Tite**: - Right! But they didn't know the exact weights. Not if what exactly they did in Turin.

The thing, however, is not important. *Brunati's comment*: "My letter dated 30/9/89, to Prof. Testore: 'I have never been able to understand, if the weighings of the single samples were made separately, far from the eyes of the envoys from the laboratories, or in the presence of everyone.'" "Letter from Prof. Testore of 28 October: 'Weighing of the samples was done in front of all those present who could see the indications of the balance on a large monitor.' "Sincerity and precision of absolutely scientific answers!"

Petrosillo: - The Shroud was not as contaminated as expected. **Tite**: - Each sample is contaminated. Everything is to see what is meant by contamination. The laboratories were very careful in the pretreatment, they made purifications [...] without sensitive variations of results. (Turning to Petrosillo.) Wasn't it you who wrote in a newspaper in Rome that I replaced the Shroud with the 4th sample? **Petrosillo**: - Oh, yes, I simply reported the news published in France. *Brunati's comment*: - "Which shows that Tite pretends not to have ever read Sox's book but is, however, quite aware, when he wants, of what is said about the Shroud, except for what the Italian newspapers publish about it."

Source: "Intervista Al Prof. Tite Del British Museum." Interview of Michael Tite on September 8, 1989 by Orazio Petrosillo and Emanuela Marinelli. Transcribed and translated with comments by Ernesto Brunati. *Collegamento Pro Sindone*," January/February 1990, pp. 38-44.

1989 *(September).* The Paris Symposium organizers published the following document:

"DECLARATION OF THE SCIENTIFIC COMMITTEE OF THE PARIS INTERNATIONAL SCIENTIFIC SYMPOSIUM"

1. The Committee was gratified that two scientists who had taken an active part in the recent dating of the Shroud by Carbon 14 attended the Symposium and presented papers:
    • Professor Luigi Gonella of Turin Polytechnic, • Doctor Mike S. Tite, director of the British Museum Research Laboratory.

It will be remembered that Dr. Tite was coordinator of the Shroud dating project and it is he who certified the samples, coordinated the results obtained by the three laboratories (Arizona, Oxford, Zurich), and interpreted them. Dr. Tite is also one of the signatories of the article that appeared in the British scientific periodical, *Nature*, 16 February 1989, the only document giving an account of the dating operations carried out on the three samples prepared from the strip removed from the Shroud; operations which led to the conclusion that the Shroud fabric dates from the XIIIth or XIVth century.

2. Even before the Symposium took place, the Scientific Committee had been informed that professional statisticians, (among them Bourcier de Carbon, Symposium moderator) had expressed strong reservations about the manner in which the results obtained by the three laboratories had been statistically analyzed.

    According to these statisticians, it would appear from the results entered in Table 2 of *Nature* (p. 613, second column) and especially the values given by the X^2 test for the three samples (6.4 in column 1) that the samples are not homogeneous in radiocarbon dates. This allows the affirmation, lacking more information, that the ensuing statistical estimates are devoid of value.*

    The hypothesis that the percentage of Carbon 14 with reference to total Carbon (from whence one deduces the radiocarbon date) is not the same in every part of the Shroud is rejected, for theoretical reasons, by almost every specialist in C14 if the object under consideration has not been heavily contaminated -- which is the case of the Shroud.

3. While the Scientific Committee leans toward the opinion of the specialists, it does not intend to take sides. But it believes that one needs only to examine the February 16 article in *Nature*, the only document relating to the results obtained by the carbon dating tests, to see the anomaly. This by itself is sufficient reason to justify the request formulated by many for a new dating of the Shroud by the same method.

4. However, because of the unparalleled value of the Shroud, historical as well as religious, one cannot go on indefinitely removing samples; therefore, the Committee believes that before proceeding to a new dating by C14, it is indispensable to elucidate the serious question raised by the divergence of the results obtained by the Oxford laboratory on the one hand and those obtained by Arizona and Zurich on the other.

The Committee believes that to accomplish the above, it would be necessary:

    a. To conduct dating tests by the C14 method on several fabrics of different ages with solid guarantees of non-contamination, and taking from each one several samples *from different areas*.

    b. If, contrary to what the Carbon 14 specialists and the Scientific Committee believe, these tests reveal, in a significant manner, a heterogeneity in the distribution of the proportion of C14 to total C within these fabrics, it would be desirable that physicists expert in nuclear physics attempt to work out a coherent explanatory theory.

The Committee recalls that numerous hypotheses have already been proposed to explain the anomalies in the distribution of C14 in an object (radiation, particle flux, etc.). But all these hypotheses have run up against grave theoretical objections.

    c. A new dating by C14 therefore should not be undertaken until after it is established in definitive manner that in the distribution of the C14 percentage, there is no heterogeneity that cannot be explained by reasonable natural causes.

The Committee recalls that every explanation given for a variation in C14 caused by the Resurrection of the Man of the Shroud can neither be demonstrated nor invalidated by Science.

    d.   Given the inestimable value of the Shroud, the Committee requests, if it is decided to proceed to a new dating, that this be done on samples on which non-destructive tests had been *previously* conducted; in particular, the chemical study of the threads that numerous scientists have called for.

5.   In case a new carbon dating test were to be decided, the Scientific Committee requests:

    a.   That a precise procedure of operations be established with the cooperation of the British Museum, the Pontifical Academy of Sciences, and STURP.

    b.   That the procedure be entirely controllable.

    c.   That the delegates of the three organizations (and eventually others to be decided upon) be charged to verify at every moment that the defined procedure is exactly followed.

    d.   That no person be granted a monopoly in the interpretation of the measurements made by the designated laboratories. For this reason, the Committee requests that the laboratories publish immediately and without any correction *all the raw results* obtained by their apparatus: that is, the percentages of C14 to the total C, according to the international standard, i.e., the standard of the atmosphere of A.D. 1950.

Of course, the laboratories would be free at the same time to publish the corrections deemed necessary and their interpretation of their results. But every physicist and every statistician ought to find in the reports *everything* he would need to make his own interpretation of the raw results obtained. No monopoly of interpretation should be conceded to anyone at all, particularly in the statistical calculations.

The present Declaration was unanimously adopted by the Scientific Committee and signed by all members at the meeting of 29 September 1989, in Paris.

* Therefore the Committee requests the publication of the comments made by Prof. Bray, of the G. Colonetti Institute, concerning this specific point. His comments are alluded to in *Nature*, p. 614, column 2. Also requested is the publication of all the raw results obtained by the three laboratories.

Source: *Shroud Spectrum International*, No. 32/33 (September/December) 1989, pp. 33-35. https://www.shroud.com/pdfs/ssi3233part5.pdf.

Comments: Notice that if another C-14 test were to be done, the organizers called for the immediate release of all the raw data from the labs -- something it took a Freedom of Information Act request and about thirty years to accomplish for the 1988 dating.

1989 *(September -- Post-Paris Conference)*. A new Shroud organization called "ASTA" (American Shroud of Turin Association) was formed by American businessman Charles Parlato. Parlato intended to hold a Shroud conference early in 1990 in Larchmont, New York. Various Shroud political issues caused a postponement, and it was rescheduled for September 15, 1990.

Shortly before the September 15th conference, Parlato sent out a letter to prospective attendees saying, "Recently I had the opportunity to see still photographs and a super-slow-motion videotape of the C-14 sample-removal from the Shroud prepared by Dr. Alan Whanger. It is of concern to me that this videotape appears to show that a seam between the side-strip and the main body of the Shroud runs through the sample which was sent to the carbon-dating labs. Tom Flaherty shares the same interpretation. If correct, this would invalidate the premise upon which the C-14 test was conducted, namely, that the sample was taken 'from the main body of the Shroud' and is representative of the cloth as a whole."

"In order to assess the meaning of this observation, it is important to me that you take the time to view this videotape very carefully and draw your conclusions from it. Please do this for me within the next two weeks. Dr. Whanger has agreed to prepare and send a set of copies of the videotape to various persons."

"In his description of the removal of the sample from the Shroud, G. Riggi states that he trimmed away extraneous threads from the sample. This resulted in the change in size from approximately 8 square cm to 7 square cm (less than 15% cut away). That report only deepens the concern I have about the sample chosen. The videotape of the sample removal, as well as photos of the sample site, indicate that the seam is so large, relative to the sample, that it would be impossible for the seam itself to be removed without losing a third or more of the sample. This would be significantly different from removing a few threads."

"The relationship of the side-strip to the main body of: the Shroud is not well-understood. Although some have speculated that it was attached to the Shroud in order to center the image, one textile expert pointed out in Paris that characteristic lines in the weave match on both the main body and the side-strip. He concluded that the side-strip was originally part of the Shroud, had been removed from the Shroud, and then re-attached in the same place. Alternately, the seam may be the point of attachment where the edge of the Shroud was folded back on itself to create the side strip. According to the best of my understanding, G. Riggi indicated in Paris that the side strip is hollow, thus making it suitable for displaying the Shroud on a pole. Presumably, this was done to create a less damaging method of display. There is no compelling evidence that the side-strip seam was part of the original weaving of the Shroud, and it may well post-date the fire of 1532, since Riggi also indicated that the seam is attached to the backing cloth."

"Others have previously criticized the choice of sample site due to its proximity to areas that may have been rewoven. What I am saying now is that Al Whanger's video seems to provide photographic evidence that the sample site is definitely a rewoven area."

Source: CD of Holy Shroud Guild archives in possession of the author. Some additional information, as mentioned in comments below, was provided via email of December 14, 2019 from Giorgio Bracaglia, as well as No. 26 (Sep/Oct 1990) of the *British Society for the Turin Shroud Newsletter*, pp. 3-4.

Comments: The September 15th conference was also postponed. According to the BSTS newsletter, "A major factor in the postponement seems to have been a pull-out by those speakers who are members of STURP, apparently due to a concern that any media mis-reporting of remarks they might make at the Symposium could jeopardize their chances of being invited to participate in any further testing work on the Shroud." We see that in the case of this group, **politics** caused the postponement of their conference two times before it was finally held in March 1991 at Columbia University in New York. I happened to organize a Shroud symposium in St. Louis that was held only a few months later. I remember that Fr. Otterbein didn't want to see either symposium held, fearful that they would possibly antagonize Church authorities. Fr. Rinaldi, on the other hand, encouraged me to proceed with mine. I don't remember what his stance was on the Columbia U. one.

The Guild archives also shows that STURP lawyer, the late Mike Minor, had sent a letter mid-Summer to both Rinaldi and Otterbein. Part of the letter seems to be missing, but it seems to refer to the question of the integrity of the C-14 corner. Minor wrote, referring to the Parlato letter, "[H]is last question is <u>impossible</u> to answer based on viewing the film clip alone. Much more information would be needed. If Adler, e.g., were to answer any or all of Parlato's questions, I believe his future usefulness insofar as Turin & Gonella are concerned would be over. The same with D'Muhala, Jackson or Schwalbe [...]. [W]hat will Parlato make of the answers he might receive to these questions? This is exactly the sort of thing the secular press will seize & sensationalize. It is also precisely the sort of things that will inflame Turin. If Luigi *[Gonella]* is to be believed Turin now regards STURP as virtually their only scientific friends, & STURP will be called on in the near future to quietly do some conservation work -- & of course much other work and *[unreadable]* at the same time, but under the mantle of conservation [...]."

Parlato's letter contains phrases that would be used a great deal after Ray Rogers authored a peer-reviewed article in 2005: "rewoven" and "from the main body of the Shroud" and "*[not]* representative of the cloth as a whole."

Unfortunately, STURP was never called on to do conservation or any other kind of work on the Shroud.

1989 *(September).* Various Shroud groups who had attended the Paris symposium had gotten together to discuss the possibility of working out "a collective approach to future research, pooling of resources, recruiting specialists for peer review, etc. There was a general consensus for this, and that this organization should consist of one delegate from each of the research-oriented Shroud groups -- effectively as represented at the present meeting. It was felt that any research project that was fully supported by all the groups could be put with greater confidence before Vatican/Turin decision-making authorities. Similarly, properly constituted, the council of Shroud groups could effectively provide an organization that the Vatican/Turin decision-making authorities could look to for advice on future Shroud matters, obviating the need for the Vatican itself to set up any such advisory body. The onus would be on the council of Shroud groups to recruit its own advisory specialists of sufficient international stature and independence to be able to carry public confidence in the interpretations and conclusions reach in the course of any future direct research on the Shroud."

This was conveyed to Otterbein, who wrote that Rinaldi had told him that he had a meeting with Gonella before coming back to the U.S. and "Gonella was talking about a Shroud commission! He was not sure what he had in mind, but he raised the possibility that he is trying to beat us and establish a 'friendly' Commission in Turin. However, several others later mentioned that Gonella has been expecting the Pontifical Academy of Sciences to appoint a Commission. In any event, it makes me think that we should move as quickly as possible to make our idea a reality, lest we be told: 'There is no need for another Commission'."

Otterbein also wrote specifically to Maloney after a phone conversation they had, "[M]y idea of the Commission was NOT a strictly peer review group. I also said it was NOT the actual hands-on participation group in science on the Shroud. I said it was an <u>Advisory Group</u> of the best informed scientists, who have years of experience in Shroud research.

I hope that I did not mislead you, for I do not exclude peer review nor actual participation in Shroud examination. Perhaps I should have said that it is PRIMARILY an advisory group of the best informed and experienced Shroud researchers. No doubt some of the members will do peer review, and also participate in research but the Commission is primarily a group of experts, who can review proposals for new tests and make recommendations to authorities, who will make decision to approve or refuse permission."

"Any group or individual can submit a proposal to the Commission for review. This would provide an opportunity for an individual, who is not able to convince his own group or who is not a member of a group, to submit a proposal and obtain an evaluation from an unbiased panel of experts."

"Each GROUP of scientists, would select a delegate, who would be a voting member of the Board of Directors. Since the number of Shroud scientific groups in the world is limited, the number of Board members would be limited. Each group would have ONE delegate, but other members of the group could be Associate members."

"Hopefully the Board of Directors would include scientists from various fields. However, since there are other important qualities necessary e.g., the delegate should have respect of his group, past experience in Shroud research, and be a good negotiator, it may be difficult to have a representative from all desired fields. But if the delegate has all the necessary qualities, he will be in touch with his group and be able to obtain the advice and assistance of experts in other fields to advise the Board, They could also make recommendations about best qualified people to do particular research and organize cooperation among the various groups."

"It is important to get moving as soon as possible, but I do not think we can move before February. I would like to see a delegate from each group represented at our last meeting, present for the Feb. meeting. I would suggest that each group be free to decide on how the delegate be chosen NOW e.g., appointment or election. The Board can then decide on the term of this organizational Board, on the term of office of future Boards, and the manner of selection. Board members could be re-appointed by their group."

"The First Board could with the advice of the other people at the meeting decide many procedural questions and organize a program for action."

It was likewise conveyed to Rinaldi, who wrote that he had discussed it in a phone call with Gonella. Rinaldi wrote, "He was very adamant in affirming that an 'Advisory Council' would be totally unacceptable to the Church authority, meaning presumably, Cardinal Ballestrero. He implied that it would be tantamount to a pressure group, a kind of lobby that would only interfere with the research."

Otterbein also said the following in another letter to a representative of the one of the groups hoping to put together the commission:

"I appreciate the urgency of a response to the C-14 dating, and I sympathize with individuals who are anxious to initiate some action as soon as possible. However, I have insisted that the next meeting must be an organizational meeting to structure the cooperation of the various groups. This must be done prior to any attempt to get agreement on a new Protocol for new testing. It would be a delay, but a necessary delay."

"I have been concerned since the appointment of the new Cardinal in Turin about the possibility of new testing. Then I heard of the possibility that Cardinal Ballestrero would be re-appointed Official Custodian of the Shroud. There was also some discussion about the possibility of the Pontifical Academy of Sciences assuming responsibility for future testing, but no official announcement had been made by the Vatican. Hence I endorsed the idea of our Advisory Council, and I hoped that we could get organized before any action by the Vatican. I hoped that we could offer the services of the Advisory Council to the Cardinal and to the Vatican before any decision and announcement was made."

"However, according to my latest information, Cardinal Ballestrero has been re-appointed Official Custodian, and judging from the actions of Prof. Gonella, Gonella will continue to advise the Cardinal Ballestrero and be the key man in future testing. This is crucial factor in our considerations."

"You will recognize this opinion of Gonella: 'I find Luigi Gonella very, very exasperating to deal with.' Another report of a telephone conservation with Gonella quoted him as saying: 'What do we need such a Council for?' A third individual quoted Gonella: 'I'll have nothing to do with that group'."

"In view of Gonella's continuation in his key position and of his attitude towards the Advisory Council that we are attempting to organize, I think there is little hope of such a Council rendering any fruitful assistance to the Cardinal or to the Vatican. Hence without some basis for hope that such a Council would be able to make an important contribution, I cannot ask individuals to contribute substantial time and money to establish such a Council. It is a great disappointment to Fr. Peter and myself, but we consider it to be an honest and realistic evaluation of the situation at this time."

"There are several other factors, which make it difficult to proceed with plans to meet and establish a Council at this time and without further information. I was surprised to learn that the Italian Gov't had protested the sample taking for the recent C-14 test, since the Shroud is located in the former Royal Chapel and hence on government property. They did not take any action over the past sample taking, but apparently want a voice in any such future action."

"We do not know what prompted the Vatican to re-appoint Cardinal Ballestrero. Several possible motivations have been suggested. One possible explanation could be that the Vatican has no intention of allowing any new tests in the near future."

"This will give you an idea of some of the reasons to doubt that an Advisory Council will have an opportunity to function in the near future. In view of these doubts and uncertainties we no longer feel that we can cooperate to organize the meeting this Spring in Turin, and we certainly could not accept the responsibility to organize such a meeting later in the summer of 1990. I personally would want more assurance about future testing and the function of the Advisory Council, and I cannot ask others to make a substantial contribution of time and money under the present circumstances."

"From a practical standpoint I doubt that anyone can organize a meeting in the Spring of 1990, especially since I strongly recommended the election or appointment of the various group delegates, who, I hope would attend the next meeting. They should then start work on By-Laws, elect officers, etc."

"We still think an Advisory Council could render great assistance, but only if the ecclesiastical authorities are interested and open to such assistance from knowledgeable and recognized Shroud experts. Hence we would recommend that the organizational meeting be postponed, and that an attempt be made to obtain factual information about the situation both in Turin and at the Vatican. At the moment I can only hope that enough assurance will be obtained to justify proceeding with what

I still consider potentially to be a most important factor in future testing on the Shroud of Turin."

Source: Memo of Fr. Adam Otterbein after the Paris symposium. CD of Holy Shroud Guild archives, in possession of the author.

Comments: The material in this entry isn't specifically just about the 1988 dating, but is included because it shows the dynamics of trying to work with the Catholic Church regarding testing of the Shroud, which has serious ramifications for the future, including any new potential C-14 test.

One can see how much time and effort was put into this proposed commission, but the Church was unwilling to give up their total control of the Shroud.

1989 *(September)*. Tite wrote to Gonella, "Dear Professor Gonella, Following or recent meeting in Paris, I am writing to put on record the fact that I myself do not consider that the result of the radiocarbon dating of the Turin Shroud shows the Shroud to be a forgery. As you have correctly pointed out, to describe the Shroud as a forgery implies a deliberate intention to defraud and the radiocarbon dating clearly provides no evidence in support of such a hypothesis. I myself have always carefully tried to avoid using the word *forgery* in discussing the radiocarbon dating of the Shroud but I fear that the description of the Shroud as a forgery has still crept into a number of newspaper articles based on interviews that I have given. I can therefore only apologize once again for any problems that such reports have caused you and others in Turin. I was very pleased to meet you and Professor Testore again in Paris."

With best wishes.

Yours sincerely,

[signed]
Michael Tite

Source: Letter of September 14 from Dr. Michael Tite to Prof. Luigi Gonella. British Society for the Turin Shroud Newsletter, No. 24, January 1990, pg. 7. https://www.shroud.com/pdfs/n24part5.pdf.

Comments: Less than two weeks before, Tite had said in a French radio interview that he believed the Shroud was medieval. Here he told Prof. Gonella that he didn't believe it was a forgery. That basically leaves an unexplained but apparently-natural medieval image-formation. Some researchers have claimed over the years that a medieval origin would be more a miracle than if it were caused by the Resurrection of Jesus.

1989 *(November)*. On the 3rd, Kersten wrote to Wolfli asking for a photograph of the unused portion of his Shroud sample. Kersten heard nothing so wrote Wolfli again about four weeks later. Wolfli wrote back saying, "[A]fter some searching among 5000 specimens we have found the remains of our Z1 sample. There are just two small pieces of about 2 mg each, and they have already been treated chemically." Kersten commented, "The photographs taken by Wolfli clearly showed that he only used a 25.9 mg. portion from the whole specimen which he had been given in Turin (52.8 mg). He had according to his own statement kept the remnant, a single 26.5 mg piece of cloth, 'in a safe place outside the lab', known only to himself and his wife. How is it then that he had to spend weeks searching for the fragment among 5,000 other samples, when he had supposedly kept the Turin cloth remnant in this safe place? And why was he left with just two meager 2 mg threads, from which nothing much could be seen, certainly nothing about the closeness of the weave? One might suppose that he could have used the four weeks to ask the 'powers that be' how he should proceed."

In the meantime, Kersten had also received from the Catholic Counter-Reformation (CRC) group in France, a photo from Zurich that showed the upper surface of the sample. Wolfli had not told Kersten about this particular photograph. Kersten wondered why Wolfli had not alerted him about this, given his seeming willingness to assist.

Source: Kersten, Holger and Elmar R. Gruber. Jesus Conspiracy: The Turin Shroud & The Truth About The Resurrection, (Rockport, MA: Element, 1994), pp. 95-97.

Comments: The CRC is another entity that believed that the Shroud samples had been switched with a medieval cloth. The CRC came out with many documents in both French and English between 1988 and 2000 and revealed many inconsistencies from the statements of the major C-14 dating participants that they interviewed.

The late Jesuit theologian, Fr. Werner Bulst, was quoted in the "Night of the Shroud" documentary saying that a photo he received from Zurich lab didn't correspond with the Shroud and even wrote a book in German in the early '90s claiming the samples were switched.

One does not have to accept the hypothesis of a sample switch or another hypothesis mentioned in a 2014 entry -- that the C-14 dates were a result of a computer hacking, to acknowledge the import of the questionable statements and/or behaviors of various individuals involved in the testing.

1989 *(November).* On the 14th, Italian engineer Brunati wrote a letter to Prof. Testore, who had been involved in the sample-taking. He wrote, "Dear Professor, I thank you first of all for having answered me. Many doubts, however, remain and lead me to return to the topic. I take the liberty of doing it publicly, first of all because your letter, as a rectification of what was said and written for a conference, can already be considered intended for the public and, then, also for the great interest with which these problems are followed, at least by the readers of this periodical."

"Please do not consider the present as the fruit of a purely polemical spirit and, if in this way my words lend themselves to being interpreted, I apologize *a priori*: polemics are not in my intentions, to me, and I believe in all of us, it is only of interest to ascertain whether the samples sent to the laboratories were the Shroud or not."

"It is also clear that I do not suspect any of you who have cooperated in taking the samples for dating. If I blame you for any fault or deficiency, they will concern entirely secondary facts: your loyalty is absolutely out of the question for me."

"On some things took place in Turin on that April 21st, therefore, I always have many questions to ask. Why, for example, did they cut 500 mg. of the Relic, knowing that only 150 should be used? Especially since these too could be reduced, taking into account that with its 50 mg. Zurich, for example, managed to do not 2 but 5 datings. I didn't even understand why a piece of fabric was cut and kept in Turin. Another possible sample could possibly have been taken at any time."

"But, especially, I did not understand why, having 300 mg of sample available, it was divided first in half. Is it not more logical to cut a first sample so that it had dimensions such as to reach the required 50 mg., weigh it for verification, and then separate the other two in conformity?"

"The cutting methods were not decided by you. And, apart from criticism, nobody today can deny that at a certain moment there were two small pieces of cloth on the table: one of around 145 and the other of around 155."

"In September, you had written and said in Paris that the first sample had been cut, that of 154.9 mg., obtaining three pieces, which according to you were 52.0, 52.8 and 53.7 but the sum of these three weights, as I already pointed out and as you also say you have noticed, is equal to 158.5, greater than the starting 154.9. So, or those weighings were wrong, or something unusual had happened. We cannot doubt about the weighings, you tell us, and I agree: the values, in part, are also seen in the television recording. As an explanation, you advance your second version. The piece of 154.9 was not cut, but the other of 144.8 mg.; you would have divided it into three parts: the first weighed 52.0, the second 52.8 and the third, of course only 39.6 mg."

"Such a small sample was not acceptable, so another 14.1 mg. were taken from the 155 sample and one of the three laboratories, instead of receiving a single piece of tissue, like the others, would have received two small separate rectangles, weighing overall of 53.7 mg."

"Allow me to express all my amazement. The whole operation was aimed at cutting three pieces of at least 50 mg. So why divide the 144 mg. sample from which the three 50 mg. pieces could never have been obtained, ignoring the other which, on the other hand, was suitable for the purpose and which was left in this way without any specific use? And, even if it all started from a trivial mistake, why didn't you cut a 50 mg. sample from the remaining piece instead of providing a two-part sample to a laboratory? It was absolutely irrelevant whether one piece or a large piece and a smaller one remained in the safe in Turin. At this point your new version intervenes. In addition, if you read your first report, you also have the feeling of a very precise description of the facts and regarding the cut, you say verbatim that the piece *'fut partagé en trois parties, qui furent presque toutes identiques'* ('it was divided into three parts which were almost all identical'). Therefore, this is not just a citation error between the first and second samples, you explicitly said that the three pieces resulting from the cut were almost identical, while today we are told that the third was so small that it needed an integration."

"Not only that, but if we read the report by Prof. Riggi in this regard, the contradiction is even more evident: the weights of the two starting pieces are assimilated to 150 mg., with the intervention of a "*circa*". But then it says: *'une d'entre elles subit une triple découpage: le hazard veut que chacune de ces trois parties soient identiques aux autres, parce que poids des trois fragments, pesé sur une balance électronique, variant d'un mile liéme de gram env. pour chaque piéce et fut équivalent à presve que 0,053 gr. en moyenne pour chaque echantillon'* ('one of them undergoes a triple cutting: it happened that each of the three parts was identical to the other because the weights of the three fragments, weighed on electronic scale, differed by a thousandth of a gram for each piece and were equivalent to about 0.053 gr. on average')."

"Here there is no escape: we speak without a shadow of doubt of only three pieces each of which weighed about 53 mg. And the merit of the equality of the three parts is also attributed to good fortune. If I am not mistaken, it was Prof. Riggi who made the cut: he no longer remembered having given a two-part sample? [...]."

Source: Brunati, Ernesto. "Lettera Aperta Al Prof. Franco Testore." *Collegamento Pro Sindone*, November/December 1989, pp. 41-45.

1989 *(November)*. Bro. Bruno sent a letter on the 14th to Wolfli, Hall and Donahue about discrepancies in the sizes and weights of the samples. Hall's secretary sent a reply on the 24th saying that he had retired a month ago and wasn't willing to engage in any correspondence regarding the Shroud.

Riggi and Testore gave a lecture in Milan on the 28th. Bro. Bruno's group CRC asked Riggi some questions:

CRC: Who holds the remaining piece (*riserva*) of 141 mg?
Riggi: I do!

CRC: Who holds the film of the sample taking?
Riggi: I do.

CRC: Can we organize a session with you to view the whole sixteen hours?
Riggi: No!

CRC: Who has the photographs of the samples?
Riggi: I have!

CRC: Can we see them?
Riggi: No!

Source: Bonnet-Eymard, Bruno and Georges de Nantes. "Holy Shroud -- The Turin Tricksters in Disarray." *Catholic Counter-Reformation in the XXth Century*, November-December 1989, No. 224, pg. 9.

Comments: One has to wonder why Riggi was not willing to share information about the materials in his possession. CRC said there sixteen hours of film. Other accounts have mentioned only ten to twelve. It's just one of the many ongoing discrepancies in the data.

1989 *(November)*. Riggi and Testore held a discussion about the sample-taking at the Rosetum in Milan.

Source: Brunati, Ernesto. "Testimoni, Non Accusati." *Collegamento Pro Sindone*, January/February 1990, pg. 45.

Comments: See Brunati's comments about this discussion under "1990 (January/February)."

Vial, who had worked closely with Testore, did not attend the Milan meeting. However, he was involved with a disagreement with Prof. Raes, regarding the cotton that had been found by the Oxford lab. Rev. Dreisbach received a letter dated November 13, 1989 from Raes *[copy in possession of author, sent by Fr. Dreisbach]*, in which he said, "In his contribution of Paris, Prof. Vial suggested that the cotton fibres mentioned in my report could be superficial fibres present on the surface of the Shroud and not belonging to the yarns of the Shroud. After the Paris meeting I was writing to Prof. Vial telling him that I could not agree with him and that in my opinion the cotton fibres were not superficial fibres. I proposed to Prof. Vial a test allowing being absolutely sure about the location of the cotton fibres. I wonder, however, if it should be possible in practice to perform such a test. Till *[sic]* now I did not have any reply of Prof. Vial. The presence of traces of cotton was confirmed by the sample tested by Oxford for the C-14 datation test. Prof. Hall found some foreign fibres in his sample identified by a laboratory of Derbyshire as being cotton fibres. As told in my contribution in Paris we may regret that this laboratory was not able to identify the variety of the cotton fibres by testing the number of reversals as told in my research of 1973 [...]."

1989 *(November and December).*

Ballestrero was due to retire in 1989. Rumors were going around that Gonella's role as scientific advisor would be ending. Apparently any future Shroud studies would be organized by the Pontifical Academy. STURP, uncertain who would be in charge in Turin, moved quickly to set dates for their proposed non-destructive tests. A meeting was called to be held in Albuquerque, New Mexico, U.S.A. to try and get things set before there were new personnel in Turin. Gonella attended; he had no new information regarding their status pertaining to their proposal, but told the attendees he would talk to Ballestrero about setting dates. According to Bracaglia, "However, Gonella's objective was not necessarily to serve STURP's interests for future testing, but rather defend his own preservation. With Gonella's reputation tarnished by his performance supervising the radiocarbon testing, Gonella was focused on clearing his name by any means."

After the meeting in Albuquerque, Gonella flew to Connecticut to try to talk Adler into participating in future testing. Adler had no intention of doing so until Gonella enticed him with something he brought with him from Turin -- miscellaneous threads that had been removed during the April 1988 sample-taking. How did Gonella come into possession? Protocol stipulated that any material taken from the Shroud was the responsibility of the Papal Custodian of the Shroud. It is believed, however, that Ballestrero had allowed Gonella and Riggi to retain peripheral samples for their own personal studies, and Turin had repossessed only the reserve.

Within two months, Adler found starch on the threads given to him by Gonella. Adler hypothesized that when labs pretreated their samples, they failed to eliminate the starch. According to Fr. Dreisbach, Adler was going to write within thirty days a report of his findings. Dreisbach sent off a letter to the Holy Shroud Guild, and wrote that it appeared that the C-14 was part of a reweave and still retained remnants of a starch. However, Adler did not end up writing a report -- perhaps because the results were ultimately deemed inconclusive.

Bracaglia wrote that he was told in confidence that Adler had sent the threads he had gotten from Gonella to the University of Toronto to have them perform a C-14 test. Reportedly, the results put the date at a time <u>before</u> the time of Jesus, reputedly due to the starch. Adler referred to this in his paper "The Nature of the Body Images on the Shroud of Turin," which was published in 1999. The dates came out BC 3000 to BC 2200. However, with scientifically-based corrections, he said that it could be a first- century date. He stated, "We still do not the provenance of the sample, we still have no measure of accuracy, the precision is poor, we have ignored all the usual corrections to such dates, and the chemical preparations of the sample are entirely inadequate." Then he said, "It does give us some evidence that the Shroud really is a first-century object and that our only problem in getting an accurate date is a chemical problem [...]."*[See my comments below about this apparent contradiction in his statements.] [More details of Adler's analysis can be found in the link listed in the "Source" below.]*

Adler's 1999 paper indicated that the threads were sent to two different labs, not just the U. of Toronto. He also mentioned that he tried to pretreat the threads with cellulose in a tris-borate buffer, before sending them to the two labs. According to Bracaglia's source, that also caused the (before corrections) erroneous dates. So, whereas Adler had not written a report after he first examined the threads in late 1989, apparently the results from the two labs enabled him to write his 1999 paper.

Source: Bracaglia, Giorgio. <u>Uncovering the Paradox within the Archives of the Holy Shroud Guild</u>. (Honeoye, N.Y.: Holy Shroud Guild), 2019, pp. 247-251. Also, https://www.shroud.com/pdfs/adler.pdf.

Comments: What I find contradictory is that Adler says the origin of the 1988 sample was suspect, but then with his analysis of the threads from that sample, he says "It does give us some evidence that the Shroud really is a first century object. If the original sample was suspect, I don't know how he could conclude there was evidence of a first-century artifact, given that he used threads from that very sample.

1989 *(December)*. Bro. Bruno wrote to French textile expert, Gabriel Vial, who had been present at the sample-taking, about the measurements discrepancies. Vial replied on December 3rd, "I too am quite disturbed by the Riggi/Testore contradictions to which, truth to tell, I had paid no attention [...]. I must say that am surprised by the certain 'lack of rigour' noticed in the result of the measurements and weighings. I wrote to Testore to express my surprise [...]."

Here are some excerpts from the letter of Vial to Testore, "Controversy has developed following the 14C test. I venture to send you a copy of the CRC review which you already certainly know. There you will see how the two reports Riggi/Testore are brought into question. It has to be admitted that a comparison of the two brings out certain contradictions. In particular, a comparison of the weights and measurements given in the two reports makes the author *[of this CRC article]* conclude that the two samples of 8.1 X 1.6 and of 7 X 1 would be of different origin and that there was perhaps 'substitution' at the time of delivering (the samples) [...]. Furthermore, the division (of the sample), which ended in two unequal pieces, one of which (the smaller or the bigger?) was then divided into three, again unequally, obliging you to make up the difference by removing a fragment from the reserve piece [...] all casts doubt on the seriousness of the operation [...]."

"[...] Have you kept details of all your weighings and were they filmed? If yes, it should perhaps be possible to discern the herringbone effect on the samples found on the scale? Where are the waste threads and the piece held in reserve? Were they photographed, and can we have a copy of the photo? Were the pieces that were cut photographed before being placed in containers?"

"All these questions follow from a reading of the article which exploits -- quite rightly -- the uncertitudes and contradictions shown up by a comparison of your two reports. Other questions will have to wait until after publication of the 'Acts' and a clarification on your part may be necessary. But where? [...]."

Source: Bonnet-Eymard, Bruno and Georges de Nantes. "Holy Shroud -- The Turin Tricksters in Disarray." *Catholic Counter-Reformation in the XXth Century*, November-December 1989, No. 224, pp. 8-9.

Comments: Given that Hall had already backed out of the Paris congress and would bow out of the Cagliari conference in April 1990 (see below), it's not surprising that Hall refused to answer.

Vial's comments about and to Testore were an indictment of the reliability of the test, and came from one who played a major role in the dating.

1989 *(December)*. CRC produced an article with questions on various aspects of the dating addressed to all the signatories of the *Nature* February 1989 report. "It is a known fact that two months before the samples were removed in Turin on 21 April 1988 and one month after the preparation meeting held in London in January, Dr Tite was looking for a 'double' of the Shroud sample, but to be dated 13th or 14th century, preferably 14th. Was this procedure allowed for in the recommendations of your January meeting? It does not figure in Tite's letter *[in Nature **337**, 611]*, which in your report constitutes the only reference to an experiment protocol. On the other hand, it forms the object of a confidential letter from Tite to Jacques Evin dated 12 February 1988."

"According to the very terms of Tite's protocol-letter, it was established that secrecy was no longer in order, the double-blind procedure having been abandoned. Furthermore, since the representatives of the labs were present at the sample removal and could observe the characteristic form of the Shroud's weave, there was no longer any secrecy, and the *mescolamento* had become meaningless."

"Why then were the samples placed in their tubes out of the witnesses' sight and in a room apart? According to your report, this operation was performed in a *tête-à-tête*. But at the Paris symposium, Luigi Gonella stated that he had also taken part. So, were two or three present? Whom are we to believe?"

French C-14 expert Jacques Evin told Bro. Bruno "that Riggi had also gone with them into the separate room so that he himself might put the samples into their containers." Bro. Bruno and de Nantes commented, "This fact was confirmed for us a few days later by Woelfli, himself an eyewitness of the sample removal. He saw Riggi enter the 'room apart,' after Ballestrero and Tite. So there were apparently not three but four, according to the testimony of Woelfli, one of the twenty-one to have signed the report affirming that only two were present. And it was neither the Cardinal nor Tite who would have mixed up the samples in the tubes destined for the laboratories, but Riggi. Riggi, therefore, was the sole guarantor of the whole operation from beginning to end, contrary to what is affirmed in your report. The gravity of these contradictions in a matter of such grave importance cannot escape you [...]."

Source: Bonnet-Eymard, Bruno and Georges de Nantes. "Appeal to the Twenty-One Co-authors of the Report on the Carbon 14 Dating of the Holy Shroud." *Catholic Counter-Reformation in the XXth Century*, November-December 1989, No. 224, pp. 12-13.

Comments: So who was in the room when the samples were put into the containers: Cardinal Ballestrero and Tite?; Cardinal Ballestrero, Tite and Gonella?; or Cardinal Ballestrero, Tite, Gonella and Riggi? The *Nature* report had stated that only Cardinal Ballestrero and Tite were there. And who actually put the samples into the containers? Video documentation could have answered all this; but, of course, that aspect wasn't recorded!! In his book, Gove (pg. 261) would only characterize this as "this mildly flawed procedure."

1989 *(December)*. Bro. Bruno recounted, "On 14 November 1989, we put to him *[Donahue of Arizona]* a very precise question, sending him a copy of page 3 of Riggi's report in its amended version, with the respective weights of the samples, of which one is in two pieces: Which one had he received? At the end of three weeks we received the answer. On 5 December 1989, Donahue wrote: 'When we arrived in Tucson with the samples from Turin, we immediately cut the Shroud sample into four pieces.' He indicated the masses of the four pieces; total 52.36 mg. He added: 'Our records do not indicate whether or not the Shroud sample was in two pieces.' He enclosed a photograph of a piece which he said measured 0.5 X 1 cm, that same colour photograph which we subsequently published as being that of the Tucson subsample."

"Hoping that he had answered our questions, he said that he would be anxious to learn of our reactions and interested in Brother Bruno's investigation. But to our surprise, when we telephoned him a few days later to ask him how the subsample in the photograph had been cut, he became evasive and did not answer our three insistent letters of 13, 21 and 29 December 89. Then for ten months there was total silence from him; he did not even acknowledge our *Appeal to the twenty-one* (January 1990)."

Source: Bonnet-Eymard, Bruno and Georges de Nantes. "The Carbon 14 Dating: In Pursuit of the Forgers." *Catholic Counter-Reformation in the XXth Century*, April 1991, No. 238, pg. 2 and also Bonnet-Eymard, Bruno. "The Crime Committed Against the Holy Shroud." *Catholic Counter-Reformation in the XXth Century*, February-March 1996, No. 283, pg. 2.

Comments: It's quite amazing that Donahue and Jull's version of the events surrounding the arrival of their sample(s) were so different from Damon's and that Donahue said that they didn't even record if the sample came as one or two pieces!

The *Appeal to the twenty-one* was an open letter sent to the twenty-one signatories of the *Nature* report.

1989. The late Prof. Jerome Lejeune, a member of the Pontifical Academy of Sciences, which was originally involved in the procedures but later eliminated, stated that the C-14 results were "invalidated by procedural defects."

Source: *The Night of the Shroud (La Notte de la Sindone)*, documentary directed by Francesca Saracino, 2011. In 2016, it was revised and retitled "Cold Case: The Shroud of Turin," which is available at amazon.com. I have a review copy of the original version, which has an English voiceover. The revised version has English subtitles. Material pertaining to the post sample-taking period starts at about the thirty-six-minute mark.

1990 *(January)*. Brunati wrote regarding the information that Riggi and Testore provided at a conference in Milan in November, "The points under discussion are as follows:"

- How come after a year and a half of conversations and conferences and after writing the Paris report, only now Prof. Testore realizes that the three weights that he mentioned are inconsistent with the total and only today we remember that a sample was made up of two particular pieces, perhaps of little practical importance, but characteristic and worthy of mention.

- How come he continued for months and months to say that a piece was cut, citing the weight of 154.9 mg, and not the other, and he went down in free details on the alleged equality between the weights of the three samples and on their origin from a single piece; all things that must then be denied.

- They had to deliver three pieces of 50 mg each at least, therefore 150 mg of material; how could they think of obtaining samples for 150 mg from a fragment of only 144.8. And they knew their weight exactly because they had weighed them a few moments before. The other 154.9 mg piece was on the table and at full disposal.

- How come even Prof. Riggi, in his report to the Paris, emphasizes the almost identity of the weight of the three samples, as they had been cut from a single fragment of about 150 mg in weight. There are two different sources, therefore, that say that from that single sample of 150 mg three have been obtained, of almost equal weight.

- How come even the Prof. Riggi, in the last few weeks, gives a new version of the facts, in which one of the samples consisting of two rectangles of linen starts. He denies, without a word of explanation, what he has said and written up to that moment.

- How come in this second version it provides completely different data from that of Prof. Testore: the sample cut, he tells us was the largest one, from 154.9 mg and the three pieces weighed 52.0, 52.8 and, presumably, 50.1 mg, so the added piece was only 3.6 mg. Unnecessary, one might say, given that the sample already exceeded the fateful limit of 50 mg. Strictly speaking, this conclusion is reached. One of the two second versions is undoubtedly false. Both are in contradiction with too many testimonies previously told, to be certainly accepted without reservation. The first versions, on the other hand, are two shared testimonies, coming from two different sources; they seem reliable to a single condition: that the three samples have been cut from a larger piece of fabric and therefore heavier than the one just cut by the Shroud.

Brunati continued, "The unit weight (in mg per cm^2) of the samples the sample cut, he tells us, was the greater one, from Prof. Testore, at the beginning of his report, says that on April 21, before taking the sample, he calculated that the cloth of the Shroud had a unit weight of about 23 mg/cm^2. He adds that he checked, after cutting and weighing, that said value could actually be considered correct. The unit weight of 23 mg/cm² is reported by numerous other scholars of the Shroud and must be confirmed. Therefore an indisputable fact is considered. With a similar unit weight, each 50 mg, each sample should have a slightly larger surface area of 2 cm^2. Prof. Testore, in his report, says that the piece as it was, just cut, could be assimilated to a rectangle of 8.1 x 1.6 cm. This piece, according to Prof. Testore was then cleaned of parts of dubious origin and finally was weighed. It is not said, therefore, if at the moment of the weight it still had its dimensions of 8.1 by 1.6; we only know that the weighing was very celebrated because it resulted in the round figure of 300.0 mg, which was then divided into two parts, etc., etc. ..."

"If Prof. Testore is not very detailed in his report, Prof. Riggi is instead very committed to it. He confirms the existence of a piece of 8.1 x 1.6, weight 540 mg, which then reduces to a 7 x 1 circa with a weight of 300.0 mg. In other words, from what Prof. Riggi, the unit weight is about 42 mg/cm^2, both for the 8.1 x 1.6 cm piece and for the other 7 x 1 piece, we are twice what it should be. And the 7 x 1 was repeated by the same Prof. Riggi on many occasions, for example in his book "Shroud Report *[Rapporto Sindone]* (1978-1987)" (Edition 3M). When I raise this objection, my interlocutors don't seem to know what to say; then Prof. Testore discovers that the 8.1 x 1.6 (12.96 cm^2) surface multiplied by 23 gives about 300 mg, so he is convinced that the piece divided into two parts for the final cut of the samples was precisely that of 8.1 x 1.6 cm. Strange that he discovered it only now, after having for so long heard and read without denying, that instead it is a 7 x 1. If we then consult the only true official report, the one signed by Prof. Tite and 20 other scientists, at least five of whom were present at the cutting of the sample, confirmed the "approximately 7 x 1". The version of Prof. Testore has its logical foundation, but, as had happened before, one is perplexed if it is found that it is advanced in retrospect. The fact that Prof. Riggi during the conference has supported it is at least strange, in that it distorts what he himself has written and said on many occasions. And, if it were really, cut the 8.1 x 1.0 piece, where would that much-proclaimed 7 x 1 come from? It is possible that if they were all invented together, Prof. Riggi and the chorus of 20 scientists, who pontificate from the journal *NATURE* under the direction of Tite? To resolve the serious doubt that the samples sent to the laboratories were not the same ones coming from the Shroud, the piece of film broadcast by RAI at the time was not helpful. I asked Prof. Riggi if he had the reels of the original and complete TV recordings, and the photographs. **He said he was personally in possession of it and also had the duly sealed sample of the Shroud with him, which remained**

unused after being collected (I thought it was in the hands of Cardinal Ballestrero). When I asked if it was possible to organize a meeting to calmly review the film and photos, **I was answered with a "NO" that did not allow replies** [...]."

Source: Brunati, Ernesto. "Testimoni, Non Accusati." *Collegamento Pro Sindone*, January/February 1990, pp. 48-51.

Comments: Again, what was the point of taking video and photographs if they couldn't be shared?

1990 (March). On the 10th, Tite gave a lecture on "Fakes," in which he presented some information on the Shroud. This was the day after the British Museum had opened an exhibition on art fakes, which prominently featured the Shroud. According to the late David Boyce of the CRC, "Dr. Tite projected onto the screen the statistical chart reproduced on page 14 of the English CRC, No. 224. It is the kind of information that no one but a trained mathematician can begin to take in, but from these arcana there emerged two plain statements: one cut of 70 sq. cms. had been taken from the Shroud, and secondly, the laboratories could not help but know which sample (supposedly) came from the Shroud. CRC readers, however, will remember that Testore's report specifically states that Riggi cut a piece measuring 81 x 16 mm. More significantly, if Dr. Tite knew in advance that the little ploy of the secret *mescolamento*, when he and Cardinal Ballestrero disappeared into a back room -- with how many others? -- to place the samples in stainless steel cylinders, could not possibly keep secret from the laboratories the identity of the sample that mattered, then why was it done at all?[...]"

"[...] Since Brother Bruno's more implacable accusations would have offended against conventional good manners in this setting, though not against the quest for truth, I had to reserve them for a private exchange with Dr. Tite at the end. First question: Had he now revoked the letter to Gonella, publicised in the *Catholic Herald* for 12 Jan. *[1990]* (CRC 224 Eng. ed., pg.6) in which he had stated that the Holy Shroud could not be regarded as a 'fake'? Answer: Had I known that my letter would be published, I would have added another paragraph. Question: Are you certain that a piece from the Shroud was really tested? <u>Answer</u>: Beyond doubt. <u>Question</u>: Have you read Bruno Bonnet-Eymard's study on the dating of the Holy Shroud? <u>Answer</u>: a big laugh. <u>Question</u>: But have you read it? <u>Answer</u>: Yes. <u>Question</u>: Are you going to answer the accusations he makes? <u>Answer</u>: But they have been answered. <u>Question</u>. No, there are accusations of grave anomalies in the size of the samples and of official reports having to be amended in the light of Bonnet- Eymard's accusations. <u>Answer</u>: But who is to answer? <u>Question</u>: You, Gonella and Riggi? <u>Answer</u>: That is for Turin to answer. Riggi must answer those questions."

Source: Boyce, David. "Dr. Tite in Surrey." *Catholic Counter-Reformation in the XXth Century*, December 1990, No. 234, pg. 24.

Comments: Tite was the coordinator. He shouldn't have been trying to pass the responsibility to the Italians. <u>But</u>, note the very next entry.

1990 *(April).* Tite wrote to Gonella, "I am writing regarding the current discussion about the size of the sample taken from the Shroud for radiocarbon dating. I find the very obvious discrepancy between the actual size of the sample taken (16mm x 81mm) and the sample size quoted in the *Nature* paper (10mm x 70mm) somewhat embarrassing."

"I myself kept no record of the size of the sample removed from the Shroud. Therefore, when I came to write the first draft of the *Nature* paper, I used the sample size that we had planned to remove rather than the size actually removed. Unfortunately I failed to highlight this problem sufficiently for any of the people reading the draft to point out and resolve the discrepancy."

"A number of my co-authors and myself believe that the actual size of the sample removed from the Shroud should be on record in a scientific journal. I wonder therefore whether you would be willing to write with me a joint letter to the Scientific Correspondence section of *Nature* pointing out the error in the sample size as publicized in *Nature* in February 1989 and indicating the actual size of the sample removed for dating."

"I hope very much that you will feel able to write such a letter which I believe would resolve the current rather embarrassing and unscientific situation. I look forward to hearing from you."

With very best wishes,

Yours sincerely,

[signed]
Michael Tite, Director

Source: Letter of April 18, 1990 from Tite to Gonella. Luigi Gonella archives, of which the author has a copy.

Comments: No joint letter by Tite and Gonella was ever printed in *Nature*. Notice that Tite's letter was the month following CRC's inquiry to Tite about the size of the sample and Tite's reply that it was Turin's responsibility to answer. Clearly, the answer to the CRC in March was an effort to buy some time.

It's striking that a letter of this *Nature* could be written nearly two years after the sample was taken for the Shroud C-14 dating and more than one year after the published results. Very striking is Tite's admission of a "rather embarrassing and unscientific situation."

1990 (April). Representatives from the three labs met in Paris on the 23rd to discuss the problem of the discrepancies in the various reports of the sizes and weights of the samples.

Wolfli distinctly remembered that one of the samples was in two pieces on April 21, 1988. His lab had received one piece, as had Oxford's. Wolfli said to Jull at the Paris meeting, "[S]ince neither Oxford nor I received a sample in two parts, it must be you?" "Well [...] yes!" was the reply of Jull.

Source: Bonnet-Eymard, Bruno. "The Crime Committed Against the Holy Shroud." *Catholic Counter-Reformation in the XXth Century*, February-March 1996, No. 283, pp. 5-6.

Comments: The fact that the representatives of the three labs met is an acknowledgment that the varying reports regarding the measurements was a serious issue.

1990 (April). As with the Paris conference, Hall was scheduled to attend the Shroud conference held in Cagliari, Italy on April 29-30, 1990 and present the paper "An attempt to answer criticism concerning the dating of the Shroud" -- but didn't show and didn't send anyone to read the paper in his place.

Source: Bonnet-Eymard, Bruno. "The Holy Shroud -- Silent Witness." *Catholic Counter-Reformation in the XXth Century*, April 1997, no.295, pg. 25.

Comments: Hall bowing out of two conferences in the space of seven months comes across as suspicious, to say the least. Surely, he could have at least sent someone to present the paper for him. The next entry specifies his reasons -- or would that be "excuses?" -- why he didn't attend.

1990 (April). According to Petrosillo and Marinelli, "Hall was invited to the 5th National Congress of Shroud, held in Cagliari *[Italy]* on 29-30 April 1990, who at first had not refused, but he made his clarifications: 'I was told by colleague Dr. Tite, who attended the recent congress in Paris, that the whole tenor of that symposium consisted in an attempt to raise doubts about the results obtained by the three dating laboratories and in an attempt to prove again that the Shroud originated in the time of Christ. Practicing it again would seem like a total waste of time for everyone involved. I mean that nobody seemed interested in the incontrovertible scientific evidence, published elsewhere, concerning the dating of the Shroud.' The British scientist added: 'After the publication of the results of the dating of the Shroud of Turin, a number of doubts were expressed regarding the negative nature of the attribution. Apart from the historical problems, there have been other sincere ones expressed mainly by non-scientists. An attempt will be made to explain the nature of the scientific evidence and the sequence of events after receiving the samples in

Turin.' Subsequently, however, he made it known: 'I won't have time. Furthermore, I feel that I would have little to offer the participants.' And with that he declined the invitation."

"Over time, after venting against the laboratories, Gonella took it out on sindonologists "whose hysterical rejection of the results did nothing but support the widespread belief that the Church accepts science only when it suits you.' The Turin engineer was shocked that the scientific debate was blocked by the sindonologists on the question of the authenticity or not of the Shroud, ruling that the only problems that must concern are the mechanism of formation of the image and the conservation. It was true that these are two problems of absolute priority, but it cannot be denied that together with them, and no less important also from a scientific point of view, there is the problem of authenticity. This observation by Riggi should not be forgotten: 'If we are unable to establish today the methods of formation of the image and which materials have interacted on this sheet, how was it possible to think of being able to obtain a dating precise and unassailable?' And then, why couldn't the era of origin of the cloth and the identity of the Man of the Shroud help in understanding how that image was created?"

"Gonella, interviewed by M. Consolata Corti, once again paradoxically affirmed: 'If we in 1978 we had found somewhere, in the folds of the fabric, a beautiful parchment with the dry stamp of Pontius Pilate and the signature of Annas and Caiaphas, who said' 'this is the funeral sheet of Jesus Nazarene, executed and crucified,' our interest would have been absolutely zero, because it would not have told us how this image was made.' While continuing to admit that the image comes completely out of the canons of medieval culture, the Turin engineer believed that an 'absurd object' remains such regardless of whether it is of medieval or first-century origin and added that 'the sheet is still safe, a large icon of the passion of Christ, comes from Christ himself or from another crucifix.' He constantly repeated his convictions publicly: 'That the Shroud is or is not of Christ is not a scientific problem. There is no proof of authenticity. The fact that it is from the first or fourteenth or twentieth century is irrelevant. The whole problem of the Shroud, of the research on the Shroud, has absolutely nothing to do with faith or with the relationship between science and faith. It is not that religious importance comes from the fact that the body of Christ was there; there is a very great religious importance even if it did not have the body of Christ inside; because it undoubtedly remains in any case a great image of the passion of Christ. We do not know how it came to be, the problem remains. There is only one thing that the Church knows and has always said: that is an image of Christ, meaning by image the broadest sense possible. The liturgy of the feast of the Shroud is not a relic liturgy but an image liturgy'[...]."

"[...] [S]cientists in the field of medicine noted how the arrangement of the blood spots was a little different from traditional iconography, but instead showed an anatomical-pathological correctness out of context with the mediaeval culture. The photograph then had an inexplicable aspect: the distribution of chiaroscuro typical of a portrait appears in the topographical negative, rather than in the positive, a concept in itself completely unknown in the Middle Ages. A medieval artefact, as such the Shroud was already denounced by a bishop in Lirey. But in conclusion the singular characteristics of the image and of the stains, revealed by modern science, clearly out of context with medieval culture. The Shroud image resembles a photographic negative. This puts it out of the technical-cultural context of the Middle Ages. The medical investigation conducted on the negatives showed an anatomical correctness extraneous to medieval culture and also deviating from the traditional iconography -- this obvious crucifixion."

Gonella was interviewed by Orazio Petrosillo:

Petrosillo: During the years of intense research, only one scientific proof against the authenticity of the Shroud has emerged? **Gonella**: None, absolutely none. It is proven that there are no pigments that suggest a painting. If it turns out that the Shroud is medieval, it would become the most interesting piece in the entire history of art, because it would be an image made with a technique unlike any other to our knowledge.

Petrosillo: Has an irrefutable or unexpected result been obtained instead? **Gonella**: In a field of materialistic investigation there is no capital proof. What counts is the consistency of the general framework of all the measures. I don't know which one to call sensational or unexpected. Unexpected is the presence of traces of serum around the blood stains as halos. They have been tested with two independent and completely different techniques: the physical fluorescence technique and the microchemical analysis of the fibril coating. The sheet was in contact with a corpse, it contains blood coming from coagulation wounds.

Petrosillo: In this sense, not even the radiocarbon dating is the decisive proof? **Gonella**: Exactly. It is one of many. It is no longer critical of those carried out in 1978 such as X-ray fluorescence. Those tests could already prove that the Shroud is medieval but they haven't done it.

Petrosillo: Is it possible that all this was blown away by the radiocarbon storm? **Gonella**: There is no scientific reason not to give credit to the data collected with the ^{14}C.

"Even Franco Testore is less categorical than he: 'From a scientific point of view it will never be possible to say that it was Jesus Christ who left that imprint, but many elements give this conclusion a good chance'."

Source: Petrosillo, Orazio and Emanuela Marinelli. La Sindone: Storia Di Un Enigma, (Milan: Rizzoli, 3rd edition, 1998), pp. 232-235.

1990 *(May)*. At a conference on the 15th at the Rosetum in Milan, Gonella said, "I wish the sindonologists and the antisindonologists would be shut up in a stadium, the key thrown away, and that they would butcher one another, so that scientists could work in piece."

The author of the Shroud periodical in which this quote appeared wrote that the "Milan conference was a strong attack against those who dared doubt the validity of the C-14 examination, and a defense of the three labs."

Gonella also commented on Italian journalist Vittorio Messori's published disappointment that Cardinal Ballestrero stopped referring to the Shroud as a "relic" after the C-14 dating results and only called it an "icon." Gonella said, "The problem whether the Shroud is authentic is extremely secondary from the scientific point of view." But he also went on to say that "the Shroud remains venerable because it bears the entire image of Christ" and "We do not know if it was fabricated or not -- we don't know anything. The image should not exist."

Source: Farkas, Ilona. "Notizie Varie" in *Collegamento Pro Sindone*, July-August 1990, pp. 53-59. Translated by Dr. Daniel Scavone.

Comments: Gonella comes across as very frustrated, but he could have saved himself a lot of the frustration if he had listened to the advice of many counselors, which he rejected. His other quoted statements from the conference, in my opinion, do not neatly fit together.

1990 *(July)*. In a letter to *Nature*, a reader noted, "SIR-Regarding the procedure for obtaining samples from the Shroud of Turin for radiocarbon testing, M. S. Tite of the British Museum (*Nature* **332**, 482; 7 April 1988) says, 'all stages will be fully documented by video film and photography'. However, the official report of the test results (*Nature* **337**, 612; 1989) says: 'All these operations, *except for the wrapping of the samples in foil and their placing in containers*, were fully documented by video film and photography' *[letter author's italics]*. The key word here is 'except'; its presence raises serious questions. Did the custodians of the Shroud agree that 'all stages' of the procedure would 'be fully documented' as Tite indicates in his letter of 7 April? If so, why did they make and suddenly break that reasonable agreement? (Note: The sampling of the Shroud took place on 21 April -- just two weeks after Tite's reassuring letter appeared in *Nature*.) Why wasn't the entire procedure documented on video film and made available to the public? Without full documentation how can one be certain that it was the Shroud -- and not another cloth -- that was carbon dated? At what point in the sampling and sealing procedure did filming end -- and why was it ended? As Tite was the independent overseer of the carbon dating test and the guarantor of its unequivocal reliability, he has an obligation to address these unanswered questions. (That obligation would be very apparent had the Shroud been carbon dated to the first century.) And it should be noted that the man of the Shroud and the person(s) who imprinted his Christlike image are, as yet, unidentified -- and not everyone is convinced that the Shroud is mediaeval as the results of carbon testing indicate."

ROBERT HALISEY, Lookout Drive, Lake Hayward, Colchester, Connecticut 06415, USA

TITE REPLY -- I confirm that, as stated in the *Nature* article, the wrapping of the samples in foil and their placing in containers was not documented by video. This was because we were continuing to follow the blind testing procedures according to which only the Cardinal, Professor Gonella and myself were to know which containers held the Shroud samples. This aspect of the procedure was, I admit, somewhat illogical, as by this time we were aware that, because of the unusual weave of the Shroud, blind testing was not feasible without unraveling the samples. However, I should emphasize that it was the Cardinal and myself who were guarantors of the samples and that the video film was intended as an *aide- memoire* rather than being meant to provide definitive proof of the identity of the samples.

M. S. TITE, Research Laboratory for Archaeology and the History of Art, University of Oxford, 6 Keble Road, Oxford OX1 3QJ, UK

Source: "More on the Shroud." [Letter], *Nature* **346**, 100, (12 July 1990).

Comments: In his April 7th letter, Tite made no mention of an exception for the video recording of "all stages." His statement in the July 12th letter flatly contradicts his previous statement. That aspect wasn't "somewhat illogical" -- it was totally illogical. It was to keep up appearances of a blind test. To have ten to sixteen hours of video of all the procedures, but to make the single most important aspect not recorded and made dependent just on the integrity of two individuals (or three or four if Gonella and Riggi were present, as in some accounts), was simply outrageous.

1990 *(July/August)*. Van Haelst wrote, "Ever since the publication of the results of the radiocarbon dating of the Shroud of Turin, I have tried in vain to obtain the complete and official report. I wrote to all the authorities involved. Without any success. In the end I wrote a letter to His Holiness the Pope. The Vatican **kindly took an interest**, but it wasn't much help for my question. I followed the advice given to me: I contacted the competent Turin authorities. I did it immediately. But I became very disenchanted by the attitude of the Turin authorities! Neither the new Archbishop, nor Prof. Barberis replied to my letters, in which I asked for the publication of the entire official report and all information about it. My letters were supported by the letter of the Pope [...] which had given me the advice to consult the competent authorities of Turin. I also asked for the opinion of the British Museum, regarding the declarations of Dr. Tite about the results from 14C. The British Museum still supported the conclusions published in *NATURE* with its full authority. The new spokesman for the British Museum, Morven Leese, replied as follows: 'Dear Mr. Van Haelst, thank you for your letter of April 4th. I regret that I cannot provide any further information on the scientific procedures used for the radiocarbon dating of the Shroud of Turin, since I did not take part in it. As for the statistical analysis, for which I was **responsible**, the methodology is explained in the *NATURE* article, where the bibliographic references are indicated. The data provided to me are also published in that article, so you are able to do an analysis yourself if you wish. Best regards, Morven Leese'."

"Mrs. Morven Leese certainly takes a clear position on this result. But since she only participated in the statistical analysis, does this mean that Mrs. Morven Leese, as a spokesman for the British Museum, cannot give her opinion on scientific procedures? The same happens for several other scientists involved [...]. In Paris, during the Symposium on the Shroud, I asked competent scientists, such as Dr. Tite, Prof. Gonella, Prof. Evin and **others**, to make the statistical analysis with the data reported on *NATURE*. Since sitting at a table, with a calculator, there is no possibility of escape. I know very well that this is not a good strategy -- it doesn't leave a way out for the antagonist. It is much easier, as has been done in Paris, to respect the point of view of others and to speak politely in neutral terms. An example of the questions I asked: 'How can we obtain the average date for (the laboratory of) Arizona, 646 +/- 31, with the sigma 13C, when the individual dates of table 1 are reported without the correction for sigma 13C??? This proves that **NONE** of the statisticians involved have checked the data for the individual Arizona dates!!! I asked them this question very clearly, but nobody gave me a clear answer. Even if I have not taken part in the statistical analysis nor in the dating with the 14C, I certainly have my opinion on the subject!!! I hope that the competent authorities of Turin will react positively to this honest question of a sindonologist, in search of the truth."

Source: Van Haelst, Remi. "Una Domando Per La Verita." *Collegamento Pro Sindone*, July/August 1990, pp. 45-47.

1990 *(August)*. Ballestrero, who had been in retirement since early in 1989, was confined to a wheelchair, and asked the Pope to remove him as Papal Custodian of the Shroud. On August 16th, Ballestrero sent a letter to Casaroli expressing regret for the accusations of "imprudence" for not having exercised controls, on the Turin archdiocese, made the recommendation that the research continue with objectivity, humility and interdisciplinarity, emphasized the urgency of a research aimed at conservation, addressed particular reservations on the International Center of Sindonology of Turin, and made the recommendation not to exclude the consultants who had worked from 1978 through the current day. Consequently, Gonella presented himself to his Ballestrero's successor, Giovanni Saldarini, who told Gonella dryly that he did not need scientific consultants, and that the question of the Shroud had been "handled very badly." From that point Gonella noted he was excluded from any question concerning the Shroud, and with him the other collaborators of Ballestrero in the operations of 1988. The successor of Saldarini, Cardinal Severino Poletto, maintained the same line, both increasingly having leaned on the Turin Sindonology Center.

Gonella was upset that the media was characterizing the Shroud as a fake, since that word implies an intent to deceive -- he believed that since it's not even known how the image was made, it's not appropriate to ascribe intent to a forger. Gonella also told Tite that the newspapers were using that term and Tite wrote to Gonella on August 14th that he agreed with the latter -- he didn't think the dating had shown that the Shroud was a "forgery" and he claimed mistakenly attributed to him having used the word.

On the other hand, Gonella also believed that pro-Shroud researchers were dismissing the dating, offering unlikely reasons why the dating could not be correct. Gonella said that the supposed proofs given for the

Shroud's authenticity were really only clues to a Middle Eastern origin, and there was also an undue emphasis on the belief that such an image could not have been the product of pre-industrial technique. Gonella also stated that they were concerned about any actions that could have been construed as pressure for the labs to come up with a first-century date.

Source: Gonella, Luigi and Riggi di Numana, Giovanni. Sindone: il mistero continua, (Milan: Fondazione 3M), 2005, pp. 84-86.

Comments: I disagree with several of Gonella's statements. I believe it is certainly possible that a forger could have produced the Shroud, even if we don't understand the method used. Regarding the assertion that some researchers were offering unlikely reasons for dismissing the dating, two of those reasons seem to me to be eminently possible: 1) the dating was altered by what we call "The Resurrection" -- after all, that was unique and the Shroud is unique, so the energy from that event could have caused an anomaly in the carbon. 2) Since the only sample taken was from a frayed area, it is possible that the area could have been repaired -- since samples weren't taken from any other area, there's nothing to compare it with -- which is the reason all the experts had recommended taking more than one sample. It was a badly-designed and badly-executed experiment -- which is why the results have always been questioned. If the test had been done properly, there would have been little or no concern that Turin was putting pressure on the labs to come up with a favorable date.

1990 *(August)*. On August 18th, the Vatican Press Office stated in its bulletin, "The result of the medieval dating became an odd point, even in contrast, compared with previous results, which were not inconsistent with a 2,000-year old dating. These are experimental data, among others, with the validity and also the limits of sectoral tests which are to be integrated in a multidisciplinary framework."

Source: Marinelli, Emanuela. "The Setting for the Radiocarbon Dating of the Shroud." Presented at 1st International Congress on the Holy Shroud in Spain -- Valencia -- Centro Español de Sindonologia (CES), April 28-30, 2012, pg. 14. www.shroud.com/pdfs/marinelliv.pdf.

Comments: It's nice to have seen the Vatican acknowledge this, but considering that before the C-14 test, it was argued that the C-14 test should have been one test among many, it's the Vatican's fault that it wasn't done in a multidisciplinary context. And since it was their fault, they should rectify it by allowing more testing.

1990 *(August)*. The Vatican made an announcement that it would consider proposals from researchers and scientists for new scientific tests on the Shroud. Meacham wrote, "The statement called the C-14 results 'strange' and pointed out that they conflicted with previous scientific findings."

Source: Meacham, William. The Rape of the Turin Shroud: How Christianity's most precious relic was wrongly condemned and violated, (Lulu.com, 2005), pg. 111.

Comments: The Vatican's statement confirms Fr. Rinaldi's statement from November 1988 that the Church had regretted saying they had no reason to doubt the results. Meacham added, "I sent a copy of my proposal already submitted in 1989, but alas this apparent openness to new research was closed as suddenly as it had appeared, for reasons known only to the inner sanctum of the Curia. The new archbishop of Turin, Cardinal Saldarini, made it known that only proposals regarding the conservation and preservation of the Shroud would be considered. In 2000, a similar call for proposals was again put out. Nothing was heard about it again until the 2005 Dallas Shroud Conference, when Monsignor Ghiberti made an announcement at a special dinner that the proposals were then being considered by The Vatican. It's hard to understand why the Church moves so slowly when they themselves have initiated requests for proposals and when positive news about the Shroud undoubtedly would provide encouragement to members of the Church and possibly even bring new converts." It's very frustrating to know that the Church doesn't trust the 1988 C-14 results, has twice asked for new test proposals, but has not moved forward to actually allow new testing.

1990 *(August)*. On the 11th, the Vice-President of the French Shroud group C.I.E.L.T. wrote a letter to the Director of the British Museum, Sir David Wilson, about an exhibition on art fakes that had opened in March and which featured the Shroud prominently. Upinsky wrote, "Mr. Director, we wish to draw your attention to a grave scientific disinformation presented in the British Museum, from 9 March 1990, in the framework of the exhibition 'False? The art of fraud.' It is unjust that a giant slide of the Turin Shroud was placed at the center of this exhibition, whose luxurious catalog presents 'the counterfeiters of the Turin Shroud' to the public's disapproval. The classification of this authentic archaeological piece as a falsification is contrary to the most basic rules of scientific ethics. So we would like to ensure that your vigilance has been taken by surprise by this use of the British Museum's scientific reputation for partisan purposes. The situation is aggravated by the central role played by the British Museum in dating to C14. In fact, you know that this sheet, the subject of a lively controversy, is the most studied archaeological piece in the world. Unique in its kind, it remains a scientific enigma for the image transfer mechanism, still unknown, that it hides. At the time of the International Scientific Symposium in Paris, on 7 and 8 September 1989, some inadmissible irregularities and scientific contradictions were noted in the C14 dating report.[1] No response to these contradictions has been given since that meeting and no member of the international scientific community responsible for the research claims more that the Shroud is a fake.[2] This means that nowadays the scientific status of the Shroud is, for everyone, that of an authentic object: the beginning of our era for almost all researchers in the international scientific community; for the small group -- and scientifically isolated -- of those who dated with the C14, of the thirteenth and fourteenth centuries. Under these conditions, the presentation of this singular archaeological piece -- *acheiropoietos [word meaning "not made by human hands"]* -- as a vulgar fake was not made possible except through the systematic concealment of the main implications of the investigations:"

a) Absolute silence on the history of the Shroud before its appearance in France;

b) Only the medieval dating to the C14 is presented to the visitor, while the considerable volume of the works which allow to deduce the authenticity is totally denied;[3]

c) No technical documentation is given on the singularity of the Shroud and on the enigma of the image transfer mechanism;

d) No mention is made of the International Symposium in Paris, which brought together most of the researchers from the international scientific community of the Shroud;

e) No complaints are made about the greatest contradictions that the medieval dating of the Shroud raises, while "the limits of the expertise" are granted to pieces of lesser importance.

"Finally, the pressure exerted on the owner of the slide to leave it in the British Museum until the end of the exhibition should be noted. And this despite a refusal had been opposed to his request to see the position of Dr Tite and despite his request for a minimum rectification was reduced to a simple personal 'opinion'.[4] Such maneuvers highlight the factious nature of this exposition, which rejects any scientific presentation loyal to the Shroud, which it targets. They disqualify their authors. In the face of such a scandal, you will understand Mr. Director, that, in order to prevent the opprobrium that is associated with such ways of acting from immediately reflecting both the reputation of the British Museum and the credibility of dating at C14, we await that you confirm:"

1. That the tendentious insertion of the Shroud of Turin in this exhibition was made without the knowledge of the British Museum Management.

2. That this exposition was not conceived under the control and scientific authority of those who dated with C14.

3. That the offensive terms of the exhibition catalog have been appropriately corrected and that you send us a copy. In the meantime, we thank you for your collaboration to prevent the consequences of this growing case.

Please accept, Mr. Director, the expression of our best feelings.

"For the President and the Commission of the C.I.E.L.T., Arnaud-Aaron Upinsky Vice-President"

[FOOTNOTES]

1. In particular, the epistemological contradiction and "nonsense" to which medieval dating leads; the statistical contradiction (confidence level of 5% and Ki2 of 6.4); the contradiction between the statements of Prof. Testore and those of Mr. Riggi di Numana on the samples and the weights of the samples.

2. Dr. Michael Tite of the British Museum, coordinator of dating at C14, kept himself from distancing himself from the thesis of the forgery, in his letter of 19 September 1989 to Prof. Gonella.

3. In particular by STURP, ASSIST, SEARCH, HOLY SHROUD GUILD and CENTRO INTERNAZIONALE DI SINDONOLOGIA.

4. See the correspondence of 8, 11, 16, 17 and 19 July 1990.

Source: Upinksy, Arnaud-Aaron. "Lettera Al Direttore Del British Museum," *Collegamento Pro Sindone*, November/December 1990, pp. 39-42.

Comments: "SEARCH" (footnote 3) was an Australian Shroud organization.

1990 *(September/October)*. Fr. Bulst wrote, "The radiocarbon dating of the Shroud, sensationally published, has not been accepted by the great majority of scientists who know the Shroud itself and previous research. Therefore the scientists meeting at the Paris congress in

September 1989 demanded a new 'controlled' test without a 'monopoly', as Dr. Tite of the British Museum usurped. In the book 'The Enigma of the Shroud: a Challenge to Science' (Rome 1990) O. Petrosillo and E. Marinelli have exposed many circumstances and the whole procedure extremely doubtful. A few weeks earlier my book in German had been published on same subject and in the same disposition, but less ample. I am very grateful to the Italian authors for letting me know many details that I did not yet know and which, in the context of everything, confirm my judgment. On the other hand, for the contact that I had with some scientists who were (directly or indirectly) involved in the radiocarbon test procedure, I could learn about some important details, but generally not yet known. Therefore my refusal of radiocarbon dating could have been even more resolute. The solution to the riddle seems to be not difficult, so here is a brief summary of the first part of my book."

"I. Prior considerations"

"There is no doubt that medieval dating is false. I propose a very brief summary of previous research:"

1. The fabric of the Shroud is ancient, it comes from the Near East. Professor Vial, director of the Lyon Textile Study Institute, confessed that he had not seen such a fabric of western origin in the fifty years of his studies.

2. The provenance is proven by the pollen found on the Shroud. Only a few of them are European plants, the vast majority correspond to the exceptional vegetation of the Jerusalem region. It is very important that two Jewish professors, the geobotanist Danin (Jerusalem) and the palynologist Horowitz (Tel Aviv), even believe that the Shroud was made in the Jerusalem region.

3. The image on the Shroud evidently is not a work of art, but was formed by a natural process: oxidation and dehydration.

4. The doctors, especially the experts in forensic medicine, agree that the Shroud is the sheet of a crucified man.

5. The existence of the Shroud is confirmed by the iconography and Byzantine documents, at least since the time of Justinian (527-565).

"Radiocarbon test examiners ignored almost all of this knowledge. All the more strange is that they refused collaboration with experts from other sciences, although this was expressly agreed. Total ignorance sometimes manifested itself in an almost incredible way. Professor Wölfli, who had seen the Shroud the first time on the day of taking the samples, wrote to me that he had understood 'immediately' that the traces of 'blood' on the Shroud were not actually blood. This was *proven [to be blood]* a few years ago. Another of the experts seeing the stain of the side asked: What is this? Professor Hall of Oxford in a conference in London the day after the publication of the results, which he himself had prepared with Tite, defining all those who believe that the Shroud is the sheet of a crucified man, 'religious fanatics' and 'pathological' especially STURP scientists. He was unaware that only four members of STURP were (or are) Catholic, three are Jews, six agnostics, the other Protestants."

"II. Additional reasons to be skeptical"

1. The collection of the Shroud samples took place on April 21, 1988 in the sacristy of the Cathedral of Turin in the presence of about twenty invited experts, among whom the representatives of the laboratories in Oxford, Zurich and Tucson. The whole procedure was filmed. However, a very important act, the **placement** of the samples in the containers, took place in a secluded room in which Tite had retired with Cardinal Ballestrero, according to later reports also Gonella and Riggi were present. But this is marginal. Effectively only Tite, the coordinator of everything, knew "which was which". All the others were excluded in this way that Gonella, the Cardinal's personal adviser, later complained of the "**mafia**" *[tendency]* of atomic scientists.

2. The laboratory representatives also received other samples, so-called "control" samples, which was totally unusual in a radiocarbon test. **According to Tite there were two such samples**, one from the first and the other from the eleventh century. The

intention, according to what was said, was a "blind" test. But such blindness already *a priori* was an illusion because the fabric of the Shroud is unmistakable. The real intention was most likely another.

3. **The key to the riddle seems to be a third control "sample"** that Tite had asked of Prof. Evin and Prof. Vial of Lyon: a sample as similar as possible to the Shroud (!): linen, of the same color, of the same texture, **almost a twin of the Shroud, but medieval**, of an age known between the XIII and XIV centuries, therefore of the period desired by the opponents of the authenticity of the Shroud. Vial procured such a fabric in the Church of Saint-Maximin-la-Saint-Baume in Provence, a fairly secluded region. He took it in an extremely hidden spot in the lining of a cope of St. Louis d'Anjou, appointed archbishop of Toulouse in 1296, but who died the following year. So this is already dated precisely from 1296-97. The removal took place **in the absence and without the knowledge of the parish priest**, who was surprised and disappointed. What was the real intention of this truly nebulous action? The doubt is strengthened for three reasons:

   a) This specimen was in the **form of threads,** why**? In M. Paolicchi's photograph you can see a larger part of this cover,** not just the threads.

   b) Tite had not prepared containers for this third sample. So he no doubt intended to hide the existence of it. But the representatives of the laboratories saw the threads brought by Vial. Therefore it was inevitable to deliver these samples too, but simply in a sachet, as seen in a photograph by Prof. Wölfli. On the contrary, Tite was photographed with a tray on which only 3 x 3 containers were placed.

   c) Tite also later attempted to hide the existence of the third control sample. Even in his publication on October 13 in London he mentioned only two control samples. It is truly strange that **in the English-speaking report there were only two samples, while in the reports in Italian, French and German reports, there were three control samples**. Only in the definitive publication, on February 16, 1989 in the journal *NATURE,* Tite, one of the authors, had to confess that this sample had also been examined in Oxford. What was the purpose of such a falsehood?

"III. The solution to the riddle"

1. Unexpectedly the truth was read in the common report of the laboratories, published in the English journal *NATURE* on February 16, 1989. The **measured values** of this mysterious sample named "Z4" **were identical** to the values of the sample "Z1", alleged to be from the Shroud. I propose the essential values for this judgment in the form of the usual calendar data:

    a) The measured values for the sample "Z1" (for the Shroud) are indicated between the years **1273 and 1288**, taking into account 68% of the measurements: between **1262 and 1312**, taking into account 95% of the measurements. 5% of the measurements are left out, because some exceptional values can falsify the result.

    b) According to the same calculation method, the age of the sample "Z4" (from the cope of St. Louis d'Anjou) is indicated between the years **1268 and 1278**, respectively **1263 and 1283**. The authors of the report had to confess that these values are very close ("very narrow"). To tell the truth, these very small differences between only statistical values are negligible, within the normal limits of values in the radiocarbon method. That is, the material of the samples "Z1" and "Z4" with the maximum probability was identical. The measured age agrees well, but not precisely, which is obvious in this method, with the true age of the cope of Louis d'Anjou. The Shroud evidently is not of this epoch. Further arguments prove this suspicion.

1. **Of decisive importance** is a comment in the 1989 yearbook by the ETH (Polytechnic) in Zurich, where Wölfli himself had examined the samples. The author of this report was certainly Wölfli himself, or at least he had approved of a text so important. In this report we read: **The analysis of the data did not indicate any contamination of the sample "Z1"** (alleged to be from the Shroud)**! Even the age of the parts of this sample that were differently cleaned and not cleaned was not distinguishable (!).** But all those who know the Shroud know that the whole Shroud is very contaminated and especially the region from which the sample was cut. This contamination was the reason why Prof. Libby, who for the creation of the radiocarbon method had received the Nobel prize, refused a Shroud test. The consequence is not doubtful: The sample "Z1" was not of the Shroud, rather then, the cope of St. Louis d'Anjou. **The lining of this cone of the interior of that liturgical dress was perfectly protected from any contamination for centuries.**

2. Also in the official report in the *NATURE* magazine, approved by all those who had collaborated in the three laboratories, it reads: After the first measurements there was no evidence of contamination ("The first set of measurements revealed no evidence of contamination", pg. 613). On page 612 of this report we read: "Sample of the Shroud had been taken from the main part away from some patches or burnt spots" (!). This claim is obviously false. **And since six of the reporting officers were present at the cutting of the samples, it's a scandalous lie or rather, a further almost incredible further proof of ignorance.** All of *NATURE*'S data must **be discarded**.

3. Therefore the sample of the Shroud had undoubtedly been changed on purpose. But the test team constituted a closed group, rightly defined by Gonella as a "mafia." Therefore it is not easy to know when and how this change has been carried

out. And certainly not everyone who participated in the test procedure was an imposter, some were probably either deceived themselves or had collaborated only in a technical way.

4. Bonnet-Eymard believed that Tite was already making the change in the secluded room in Turin. Undoubtedly this procedure was quite obscure, significant in the atmosphere that existed from the beginning. But change, it seems to me, was not possible. **Much more likely that the change was made in Zurich, and shortly thereafter, in Tucson**. Wölfli had waited three weeks and only in the presence of David Sox, a well-known fanatic adversary of the authenticity of the Shroud, did he open the containers. Sox spent two days in Zurich, **even in the private home of Wölfli**. Here certainly there was an opportunity to exchange the containers, probably without Wölfli's knowledge. The containers had been procured by Tite, a larger number of them were no problem. These days, Sox cites the significant affirmation that Zurich, where Zwingli at the time fought against the cult of relics, was the right place to unmask the Shroud, The title of the book of Sox: "The Shroud unmasked, the unmasking of the greatest forgery of all time" unmasks the intention in itself. It is noteworthy that this defamatory book had already been printed before the results were published in London.

"A few days after the arrival of Sox in Zurich, another militant opponent of the Shroud, Prof. Gove visited Tucson, which irritated Prof. Gonella. This is further proof that it is not true that institutions did not have contact with each other. Sox said that Gove had arranged for 'suitable' characters to be hired for the radiocarbon test procedure."

"IV. Epilogue"

1. This radiocarbon test is undoubtedly of absolutely zero value. Therefore it is understood that scientists gathered in Paris by mutual agreement have not accepted the results (with the exception of very few atomic scientists). **A new test under "control" is required without the "monopolistic" position which**

Tite claimed.

2. But I see serious resistance against such a repetition:

   a) A new test would inevitably be a serious affront to the three world-renowned institutes;

   b) The test itself, in the case of such a contaminated object, is very problematic;

   c) There are solid arguments for much more precise dating for the Shroud. A summary is found in the fourth part of my book "The Shroud Deception. The manipulated radiocarbon test".

Source: Bulst, Werner, S.J. "L'Enigma Del Test radiocarbonico Della Sindone, Proposta Della Soluzione." *Collegamento Pro Sindone*, September/October 1990, pp. 37-44.

1990 *(October and November)*. Bonnet-Eymard and de Nantes noted that after repeated questioning of Testore and Riggi, Bro. Bruno had discovered that one of the labs had received their sample in two pieces.

Bro. Bruno and a colleague requested to be received at all three labs and the labs did grant permission for on-site interviews. They mainly wanted to resolve the issue of which lab had received their Shroud sample in two parts. They first traveled to Arizona.

On October 26th, Bro. Bruno, following up on Jull's reply to Wolfli in April that Arizona had received their sample in two pieces, asked Donahue about it, he said, "All right, I don't know. I think we received [...] I think -- but I'm not sure -- that we received two pieces, two fragments. But I have no record of that; it's only my memory." But Arizona had reported that it was Donahue who had cut the sample into four pieces before giving them to Toolin, the team chemist, for cleaning. Toolin was listening to the conversation between Donahue and Bro. Bruno. Toolin whispered to Bro. Bruno, "As far as I'm concerned, it was in one single piece." But as he said it while going out of the room and only in a whisper, it was not picked up by Bro. Bruno's tape recorder. Bro. Bruno wanted to hear him say it again so asked Jull to get Toolin to come back in. Bro. Bruno asked Toolin if he had been present when the piece was cut.

Toolin: I was there, yes, when we opened it.
Bro. Bruno: Dr. Jull is not sure whether the sample was one or two pieces when you opened the metal tube. Did you notice whether one of the samples was in two pieces?
Toolin: I don't remember that. No, no.

Donahue, Jull, and Toolin answered all questions, but there were discrepancies in their answers pertaining to several aspects, especially regarding the appearance of the sample when the steel tube was opened.

Damon wasn't at the lab when Bro. Bruno had visited so he phoned Damon on November 1st. Damon told Bro. Bruno that only he and Donahue had been present -- Jull and Toolin were not. Damon claimed they examined the sample microscopically and took photos. When asked if Toolin was there, Damon said, "No, Toolin was not there. Toolin came the next day. The opening took place on a Sunday, and Toolin did not come till Monday, when the others came into work." But both Jull and Toolin certified in the laboratory notebook that the container seals were unbroken when they were opened.

Bonnet-Eymard and de Nantes sum up the conversations at University of Arizona: "If we weigh up the conversations at Tucson, we are bound to conclude that the testimonies of Doug Donahue and of his assistant Jull, on the one hand, and of Paul Damon, on the other, are absolutely contradictory. On the one side: *no records and no photographs*, and on the other, *photographs and TV record*."

When they traveled to Zurich, Wolfli admitted there was an error in the *Nature* report regarding the 1 X 7 cm strip. When Bro. Bruno asked how they could have made an error like that, Wolfli replied, "Well, we were [...] for publication, we were put under heavy pressure in February 1989 at the time of publishing the report in *Nature*. And we didn't have time to check. Tite wrote the dimension in his report, and this was the dimension we agreed in London at the meeting in January 1988 [...]. Nobody saw this mistake, not even those who are not signatories (to the report)." Bro. Bruno commented, "An allusion to the Italians responsible for taking the samples. Today, they give measurements different from the official ones, which they themselves accepted for seventeen months." Wolfli continued, "As a matter of, it was the job, I mean [...]. But I don't want to blame anybody because I should have rechecked with Turin, asking them: are you sure this number is correct? I didn't do this. The only excuse is: we are under pressure, but that is no excuse." Bro. Bruno remarked on Wolfli's incomplete sentence, "Wolfli had Tite's name on the tip of his tongue but he was careful not to pronounce it [...]."

On November 16th, Bro. Bruno and his colleague arrived at Oxford. Hall, who retired in October 1989, apparently took his records related to the Shroud with him. They first met with Robert Hedges, who had been present in Turin, and actually performed the test at Oxford. There was this exchange with Tite, who had replaced Hall: "**First question**: 'Did you yourself precisely measure the samples removed from the Shroud, the dimensions of which are reported in your *Nature* article?" **Answer**: "No, I didn't measure it at all. I was watching the sample being taken. I saw the sample being taken and cut. But I didn't actually take any measurement of it. And therefore that 1 by 7 cm is just a sort of approximate estimate." "**Objection**: But, if you will excuse me, you wrote 70 by 10 mm. An approximation is all right for give or take a millimeter, but not whole centimeters!" **Answer**: "It's a mistake? Right, yet. It was a very rough figure based on my memory, but I mean [...] yes I couldn't say [...]. If Riggi's figures come out with something which is shorter or bigger, then I would tell you: you can go by those!"

Bro. Bruno was still trying to figure out which lab got their sample in two pieces. When they asked Hedges if he remembers, he replied, "Professor Tite will certainly provide you with facts about that. As guarantor and overseer of the whole experiment, as we have said, it was he who put the samples into an envelope, in a separate room in Turin with the Cardinal, so as then to place them in their little steel containers. He will surely remember [...]." *[Or maybe not.]*

Bro. Bruno commented, "Incredible to relate: Tite has no memory of this detail." "I can't remember' [...]. I simply can't remember that at all [...]. I cannot remember whether there were one piece or two, I just can't [...]." We put this question to Tite: "When everything was put on the table, it was yourself who placed everything in these metal foil containers [...]?" He cut in, "That's right. Inside!" "You didn't notice that one of the Shroud samples was in two pieces?" Answer: "I can't remember." "Astonished and doubtful, we reminded him that he alone had packed the samples that were on a plate, in a room apart with the Cardinal." He interrupted: "The Cardinal and Professor Gonella [...] he was there the whole time, as interpreter." "Ah, all right! and Riggi too perhaps?" Tite answered: "He wasn't there. I have a feeling that he may have come in to ask: 'Is everything all right?', when the samples were brought."

Because the spread on the Shroud samples was wider than the control samples, Bro. Bruno asked about the possibility of doing additional tests. Tite said, "In my opinion, there were no other measurements we could take under the circumstances we were in."

Bro. Bruno: But Arizona and Zurich still have a reserve piece available.
Tite: I think [...] I don't know.
Bro. Bruno: But [...] they told us so.
Tite: All right! Fine! OK!
Bro. Bruno: Arizona sent you a letter stating that they still had material left for further measurements.
Tite: I don't remember that.
Bro. Bruno: You were informed of that in time.
Tite: OK. All right. I don't remember.

On November 22nd, Bro. Bruno called Paul Damon to see if he could get any clarifications about the contradictory evidence that the Arizona lab had given them. "We have the photographs now; we've found them! It's Doug Donahue's wife who took them on the Sunday when we took the samples out of their steel tubes. It's the answer to your question. She had given to him in an envelope and she had forgotten to tell him where they were, and that's where we discovered them." Bro. Bruno asked, "But does this photograph show a sample in two pieces?" Damon, seemingly embarrassed, replied, "No, it's we who divided it...er.. in ...er...er.. into several pieces, yes. We have the photographs of these operations." He added, with further embarrassment according to Bro. Bruno, "There was no video because it was a Sunday. We came back on the Saturday and we did that with Jull, Donahue and I, but we took bad photographs." And the video? "We made a video recording after having begun to prepare the samples, on Monday [...]."

Source: Bonnet-Eymard, Bruno and Georges de Nantes. "The Carbon 14 Dating: In Pursuit of the Forgers." *Catholic Counter-Reformation in the XXth Century*, April 1991, No. 238, pp. 3-7 and also Bonnet-Eymard, Bruno. "The Crime Committed Against the Holy Shroud." *Catholic Counter-Reformation in the XXth Century*, Feb-March 1996, No. 283, pp. 6-9.

Comments: It was suspicious that Hall took the Shroud-related records with him into retirement. One certainly is not able to rely on the memory of Tite! It's difficult to try and reconstruct what exactly happened at Arizona. The people in Tucson had bad memories, bad photographs, bad record-keeping -- but we're still supposed to take at face value their results?? Georges Bonani, one of the Zurich C-14 lab scientists, emailed my late wife, Sue, on September 9, 2008 saying, "We have absolutely no spare material. We have used the whole sample for the tests in 1988," which contradicts the fact that someone at Zurich had told Bro. Bruno that they had preserved a sample and Wolfli had also admitted that to Kersten. The representatives of the labs seemed neither confident nor consistent in their answers.

1990 *(December)*. In an interview, STURP member Kenneth Stevenson said, "I was appalled that they were able to get away with calling this standard operating rules: dropping the double-blind controls; refusing to provide their raw data for peer review; failing to have the samples notarized; arbitrarily throwing out data that conflicted with their announced prejudice. According to their own published reports, they *[C-14 scientists]* discarded readings that didn't fit what they wanted [...]. Several of the C-14 team members made public statements before the testing to the effect that 'it's a fake and we're going to prove it,' which tends to taint their credibility from the start."

In the same interview, the author highlighted some of the research by Bro. Bruno Bonnet-Eymard. "Right from the start major discrepancies appeared. The notes of Franco Testore strongly disagree with those of Michael Tite and Giovanni Riggi as to the size of the sample cut from the

Shroud. According to Tite and Riggi, it was seven square centimeters. Testore says it was almost double that: 12.96 centimeters. The participants contradict each other and themselves concerning the number, size, and weight of the pieces cut from the sample. 'Since nothing was checked or certified by notarized acts,' observes Bonnet-Eymard, 'everything is possible and everything is plausible.' Including outright fraud perpetrated by Tite and one or more of his associates. Bonnet-Eymard cites a string of glaring inconsistencies, contradictions, and blatant violations of scientific methodology, not to mention a militant blast against the Shroud, to support his hypothesis of a conspiracy to discredit the linen relic."

"'The most critical point in the whole matter was the placing of the samples into the containers,' says Bruno. 'Tite, with the Cardinal, retired into a separate room. There exists no documentation, no independent witnesses'."

The author also brought up the fact that the Oxford lab had received a one-million-pound donation and that Tite took Hall's place after the latter retired. "So, besides his undisguised animus toward the Shroud, Tite had ample motive to fudge on the C-14 testing: money, prestige, status, security. It is worth noting that all three of the principal characters in the radiocarbon testing -- Tite, Hall, and Ballestrero -- resigned from their positions shortly after the Shroud dating results were made public, thus making it even more difficult to question them and resolve the many serious problems surrounding the whole Shroud/Carbon 14 affair."

Source: Jasper, William. "Science and Faith," *The New American*, **6** (26), December 17, 1990, pg. 28.

Comments: As was previously noted, not long after retiring, Hall refused to answer questions related to the dating.

1991 *(January/February).* Italian journalist Petrosillo reacted to a negative quote by a French monsignor with a mock summary, "It is a medieval fake to be removed forever from the veneration of the faithful. Like a vulgar fake or a banal decoration." The quote from the monsignor: "And the ecclesiastical authorities, in order to cut short any misplaced nostalgia, should have the courage to take for the

Shroud of Turin the decision adopted for the Cadouin sheet". Petrosillo continued, "That is, to hide it, as happened for that cloth served to a twelfth-century crusader to transport some relics from the Holy Land and then exchanged for a relic himself. To express these dry judgments and also some criticisms of the ecclesiastical authorities '**which often shine the most in the art of compromise**', allowing devotion to a false relic, is not an anticlerical or one of those agnostics convinced that the Catholic Church proposes meekness, but he is a prelate, an appreciated historian and archaeologist with two official positions in the Vatican; this is the French Monsignor Victor Saxer, rector of the Pontifical Institute of Christian Archaeology and for just over a year also president of the Pontifical Committee for Historical Sciences, an organism that represents the Holy See as a member of the 'International Committee of Historical Sciences.' This anti-Shroud indictment, supported by the titles of Saxer, was published last year in the 'Journal of Church History in Italy', (see *Collegamento Pro Sindone*, November-December 1989, pp. 46-49) whose management is based in the Vatican, and recently on the no less prestigious '*Revue d'Histoire de l'Eglise en France*'. In the first article, the author declares himself completely sure of the non-authenticity of the sheet -- which appeared in Lirey in France shortly after 1350 -- because it is not mentioned in previous historical documents and because the local bishop, in a reminder of 1389 to the Avignon antipope Clement VII, referring to the judgment of one of his predecessors, defined the Shroud as 'an artificially-produced work whose author was known, who had confessed the wrongdoing'."

"Saxer, without concealing his annoyance at the 'intense and systematic publicity' around the Shroud and its ostensions (in 1978 three and a half million people went to Turin, Cardinal Wojtyla *[the future Pope John Paul II]* at the head), concludes his essay with the sentences proclaimed on 13 October 1988 by the archbishop of Turin, Cardinal Ballestrero. Announcing the response of the radiocarbon dating of the Shroud that placed it in a period of time between 1260 and 1390, the Cardinal had reiterated 'the respect and veneration of the Church for this venerable icon of Christ who remains today -- casting of the cult of the faithful in line with the attitude always adopted'. In reality, the Cardinal had arbitrarily classified devotion to the Shroud as a relic -- attributed by many Popes -- to devotion to the Shroud as an image. In the article for the

French magazine, Monsignor Saxer suppressed Ballestrero's statement by adding a long conclusion to it with a judgment without appeal: the Shroud is to be hidden. This is in striking contrast to the recent position taken by the Holy See on 18 August with which new expert opinions are authorized. Saxer relies on convergence, in his view irrefutable, between the historical documents and the 'verdict of Science' expressed by radiocarbon dating performed by the laboratories of Oxford, Tucson and Zurich. The prelate admits that he has no other competence than the historical one (and in fact speaks of 'forty Shrouds in competition', which is completely false) based on the dossier collected by the canon Ulysse Chevalier after, in 1898, the first photograph of the Shroud had rekindled interest in the sheet by revealing a perfect photographic negative and making the clear features of the face and body of a tortured and crucified man emerge, not perceptible by direct observation -- memorandum without date or signature -- by Bishop Pierre d'Arcis who, in controversy with the holders of the relic, expresses his conviction, supported by 'they said' (*'ut dicitur'*), that the image on the sheet has been painted."

"And Clement VII, changing his opinion several times, gives a little reason to one and a little to the other of the two parts. The foundation of this dossier, already dismantled by the historian Luigi Fossati, (see *Collegamento Pro Sindone*, March-April 1990, pp. 34-41) has collapsed since it has been scientifically proven that the Shroud is not a painting and the hypothesis of the forger who would have wrapped a body with wounds similar to those told in the Gospels can only be accepted by those who, like Saxer, **know very little about the Shroud**. Scientists have already answered negatively to the naive question about the possibility or not of making such a perfect impression. All the wounds of the man of the Shroud were inflicted during life and only the wound at the side was post-mortem with separate blood and serum. How would the manufacturer of that image have calculated the exact time of laying the corpse on the sheet (from 30 to 36 hours) and how could he make it without making smudges on the cloth? The forger should have known the flagellation and crucifixion techniques used by the Romans in the first century and whose memory was then completely lost. He should have found a victim whose face was congruent in dozens of points with the icons of Christ spread in Byzantine art (and not of 'medieval' iconography, as Saxer surprisingly writes). As for the 'verdict of Science' given to the radiocarbon examination, in contrast to other scientific verdicts, I have documented in a written book for Rizzoli together with Emanuela Marinelli, the many perplexities on the applicability of the method to the Shroud, on the scientific reliability of the test and on the unclear aspects of its development to push the Holy See, to new interdisciplinary examinations."

Source: Petrosillo, Orazio. "La Sindone? Un Falso Da Nascondere." *Collegamento Pro Sindone*, January/February 1991, pp. 4-7.

Comments: The "Cadouin sheet," held in the Abbey of Cadouin, in Dordogne, was long considered as the cloth believed to have covered the face of Jesus when he was placed in the tomb. In the 1930s, it was discovered to be an eleventh-century Islamic cloth.

1991 *(February)*. In a letter to *Nature* on February 14th, a French physician wrote "SIR-Damon *et al.* (*Nature* **337**, 611-615; 1989) asserted that radiocarbon dating performed in 1988 provided conclusive evidence that the linen of the Shroud of Turin was mediaeval. However, most of the scientists involved in the studies on the Shroud clearly showed at the Paris International Symposium (7-8 September 1989) that they utterly disagreed with the conclusions of this article; the main reason was the lack of reliability of the results due to several methodological inadequacies. On a matter of such wide interest, it is important that the scientific community should be seen to come to a definitive conclusion about the value of the published dates. For this reason I would like to re-open the debate. As a matter of fact, survey protocols have to be performed according to a method capable of avoiding investigator bias, in order to achieve relevant and accepted conclusions (F. Ederer, *Amer. J. Med.* **58**, 295-299; 1975). In the case of the radiocarbon dating of the Shroud, the procedures were neither blind nor controlled, contrary to the assertions of M. S. Tite (*Nature* **332**, 482; 1988). As a result, the following questions have to be asked: (1) What were the scientific reasons for abandoning the blind procedure and the full documentation by video film and photography? (I showed at the Cagliari conference that a true blind radiocarbon dating was feasible.) (2) What was the methodological need for giving the ages of the control samples to the laboratories before the radiocarbon dating procedure? (3) What were the detailed data of the carbon measurements in. each series of analysis? (4) What were the detailed data of the statistical analysis supervised by the 'G. Colonetti' Institute? (5) What was the scientific reason for asserting without any discussion that the results obtained provided conclusive evidence that the linen of the Shroud was mediaeval, whereas it is in complete disagreement with every result obtained previously by scientists in the past 90 years? All these important questions should have been discussed at the scientific conference held in Cagliari (29-30 April 1990), since its topic was precisely the Shroud dating. It was reasonable to expect some of the scientists involved in the dating performed in 1988 to attend: And above all, the paper to be presented by Professor Hall on 'An attempt to answer criticisms concerning the dating of the Shroud' was eagerly anticipated. Unfortunately, none of the 21 authors of the article quoted previously was present in Cagliari, including Hall. So, the international community of scientists interested in research on the Turin Shroud is still awaiting answers from Hall and Tite to these questions."

OLIVIER POURRAT Clinique Medicate, Univetstte de Poitiers, Hôpital Jean-Bernard, BP577, 86021 Poiters, France."

Source: "Shroud dating still questioned." [Letter], *Nature*, **349**, 558 (14 February 1991).

Comments: Pourrat raised legitimate questions but no satisfactory answers were ever given.

1991 *(March)*. At a conference held at Columbia University in New York, Donahue told the attendees that the sample received by his lab was in two pieces, one weighing 14 mg. and the other 40 mg.

Source: Bonnet-Eymard, Bruno. "The Holy Shroud, Silent Witness." *Catholic Counter-Reformation in the XXth Century*, April 1997, no. 295, pg. 26.

Comments: But the weights given there don't match what Bro. Bruno had noted in Arizona's own notebook that he observed in his visit of October 26, 1990!

1991 *(July)*. In a letter to *Nature* published on July 18th, a U.S. physician wrote, "SIR - Others have drawn attention to serious anomalies in the procedure undertaken for radiocarbon dating the Shroud of Turin (Damon *et al. Nature* **337**, 611-615; 1989). Your readers should know that anxiety about the procedures followed has been heightened by a recent declaration of Professor Wolfli, one of the 21 co-authors of Damon's report. In a short interview published in the French monthly journal *Contre-Reforme Catholique*, Wolfli asserts that the size and weight of the shroud samples mentioned in Damon's paper were erroneous. According to the French journal, he declared: 'Nobody (among the authors) has seen this error. We were under pressure, but that is not an excuse.' So far, this statement has not been challenged in any way by the first author, Damon. Because sampling procedures have always been regarded as critical in the dating of the shroud, this situation is most disturbing. This unique archaeological artefact deserves more serious attention. Logically, scientists who question this procedure should be allowed to review the original records, including the videotapes recorded during sample collection in Turin in April 1988. How can this be achieved?"

PIERRE BUSSON, Lineberger Cancer Research Center, University of North Carolina, Campus Box 7295, Chapel Hill, North Carolina 27516, USA."

Source: "Sampling error?" [Letter], *Nature*, **352**, 187 (18 July 1991).

1991 *(July/August)*. Van Haelst wrote, "Ever since the development of radiocarbon dating by Prof. Libby, experts in the field have tried to prove that their results were beyond doubt, 'Four-star' results with the accuracy of a stopwatch. Despite many 'aberrant data', in conflict with archaeological and/or statistical evidence. As clearly happened with the radiocarbon dating of the Shroud of Turin. Here specialists of dating at C14 are jumping from a somewhat arbitrary 'statistical reliability of 95%' to a 'definitive proof' without any justification [...]. The question remains: can an **objective** scientist rightly **doubt** radiocarbon dating as it has been given by the three laboratories???"

"**International Collaboration Program (1990).** During the Trondheim Radiocarbon Conference, an 'International Collaboration Program' was launched in three **successive** stages (counting, analysis and pre-treatment). The labs received a series of duplicate samples, which they did **not** know beforehand about their identity. Difference values greater than **two** would have indicated that a laboratory was unable to detect the 'presence at duplicates within the limits of their declared precision' of Glasgow, NERC Radiocarbon Laboratory and Scottish Universities Research and Reactor Center. Despite the almost universal NBS standard (the 'oxalic acid standard SRM 4990'), deviations still exist. The program was conducted by 'Scott and others' (University systematic within and between laboratories for the C14 part of Otlet and others (1980) and ISG (1982) identified the existence of systematic deviations and unexplained variability. This failure leads to a loss of resolution and reduced user confidence in C14. The previous practice of an **ad hoc** exchange of samples between laboratories is now rightly **perceived** as **inadequate**. A relevant question is: how do AMS laboratories choose to define and direct representative sampling. Deliberate small-scale pre-selection from a typical natural deposit can reinforce the possibility of a significant deviation from the most representative age. There is clearly a need for particular awareness when dating micro-samples. The quality and homogeneity of these samples is of **utmost** importance. The final report was produced in the early 1990s. The entire report can be obtained from Glasgow Unv. Dept. of Statistics."

"**CONCLUSIONS.** It seems reasonable to consider that a laboratory operates adequately, if it does not have a significant systematic deviation (which should include zero and evaluate internal and external variability adequately, i.e., with error multipliers not significantly different from 1). Of the 80 invited workshops, 58 accepted, Phase No. 1 was completed by 52 laboratories, Phase No. 2 by 37 and Phase No. 3 by 38. In total, 23 of these 38 laboratories **failed** to meet the three criteria basic for an adequate operation in the production of radiocarbon data. This is clearly a cause for much concern. Dr. Baxter said in an interview: 'The AMS used to date the Shroud of Turin came out of the overview very badly' (*New*

Scientist, 30/9/1989, p. 29). The geologists at Lamont-Doherty offered solid proof, published in *NATURE*, of how uncertain radiocarbon dating is!!! (*Time* Magazine, June 1990). Now the question is: do the results of the Turin Shroud dating to C14 meet the new criteria proposed in the conclusions of the 'Collaboration Program'??? I asked this question directly to Dr. Scott, to Dr. Baxter, to Dr. Garwin (*NATURE*) and Dr. Leese (British Museum), sending them my critical analysis and asking them to indicate if and where I was wrong and if the statements made in *NATURE* were still to be considered 'decisive evidence'. Dr. Scott and Dr. Baxter did not reply. Dr Leese avoided a direct response. Dr Garwin, editor of the Physical Sciences of *NATURE*, wrote: 'You are asking me questions that go beyond my ability to answer [...]'. Dr Leese co-authored a recent article in *Radiocarbon*, explaining that evaporation losses could be one possible reason for errors in AMS radiocarbon dating. In view of all these tests, I asked the British Museum if they could still support the claims made in *NATURE*."

"The British Museum replied that the matter was now closed for them and that they could not devote any more time to their staff for this matter and that their part in the correspondence was over. In view of the Symposium on the Shroud of New York, Paul Maloney asked for my 'Critical Analysis'. I sent him some written questions concerning the same questions that remained unanswered that I had addressed to Dr. Tite and Prof. Evin at the Paris Symposium. Coincidentally, Dr. Scott was a member of the C14 commission!!! And therefore Maloney submitted my question and my 'Critical Analysis' to her. Since Dr. Scott did not wish to answer in public, she promised to answer my questions by letter. Furthermore, it takes only a few minutes to perform the statistical analysis following the new proposed criteria. The result will be the exit of Oxford and Arizona slightly out of line. I'm still waiting. The interpretation of the results of radiocarbon dating is a very complicated matter, but the statistical analysis can be followed by any mathematician. Any statistician **must disagree** with the claims '95% reliability' and 'definitive proof'. To conclude, I quote from a personal letter from a professor (Geneva): 'Dear Sir, I have happily received your booklet on dating the C14 of the Shroud. First of all, I would like to congratulate you on the quality of your work. Not only have you confirmed the work of others, but, with a different approach, you have established definitive proof that the measurements of C14 made on the linen of the Shroud are **not** homogeneous and should be rejected [...].' P.S. 'The Statistical Analysis' is mentioned in many new books on the Shroud written by Dr. Baima Bollone, by Petrosillo-Marinelli and by Dr. M. C. Van Oosterwyck (France)."

Source: Van Haelst, Remi. "La Datazione: Al Radiocarbonio Rivista Dagli Esperti." *Collegamento Pro Sindone*, July/August 1991 pp. 44-48.

1991 *(September/October)*. Van Haelst wrote, "In the 1988 yearbook of ETH (Zurich) the radiocarbon dating of the Shroud is described by the Zurich laboratory. The only reference to other studies on the Shroud is an article published in *National Geographic*. It is interesting that a photograph of the Zurich sample is published with the measures 1.8 x 1.4 cm, when it clearly states in *Nature*: 'a strip was cut (10 x 70 mm) [...]' from which 'three samples were prepared, each weighing 50 mg ...'. In fact, the Zurich sample weighed 52.8 mg. One may ask: is it possible to cut the Zurich sample from a 1 x 7 cm strip???? ? -- The photograph of the Zurich sample caused a lot of confusion, since the structure of the linen does not seem identical to that of the flax of the Shroud. The interpretation of the Chi square test is interesting!!! First of all the Chi square test is used to show how valid the results of Zurich are. In fact, here the value of this test is well below critical value. But when the Chi square test for the general results of the Shroud dating was **greater** than the critical value, the report states that this is not **forbidden**!!! In *Nature* it was thus stated: 'The Chi test framework shows that the errors mentioned are unlikely to fully reflect the global dispersion' [...]. When it should have been written: 'The NEGATIVE Chi square test shows that the results obtained by the three laboratories are not HOMOGENEOUS'!"

"We preferred to play with the coefficients [...]. *Nature's* relationship cannot be incomplete, since the doubts were not clearly stated in the 'significance level' of 5% for the Shroud. In fact, there is only a **5%** probability that a dispersion between dates as high as the one observed can be obtained by chance. This means that dispersion is caused by real problems such as sample contamination, equipment setup, etc. If you think of protocol violations when cutting, measuring, weighing and handling specimens! EXPERTS gave five different reports on this. ONE of the labs received **a two-part sample**. It is clear that the whole experiment was conducted in an unscientific way that renders the results **worthless**."

Source: Van Haelst, Remi. "Onesta' Scientifica." *Collegemento Pro Sindone*, September/October 1991, pp. 52-53.

1992 *(January/February)*. Petrosillo and Marinelli recounted, "The pickaxes left and right are now in fashion. And the engineer Gonella, scientific consultant of the then custodian of the Shroud, Cardinal Ballestrero, is no exception. Except that he is exclusively dedicated to sindonologists. His bitterness is understandable. The result of the radiocarbon test was certainly not what he hoped for and expected. The medieval date turned out -- we imagine -- a hard blow for him too. To the disappointment was added the annoyance at being teased or in any case not respected by Tite and by the heads of the three laboratories who have committed more than one impropriety towards him. Gonella's anger is widely recorded in the declarations of August-October '88 when he said it of all colors against the scientists who had not behaved as such, indeed had acted as real 'mafia', ordering an 'anti-Catholic conspiracy'. Even worse, however, the accusations made by sindonologists and of those experts who, having no reverential fears for the carbonists, went on the counterattack by detecting everything that raised doubts before, during and after the test. Gonella soon forgot the violent accusations against Tite and the laboratories for not making a bad impression of rejecting a 'scientific' result (despite all the accusations of unscientific behavior to its authors). And he got angry with the sindonologists. Also because it is easier for him, sindonologists being almost all non-scientists, not versed in the so-called 'hard sciences'."

"Almost all the public servants of the engineer [...], have made violent attacks against the sindonologists and to those who have written on the story of radiocarbon dating. It is understandable that Gonella wants to vent his bitterness and wants to respond to the accusations against him and Cardinal Ballestrero [...]. But from here to take back sindonological convictions proclaimed by himself until recently and, what is worse, to spread falsehood and in turn slander, this is not justifiable. Is it really irrelevant whether the Shroud is or is not of Christ? Gonella is shocked that the scientific debate is blocked by the sindonologists on the question of the authenticity or otherwise of the Turin cloth. And from the top of his 'scientific' chair he ruled, with ill-concealed contempt for the uninitiated, that the only problems that must concern are the mechanism of image formation and the preservation of the Shroud image. It is true that these are two problems of absolute priority but it cannot be denied that together with them, and no less important also from a scientific point of view, there is the problem of authenticity. 'That the Shroud is or is not of Christ is not a scientific problem,' Gonella repeated with confidence and up to boredom. Allow us to reply softly that this is a colossal nonsense. Unless only physics and chemistry are to be understood by science. And where do we put archaeology and history and other 'scientific' disciplines that come into play in the discourse of authenticity? And if it is just like Gonella says, why did he and Cardinal Ballestrero allow the radiocarbon dating that served only to demonstrate the (non) authenticity of the sheet? And then that the Shroud is from the first century or the fourteenth is not exactly the same thing. That 'impossible' object, as he himself declared, is absolutely out of the medieval context. It becomes 'possible' only if it belonged to Christ. Otherwise we should resort to the miracle hypothesis which, paradoxically, in order not to question the 'scientific' result, the Turin physicist tends to prefer."

"The Shroud a dogma? Is the Christian faith at stake? In his reply, not serene, Gonella accuses us of considering the authenticity of the Shroud a dogma. In the heat of statements he does not realize that what he says is crazy and not what he has read in our book. Unfortunately, he did not understand the irony against those who, with a completely illegitimate syllogism, ideologized the medieval result by interpreting the test as 'the plaintive unmasking of an unauthentic relic' and drawing the undue consequence that 'science assigns to the Church the kingdom of myths'. 'As if -- it was our ironic deduction -- the dogmas were attackable by radioactivity'. That is -- for those who have not understood the joke -- any laboratory response about the non-authenticity of a relic remains irrelevant to faith. But not that the Shroud is an object of faith. Not even Bonnet-Eymard, who was considered a fanatic by Gonella, ever thought of it. It is enough to know a little about theology or to be sure of one's own catechism in order not to hypothesize such a slander."

"The Pope and Cardinal Ballestrero: a denial or a 'rascal journalism'? On this subject, Gonella plays the three-card game and misleads the listener. We specify. One thing is the communiqué read by Cardinal Ballestrero on 13 October 1988 and whose text was prepared in Turin and adjusted or revised by the Secretary of State, where it is said that 'the Church leaves the evaluation of these results to science and reaffirms its respect and reverence for this venerable icon of Christ'. Another thing is the 'witticism' of the same Cardinal during the press conference, of the type: 'I would not want anyone to think that the Church is dismayed by the results of the exams. It would be good if the consternations were all there'. Or the joke about the 'miracle' consisted in the fact that the operation did not cost anything, even with the addition that the Shroud 'did miracles and still does'. It is obvious that the Pope's authority is implicated in the first case and not in the second. As for John Paul II's 'denial' to Cardinal 'that's enough', calmly compare their statements. Ballestrero that October 13 boasted like this: 'I never considered the Shroud a relic. And, given the results of the research, I was prophetic'. The Pope, whom I questioned, reiterated: 'Relic certainly is, relic certainly is, (he repeated twice, editor's note) it cannot be changed [...]'. And then, after considering the Shroud as an image, he went on: 'But is it a relic, I think it is. And so also the guardian of the Shroud, Cardinal Ballestrero. He is a closer expert, a witness more near this Shroud'. One can perceive a mile away that the Pope makes a gesture of exquisite kindness in not wanting to underline the difference and in calling the elder, and in many other respects archbishop, to witness something that Ballestrero is not convinced of. And always. As for the 'scoundrel journalism,' which consisted in asking a question like this 'point-blank' (in a press conference) to the Pope, already on his own 'very worried about a trip to South Africa' (actually the South Africa trip had nothing to do with it because we were going to Madagascar) *[Petrosillo was a Vatican correspondent for an Italian newspaper and often travelled with the Pope]*, this is part of Gonella's dramatization. In hindsight, not very respectful of the Pope who is up to having a clear distinction between icon and relic and having had a precise idea of the Shroud after having venerated it twice. One last note -- the entire interview was published by *Collegamento Pro Sindone* five months before 'Newsletter' where the good Gonella read it -- with inevitable adjustments of a translation from

Italian to English -- who imprudently complains that no Italian reported the text of the interview."

"Relic or image? In order to better attack the Shroud scientists and to fully align with the positions of the Cardinal whom *Repubblica* wrongly called 'a rationalist with the purple' *[vestment],* the engineer of the Polytechnic discusses relic and icon, speaking of 'relic in the broad sense', of 'image in the widest possible sense' in his conference of 14 May at the 'Rosetum', a true synthesis of his anti-Shroudologist statements. For Gonella there is no doubt: the Shroud is an image of Christ. Indeed: 'there is only one thing that the Church knows and has always said: that is an image of Christ, meaning by image the widest possible sense'. And who does not think so, according to Gonella, is a 'Lefebvrian'. *[Marcel Lefebvre was a French dissident archbishop who was unhappy with Second Vatican Council reforms.]* We leave aside every comment and remember, with a touch of regret, these statements by the same professor, made to the monthly 'Jesus' (June '87): 'When I approached the Shroud for the first time I was very excited, much more than doing the communion. And that made me reflect. Why did I take that host almost as a matter of course, while that Face made me tremble? Why that difference, between one body of Christ and another? In reality, like all relics, the Shroud is a sign of faith that puts us in direct contact with the past'. Was a test enough, on whose infallibility we at least have the right not to be blindly fideistic, were anti-Shroud controversies enough to throw these beliefs overboard? Hopefully not [...]."

"'There is no proof of authenticity on the Shroud!' In his remote controversy against the sindonologists, at least perplexed about the result of carbon 14, as it was the only one to go in the opposite direction compared to the agreed tests about the authenticity of the cloth, Gonella surprises us for his 'extreme frankness' and declares in no uncertain terms: 'There is no proof of authenticity'. So, with the sufficiency of those who explain almost elementary things to the uninitiated, he makes us aware of the 'big difference' between 'proofs of non-contradiction' and 'proofs of existence', between 'sufficient conditions' and 'necessary conditions'. Very well. We will not endeavor to reply with our own arguments, as non-scientists. We will quote the same scientist Luigi Gonella, who, no later than five years ago, drew these conclusions on all the scientific work carried out on the Shroud. 'What is the result?'

the reporter asked. And he: 'Only agreed results. One always expects the single, striking experience, the revelatory proof from science. While the concordance of the elements, the interdisciplinary mosaic is more probative than the sensational discovery'. *Et de hoc satis*. Let's forget the rest: the fact that nobody had indicated another site for the withdrawal (ask Meacham); that there are no objections to the radiocarbon test (see page 140 of our book *[La Sindone ...]*, ask statisticians and many others); that there are no proposals for new exams (incredible distraction: STURP has filled 177 pages to propose a project of 26 exams for 85 questions)."

Source: Petrosillo Orazio and Emanuela Marinelli. "Le Esternazioni Contro I Sindonologi." *Collegamento Pro Sindone*, January/February 1992, pp. 17-22.

Comments: The authors call Gonella an engineer and also a physicist.

1992 *(January/February)*. Van Haelst wrote, "Since the publication of the report on the radiocarbon dating of the Shroud of Turin (*Nature* **337**, 16 February 1989) many scientists have raised serious doubts about the scientific value of the whole experiment, which was conducted under the supervision of Dr. M. Tite of the British Museum.

"1. Measurements and weights following Riggi - Testore – *Nature*"

Source	Dimens	Raw	Net	Res.	A	B	C	D
Nature	10 x 70				+-50	+-50	+-50	
Riggi 1	16 x 81	540	497	197				
	10 x 70		300		+-53	+-53	+-53	150
Testore 1	16 x 81		300		52	52.8	53.7	144.8
Testore 2			300		52	52.8	39.6+14.1	154.9-14.1
Riggi 2					52	52.8	50.1+3.6	144.8-3.6

"Note: The weights given by Riggi 1 are not in accordance to the established dimensions. 540 should be 300 and 300 should be 159 g. As noted on the scheme, made by Riggi and published by Testore. Total weight 0.497 g, weight (A + B + C + D) = 0.3 g. D = 0.141 g."

"It is physically impossible to divide a 1.6 x 8.1 cm strip into approximately equal parts and obtain a 1 X 7 cm strip!!! None of the laboratories and also the French textile expert Vial, present during the cut, notified of a compound sample. The Zurich laboratory published a photo of his sample (ETH Yearbook 1988). The dimensions are approximately 1.7 x 3 cm, weight 52.8 mg. The textile specialist Prof. Raes (Belgium) examined this photo and was unable to identify the particular texture of the Shroud. The Arizona lab also publishes a photo of a subsample. The French textile specialist Vial agreed that the weaving could be identical to the Shroud here. But Dr. Donahue (Arizona) declared that he was surprised by the brightest color of the sample. The 52.36 mg weight also does not comply with the weight given by Testore. At the symposium of New York (March 1991) Dr. Donahue admitted that the Arizona lab actually received a compound sample, with a total weight of 53.7 mg. (The information on Arizona was given by Br. Bruno Bonnet-Eymard). The Oxford laboratory published a photo of the sample, with no detail on the size or weight. To quote Prof. Gove: 'A process that becomes a rather poor ["shoddy" in Gove's letter to the British Museum, which Van Haelst is attempting to quote] enterprise'."

"2. None of the laboratories reported contamination of the samples."

"This contradicts the observations made by Riggi (1978) also affirmed in his Paris report (p.33). 'The Shroud is rich in biological residues also coming from the animal type (orabatei), about 4-5 units per cm^2.' On page 40: 'Under the microscope you could see dust of all kinds, but no parasites' ... Prof. Testore (p. 61): 'Traces of humidity and the presence of parasites in the dust could be observed'."

"Other scientists, such as Prof. Frei, Maloney, Prof. Raes, Prof. Morano, not to mention STURP [...] found a significant contamination:

pollen, dust, etc., etc. And look at what Dr. Wölfli (private communication) said: 'Therefore all the samples were examined under the microscope in order to identify and remove the foreign material present. In our case no material of this type was discovered'. Oxford found some cotton on the sample, but this was not stated in the *Nature* report. Following also P.H. South of Derbyshire, this cotton was of Egyptian origin and quite ancient (*Textile Horizons*). From points 1 and 2 one wonders if the laboratories really dated the samples cut by Riggi and weighed by Testore."

"3. The blind-test"

"Collaborative exams are done blindly. Following Riggi, for the Shroud even a 'trio of blind tests' (p. 36 of Paris report). But everyone agreed that it would be a 'fake blind test trio'. And therefore, on the advice of the laboratories, the samples were not disintegrated [...] Also Prof. Gove (Rochester) recommended this procedure, to ensure not only the homogeneity of the samples but also a true blind examination. The proof is the dating of the fourth 'unknown' sample which was in the form of threads. One could be sure that such a violation of the protocol will make any legal cause insignificant."

"4. Quality assurance with radiocarbon."

"The results of a collaborative study for the intercalibration of dating laboratories with the ^{14}C (*Radiocarbon*, vol. **28**, n ° 1, 1986) are given in *Antiquity* vol. **64**, No. 243, June 1990, pg. 319. The international community of laboratories dating with radiocarbon has recognized the possibility of significant analytical deficiencies that have influenced some age measurements with ^{14}C reports. This report is written by Dr. Scott, by Dr. Baxter and others [...]. Dr Baxter has even stated[1] that the AMS used last year by a laboratory at the University of Oxford to date the Turin Shroud came out of the investigation badly. The *Time* Magazine investigation (11 June 1990) has published a recent report by scientists from the Lamont Geological Laboratory (USA). I quote: 'Writing in *Nature* they have shown that some dates with radiocarbon can be even 3500 years old -- perhaps it is enough to force a change in current opinion.' Is the same also probable on the value of the radiocarbon dating of the Shroud?"

"5. Statistical analysis"

"The British Museum ensured that statistical analysis was done with data presented in *Nature*. I asked Dr. Leese (British Museum) and Prof. Bray (Turin) **to show me only with figures on the card, how can the Arizona average be verified with the correction of ^{13}C (table 2) from table 1 with the data formed without correction with the ^{13}C.** This is IMPOSSIBLE, although one can make an assessment. The 'weak' statistical analysis, proven by a NEGATIVE X^2 exam, as provided in *Nature*, **cannot be an argument to support the vaunted claims of '95% reliability' and 'conclusive evidence'.** The Oxford data are too far from other data for such small samples taken in the same place. The EEM criteria (External Error Multiplier), recently proposed, for collaborative comparisons, are a good tool to prove this point of view, Only Zurich has given results lower than 1, the maximum EEM. Oxford is out, Arizona to the limit."

"**Conclusion**: despite the fact that the individual radiocarbon data provided by Oxford, Arizona and Zurich indicate a medieval age for the Shroud of Turin, these results are statistically uneven and should not be used further. Radiocarbon data are in contradiction with all the rest of scientific research and historical data on the Shroud. Due to the protocol violations previously declared here, the uncertainty about the sampling, the weight and the measurements, the contamination and the identity of the samples, the X^2 NEGATIVE test, these results cannot be considered decisive tests of a medieval age of the Shroud of Turin.

"I asked Dr. Hedge of the Oxford Radiocarbon Accelerator Unit, to comment on this text. Here is his answer:"

6 August 1991

"Please understand that this constant quibble on the part of those who wish to discredit the dates of the Shroud of Turin means that my answer can only be short."

1. Since we did not participate in the particular comparison to which the observations of Dr. Baxter[1], these are irrelevant. A subsequent letter from my opinion in the *New Scientist* makes it very clear. In any case, the AMS laboratories that actually participated in it came out better than many other radiocarbon laboratories (apart from the "high precision" laboratories. The results of the second comparisons are now available and show that the Oxford measurements are perfectly satisfactory. We did not participate in the first comparison because at that time we were engaged in a research and development program of our system that prevented us from using our typical dating methods)."

2. I really don't follow the details of your argument about the validity of the real samples that have been examined. But I would like to clarify two points: first of all, the statement in *Nature* that the sample was 10 x 70 mm meant only to give a general indication of the quantity taken and was based on a rough visual estimate made by Prof. Tite. Looking back, we realize that it would have been preferable to have given a description that would have identified the cut much more precisely. It is arguing about the differences in weights a few milligrams. None of the weights were made for analytical or identification purposes, but simply to establish the general dimensions of the pieces. Actually, given the circumstances, I think that the weights of the individual samples are in reasonable agreement. I would like to clarify one last point concerning the fact that, as you know, the Shroud has such a characteristic texture that it is perfectly obvious to anyone who works on it whether it is having to do with the Shroud or not. I don't think this will lessen your suspicion that someone has mysteriously replaced another cleverly prepared piece of fabric having an identical texture and an adequate medieval date, but frankly, I doubt there is anything I can say will lessen that suspicion.

Sincerely,

Dr. R.E.M. Hedges, Director, Radiocarbon Accelerator Unit

cc. M.S. Tite

"Note *[by Van Haelst]:* Any scientist, archaeologist or chemist involved in the analysis of specimens would not consider any unauthorized sample. In my profession, where one deals with millions, any discussion between parties is resolved by NEUTRAL laboratories that analyze AUTHENTICATED samples. And believe me, it is not easy to perform ppm-acceptable range to all concerned. If one follows Dr. Hedges, the only important part, remains the result of the analysis with radiocarbon. Also performed on a NON-AUTHENTICATED sample. Calling the observations of other researchers on the Shroud 'quibbles' seems rather 'a desperate act'. Most of this 'quibble' could be solved by submitting full reports to the judgment of a neutral group of qualified referees. But so far these relationships are held in great secrecy. From Oxford, Arizona, Zurich, the British Museum and in the Vatican. In fact, to quote Prof. Gove, Dr. Tite has taken on a responsibility that he and the British Museum could later regret."

"Note 1) Article published in *New Scientist* on 30-9-1989. Baxter says that accelerator mass spectrometry, used last year by a laboratory at the University of Oxford to date the Turin Shroud, presumably the funeral sheet of Jesus Christ, came out of the investigation badly. Five of the 38 participating laboratories used this technique, for which samples weighing a few milligrams are acceptable. The other techniques require grams of the sample. Baxter claims that some of the accelerator labs were wrong when they dated samples that were only 200 years old. Since so little material is used in accelerator mass spectrometry, the effects of chemical pre-treatment are likely to be more serious, says Baxter. 'The samples are probably more prone to atmospheric dust or dandruff,' he said. In light of the results, researchers need to adopt new practices to improve quality control. The frequency with which laboratories have samples blindly examined by others should be increased."

Source: Van Haelst, Remi. "UNA RECENSIONE DELLE RELAZIONI DI PARIGI Sul. Prelievo Di Campione E LA Datazione Al Radiocarbono Della Sindone Di Torino." *Collegamento Pro Sindone*, January/February 1992, pp. 31-38.

Comments: Some of the material in this article was already covered by Van Haelst in one of his previous articles cited, but I felt it was still worth reproducing all of the material in this article, which does include some points not covered previously.

1992 (January/February). In another article in the same issue as the above entry, Van Haelst wrote, "In my first contacts (just after 13 October 1988) with Dr. Hedges and Prof. Hall (Oxford), I came to know that only the carbonists of Oxford can judge their date of the Holy Shroud by radiocarbon. This is the reason why Dr. Hedges refused me and even any other radiocarbon specialist permission to examine Oxford's 'raw data'. Professor Hall had no time to waste on such nonsense. The ^{14}C verdict is absolute. All the observations made by scientists from other disciplines about the lack of data were considered 'quibbles.' To illustrate, **none** of the laboratories even posed the problem that NOT finding any contamination was inconsistent with previous reports. The justified criticism on the arbitrary statistical analysis and the non-homogeneity of the average data was caused only by a 'too optimistic estimate of the errors' quoted by the laboratories involved. After the publication of the Proceedings of the Paris Symposium it became clear that the claims in *Nature* were not in agreement with Riggi-Testore's reports on the cutting and weighing of samples. How could you be sure that the samples released for radiocarbon dating are really the samples cut by Prof. Riggi and weighed from Dr. Testore? Here from a 'distraction' from Dr. Tite while writing the report for *Nature* [...] He could not see the difference between three oblong pieces of + - 1 x 2.3 cm and three almost square pieces of + - 1.6 x 1.4 cm, of which ONE was made up of TWO parts [...]. None of the 21 co-authors of the *Nature* report made any observations. Now the Arizona lab, about **two** years after the facts and **three** different versions, has admitted that they actually received a compound sample."

"On August 6, 1991, Dr. Hedges of Oxford wrote me the letter published in my first article. Let's see what the French carbonist Prof. J. Evin says about this: 'I want to emphasize that any dating with radiocarbon made on NON-AUTHENTICATED samples cannot be taken into consideration'. (Private letter regarding a "secret" ^{14}C exam of samples taken from the Shroud in 1973). Let's go back to Dr. Hedges: 'I would like to clarify one last point. Which, as you know, is the fact that the Shroud has such a characteristic texture that it is perfectly obvious to anyone who works on it if it is a question of the Shroud.' But is this really so obvious for a NON-EXPERT textile, to whom the Shroud is shown only once, that is to identify the Shroud on a small sample of + - 2 cm²? Make a comparison with the text of the report by Prof. G. Raes, a textile expert who examined the Shroud in 1973. 'Due to the limited length and irregularity of the threads I was unable to determine the textile type and consequently it was impossible for me to establish whether the pieces n ° 1 (the Shroud) and n ° 2 (the side strip) are made of different materials '. (Sup. Riv. Torinese, 1976, Ap. B. Rapport d'Analise PL II-II). Therefore a world-renowned textile expert like Prof. Raes (Univ. Ghent - Belgium) was unable to establish the difference between the Shroud and the lateral strip, which certainly are not from the same era. Professor Raes did the tests in his laboratory with the two pieces under his microscope. The dimensions of the pieces were about 1.3 x 4 (Shroud) and 1 x 4 cm (lateral strip). Dr. Hedges assumes that 'a 3/1 diagonal fabric' identifies the Shroud, but Prof. Raes wrote in his report: 'The diagonal weave of the two pieces is similar. It is a 3/1 twill. The type of the two threads seems to be different, especially the threads of the weave'. Gove, you fear Dr. Tite and the British Museum may regret the examination performed on the Shroud."

Source: Van Haelst, Remi. "Quando Gli Esperti Del Radiocarbonio Diventano Esperti Tessli." *Collegamento Pro Sindone*, January/February 1992, pp. 39-41.

1993 *(May)*. Petrosillo and Marinelli wrote, "Gonella permitted a challenge to the official results from 1988: 'the studies of Leoncio Garza-Valdes.' This doctor and microbiologist from the University of San Antonio (Texas), on 18 May 1993 obtained some samples of the Shroud from Riggi in an unofficial way. Faustino Cervantes Ibarrola, ecclesiastical assistant of the Mexican Center of Sindonology, related that Garza-Valdes asked him to ask Cardinal Saldarini for the possibility of examining the Shroud under a microscope and making a small sample in an area certainly stained with blood. On April 21, 1993 the Mexican priest sent a fax to the Cardinal, 'absolutely convinced of its uselessness.'

In fact, on April 30, the answer, signed by *[Msgr.]* Giuseppe Ghiberti on behalf of Saldarini, left from Turin: 'a negative response, as was logical,' says Cervantes Ibarrola. Garza-Valdes, however, does not give up: 'I left for Turin. I could see the Shroud but the new custodian, Cardinal Giovanni Saldarini, let me know that it was not possible. So I contacted Professor Riggi, I went to his house and there I was able to examine under the microscope some fragments of the Shroud taken e.g., 1988. This time Riggi wore a mask, imitated (or invited?) by Garza-Valdes. The Mexican microbiologist found what he was looking for: a conspicuous patina of fungi and bacteria that polluted the Shroud. Cervantes Ibarrola, also present, rejoiced: 'We invite Professor Riggi to dinner and with a magnificent Piedmontese wine we celebrate the result.' And he added: 'We were returning home with a very rich sample of raw materials.' Garza-Valdes can thus announce his discovery at the international symposium organized by C.I.E.L.T. in Rome from 10 to 12 June 1993. He does not name in the report Riggi as the source of the material; he simply says that he has studied 'a fragment of blood and pieces of cloth from the Shroud of Turin, coming from the area where the samples for the 1988 radiocarbon analysis were taken.' Riggi and Gonella minimized the size of the material: only 'waste and threads' ... Riggi, for his part, found it perfectly legitimate to give samples to the laboratories of San Antonio: 'When this new idea arrives, a quick consultation occurs between myself and the Polytechnic and I decide to analyse this, because the idea could seem reliable.' Interesting is the fact that Riggi uses the euphemism of calling Gonella 'Polytechnic'. And anyway, shouldn't he have consulted the new caretaker instead? Not at all: Riggi had no doubts. 'Cardinal Ballestrero had entrusted me with material to be able to do some useful research and therefore I embarked for San Antonio'."

Source: Petrosillo, Orazio and Emanuela Marinelli. La Sindone: Storia Di Un Enigma, (Milan: Rizzoli, 3rd edition, 1998), pg. 236.

Comments: To continue this thread, go to the Petrosillo/Marinelli entry under "1994 *(Fall)*."

1993 *(June)*. At an international Shroud symposium held in Rome, statistician Philippe Bourcier de Carbon listed fifteen failures in the Shroud C-14 procedures: "1) absence of a formal report of the sampling; 2) absence of a video archive on the final steps of the samples packaging; 3) in the official reports, contradictions about the cutting and the weight of the samples by people in charge of sampling; 4) breaches of the protocols initially planned for the operation of dating; 5) rejection of the usual procedure of double-blind test; 6) refusal of the interdisciplinary documentation, which is usual in the procedures for radiocarbon dating; 7) exclusion of acknowledged specialists in the Shroud, particularly American scientists who participated in previous works of STURP; 8) communication to the laboratories, most unusual, of the dates of the control samples prior to testing; 9) intercommunication of results among the three laboratories during the job; 10) disclosure to the media of the first results before the delivering of the findings; 11) refusal to publish raw results of the measurements (requested also with insistence in its official statement by the Scientific Committee which prepared the Symposium in Paris in 1989); 12) non-explanation of the unique isolation of the confidence interval of the measures performed by the Oxford laboratory compared to those made by other laboratories; 13) unacceptable value of 6.4 published in the journal *Nature* for the chi-squared statistical test on the results of the radiocarbon dosage on the Shroud; 14) rejection of any cross-debate on the statistical measures performed; 15) rejection, absolutely uncommon, of the publication of the statistical expertise of this operation, officially entrusted to professor Bray of 'G. Colonetti' Institute of Turin (requested also with insistence in its official statement by the Scientific Committee which prepared the Symposium in Paris in 1989)." Bourcier de Carbon concluded: "Such a remark of deficiencies remains completely unusual in the context of a truly scientific debate, and one can only deplore this exception to the usual ethics."

Source: Marinelli, Emanuela. "The Setting for the Radiocarbon Dating of the Shroud." Presented at 1st International Congress on the Holy Shroud in Spain -- Valencia -- Centro Español de Sindonologia (CES), April 28-30, 2012, pg. 16. www.shroud.com/pdfs/marinelliv.pdf.

1993 (July/August). Van Haest wrote "Ever since the results of radiocarbon dating on the Shroud were announced, the sindonologists have tried in vain to obtain the complete set of data provided by the laboratories at the British Museum. It is important to know that AMS radiocarbon dating is not based on a **single** measurement, but on a number of tests (up to 4), with a certain sequence (between 10 and 20). This means that the Shroud has been dated not 12 times (*Nature's* table 1) but between 120 and 1280 times! This indicates the importance of the data used in the statistical analysis, because the standard error is a function of the number of measurements. The more the measurements, the lower the error, the greater the dispersion. Example: a standard error of 61 based on four tests, sequence 20, indicates a dispersion of results of + - 545 years, assuming a normal distribution -- which will place some of the results in modern times [...]. In the case of a non-normal distribution, some data that falls out may even indicate a date in the era of Christ. This is the reason why the complete raw data should be published by the British Museum and laboratories. Dr. Hedges refused to disclose raw Oxford data, even to other radiocarbon experts [....]. I reworked the statistical analysis, following the data presented in *Nature*. I asked Dr. Tite (British Museum), who declared himself incompetent in the matter during the Paris Symposium 1989. I must say that he kept his promise to seek the opinion of Dr. Morven Leese (British Museum), author of the statistical analysis, who agreed with the results of my work. Part of my work has been used in recent books on the Shroud of eminent sindonologists such as Dr. Van Oosterwyck, Dr. Baima Bollone, Prof. Marinelli *et al*."

"Subsequently, having received some confidential information regarding the shape of the Arizona sample and the combination of the **eight** original data in **four**, I asked the British Museum and Dr. Donahue or some explanations. Dr. Donahue (Tucson) did not reply. The answer of Dr. Tite was rather strange: 'From a strip of about 10 X 70 mm **three** parts of about 50 mg were cut was written on the basis of what was remembered in mind.' Dr Bowman (British Museum) made the consideration that 'approximately 10 X 70 mm' indicates that the measurements are not exact.' Of course I know that about 10 X 70 mm means + - 1 mm, but nevertheless it remains **impossible** to cut **three** samples of about 1.3 X 1.6 cm from a strip of 10 X 70 mm. Oxford and Zurich took pictures of samples of these sizes! Dr. Bowman explained that the difference in measurements and weight could be caused by differences in temperature and humidity! He did not comment on the fact that Arizona actually received a **two-part** sample. Dr Hedges (Oxford) agrees, and now realizes it, that it would have been better to have provided a more accurate description of the samples. In a subsequent letter he asked me not to publish this statement [...] letting me know that our correspondence was over! Dr. Bowman and Dr. Leese provided a different reading regarding the application of the ^{13}C correction and how the result 646 (table 2) was obtained from Arizona with the ^{13}C, based on data presented without the ^{13}C in table 1. Based on the data in table 1, the same data 646 is obtained, it cannot be corrected. You should get around 639. The easiest way to explain it was to show in writing how it was done. Dr. Leese also explained that **four** Arizona data (*Nature's* table 1) are actually independent data, obtained from the combination of **four pairs** of dependent data. (The same standard samples were used). But she didn't explain why this wasn't noted in *Nature* [...] and kept secret until an Arizona scientist revealed it [...]. To demonstrate the importance of publishing all the data, I'll do the basic t-test, which indicates the data. In my Perry 'Chemical Engineers' Handbook, my professional bible for many years, it is clearly written: 'Before a statistical test is applied, the level of significance must be selected.' Original Arizona data (not published in *Nature*); all calculations are based on dispersion."

Maximum t value for 8 data and 97.5% confidence = 2.365

Maximum t value for 4 data and 97.5% confidence = 3.182

606	574	753	632	676	540	701	7.01	Average 648 S.E. 25.56
t 1.64	2.89	4.11	0.62	1.10	4.22	2.08	2.08	
591	690	606	701					Average 647 SE 28.26
t 1.98	1.52	1.45	2.08					

Example (648-540) / 25.56 = 4.22

"Since the calculated value of t, based on 8 data, is **greater** than 2,365, the assumption that the true average is equal at 648 it is rejected. Based on 4 data, the calculated value of t is **less** than 3,182 and therefore the hypothesis that the true average is equal to 647 is not rejected. This clearly shows why the original Arizona data was not published! The same applies to the comparison of the final results of each laboratory with the general average. Without any scientific justification one jumps from 16, the error obtained on the basis of the cited errors (Wilson-Ward) to 31, calculated on the basis of the dispersion. Notwithstanding this mixture of 'selective' methods, the value 6.4 of the chi-squared test is even greater than the critical value 5.99 for 95% confidence. A value of 6.4 of the chi square test is not in agreement with a significance of 5%, the minimum acceptable level (Table 2 of *Nature*). Dr. Leese doesn't comment on this. On each occasion I asked Dr. Leese to show me her calculations in writing. Instead of this, I only received some undefined information. Eventually I received a letter from the British Museum Council stating, to my surprise, that I had received complete answers to my questions and that our correspondence had ended. In view of a conference to be held before the International Symposium on the Shroud (Rome 1993), I asked Dr. Leese to allow me access to the British Museum documentation. Dr. Leese agreed on the condition that I get permission from the labs. I contacted all parties involved. None of the three laboratories, nor Msgr. Saldarini of Turin, nor the Vatican replied to my letters. The only positive reaction came from Prof. Bray of Turin. He gave the British Museum permission to publish their correspondence regarding his review of the statistical analysis made by Dr. Leese. Professor Bray was in fact one of the few who saw the report made by the British Museum. But despite its positive reaction, the British Museum has so far not published correspondence. To conclude, I want to state my questions to the British Museum."

    A. The reason why the blind test was abandoned. Blind testing is the basis of all comparative investigations, as it "limits" unconscious and/or conscious influences. In fact, identification of the samples should be done **only after** the publication of the results. The reason for abandoning the test as blind provided in *Nature's* report, "anyone could recognize the Shroud" is not serious. Prof. Raes, a textile expert, who examined the Shroud in his laboratory, could only establish small differences between the Shroud and the lateral strip.

B. An explanation of the differences in size, weight and shape of the samples. It seems that we have forgotten that the **certification** of the collection and treatment of the samples is the most important part of any analysis.

C. An explanation of the fact that none of the laboratories reported a contamination of the samples, while **all** the other reports (Raes, Frei, Maloney, and also Riggi-Testore) mention the contamination (pollen, fungi, ...). Maloney spent more than 500 hours of research on Frei's samples. Dr. Wölfli assured me that he found no contamination on the Shroud sample of Zurich at the microscope. One may wonder why he made absolutely no reservations. The presence of an "ancient Egyptian cotton" thread in Oxford and a red thread in Arizona are also **not** mentioned in the *Nature* report.

D. The publication of all the **raw** data of each laboratory. One can only hope that the British Museum and the laboratories involved will eventually work together once and for all to establish the truth.

Source: Van Haelst, Remi. "Il British Museum All Fine Aprira' I Suoi Archivi Sindonici?" *Collegamento Pro Sindone*, July/August 1993, pp. 29-33.

Comments: Although the results were finally released in 2017 (but only because of a Freedom of Information Act request) and finally publicized in March 2019, the plea for this by Van Haelst is still worth producing to give pertinent facts that were known back in 1993. In 2018, Italian physicist Paolo Di Lazzaro noted that the three labs, "have always refused to provide the exact distribution of raw data. This is the only case I know of authors of an article refusing to provide data that would allow other scientists to repeat the calculation and verify whether it was done correctly" (https://www.lastampa.it/vatican-insider/en/2018/05/03/news/doubts-about-the-age-of-the-shroud-experts-re-open-case-1.34012898).

Several years before Hedges ended their correspondence, Van Haelst had requested to see Oxford's notes regarding their samples. Despite the fact that it's normal for labs to release their raw data, Hedges replied to Van Haelst "that I will not hand over my notes, even to a radiocarbon specialist [...] because it would take too much of my time" (pg. 2 of the Van Haelst draft document referred to earlier).

1993. The late Prof. Jerome Lejeune of the Pontifical Academy of Sciences was interviewed by Italian journalist Stefano Paci.

Paci: So the British Museum and the other scientists were wrong [...].
Lejeune: There is no doubt about it. The Carbon 14 dating by the three Laboratories does not give the age of the Shroud of Turin. Their dating (1260-1390) is in disaccord with the historic certainty that between 1100 and 1200 a painter saw all the details of the Shroud today kept in Turin, including the burn holes which are not at all interesting from the artistic point of view.

Paci: The C14 dating was authenticated by the British Museum authorities. Your criticism of it is likely to cause controversy.
Lejeune: That wouldn't be surprising -- it happens often in the scientific world. Something gets published and then they realize it is not true. The errors of science can also be made in good faith. But sometimes certain tricks are used. The British Museum itself fell afoul of them. For 20 years it exhibited the so-called "Pildaur Man" *[this was actually "Piltdown Man"]* whose image appeared on every book on evolution. But in the 1950s they realized it was a fake. It had been covered up by British Museum authorities who had attested to its authenticity and that error spread throughout the world.

Paci: Many Carbon 14 tests, including some by the same laboratories commissioned to date the Shroud, have given absurd results. There has been some criticism of the excessive weight ecclesiastical authorities gave to that one experiment, instead of integrating it within a series of inter-disciplinary examinations. Do you share that view?

Lejeune: That poor custodian of the Shroud at the time, Cardinal Anastasio Ballestrero! He knew nothing at all about Carbon 14. He was obviously not an expert on it. What he said of the Shroud before or after that experiment is not important. It is with respect that I say that because a Cardinal is not an expert in Carbon 14.

Source: Paci, Stefano M. "All Those Carbon Errors." *30 Days in the Church and in the World*, No. 9, 1993, pp. 60-63, on pg. 63.

Comments: If the labs had done the appropriate chemical analysis and found that the sample was spurious, what were the odds that they would have made it known instead of just proceeding?

1994 *(Fall)*. "It was the autumn of 1994. Riggi arrived in San Antonio with a supply of samples of the Shroud, including particles of blood. Was Saldarini aware of the delivery of Shroud material to San Antonio? 'It appears to me,' says Gonella, 'that Riggi advised Cardinals Saldarini and Ballestrero of his journey.' From Saldarini's statements, though, it didn't seem so. Referring to those samples, he says firmly: 'I have no documentary evidence that they took them'."

Source: Petrosillo, Orazio and Emanuela Marinelli. La Sindone: Storia Di Un Enigma, (Milan: Rizzoli, 3rd edition, 1998), pg. 237.

Comments: To continue this thread, go to the Petrosillo/Marinelli entry under "1995 *(September)*."
 This is another good example of the Church's left hand not knowing what the right hand was doing.

1994 *(November/December)*. Van Haelst wrote, "To judge the measurements made in the laboratory, any prejudice must be abandoned. Theoretically, there should have been no difference if the Shroud's radiocarbon dating had been coordinated by the Pontifical Academy of Sciences instead of the British Museum. **Any** scholar, evaluating the results in a scientific manner, must reach the conclusion that the data reported in the semi-official report of *Nature* (February 16, 1989, vol. **337** No. 6200) are **not consistent**. According to the code of scientific ethics, the full expert report with the **raw** data was to be published. This means that the radiocarbon data for the Shroud had to be examined in the **same** way that British Museum experts assessed the 'intercomparison test' in 1983 (*Radiocarbon* **28**, pp. 571-577, 1986. Authors: Burleigh, Tite and Leese). In this examination, sample 2, a piece of Peruvian cotton dating back to 1200 AD, was dated by radiocarbon only to 362 + 94 RCY (radiocarbon years). Despite the fact that **no** points were **reported outside**, the result was **rejected**. The British Museum scientists themselves did not reject radiocarbon data for the Shroud, despite the fact that statistical analysis shows that the data for the Shroud are **incompatible**. The statistical analysis of *Nature* data is based on a small number of measurements with a large dispersion 7.13 which is greater than the value of *Nature* 6.4, (both above the critical value 5.99). Due to the limited number of data, it was decided not to reject any results (letter from Dr. Tite)."

"Scientifically it had to be that the results were the 'definitive proof' of the medieval age of the Shroud. In *Nature* we see how radiocarbon data for the Shroud are influenced to obtain the declared levels of 95% more cautious in declaring confidence and 5% of significance. During the C.I.E.L.T. Shroud Symposium held in Rome in 1993 I demonstrated this using a computer program. Any statistician will come to the same conclusion. Even Dr. Leese of the British Museum agreed that my calculations were correct. But she refuses any public admission, she is being a chemist. I checked every step of the radiocarbon date according to standard procedures (ASTM and API) to ensure good sampling and compatibility of the results between the different laboratories. There is clearly a contradiction between the descriptions of the samples found in *Nature* and in the reports by Riggi and Testore. Even Oxford agrees on this point. The justification: 'The *Nature* report was written from memory [...]'*. This means that radiocarbon dating was done on uncertified samples, which is scientifically unacceptable. About three years ago Oxford dated an artifact found in Natal (South Africa) as having 750 radiocarbon years. In the report, Oxford stressed that this could be considered as dating of the painting only if the composition of the pigments were known and thus the radiocarbon date could be related with the carbon source in the pigments themselves. The Natal museum published the date without taking into account the 'caveat' (warning) of the Oxford experts. Later it became known that the artifact had been created in recent times by an old lady, Joan Ahrens, from whom it had been stolen [...] and subsequently thrown into the bush. This was a topic of considerable irritation for Oxford [...]. The story on Ahrens appeared in the *Oxford Star* on April 11, 1991."

"Dr. Hedges commented, 'We dated the material on the rock. It was all a bit unfortunate.' Dr. Tite said, 'We warn the client that the date has no meaning until the material has been identified.' For Oxford experts, the pigments used were 40,000 years old. According to the *BSTS Newsletter* (N. 28, 1991) the paint used by the seventy-two year old lady was based on wheat [...] certainly not reaped 40,000 years ago! Since the pigments were of petrochemical origin, for which the radiocarbon age is at least 40,000 years, a mixture of these with modern oils and paints will provide a radiocarbon date in any era. If you recalculate the radiocarbon dating from Natal, you will find that about 8% of 'ancient' pigments are mixed in modern oils and paints. Oxford has learned from this 'embarrassing' episode to give greater prominence, when communicating a date, to the warning regarding the possible problems that exist in linking the radiocarbon measurement, especially on unknown material, with the 'real date'. This is the spirit of true science. One may wonder why Oxford does not publish such a 'caveat' for the dating of the Shroud, but still continues to affirm that this dating is absolutely accurate..."

Source: Van Haelst, Remi. "Un Caveat Riguardante La Datazione Radiocarbonica Della Sindone," *Collegamento Pro Sindone*, November/December 1994, pp. 41-43.

Comments: With the eyes of the world on them, there was no way the labs were going to throw out the Shroud date despite the anomalies as they had with the Peruvian cotton. It paid off for them -- radiocarbon dating is now an approximately four-billion dollar-per-year industry. There is no doubt it would not be worth as much if they hadn't come up with a firm date for the Shroud.

*One of the most important scientific reports of all time – and it was written from memory????

1995 *(September)*. Petrosillo and Marinelli wrote, "During a press conference held in Turin on 5 September 1995, the Cardinal officially denied that they are in new experiments on the sacred linen, specifying that 'no new material has taken place on the Holy Shroud after April 21, 1988 and the Shroud Custodian does not know that residual material from that sample may be in the hands of third parties.' If this material existed,' added Saldarini, 'the Custodian remembers that the Holy See has not given anyone permission to keep it and make any use of it and asks those concerned to put it back in its hands. Since there is no degree of security regarding the belonging of the materials on which said experiments on the Shroud sheet would have been carried out, the Holy See and the Custody declare that they cannot recognize any serious value to the results of the alleged experiments; this obviously does not apply to research undertaken with material taken with explicit authorization from the Custodian during the October 1978 exams. Gonella was surprised: 'We are talking about unauthorized investigations. I specify that the protocol of examinations established in '78 and '88 provided for the total freedom of research.' For his part Steve Mattingly, director of the microbiology laboratory of the University of San Antonio, declared himself willing to return what remained after the tests conducted there in Texas. But there must not be much left, since the return did not take place..."

"Nine years after the stormy dating, Ballestrero returned to the aforementioned interview with Cavaglia. 'I don't think,' the elderly Cardinal said, 'that there may have been a fraud in the three analyses that were made by the three chosen institutes. I am rather convinced that the necessary diligence in the procedure that had been agreed upon was not observed.' These analyses, in fact, are terribly conditioned by all the treatments that the finding submitted to the examination has undergone over time. If I submit to the examination an artifact that has just been discovered, which has been buried for centuries, the artifact has no accessory pollution: it is as it is. But when a find has undergone so many events over the centuries as in the case of the Shroud, it has experienced travel and transfers, it has been burned in a fire, it has been boiled, it has been exposed to worship and therefore exposed to the smoke of candles and incense, at atmospheric conditions (humidity, heat, light...), it is clear how all these events have unquestionably left traces on the Shroud relic. Then it was necessary that, before proceeding to the analysis of the sample it was carefully decontaminated from all subsequent manipulations that it had undergone, with analytical procedures, which are also possible. But perhaps these scholars have proceeded to do this with a little lightness, for an excessive trust in their techniques. And this, in the opinion of not a few, would make the results of the analyses unreliable. The Cardinal, who in 1988 boasted that he had never used the term 'relic' on the Shroud, now made an intriguing statement: 'I, instinctively, think that it is authentic and therefore I understand well how science tries to realize how. The events around the Shroud are such and many and most unknown. Today many bring as an explanation the physical fact of the Lord's resurrection, this renewal of life in a corpse, was certainly a physical phenomenon, of transcendent origin, but physical! Because the flesh of a dead man has become the flesh of a living person. Then they say: the power to revive the body has also rejuvenated the fabric in which it was wrapped. The idea is fascinating, isn't it? The technical analysis of the fabric is impressive. There is no doubt that fabric as a weaving and spinning technique is contemporary with the time of Christ. The fact of the negative is incontrovertible. Explain it though! Those scientists who are right: those who deny the Shroud's authenticity must explain to us how this hypothetical forger managed to obtain a negative, in times when nothing was known about the negative. And more so still three-dimensionality! But then the striking correspondence between the Gospel description of the Passion, of the torture, of the crown of thorns, of the crucifixion, of the piercing and of the Shroud data: it is a great miracle'."

"Regarding the origin of the Shroud, Giuseppe Ghiberti, professor of New Testament Philology at the Catholic University of Milan, asked Ballestrero how he imagined the Shroud was created, the ^{14}C verdict had proved final. The Cardinal replied: 'Why do we not want to include the supernatural intervention of God in the list of possible causes? Nor would it be the first time that we have to seriously resort to this hypothesis: we also think only of the origin of the image of the Madonna of Guadalupe.' It is useless to resort to similar hypotheses of miracles, even in medieval times, when the study of the sheet brings with reasonable certainty to tomb of Joseph of Arimathea where the body of Jesus of Nazareth was wrapped in a 'linen shroud' (Mt 27:59); even if the type of footprint postulates the occurrence of a scientifically-unlikely and even less-verifiable event."

Source: Petrosillo, Orazio and Emanuela Marinelli. La Sindone: Storia Di Un Enigma, (Milan: Rizzoli, 3rd edition, 1998), pp. 237-239.

Comments: In addition to Ballestrero's remarks from above, he would soon be saying he believed there was a Masonic plot involved, which will be detailed further below.

The "image of the Madonna of Guadalupe," also known as "Image of Our Lady of Guadalupe," is believed by many Catholics to be the miraculous imprint of the mother of Jesus, formed on a cloak in 1531 in Mexico.

1996. Author T.W. Case wrote, "The widely reported '95% chance that the Shroud was made between 1260 and 1390 A.D.' sounds impressive, but it is the result of statistical sleight of hand [...]. It all amounts to internal massaging of numbers which hides certain warning signals. In fact the wide range of dates among the three labs obtained in the Shroud sample as compared to the much narrower range in the three control samples indicates that the Shroud test gave an anomalous result. The report in Nature hints at the problem when it notes (in table 2) that there is only a 5% probability of attaining by chance -- a scatter among the three dates as high as that observed, under the assumption that the quoted errors reflect all sources of random variation. In plain English this means that all the statistical manipulation in the world can't get rid of the fact that the range of dates is much too large to be accounted for by the expected errors built into radiocarbon dating [...]. And since the samples were taken from the same tiny area, the range of dates most probably means that all you have to do is go one or two millimeters up the sample, closer to a scorch mark, or perhaps within an area containing a restoration thread or two, to throw off your results a couple of hundred years or more -- perhaps much more."

Source: Case, T.W. The Shroud of Turin And The C-14 Dating Fiasco: A Scientific Detective Story. (Cincinnati: White Horse Press), 1996, pp. 32-33.

Comments: Although it's likely that a restoration thread or two would not have accounted for the approximate 1,200-year difference needed to bring the dating to a 1st-century range, a larger repair certainly would have.

1996. Even though Riggi had given assurances that the excised C-14 samples given to the labs were free of foreign threads, The University of Arizona documented both red silk and blue satin in its sample.

Source: Petrosillo, Orazio and Marinelli, Emanuela. The Enigma of the Shroud: A Challenge to Science, (San Gwann, Malta: Publishers Enterprises Group), 1996, pg. 86.

Comments: The University of Arizona lab had conducted eight separate C-14 tests on the Shroud samples they had been given. But there was such a wide variance in the computed dates, the team in Arizona combined the data to produce four results, thus eliminating the more outlying dates (possibly they did so at the request of the British Museum, which was overseeing the tests). As noted above, Van Haelst documented that the results failed to meet the minimum statistical standards of the chi-square test. Questions to ask about the Arizona results are: Why the wide variance in the dates? Was it because of testing errors? Or was it because the sample was not sufficiently homogeneous?

1996. STURP chemist Adler, in discussing a graph that illustrates the absorbance patterns of image, nonimage, radiocarbon warp, waterstain, scorch, and serum single fiber samples, wrote, "The patterns [...] are all distinguishably different from one another, clearly indicating differences in their chemical composition. In particular the radiocarbon samples are not representative of the non-image samples that comprise the bulk of the cloth. In fact, the radiocarbon fibers appear to be an exaggerated composite of the waterstain and scorch fibers, thus confirming the physical location of the suspect radiosample site and demonstrating that it is not typical of the non-image sections of the main cloth [...]."

Source: Adler, Alan D. Updating Recent Studies on the Shroud of Turin. In M.V. Orna (Ed.), *Archaeological Chemistry: Organic, inorganic and biochemical analysis* (pg. 225) *ACS Symposium Series*, vol. **625**, 1996. Washington, DC: American Chemical Society.

Comments: Adler was very clear that the C-14 area of the Shroud was different than the main part of the Shroud. This is another example of why the lack of chemical characterization by the labs was so important -- and should not have been ignored.

1996. Adler also stated, "So you can talk all you want about how reproducible the date is, but you can't talk about how accurate it is. You have no way of knowing if the area you took the C14 sample from represents the whole cloth. That's an area which has obviously been repaired. There's cloth missing there. It's been rewoven on the edge. They even cut part of it off, because it was obviously rewoven on the edge. The simplest explanation why the date may be off is that it's rewoven cloth there. And that's not been tested."

Source: Case, T.W. The Shroud of Turin And The C-14 Dating Fiasco: A Scientific Detective Story. (Cincinnati: White Horse Press), 1996, pg. 73.

Comments: If all possibilities for a data point are not explored, it is unscientific to make conclusions. Evin (at Paris Symposium in 1989) and Adler both firmly asserted that the labs didn't consider the possibility that their sample had been a repair. Ray Rogers would later say that all of the scientific data needed to make a conclusion simply wasn't gathered.

1997. Italian author Ernesto Brunati put forth some questions regarding the labs' statistical analysis to both the labs and the British Museum but did not receive satisfactory answers despite various letters sent to them as well as numerous publications on the subject by Brunati, who maintained that the Shroud samples were not homogeneous and who suspected a deliberate manipulation of the data. Brunati's calculations regarding the non-homogeneity of the samples were confirmed by two professors of statistics at La Sapienza University of Rome, Livia De Giovanni and Pierluigi Conti.

Source: Marinelli, Emanuela. "The Setting for the Radiocarbon Dating of the Shroud." Presented at 1st International Congress on the Holy Shroud in Spain -- Valencia -- Centro Español de Sindonologia (CES), April 28-30, 2012, pg. 27. www.shroud.com/pdfs/marinelliv.pdf.

Comments: Belgian chemist Remi Van Haelst was another who consistently questioned the statistical analysis, and wrote the labs as well as published many papers between 1990 and 2002 about the dating. Like Brunati, he also did not receive satisfactory answers. But in an Italian book published in 2011, Timothy Jull of Arizona admitted, "This is a bad level. Normally, with such a result, I make the measures again." When a rogue date is dismissed on a normal object, no one cares, but a rogue date on the Shroud with skewed results would garner worldwide attention. There is absolutely no doubt that the labs would not have dismissed their Shroud findings as rogue because of the negative publicity they would have received.

1997 *(September)*. In an article published in the German newspaper *Die Welt*, Ballestrero made some surprising comments in light of what he had said right after the dating in 1988. The article stated, "The Turin Shroud is, in Cardinal Anastasio Ballestrero's opinion, authentic. The laboratory tests conducted in the 80s, which dated the cloth back to the Middle Ages, would appear to have been performed without due care, declared Ballestrero in an interview. At the time, the Cardinal had himself published the results of this research. After the publication of these results, criticism was swift to follow. Under the title (translated from the Italian) 'A masonic plot against the Holy Shroud?' *Corrispondenza Romana,* dated September 20, provided additional information. The main outline of this article is given below."

"Cardinal Anastasio Ballestrero, the former archbishop of Turin, in an interview with his private secretary, Father Giuseppe Caviglia, which appeared in the latest issue of the Carmelite review and was reproduced by the paper *Avvenire* (September 4, 1997), declared that he strongly suspects free-masonry of playing an important role during the scientific research which led to the surprising announcement of October 13, 1988 which denied the authenticity of the Holy Shroud. The Cardinal declares that he is 'convinced' that at the time 'proper care was not taken in the set procedure' (....) 'With the examinations that I had myself authorised, as soon as the solemn exposition (of 1978) was over, science became unleashed and centres for study of the Shroud shot up everywhere (...) for the most part in protestant countries (...), and this context gave rise to the most insistent requests for an examination to be conducted using carbon 14. At the same time, vicious calumny about the Church was purposely being spread around, accusing it of being the enemy of Science because it feared the truth and was frightened of losing the relics from which it made money (...)'. At this point, Father Cavaglia asked. Cardinal Ballestrero whether free-masonry had not played a certain role in all this campaign. 'Without question', came the Cardinal's reply."

Source: "Cardinal Ballestrero challenges the carbon 14 dating test of 1988." *Revue Internationale Du Linceul De Turin*, No. 6 (Autumn 1997), pg. 28.

Comments: Ballestrero's opinion about the lack of due care echoes Jull's statement in the previous entry.

1998. Piero Savarino, the scientific advisor to a successor of Cardinal Ballestrero, stated, "The reported considerations allow us to affirm that the problem connected with the radiodating of the Shroud is open and that the results of the 1988 exams, although representing a step in the complex scientific and historical history, cannot be considered axiomatically conclusive. The wide margins of uncertainty, the lack of a precise knowledge of the aging behavior of cellulosic fabrics stored in historically uncontrolled conditions, suggest a necessary prudence; it is also necessary to carry out a serious, systematic and multidisciplinary study of all the complex parameters capable of giving rise to reactions and modifications to the system before issuing definitive sentences. The controversies against authenticity or the passionate defenses in its favor are and have been numerous. As we have seen in this communication, the studies carried out on the Shroud document a serious examination of a complex problem. Many topics have been widely debated and are sufficiently clear. Still others present margins of uncertainty and need further future investigation."

Source: Savarino, Piero. "La radiodatazione dell Sindone." In Sindone E Scienza -- Atti III Congresso Internazionale di studi sulla Sindone -- a cura di Pier Luigi Baima Bollone -- Maruizio Lazzero -- Carolina Marino (CD-ROM).

Comments: So why does the Vatican continue to allow the old results to stand and not permit new testing??? Perhaps they're content to leave it a mystery, but millions of people no doubt want to see further analysis.

1998. Savarino also made a startling statement in a booklet that he co-authored, in which he stated that the C-14 results may have been erroneous due to "extraneous thread left over from 'invisible mending' routinely carried out in the past on parts of the cloth in poor repair." He went on to emphasize, "[I]f the sample taken had been the subject of 'invisible mending' the carbon-dating results would not be reliable. What is more, the site from which samples actually were taken does not preclude this hypothesis."

Source: Savarino, Piero and Bruno Barberis. Shroud, Carbon Dating and Calculus of Probabilities, London: St. Paul's, 1998, pp. 21-22.

Comments: Savarino's use of the phrase "invisible mending" shows that the Turin authorities, at the very least, should have been cognizant that the Shroud has undergone many repairs. Again, the Vatican could open the door to a solution by allowing new testing.

1998. Adler wrote, "A recent investigation comparing STURP sticky tape sample fibers with those of the radiocarbon sample by Fourier Transform Infrared Microspectrophotometry and also Scanning Electron Microprobe Spectroscopy demonstrated a clear difference in the chemical composition of the radiocarbon fibers from those of the various types of Shroud fibers." Adler also found "large amounts of aluminum in yarn segments from the radiocarbon sample, up to 2%, by energy-dispersive X-ray analysis."

Source: Adler, Alan D. and Alan and Mary Whanger, "Concerning the Side Strip on the Shroud of Turin," http://www.shroud.com/adler2.htm.

Comments: The aluminum finding is significant because it has not been found anywhere else on the Shroud and was later confirmed by chemist Ray Rogers in 2002. This is additional evidence that the labs did not chemically characterize the samples.

2000 *(March)*. In the first paragraph of a published Proceedings paper, Meacham asserted, "[T]he sampling, testing and interpretation done in 1988 were certainly very badly designed and executed." Meacham had plenty more to say:

"As an archaeologist who had used radiocarbon dating on a regular basis for the last 30 years, my own position is that nothing has been proved about the age of the Shroud. However, as someone who has debated the Shroud ^{14}C results with colleagues and scientists of various backgrounds (see for example my debate with a practicing radiocarbon physicist at http://www.shroud.com/c14debat.htm), it is clear that, at this stage, further debate on the subject is largely a waste of time and energy [...]."

"The first point I would like to make is that the ^{14}C dating of the Shroud, like the relic itself, is unique. This is a point that I harped on repeatedly in the years preceding the actual test, but alas, to no avail. As there is no provenance for this extraordinary relic, and there are no similar or associated objects that can be dated, there is no way to see confirmation or refutation of the ^{14}C dates in the normal archaeological manner. Furthermore, if it is really 2000 years old, it has been handed down over most of that time and kept in extremely different environments. And it was partly burnt several centuries ago. I know of no other object ever dated by ^{14}C that has such a history."

"Another aspect of the Shroud dating that was unique, highly irregular and unfortunate, was the fact that the person or group (STURP) studying the object were not the submitters and interpreters of the result. The British Museum stated flatly at the outset that it was not undertaking a study of the object, and its official role was that of coordinator and notary -- a role which, I might add, it carried out spectacularly poorly. It was thus an oddity and a fiasco to see Cardinal Ballestrero and the directors of the ^{14}C labs vying for the interpretive role when the results were announced. It was clear that this was going to happen, and in a 1987 letter to Luigi Gonella I wrote:"

"'The major problem with the entire Shroud ^{14}C issue seems to me to be that, unlike all other archaeological and museum ^{14}C dating, there is no person or body officially collecting and submitting the samples [...]. The labs seem to have put themselves in charge of the entire operation'."

"I had made a similar warning in my submission to Cardinal Ballestrero in 1985. It may not have seemed so important to them at the time, but many of the errors that were made in 1988 were brought on by this highly irregular testing programme."

"The ^{14}C results on the one sample that was taken from the Shroud and split into several pieces are not conclusive proof of anything, and this fact should have been stressed at the time the results were announced. 'Rogue dates' are common in archaeology and geology (see http://www.shroud.com/ meacham.htm for a discussion), and they are usually not subjected to any further detailed study. Instead, the normal practice would be to seek more and better samples, obtain new ^{14}C dates and review the overall clustering pattern indicated by the dates. Such has been my experience as an archaeologist who has excavated, submitted and interpreted more than one hundred ^{14}C samples from Neolithic, Bronze Age and Early Historical sites. Of these dates obtained, 78 were considered credible, 26 were rejected as unreliable and 11 were problematic. (This data is published on my website at the University of Hong Kong -- http://www.hku.hk/hkprehis). I mention this merely to inform the non-specialist that rogue dates are quite common in the general use of ^{14}C in archaeology. Willi Wolfli, director of the Swiss lab which dated a Shroud piece, co-authored (Johnson et. al 1985) the following in a similar vein, after a set of interlaboratory comparisons on freshly excavated samples:"

"'The existence of significant indeterminant errors can never be excluded from any age determination. No method is immune from giving grossly incorrect datings when there are non-apparent problems with the samples originating in the field. The results illustrated (in this paper) show that this situation occurs frequently'."

"It is important for anyone wishing to understand the normal use of ^{14}C to know that a single date or even a series of dates on a single object or feature is seldom if ever cited to answer important questions about the age of a culture or a site. To put the radiocarbon method in the position of being the ultimate arbiter of the age of the Turin Shroud is a blatant departure from the way ^{14}C is normally used. Unfortunately, the blame for this fiasco lies mainly on the shoulders of the extremely over-confident, over-bearing and haughty attitudes on the part of most of the ^{14}C lab directors who were involved [...]."

"[...] In the scientific community, the debate centered around how the testing would be done, and in particular on the reliability and statistical accuracy of the results. There was huge amount of acrimonious debate over the number of labs to be involved, whether the tests would be truly blind, etc. These were certainly valid concerns, but it was focused too much on the wood and not the trees! Unfortunately, despite my harping on the subject, no attention was devoted to the possibility of contamination which might escape normal pretreatment. Equally unfortunately, no significance was being accorded to the fact that the Shroud had been through a fire and a series of other events that could conceivably affect its radiocarbon content. When I first became involved in Shroud research in 1981, I was appalled by many of the things being written by STURP and by the radiocarbon specialists about the Shroud's eventual ^{14}C dating. These concerns were summarized in my *Current Anthropology* article (Meacham 1983) and further elaborated on in an article in *Shroud Spectrum* (Meacham 1986)."

"During the mid-1980s, I was extremely critical of the two proposals then being formulated and discussed: that of STURP and that of Gove/Harbottle. Neither of these proposals took seriously the possibility of contamination and heat-induced isotope exchange. In a submission to

Cardinal Ballestrero made jointly by me and two Italian archaeologists (Maurizio Tosi and Roberto Ciarla) of the Institute for Near and Far East (IsMeo), it was argued that 'in this crucial test awaited by millions of people, it is necessary to proceed with great caution so that the eventual result is the best that modern science can produce.' Specific proposals were made to insure that extensive chemical screening would be carried out prior to testing. Sadly, this did not happen. But at least the debate was beginning to focus on some of the crucial issues, namely the number and location of the samples to be tested."

"In the run-up to 1988, a major debate took place over the choice of samples to be dated. From the earliest discussions on the possibility of ^{14}C dating of the Shroud, it was generally assumed that any fragment of the cloth would suffice. The Gove/Harbottle proposal of 1979 (at first supported by STURP) called for the Raes piece to be used -- probably the most ridiculous idea of the entire saga. McCrone and Sox attempted to obtain from Raes the sample 'which was kept in what looked like an old scrapbook for postage stamps.' Eventually Gove and Harbottle accepted that credibility and chain of evidence required a fresh sample to be taken from the Shroud, but they proposed, and STURP concurred (at first), to use charred material under one of the patches as the sole sample to be divided amongst the labs. My strenuous opposition to the use of charred cloth led Harbottle to write to more than 40 practicing radiocarbon physicists, seeking their opinions on the proposal. Although Harbottle misrepresented somewhat my major concern (which was not with the possibility that carboxyl groups present in the linen could have exchanged carbon with CO_2 of the atmosphere but rather that they might have exchanged carbon with contaminants then on the cloth), the responses he obtained were interesting. He wrote that 'no one had any data directly testing the Meacham hypothesis' nor did any of the respondents know of any case in which a sample had been charred long after its lifetime, but well before being dated. Comments were obtained such as 'the use of the charred material would pose problems,' 'there is the possibility of isotopic exchange with volatile or gaseous combustion products,' 'why take the chance?,' etc. After lengthy discussion at the Turin conference in 1986, it was agreed that the charred material would not be used. This was achieved with the strong support of Alan Adler and

Bob Otlet, both of whom were well aware of the problems that might be involved in ^{14}C dating the Shroud."

"One of the main points of debate was the number of samples. A major divergence of views occurred over the sampling strategy. Strangely, the ^{14}C specialists insisted on having splits of the same single sample. It appeared as if they wanted above all else to achieve harmonious results amongst themselves, as opposed to any results that might indicate a variation of the Shroud's radiocarbon content. It was said that the reason for this was 'to maximize the credibility of the enterprise to the public.' This led to the most unfortunate and unscientific sampling of only one location on the corner of the Shroud (and a terrible choice of site at that!). It seemed to me that, if isotope exchange had occurred during the 1532 fire, it would most likely not be uniform over the entire cloth, and three samples from different sites would provide the best evidence about this possibility, and also for inter-corroboration of the results obtained. At the 1986 Turin conference which was convened to draw up a protocol for ^{14}C dating of the Shroud, no amount of pleading and cajoling by me and Adler could persuade the assembled radiocarbon luminaries that a minimum of two sampling sites should be proposed. They were supported by the Church representatives who naturally wanted to limit the disturbance to the relic to the barest minimum. Only Otlet and Hedges supported the proposal. In exasperation, I suggested using the charred material as a second sample, only to be generally heckled about a seeming reversal of position. As a result of the sampling strategy which was adopted, no hard data is available on the radiocarbon content of the rest of the cloth. An extrapolation from the corner piece to the rest of the cloth is only that -- i.e., it is little more than conjecture. Ironically, even though the labs did finally all obtain and date the same sample, and did produce reasonably harmonious results, it can be argued that a possible thermal gradient can be observed from their data […]."

"[…] When the conference was convened in Turin in 1986, there was great hope that a thorough and rigorous plan would be adopted to insure that the ^{14}C date on the Shroud was indeed the best that science in the 1980s could offer. This hope was dashed very quickly. At the meeting itself, most of the ^{14}C laboratory directors were adamant, and rather arrogant, in their claim that a totally reliable date, to within one or two percent accuracy, could be obtained if they could just get their hands on a tiny piece of the cloth. Their attitude toward the question of possible contamination, which I brought up several times, was highly and haughtily dismissive. Gove and Harbottle were particularly dismissive of the possibility that any contamination might survive the standard pretreatment, even though I pointed out to them as forcefully as I could without shouting, that a simple SEM screening of the Shroud by Marano (1978:**202**,381) had shown: *'la superficie delle singole fibre presenta un aspetto 'sporco' con abbondante deposito di materiale estraneo inquinante ma intimamente conesso con le singole fibre del tessuto'* ('the surface of the fibers presented a 'filthy' appearance with abundant deposits of pollutant material extraneous to but intimately connected with the individual fibers of the cloth'). Later, Garza-Valdes looked for possible microbiological contamination, and found it with embarrassing ease: 'Even the untrained viewer could see that the fibers of the thread were completely covered with a bio-plastic coating' (Garza-Valdes 1999:27)."

"During and immediately after the conference, STURP allowed themselves to be gradually pushed aside, unfortunately, in order for the dating to take place. Among the Church representatives there were various cliques and rivalries, and it was difficult to understand what their motives or reasoning was. The minutes of this conference will reveal that Alan Adler and I urged, pleaded, cajoled, and literally begged for extensive chemical screening of the samples before being dated, and for at least two sites on the Shroud to be sampled. One or two others at the conference were supportive, but these were voices crying in the wilderness. In March 1987, I circulated to all who had attended the Turin conference a long paper on the problems and pitfalls that should be considered before the Shroud samples were dated. A major emphasis was on possible contamination due to '[...] mold, mildew and fungal growths which are encouraged on linen [...] organic materials [such as] bacterial or insect residues and fine particulates [...] locked in the cellulose structure.' To counter this possibility I suggested that 'all samples be subjected to elaborate pretreatment, SEM screening and testing (microchemical, mass spectrometry, micro-Raman) for impurities and intrusive substances.' The reaction of the ^{14}C specialists was precisely the same as it had been in Turin the year before -- marked for the most part by arrogance and disdain. A year later, several of them did finally succeed in getting Shroud samples, which were run with only standard pretreatment, the results announced to all the world, and the rest is history. The Shroud was relegated in the public mind to a medieval forgery, or at best a medieval oddity. This was a disaster that could have been averted […]."

"[…] Could it be that some people in authority believe that the Shroud can somehow stand aloof from the ^{14}C dating? Three years ago, a prominent Shroud researcher, who has a background in science, wrote the following to me: '[T]he ^{14}C test results are no longer a hot issue. No one, including Harry Gove, takes them seriously. We all know and accept that a whole number of things went wrong with that test, including the fact that the very method is faulty in the case of old linen. In addition to that, new and disturbing facts are emerging about cosmic radiation […].' I am afraid that this attitude is profoundly in error, and represents a tiny percentage of people. To the vast majority people who know what the Shroud of Turin is, the matter was settled by the ^{14}C dates -- for them it is a curious forgery or oddity from the medieval period. There is absolutely no point and no hope in attempting to change this broad perception by debate, or even by experiment. There is one and only one thing that will reignite wide interest in the Shroud and revitalize the possibility in the public mind that this really is or could be the burial cloth of Christ, and that one thing is NEW ^{14}C TESTING! Why this should be a problem, and why 12 years have elapsed without this happening, is a great mystery to me. For it is in the best interest not only of the Church, but also of the intense public interest in the Shroud, and of science, that we focus the highest level of technology that can be mustered on the issue. What is required is merely a few grams of the cloth from three new locations. This would cause no damage or disfigurement to the relic and is a negligible sum to pay for the potential significance of what can be learned. Let us hope that the futile ^{14}C debate of the last twenty years can soon be left behind and Shroud studies can begin the new millennium with new data and a fresh approach to the true age of this fascinating object."

Source: Meacham, William. "Thoughts on the Shroud ^{14}C Debate." In <u>The Turin Shroud: past, present and future.</u> International Scientific Symposium. Torino, 2-5 March 2000. Silvano Scannerini and Piero Savarino, editors.

Published jointly by *Sindon*, journal of the Centro Internazionale di Sindonologia and Effata Editrice, pp. 441-453.

Comments: We are now an additional twenty years beyond when Meacham wrote those words about trying to get new data and a fresh approach, but still nothing has been done. Especially significant is the statement by the unnamed researcher who wrote to Meacham, "No one, including Harry Gove, takes them seriously. We all know and accept that a whole number of things went wrong with that test, including the fact that the very method is faulty in the case of old linen." It's too bad the researcher was apparently unwilling to put his/her name to the quote.

2000 (May). During the occasion of a public exposition in Turin, Pope John Paul II asserted, "The Shroud is a challenge to our intelligence. It first of all requires of every person, particularly the researcher, that he humbly grasp the profound message it sends to his reason and his life. The mysterious fascination of the Shroud forces questions to be raised about the sacred Linen and the historical life of Jesus. Since it is not a matter of faith, the Church has no specific competence to pronounce on these questions. She entrusts to scientists the task of continuing to investigate, so that satisfactory answers may be found to the questions connected with this Sheet which, according to tradition, wrapped the body of our Redeemer after he had been taken down from the Cross. The Church urges that the Shroud be studied without pre-established positions that take for granted results that are not such; she invites them to act with interior freedom and attentive respect for both scientific methodology and the sensibilities of believers."

Source: Address by Pope John Paul II at Mass in Turin. Reproduced in full in Cassanelli, Antonio. The Holy Shroud: A comparison between the Gospel narrative of the five stages of the Passion (Flagellation, Crowning with Thorns, Way of the Cross, Crucifixion and Burial), and the Shroud as Evidence, Published jointly by Gracewing (Herefordshire, ENGLAND) and Nova Millennium Romae (Rome), 2002, pp. 7-11.

Comments: The key phrase seems to be, "The Church urges that the Shroud be studied without pre-established positions that take for granted results that are not such [...]." The Pope seemed to be pointing there to the 1988 C-14 dating results. But frankly, if he himself had made some better decisions, the results might not have been in doubt.

2002. French sindonologist Raymond Souverain wrote in regard to the 1988 dating, "There is an obligation to apply all good practices required by the Scientific Community. These good practices are established either by international organizations such as the ISO (International Organization for Standardization) or by national or professional organizations [...]."

"The 1988 experiment is full of uncertitudes. For my part, after a 43-year career in inspection services -- I was general inspector of fraud repression laboratories -- I denounce its weaknesses. I would never have dared submit to a Tribunal something founded on such a deficient procedure as that adopted in 1988. Never, at meetings of international organizations on analytical methods, have I omitted to uphold the utmost importance that should be given to sampling. The dating test was a failure. It should be redone."

"Finally, the procedure of this test shows that methodological rules should not be transgressed, that an experiment should not be launched if it has not matured in a multidisciplinary atmosphere and if it is not carried out transparently."

Source: Souverain, Raymond. "Dating of the Turin Shroud: situation in 2001," *Revue Internationale Du Linceul De Turin*, no. 22 (February 2002), pp. 16 and 18.

2002. Ray Rogers and Anna Arnoldi revealed that ultraviolet photography and spectral analysis showed that the area from which the samples were taken was chemically unlike the rest of the cloth. In that area, madder root dye and an aluminum oxide mordant (a reagent that fixes dyes to textiles) were found, but these do not appear to be present elsewhere on the Shroud. Rogers also revealed the existence of a splice in one of the Raes threads, which comes from an area right next to the

C-14 sample area. He wrote, "Raes thread #1 shows distinct encrustation and color on one end, but the other end is nearly white. The photograph was taken on a 50% gray card for color comparison. Fibers have popped out of the central part of the thread, and the fibers from the two ends point in opposite directions. This section of yarn is obviously an end-to-end splice of two different batches of yarn. No splices of this type were observed in the main part of the Shroud."

Source: "Scientific Method Applied To The Shroud of Turin: A Review" by Raymond N. Rogers and Anna Arnoldi. www.shroud.com/pdfs/rogers2.pdf.

Comments: The finding of the madder root dye and the aluminum oxide mordant that fixes dyes to textiles is consistent with Adler's finding of aluminum. These findings were the preliminary work by Rogers, which culminated in a 2005 peer-reviewed paper published in *Thermochimica Acta*.

2002. I posted the following to the Shroud blog at www.shroudstory.com. "In late 2001, Sue and I submitted to *Radiocarbon* our Orvieto paper (http://www.shroud.com/pdfs/marben.pdf). In a letter dated January 1, 2002, Dr. Timothy Jull, editor of the journal *Radiocarbon*, and one of the scientists from the University of Arizona laboratory that dated the Shroud in 1988, sent Sue and me a reply regarding the submission of our C-14 paper. For those not familiar with the process by which papers are published in scientific journals, the editor chooses various reviewers, usually anonymous to the author and supposedly objective, who then make suggestions to the author(s) on how to make the paper better. After changes are made, the reviewers read the paper again, and make their recommendations to the editor as to whether the paper should be published or not. However, the final decision is in the hands of the editor. The review of our paper was out of the ordinary insofar as the reviewers were revealed to us, something that normally doesn't occur. They were all actually directly involved in the specific topic of our paper, the 1988 Shroud C-14 dating. It was our contention that the C-14 dating was skewed due to the presence of a sixteenth century repair. Here is a list of the reviewers of our paper:"

* The late Paul Damon, head of the Arizona laboratory that participated in the 1988 Shroud dating
* Jacques Evin, French C-14 expert present at the 1988 sample-taking
* The late Gabriel Vial, French textile expert present at the 1988 sample-taking
* The late Franco Testore, Italian French textile expert present at the 1988 sample-taking
* The late Harry Gove, inventor of the AMS radiocarbon dating method, who had literally bet a companion that the Shroud was medieval and was heavily involved in various aspects of the dating

"What were the chances that any of these men, each of whom would publicly look bad if our theory were correct, would want to see our paper published? The answer was obvious. Needless to say, our paper was not accepted. Most interesting was a comment by Evin, who wrote in the review sent by the editor to Sue and me:"

"'The authors, who, for several reasons, are convinced that the shroud is authentic, want to publish an article in *Radiocarbon* only to introduce a doubt about the dating. All people involved in the sampling and in laboratory analyses, will be very angry with these suspicions turning on so an important mistake or a misconduct' [...]."

Source: Correspondence sent by editor of *Radiocarbon* to author and Sue Benford.

Comments: Orvieto is in Italy. My late wife and I presented our hypothesis at a conference there in August 2000, that the area where the sample had been taken had been rewoven. I also said on the blog posting: "Enigmatic comment by Evin, is it not? How fair or ethical was of it of *Radiocarbon* to use reviewers who were directly or closely involved with the Shroud C-14 dating?"

Ray Rogers emailed various sindonologists, including me and Sue, on October 20, 2001 and said, "*Radiocarbon* sent Sue and Joe's paper to the referees with the most to lose by its publication. That is the serious problem with peer review. Science presumes honesty. The information we now have is important in Shroud studies, and we should discuss the best way to use the information within the constraints of the real practice of 'science.' Being nasty to the core, I would also like to rub some noses in objectivity and honesty." Barrie Schwortz has told me that Rogers was the most ethical scientist he ever knew. Also, on August 19, 2002, Ray Rogers sent an email to Sue, and me, saying, "It was a good thing to question the radiocarbon sample. The most common problem with 'dating' is the use of invalid samples. I thought Damon would have been more careful, because I have seen him question 'association' many times. Vance Haynes of AZ *[Arizona]* almost sings association as a theme song. Damon *et. al*, fell into the trap of getting anxious and not doing a good job of characterization, probably the only time he has done that. They should have done some background study, and now they are trying to cover up. They are very successful at it, but the 'truth will out'."

2003 *(July)*. On the 20th, I sent a long letter by snail mail to Cardinal Poletto of Turin, who was custodian of the Shroud at that time, putting forth various pieces of evidence showing the anomalous nature of the C-14 sample. I received a letter back from him on September 2nd. It was in Italian, so I asked Bill Meacham, who is fluent in Italian, to translate. The Cardinal stated he was not in a position to make judgments on the scientific matters raised in my letter. He did say that proposals for new testing were being evaluated by a group of scientists chosen by the Turin Centro. When they had made a judgment, he would then refer it to Pope John Paul II. Meacham commented to me in an email, "He mentions the jury of scientists. *[Monsignor]* Ghiberti wrote me a few days ago saying the proposals have *already* been sent to the jury for review. I am very suspicious of this, since I heard just a few weeks ago that none of the suggested international peer reviewers had been contacted yet. They could of course have gathered a bunch of local cronies, plus one or two non-Italian scientists who know little or nothing about the Shroud. This would be wide open to manipulation and facade, which I suspect is what this exercise is all about."

Source: My book, Wrapped Up in the Shroud: Chronicle of a Passion (Bowker Identifier Services), 2020, pp. 137-143.

Comments: Monsignor Ghiberti was president of the Shroud Commission of the Archdiocese of Turin and part of the Turin "Centro." Meacham and Rogers sent a follow up letter to Cardinal Poletto on August 5th. Rogers also sent a letter to Piero Savarino, Cardinal Poletto's scientific advisor. I then sent another letter to Cardinal Poletto on September 29th. We received no replies. Meacham and I then sent a letter on May 17th to the American prelate, Cardinal McCarrick, of Washington, D.C. No reply was received. A group that Meacham and I belong to, the Shroud Science Group, sent another letter to Cardinal Poletto in July. Once again, there was no reply (my Wrapped Up In The Shroud book, pp. 146-172).

2003. In 1978, STURP had taken some pictures now known as the "Quad Mosaics," which included the area from which the C-14 sample had been taken. They surfaced again in 2003 in light of all the research that was going on regarding that corner of the cloth. STURP member imaging specialist, the late Jean Lorre, was asked about these photos, on which Lorre said different colors represent different chemical compositions. Barrie Schwortz wrote "[N]otice that the area adjoining the patch (where the C14 sample was taken from, and ostensibly part of the actual Shroud) is also mostly the same color of green. This is further convincing, supportive, scientific evidence that this area is inherently different in composition than the rest of the Shroud." Archaeologist Paul Maloney noted in a presentation at a conference in Columbus, Ohio in 2008, that he wished he had been aware of the Quad Mosaic photos back in 1987, when he gave Luigi Gonella advice regarding the sample-taking location. Maloney felt that the pictures could have helped convince Gonella not to take the sample from that location.

Sources: (1) Schwortz, Barrie: (2003-2011). "Some Details About The STURP Quad Mosaic Images." http://www.shroud.com/pdfs/quad.pdf. (2) Maloney, Paul: 2008. "What Went Wrong With the Shroud's Radiocarbon Date? Setting it all in Context. Presented at "The Shroud of Turin: Perspectives on a Multi-Faceted Enigma." August 14-17, 2008 at Blackwell Hotel in Columbus, Ohio. https://www.shroud.com/pdfs/ohiomaloneypaper.pdf.

Comments: It is likely that if STURP had been allowed to be a part of the C-14 test, the area ultimately chosen for the sample-taking would have been avoided. For three different color photos of the quad mosaics, see pg. 6 of the Benford/Marino peer-reviewed paper "Discrepancies in the radiocarbon dating area of the Turin shroud" at: https://www.shroud.com/pdfs/benfordmarino2008.pdf.

2004 *(February)*. Despite us (Rogers, Meacham and myself) having not received replies to letters in late 2003 from Cardinal Poletto, his advisor Piero Savarino and Cardinal McCarrick, Meacham was still determined to try to reach Church authorities. I suggested trying to contact Cardinal Justin Rigali, at one point a high-ranking Curia member, who was Archbishop of St. Louis for part of the time I had been a monk in St. Louis. We sent him a letter, signed by various American sindonologists, including Ray Rogers, which in part read, "Dear Cardinal Rigali, We the undersigned are Americans from Catholic, Protestant and Jewish backgrounds who are deeply concerned over the future of the Turin Shroud. We are writing to you in the hope that you would be willing to raise certain issues with Cardinal Severino Poletto, Archbishop of Turin and Pontifical Custodian of the Shroud. We feel that much closer and more effective communication with Cardinal Poletto would transpire if conducted by a fellow Cardinal."

"The issues that we feel very strongly about are:"

1. the need for a proper international commission of prominent scholars, whose names would be published, and the proceedings of which would be open. Currently, an "international jury" has supposedly been appointed by Cardinal Poletto to review proposals for scientific testing of the precious relic. Nothing is known of the membership of this jury, how it was selected, or what its terms of reference are. We fear that it could be subject to manipulation, that it may be packed with cronies, and that it may not be fair and objective. The international commission we would like to see was requested in a petition to the Pope signed by 52 Shroud researchers from around the world (see Annex I) *[not reproduced here]*. It was suggested to have a much wider remit,

not only to oversee future scientific testing but also conservation of the relic. There has been no reply to this petition, and we would be most grateful if you could also make discrete inquiries with the Vatican on the matter.

2. the need for C14 dating, without further delay, of the charred fiber bits and carbon particles removed from the Shroud in 2002, as proposed by Ray Rogers and William Meacham to Cardinal Poletto last year (see Annex II) *[not reproduced here]*. This proposal has wide support in the community of Shroud researchers (see Annex III) *[not reproduced here]*. Much future scientific testing (both research design and fund raising} would hinge on the outcome of this test, and the result if significantly different from the medieval date announced in 1988 would have an electrifying effect on the world. Cardinal Poletto's response has been that the decision of "the "jury" must be awaited, then a recommendation made to the Pope, and then a final decision awaited from the Vatican. The proposal of Rogers and Meacham is different from the other proposals in that it does not require access to or direct testing of the Shroud itself, but rather of the minute samples removed already [...].

"[...] In the light of the considerable media interest in the crucifixion of Christ generated by the movie by Mel Gibson, we feel that the time is ripe for a proper re-dating of the Shroud. All of us believe that the C14 dating done in 1988 was not conducted properly, the medieval result may not be the true age of the Shroud, and the impact of a new dating will be tremendous. What is most compelling is that the amount of material required to find out the truth is minuscule -- about two spoonfuls!

Source: Email of February 14, 2014 from Bill Meacham to author.

Comments: No reply was received. One would have thought we could have been extended the courtesy of a reply.

2004. Ray Rogers was interviewed by the BSTS for their newsletter. The last questions asked to him were: "What are your views on the Shroud world at the beginning of the twenty-first century? Where do you see Shroud studies going from here?" He responded, "Future operations must be carefully planned and executed, and they cannot involve management by dilettantes. The secretive 'restoration' illustrates the problem. It was one of the most poorly planned and executed disasters that has ever befallen the Shroud. It did much more damage for future scientific studies than did the fire of 1532. Actually the fire of 1532 provided an excellent chemical test that was easy to read. That information is now largely gone. It is a good illustration of why rigorous, competent scientists open their plans and preliminary results to comment. All existing knowledge should be assembled before pursuing such a project: The persons involved with the restoration hid their plans from peer review. The 'restoration' removed many opportunities for cogent chemical analyses in the future. The thymol application in 1988 may have done similar damage: it certainly eliminates cloth samples from consideration for dating work. Charred samples now exist that could be used to provide an accurate age for the cloth, but I doubt anyone who knows how will be allowed to prepare them for radiocarbon analyses. Too many people who know nothing about kinetic isotope effects believe that heating changed the age of the cloth, further confusing the issues. As long as such doubts exist, and they cannot be discussed in a scientific forum, it will be impossible to get Turin's approval for new radiocarbon analyses. Poorly advised officials are afraid to move. Nothing new may ever be learned about the Shroud, and the officials in Turin take no cognizance of previous scientific work. I cannot even get responses from the Cardinal's scientific advisor. I cannot communicate with Luigi Gonella or Giovanni Riggi, persons I thought were my friends. I believe that competent scientific efforts to understand the Shroud have a bleak future."

Source: [Wilson, Ian.] "Who's Who in the Shroud World Ray Rogers." *British Society for the Turin Shroud Newsletter*, No. 60 -- December 2004, pp. 16-23. https://www.shroud.com/pdfs/n60part3.pdf.

Comments: Rogers gave a rather depressing view of any future studies. Since he gave this response, there have been two public expositions, but no new studies on the cloth. For further information regarding the controversial (could it have been anything but?) 2002 restoration of the Shroud, see https://www.shroud.com/restored.htm. Once again, it must be emphasized that the Church seems to be wasting a wonderful opportunity to spread its own message.

2005 *(January)*. Ray Rogers' peer-reviewed paper in the world-renowned journal *Thermochimica Acta* was published. Rogers wrote, "The presence of alizarin dye and red lakes in the Raes and radiocarbon samples indicates that the color has been manipulated. Specifically, the color and distribution of the coating implies that repairs were made at an unknown time with foreign linen dyed to match the older original material. Such repairs were suggested by Benford and Marino." Rogers concluded, "Pyrolysis-mass-spectrometry results from the sample area coupled with microscopic and microchemical observations prove that the radiocarbon sample was not part of the original cloth of the Shroud of Turin. The radiocarbon date was thus not valid for determining the true age of the shroud." Rogers also noted regarding one of his chemical analyses, "The Raes threads, the Holland cloth, and all other medieval linens gave the test for vanillin wherever lignin could be observed on growth nodes. The disappearance of all traces of vanillin from the lignin in the shroud indicates a much older age than the radiocarbon laboratories reported."

Source: Rogers, R.N. Studies on the Radiocarbon Sample of the Shroud of Turin. *Thermochimica Acta*, Vol. **425**, No. 1/ 2, 20 January 2005, pp. 192-193. Accessible at http://www.shroud.it/ROGERS-3.PDF.

Comments: It's important to note that Rogers' findings were based on an actual leftover sample from the 1988 dating. Some have questioned the provenance of the material he used, but I showed in my "The Invisible Reweave and Other Challenges to the Turin Shroud's C-14 Medieval Dating: A Review" paper, that there is actual documentation proving that it was definitely from the 1988 sample.

Regarding the findings by Sue and me, Rogers had said "I believed that it would be easy to completely refute them. It is highly embarrassing that

I could not. This is the first time I have had to present information that seemed to support what I consider to be the 'lunatic fringe.' However, an ethical scientist absolutely must publish accurate information no matter what the emotional implications" (as cited in his article "Ghiberti's pronouncement on my analyses," http://www.shroud.it/ROGERS-5.PDF).

The late monumental artist Isabel Piczek wrote, "It is not good enough just to look (with the naked eye) for a re-woven patch. It is an invisible reweave, which requires microscopic and microchemical analysis (to discover). Rogers' paper has to be accepted. New discoveries always cause lots of controversy, but (Roger's report) should be trusted because it was published in a peer-reviewed journal (as cited by Muldoon, Shena. "Was the Dating a Hoax?" *Inside the Vatican*, **13**:2 [March 2005], pg. 25). Even after this data was released, Turin's Monsignor Giuseppe Ghiberti told an Italian newspaper, "I am astonished that an expert like Rogers could fall into so many inaccuracies in his article." However, a short time after that, The Diocesan Commission for the Holy Shroud released another statement, saying that the study of Rogers was "very interesting and would be the basis for a future study on the chemical characteristics of the cloth and its possible inhomogeneity" (as cited in Muldoon, Shena. "Was the Dating a Hoax?" *Inside the Vatican*, **13**:2 [March 2005], pg. 25.) It cannot be emphasized enough that insofar as he was the only person to have access to main Shroud samples and samples from the C-14 sample area, Rogers' judgment should carry enormous weight. To see Rogers' impressive resume, go to: http://www.shroud.com/pdfs/rogersresume.pdf.

In the Muldoon article cited above, Rogers stated (pg. 24), "The sampling operation should have involved many persons from different fields before cutting anything. And if you really want to get a radiocarbon data, take a lot of samples." Asked if he thought the authorities had been aware that some of the 1978 STURP photos indicated that the corner from which they took the sample was unlike the rest of the cloth, Rogers replied, "It doesn't matter if they ignored it or were unaware of it. Part of science is to assemble all the pertinent data. They didn't even try."

Sadly, Rogers died in March 2005, less than two months after his paper was published.

2005 (March). Meacham was still trying to find ways to get Turin to reconsider doing another C-14 test. Maloney wrote to Meacham, "[...] But I think I know why Turin has resisted and will continue to resist any and all overatures *[sic]* to have the Shroud radiocarbon dated again. Their science advisor, Luigi Gonella, was dead wrong in 1985 through early 1988 when he repeatedly refused to consider advice from many of us about taking samples from anywhere but the Raes Corner. Authoritarian groups -- and the Catholic Hierarchy is certainly to be included in this category -- do not like to be made publically wrong. Turin was stringently criticized in 2002 and later for their 'restoration' of the Shroud. That's why Ghiberti and others representing Turin are on the defensive. Because they don't know or understand the nature of science they are fearful of being proven that their decisions in 1988 and in 2002 were wrong and they are very sensitive to this. Is Poletto one of those due for an advancement in the hierarchy if the Pope dies? *[Pope John Paul II, in fact, died about two weeks later.]* He may feel he will 'look bad' if he rocks the boat now. I suspect we will hear a very loud silence from any and all future requests for further work on the Shroud. If I am wrong I will be pleasantly surprised and absolutely pleased! But it sure doesn't hurt to try.

Source: Email of March 17, 2005 from Paul Maloney to Bill Meacham, author copied.

Comments: Although Ballestrero in one of his communications had mentioned the Church's "love of truth," there is at least a perception among many, that the Church hierarchy's "do not like to be made publically wrong," in the words of Maloney, is a more dominant pursuit.

2005 (May). Brunati wrote, "[A] complex of circumstances convinced me of the opportunity to rewrite this story once again, also because, today, we can consider it from new points of view, which accentuate its aspect of a truly malicious event. And because I hope that going back to talking about the falsity of that test will convince those who have the duty to openly declare that that response that they put, in 1988, on the lips of the Cardinal of Turin, must no longer be considered of any value. What do you expect to do? In short, we cannot admit that we continue to consider medieval and, therefore, what is not medieval at all a fake, that Jesus himself left us [...]. It is an unacceptable behavior that puts us on the same level as those who are responsible for that fake."

"I begin by taking into consideration what Prof. Franco Testore in the report he wrote on the occasion of the 1989 Sindonology Congress in Paris. It is an important document in that it allows you to understand exactly what happened then, so much so that, in its time, I also sent a copy to Ray Rogers who asked me for it. Frankly I do not know, however, to what extent, he has taken this into account. Before then, Testore had never dealt with the Shroud. He taught textile technology at the Polytechnic of Turin, when he was summoned as technical consultant on the occasion of the dating. And, on the occasion, he had borrowed from the Polytechnic himself an electronic precision scale, with which he made all the weighing that the dating required and the results of which he then reported on the report. Since no minutes were drawn up (which is rather surprising), those data, even if in practice they were only personal notes, have therefore become the only reliable source of information on what has actually been taken. And it was the same professor, that morning, who made the proposal not to take only 150, but 300 mg of cloth. By changing, with this, it changed the withdrawal procedure that the laboratories had planned to adopt [...]."

"[T]hat morning, a sample of about 150 mg of cloth was closed and sealed with the Cardinal's signature in one of the stainless steel cylinders intended for the transport of the samples to the laboratories. I said that Testore's proposal changed the procedure that had been planned for the project. Only after a couple of years did we know what it was. And it was the French friend, Fr. Bruno Bonnet-Eymard, who learned about it, when, on a visit to the laboratory in Zurich, he had the news from the director. So we learned that, well before going to Turin for dating, the directors of the laboratories, given that 150 mg of cloth would have been sufficient for dating, had decided to take a 7 x 1 cm rectangle of material, a piece that is, it should have had that weight. Then, evidently, they did not know what Prof. Testore on the day of collection, that is, that the unit weight of that flax was, in reality, about 23 mg/cm^2.

Therefore, a strip of that size would have made available not 150 but about 165 mg of cloth to the laboratories. In theory, in addition to the Testore report, we should also consider the one presented on the same occasion by Prof. Riggi di Numana, who, as is well known, was in charge of the material execution of the project. However, that report contains data so obviously wrong as to make it unreliable *a priori*. Not even the first edition of the Testore report was actually correct. All because of a wrong sum and, when we were able to examine the revised version, we noticed a strange fact. While the initial withdrawal of a piece of 300 mg was confirmed and it was added that, as was already known, 154.9 mg of these would be made available to the Cardinal, it was also specified that, from the remaining 144.8 mg (0.3 mg had been loss of cut) a first sample of 52.8 mg and a second of 52.0 mg were removed; therefore, for the third sample there remained only 39.6 mg. Too few, compared to those 50 mg they had pledged to give to each laboratory. It was then decided to remove further 14.1 mg, cutting them from 154.9 mg intended to remain in Turin for the Cardinal and this meant that one of the three laboratories, to be exact the American one in Tucson-Arizona, received a sample of 53.7 mg which was in two parts. A strange and surprising fact, even if, after all, it would not have compromised the regularity of the test [...]."

[Brunati saw on RAI TV some clips of the sample taking and took some photos and made some diagrams not shown here. Brunati believed he found proof that the sample taken for dating was not a long and narrow divided strip of cloth, in three parts with two transversal cuts as indicated in the "Nature" report. Brunati said the display on the electronic scale indicated exactly 300.0 mg., (and later asserted by both Riggi and Testore,) which "left everyone amazed."]

Brunati continued, "A purely coincidental fact, but one that interests me in that, it too, disproves what the report in *Nature* will say, when it attributes to the amount taken for dating a weight of only 150 mg. Wanting to conclude by making a summary of what has been said up to this point, I believe we can be sure that 300 mg of cloth were taken from the Shroud that morning, 154.9 of which were used to make those three samples (of 52.8 mg for Zurich, 52.0 mg for Oxford and 39.6 plus

14.1 mg for Tucson) personally delivered by Cardinal Ballestrero to the directors of the laboratories, while the rest remained in Turin. Even if we do not know the exact dimensions of these samples, I think it is not wrong to say that the 300 mg piece was almost rectangular and measured approximately 8.2 x 1.7 cm, while the four fragments to be dated, placed side by side, also had to form a rectangle, 3.6 - 3.7 long and 1.7 cm wide, approximately."

"It should be noted that Marino and Benford, in their report, also give the sample they consider mended, similar shape and size. According to the official report, however, the one prepared by the British Museum and countersigned by the twenty professors, it would appear that the laboratories, in reality, what was subjected to dating consisted of three samples obtained by dividing a strip 7 cm long and 1 cm wide into three parts. A strip that, therefore, could have no reference whatsoever with what appears to us to have been really removed from the Shroud. So, we must take note that what was subjected to dating must have had nothing to do with the Shroud. This is an extremely serious fact which confirms the replacement of the samples. Other than mending! An official declaration, however, to which, however, normally very little importance is given. It often happens to read learned dissertations on the possibility that the result of the test has been distorted by the strangest phenomena, such as, in fact, the mending, or, perhaps, a fire while it is not taken in the slightest account that, officially, they claimed having sent to the laboratories a piece of cloth that did not come from the Shroud."

"That they have not dated the samples received in Turin by the Cardinal at all was also confirmed by a letter sent on December 5, 1989 by the Tucson laboratory to the French friend Claude De Cointet, a collaborator of the aforementioned Fr. Bruno Bonnet-Eymard. That, knowing this professor personally, he had written to him to confirm that the sample that had happened to them had been in two parts. Rather disconcerting answer. The professor, who had gone on purpose to Turin for seeing that piece cut and withdrawn, he was unable to say whether it was in two parts or not. Then he adds that, in any case, he weighed 52.36 mg (and therefore not 53.7 mg as the authentic sample). Evidently, that gentleman did not imagine that, by this side of the ocean, we were now perfectly aware of the weight of the samples submitted to that test. It was logical that, at that point, one had become curious about the data also relating to what the other laboratories had dated. Which prompted me to write to them to ask for the weights of the respective samples. Very embarrassed answers and no clarification regarding my precise questions. It would seem that those laboratories did not date at all the samples of the Shroud material that had been taken for that purpose by the Shroud that morning [...]."

"[...] [I]n *Nature*, the slightest mention was never made of the 150 mg left in Turin for the Cardinal. Was this the reason that led to the disappearance of that sample, delivering it to the Cardinal only after two years? Strange, too, that the dimensions attributed to the dated sample (that 7 x 1 cm), while not in complete agreement with those of the pieces actually taken from the Shroud, instead fully agreed with those of the sample that should have been taken, if had not been accepted the proposal of Prof. Testore I mentioned. It was not surprising, among other things, that the 7 x 1 cm appeared for the first time on the press bulletin issued by the Curia on the evening of the day of the sampling and that, consequently, it was read the next morning in most newspapers. There is no doubt that the author or inspirer of that release was who had, if not direct, at least born all that work: Prof. Michael Tite, of the British Museum. The same who, then, must also have been the author of the report in *Nature* (I'm sure, as he was also the last signatory). Moreover, he was the only person who, apart from the directors of the laboratories who, however, could not intervene, as well as having seen what the samples were like, knew, also how they should have been if the collection procedure had not been changed by the proposal of Prof. Testore. He was the only person, that is, who knew that, if there had not been the Testore intervention, the pieces to be dated would have had to come from a strip of cloth of 7 x 1 cm."

"And, having made mention of those dimensions from the first moment was not a minor thing. After reading them in the newspapers, we too were sure that these were the actual dimensions of the samples, and when we saw them appear in *Nature* three months later, this security became absolute. Only after a year, or almost, after we got hold of the Testore report and the relative photographic documentation, did we realize that the 7 x 1 could not be correct and we started asking ourselves where it came from. Also because there hadn't been any specimens of that size that day. No possibility of data exchange, therefore. At first, we had also thought of an involuntary error by Prof. Tite. But it was not possible that, that evening, and just when he was writing that press release, he had suddenly forgotten all that had happened during that day. He had forgotten what he had discussed in the morning about the Testore proposal and, therefore, just of that size, he had forgotten what he had seen during the afternoon. When he had personally put them, one by one, only he and the Cardinal, in those stainless- steel cylinders. It could not therefore have been an involuntary error. It should not be forgotten, among other things, how it was found that, before the Testore proposal revolutionized the collection procedure, it was expected that the three samples for dating should be obtained by dividing a rectangle of 7 x 1 cm into three equal parts. So it is also conceivable that those were not only the dimensions of the pieces that should have been removed from the Shroud. If they had really wanted to replace the samples, even the pieces of medieval cloths that would have had to replace them, would have had, in all probability, those dimensions. This, Prof. Tite must have known it perfectly, so that quote could have been made to facilitate the introduction into the laboratories of those substitute pieces of medieval cloths: if Prof. Tite had not mentioned that in the press release, some journalists might have been able to investigate on behalf and perhaps have ended up knowing how the samples were taken, then making it known, in some newspaper. So, if those fake samples of medieval cloths had been delivered to the laboratories, difficulties could have occurred. Difficulties that would not have been met if the newspapers had said from the first moment that the dimensions of the pieces to be dated were precisely those of the fake medieval specimens."

"This does not mean that it remains always difficult to hypothesize that a British Museum and three very serious universities may have resorted to a device, such as the replacement of samples, decidedly contrary to professional ethics. Although, then, I realized that such perplexities were not entirely justified. To begin with, I was rather surprised at the way in which the three laboratories and the British Museum highlighted, when they claimed to be present at the collection of the samples, the suspicion that the Cardinal, acting on behalf of the Vatican, could deliver them samples that did not come from the Shroud. Of the samples, for example, of 1st century cloth, although knowing, perhaps, that the Shroud was more modern. Surprise and even disappointment in that it was considered impossible that one could doubt the honesty of that person. Although such an attitude, given the attitude of many groups or especially British associations, should not have surprised us very much. Typical, in this regard, a letter from a university professor from New Zealand, published in *Nature* (in *Nature*, I stress) well before dating began, a letter that ended with the question *'how will independent observers be sure that those samples that laboratories they receive are from the Shroud? Should we take the Vatican's word for it?'*. An environment, therefore, decidedly hostile."

"I also had the feeling that the prompt acceptance of the request to assist with the excision made the situation worse [...]. [N]o representative of the Vatican was actually admitted to the laboratories and therefore no one would have been able to verify for themselves what they actually would have done. Furthermore, radiocarbon dating is a destructive test and, from what it would have left, it would never have been possible to trace what they would have actually undergone that test. It is true that, from an ethical point of view, a replacement would have been highly unbecoming. But I do not know to what extent they judged it as such, given that it was proposed to expose what, in their opinion, was certainly already a fraud. And it was also a unique occasion, which would never come again. And the behavior, in Turin, of the directors who witnessed the removal was, if you think about it, such as to confirm that suspicion. They never moved from the bench on which they had been placed, they never appear to have asked a question or made a check. They just watched what others did from afar. Just as if they were no longer interested in what was being taken and knew that, in any case, they would receive what they wanted to be dated. And it could have been the replacement of the samples that put the laboratories in trouble. If they had, for example, replaced the Shroud with a too young, too recent cloth. And, to get out of the hindrances, they could have resorted to those messes that, in the end, exposed them."

"I try to explain myself. It may or may not be believed that the Shroud was the Shroud of Jesus, but one could not ignore its history. Of a fact documented as when that relic, already well known and the subject of discussions, in the years around 1350, had been displayed in Lirey, France. In one of its first lines, the report in *Nature* also mentions it. So the dating would have necessarily had to give, as a result, an age that would make it considered done before 1350. What was the result of that analysis instead? To realize this, we must try to translate into years of normal calendar the age that carbonists, as is their norm, conventionally express in BP years (Before Present). Table 3 of the report can help us in this regard. Given that the BP ages always refer, by convention, to 1950, we note that at an age 691 minus 31 years, then 629 years, it was said to correspond, as the calendar year of manufacture, to 1384. While at an age of 691 plus 31 (then 722) years, 1288 was taken as the corresponding figure. This meant that at those ages of 646 and 676 years, which Tucson-Arizona and Zurich had attributed to the samples they had dated, a manufacture in years very close to 1350 in Lirey."

"If the age obtained by the third laboratory, that of Oxford, had been even slightly less than that of the other two, it would have come that the sheet that we consider as the Shroud was surely made after 1350, after being, therefore, displayed at Lirey. It would evidently have been a disaster and nobody would have believed that dating reliable. This, then, explains that age of 750 from Oxford, which they probably had to resort to precisely to avoid such a possible trouble. Even if, then, those 750 years (1 century out of 6 more than the other two) would have proved too different compared to the 646 and 676 obtained by Tucson and Zurich. Apart from the fact that, always considering that 1350, the result to which they would have landed would not have been exciting anyway. It is true that the 95% probability placed in the Cardinal's mouth made a great impression, but it also referred to two date ranges that were not exactly distant from Lirey. They ascribed (see Table 3 of *Nature*), in fact, the Shroud as having been made between 1262 and 1312, or, between 1353 and 1384.The second period was definitely incompatible with Lirey, while the first assumed that it was it was displayed as a shroud of Christ, and therefore as a 1300 year old cloth, despite having been made a few years earlier. In any case, the Oxford result would have been a problem, since the statistics require that the averages are made only between homogeneous quantities. This, the British Museum knew very well, if only because the dating process that had been agreed with the Shroud history, that suggested by Ward and Wilson, would have required, in this regard, the passing of a very severe test, the chi square test. But that, evidently, must have been considered an obstacle of secondary importance, also because few could have noticed the messes that they would have had to resort to overcome it. That was the reality."

"Funny, among other things, is that it was a supreme ignorant in statistical matters, such as the undersigned, who noticed that something was wrong. It happened when I checked the averages, which are purely arithmetic calculations and which, therefore, were within my reach. The dating procedure provided that each laboratory, having received the samples, divided them into several parts, each of which would have undergone a separate dating. Obtaining, as a result and as we have seen, an age (in BP years) with relative tolerance (in so many years). A procedure that required, therefore, to arrive at a final result, to average, first of all the partial results obtained by a single laboratory, and then among those of the three laboratories. All mediums to which the British Museum was to provide. When I checked if the calculations of these averages were done correctly, at least at first, everything seemed perfect. The report reported the results made by each laboratory for the four samples. Even if I have checked more than one, here I report only what concerns, at first, the dating made by Oxford on sample 1 (the Shroud). Having divided that piece of cloth into three parts, the results achieved were as follows: 795 ± 65, 730 ± 45, 745 ± 55 [...]. So, the (weighted) average of the ages (the average, that is, between 795, 730, and 745) [...] = 749.13 years, corresponding, practically, to the 750 years of table 2. While the (weighted) average of the tolerances (between 65, 45, and 55) [...] = 30.6, practically identical to 30 in table 2. The differences were negligible and are likely to be due to some rounding not shown in the table 1. With that I also had the demonstration that, to make those averages, I had used the same formula and I started from the same data, those taken from table 1, from which the British Museum had started. Everything, therefore, regular. Tucson-Arizona, however, had divided its sample into four parts (each laboratory was free to divide the sample as it wished) so the four partial results that referred to sample 1 (the Shroud), were: 591 ± 30, 690 ± 35, 606 ± 41, and 701 ± 33 and the average age was [...] = 646.67, also practically corresponding to 646 years of table 2. When, on the other hand, I checked the average of the tolerances the result was [...] = 17.04, which was completely different from the 31 that appears in table 2."

"It was a great surprise and, when I found out, I pointed it out in the November-December issue of the magazine *Il Telo* in 1997. And nobody denied or corrected me. However, we all considered it a miscalculation. Inadmissible but which, at first glance, did not seem to change much. If the age of the Shroud had really been 646 ± 17 instead of being 646 ± 31 years, this would have meant that instead of having an age between 677 and 615 years, that cloth would have had an age between 663 and 629 years. We would have been, however, far from those 2000 years we were aiming for. In reality, however, the consequences of that diversity were very different, so much so that we could no longer consider it a simple mistake."

"I said, above, how, between the Shroud's custodians and the laboratories, it had been agreed to do the tests according to the methods established by Ward and Wilson, and that these required the execution of a verification test with the chi square system. The fact that this test took into account the values of the tolerances immediately gives an idea of the importance of that change. It should be noted that the chi square test concerned only the final average that between the results of Oxford, Tucson and Zurich. It was not required for the averages among the partial results, since these, being the work of a single laboratory, were always believed to be homogeneous with each other. The official report in *Nature*, however, was not limited to the clarification detectable by the text, that this test was mandatory. Table 2 also provided the results -- both those that concerned the Shroud and those of the feedback samples. The check with the chi square required, first, the calculation of the value of the same who square from which we passed, then, to the so-called significance level. The test was considered passed if the latter, a percentage, was at least 5%. The results reported on the report table had obviously been obtained starting from the wrong response from Tucson

Arizona, 646 ± 31. They gave a chi square of 6.4 and a significance level of 5%, the minimum required. Such a value, however, to be considered the passed test."

"When, however, I redone the calculation starting from the correct average, 646-17, the *chi square* became 9.8 while the significance level was reduced to a very modest 1.04%, far too low for the test to be considered acceptable. This finally made me understand what was the real reason that led them to change that result. They had to resort to that fake to pass, at least on paper, that test that would allow them to make that average between the 646 years of Tucson, the 750 years of Oxford, and the 676 years of Zurich, the average that would have concluded the dating. Therefore, it was clear that this was not an error. I speak of sure alteration of the result excluding a possible accidental error also for the following reason. If you consider the formula that leads to the average of the ages (the one that, in the particular case, led to 646) you will notice that it practically consists of a fraction with the expression $1/30^2 + 1/35^2 + 1/41^2 + 1/33^2$, which performs the function of denominator. It gives, as a result, 0.003440595 and it is the same that leads to the determination of the average of the tolerances. It follows that, if the British Museum has obtained the correct value of the average age (646), it must necessarily have obtained, as a result of the calculation of that expression, the number 0.003440595, the square root of the inverse of which is none other than that 17 which represents the correct value of the average of the tolerances. Giving as an average of the ages that 646, therefore, implicitly, it is admitted that the average of the tolerances is 17 and not 31. That result that we find in table 2 of the official report, that 646 ± 31, therefore, cannot exist. And it cannot be a calculation error, but it is a deliberately-altered result, it is a real malice. And the same conclusion is reached by considering another fact. In table 2 of *Nature* it is noted that not only is the average tolerance of the value of sample 1, the Shroud one, but also that of the averages of the results obtained by that same laboratory for samples 2, 3 and 4, the control ones. In fact, in table 2 the results are as follows:"

Sample	1	2	3	4
Averages indicated:	646 ± 31	927 ± 32	1995 ± 46	722 ±43
[but] the figures should have been				
	646 ±17	927 ± 20	1995 ± 20	722 ±20

"But even here, it is clear that they must have done it on purpose. If, for the second sample, for example, the average age was 927 years, which is correct, they must also have obtained, an average of the tolerances of 20 and not of 32. Otherwise, I repeat, they would not have reached 927. And the same can be said for 1995 ± 46 of sample 3 and for 722 ± 43 of *[sample]* 4. But an observation should be made. We have seen how they were forced to replace the ± 17 of sample 1, the Shroud one, in ± 31 for the final average to be allowed. An incorrect change, therefore, but, at least, justified. There was no reason, however, that it was necessary to transform the various ± 20 into ± 32, ± 46, and ± 43. If, I repeat, the transformation of 17 into 31, it could have been necessary to make that unacceptable *significance level* equal to 1.04% equal to 5%, what advantage would have been obtained by transforming *a significance level* equal to 88%, like what you had if you adopted the correct ±20, in 90% which, instead, corresponded to ± 32? Both 88 and 90% were well above that 5% which was the minimum. Those three transformations, therefore, must have been carried out only to confuse the waters a little, to raise suspicion in any inspectors that the results obtained in Tucson required a treatment different from that required by those of Oxford and Zurich. Further fraud, therefore, which further exacerbates the position of those gentlemen."

"Regarding Oxford's result and its dubious sincerity, it is interesting what, the American professor Harry E. Gove in his recent book. He was the inventor of the AMS dating process, that process which, by requesting very small samples, had made it possible to carry out that test also for the Shroud. This did not prevent that, when the Vatican reduced the laboratories to which to entrust the test from 7 to 3, the laboratory of Prof. Gove of the United States University of Rochester, was left out. You can imagine how sorry this is to our Professor. And the fact that the dating of the Shroud has continued, more than before, to deal with it, has not failed to amaze. To realize this concern, I think that the fact that Gove was, in spite of all the confidentiality commitments of the laboratories, was the only one to be invited to attend the first final radiocarbon analysis carried out by Tucson on the first sample fragment that they had subjected to dating. He was among the very first, therefore, to know a result, albeit partial, of that operation. Perhaps to follow the tests and keep in touch with the laboratories, he flew to Europe twice in a couple of months. Obviously, when they waited for Oxford's response and this never came, Gove was in great anxiety. And he talks about it, in fact, in his book. It tells us, for example, of all telephone exchanges. It was evident that Oxford had difficulties and everyone imagined what they were. And, when in the end, on 9 September (a month and a half of waiting, five months after receiving the sample in Turin) they telephoned him that, by joint decision of the three laboratories, that result of Oxford (the well-known 750 ± 30) had been accepted, comments: *I was considerably relieved*, I really took a breath. Obviously nobody explains why it was necessary to interrogate the three laboratories although it is not difficult to imagine how the acceptance of that result entailed the use of all those messes that would have involved the responsibility of three Institutes. Interestingly, rather, as the only ones to wait quietly, were, at that moment, the unsuspecting Turinese."

"Finally, I certainly cannot neglect the letter sent to me by the British Museum in response to my request for explanations on the error found in *Nature*. Dated September 5, 1997, it was written by Dr. S. Bowman, then director of the Research Department of that institute. Bowman had replaced Prof. Tite, who in the meantime had been appointed director of the similar department of the University of Oxford. It is true that Prof. Tite had been the coordinator of the whole Shroud dating operation, but in this work he had always been followed by Dr. Bowman, so much so that she was one of the three signatories, on behalf of the British

Museum, also of the relationship in *Nature*. That letter said: *You refer to an error in calculating the average of the variables of the Tucson (Arizona) results. I am not a statistician and I am therefore only able to say that, whatever the mistake may have been, it is unpleasant. But a much more important mistake should have been made to significantly change the results obtained.* Really irritating. To begin with the fact that Bowman does not consider herself a statistician. I understand that she has written at least one book on the subject, the one I have a copy of. But I think it inadmissible that the director of that institute to whom the statistical processing of the dating results were entrusted could be classified as ignorant in statistics. If she is truly so ignorant, resign or, at least, let some of her more expert employees answer it. Apart from the fact that the problem in question did not concern statistics at all, but the most modest of arithmetics."

"Second point. She does not want to deny the error, but agrees that the error exists, she should at least feel the duty to correct it. Third and most important point. Although she is also, as I said, a signatory of that report on which it is explained how dating requires that verification with the chi square, she tries to minimize the problem by pretending that the tolerances have the same function as the decimals. I don't admit that, make fun of me this way. It completely ignores the fact that she shouldn't have finished the dating. Such a letter definitively disqualifies the British Museum, demonstrating how they only intended to falsify the result and, if such deliberate misconduct is admitted, it must be admitted that they may also have replaced the samples and everything that follows. If I insist on highlighting the lack of honesty of those who made that test, I do not do it, of course, to stand as judge against those who committed the fraud, but because I would like to point out that dating is certainly not the indisputable scientific proof that was said at the time of its presentation. Sometimes, faced with such a situation, there are those who ask themselves: but why should they have done so? Certainly the Shroud, the more you know it, the more mysterious it appears. It is not possible to justify all its strange characteristics, starting from being a negative, ending with having to recognize it only as an orthogonal projection of a figure and not as an imprint due to a contact."

"This has always irritated especially those who, considering themselves a scientist, cannot admit that this may have been the physically explainable effect of a miraculous phenomenon such as the Resurrection. It is probable that they hoped to clarify the situation with the dating, but we have seen how, trying to avoid the absolutely unjustified fear of being cheated, they ended up becoming the cheaters themselves, putting themselves in a sea of trouble from which have never been able to save themselves. What seems indisputable to me is that there has been a fraud. And this is the important thing, rather than understanding why they did it. But there is another aspect of the story that I don't think should be overlooked. The dating was a test held in front of the whole world and who carried it out did not work privately, but on behalf of illustrious universities and, especially, on behalf of a famous institution such as the British Museum. Is it possible that such academic authorities would accept that their employees alleged, in their name, that the result of that famous elementary arithmetic expression was equal to 31 instead of 17? And that, moreover, they also published it, always in their name, in a magazine considered famous and scientific as *Nature*? And, since that mistake involves the falsification of a test that everyone considers important, is it possible that those academic authorities do not intervene?"

Source: Brunati, Ernesto. "Altro che rammendi! La datazione della Sindone e tutta un falso." *Collegamento Pro Sindone*, May 2005. http://www.sindone.info/BRUNATI1.PDF.

Comments: Brunati referred to "Fr." Bruno Bonnet-Eymard, whereas most of the time he was usually referred to as "Bro." In some monastic communities, priests sometimes are referred to as "Bro."

2005. Meacham wrote, "I doubt that anyone with significant experience in the dating of excavated samples would dismiss for one moment the potential danger of contamination and other sources of error. No responsible field archaeologist would trust a single date, or a series of dates on a single feature, to settle a major historical issue, establish a site or cultural chronology, etc. No responsible radiocarbon scientist would claim that it was certain that all contaminants had been removed and that the dating range produced for a sample was without doubt its actual calendar age. The public and many non-specialist academics do seem to share the misconception that C-14 dates are absolute."

Source: Meacham, William. The Rape of the Turin Shroud: How Christianity's most precious relic was wrongly condemned and violated, (Lulu.com, 2005), pg. 55.

Comments: Unfortunately, Meacham was right that many in the public and academia share the misconception the 1988 C-14 dating for the Shroud was somehow absolute, and the cloth is a medieval fake. Without a doubt, much of this misconception began as a result of the "case closed" attitude of Professor Edward Hall of Oxford and his colleagues when the C-14 results were announced -- and Hall's encouragement to any who doubted Oxford's results that they should also join the Flat Earth Society. More than thirty years have gone by since this announcement, with the media spotlight primarily on the "we told you so" professional Shroud debunkers. The labs themselves gave the impression that there was virtually no chance they could have been in error about the results. However, in recent years, those with experience in Shroud research, textile experts, chemists, physicists, and even some of the participants in the original C-14 dating have been publicly airing growing doubts about the 1988 results.

2006 *(February)*. Brunati wrote, "[...] I had tried to explain that, by now, one could be mathematically sure that the dating of the Shroud of 1988 had never had any value. And I also tried to make it clear that I had not arrived at this conclusion by formulating vague, less reliable hypotheses or theories, but, verifying, mathematics in hand, the text of the official report, published at the time in the English scientific journal *Nature*. A couple of topics were considered decisive. The first was that the report attributed to the samples, which via the laboratories had undergone dating, had dimensions and shapes were completely different from those of the pieces of the Shroud that Cardinal Ballestrero had given them to make the dating. The second was that the result achieved by dating those pieces of cloth of unknown origin had also been grossly altered. Probably, to be made more credible. That being the case, what is the point of those discussions? All of this I had exposed using formulas and, obviously, numbers. Trusting in their power of conviction but ignoring that most, perhaps because of an old dislike dating back to school, refused to take them into consideration. Therefore, very few must have been those who let themselves be persuaded by those data. Even though most of those who are interested in the Shroud continue not to believe that that cloth is truly medieval, they are always waiting for an explicit message, a message that must, however, come from absolutely credible sources. At the limit, from the same laboratories. After all, I too have always considered such a message more than appropriate. Also because I hoped that those indisputable inconsistencies were due to involuntary and, as such, admissible errors."

"So, in my time, I also wrote first to the British Museum and then to *Nature*. Receiving, however, as I have reported, truly disappointing answers. However, I did not give up, I continued in my attempts and, recently, I contacted the American journal *Radiocarbon*, a publication specialized in the field of dating which, among other things, in due course, had already dealt with the Shroud case. It should be noted that this is published by that same University of Arizona which also owns the Tucson laboratory which participated in our dating. This time, finally, I believe I have achieved, at least in part, a certain result and this is the reason for this article. It should be noted that, since *Radiocarbon* specializes in dating, writing them, I did not mention the differences between dated samples and samples taken, but I only focused on the alteration of the result. Trying to present it, however, in a slightly different way from that used until then, in such a way, that is, to make its voluntariness evident. So I set up my report remembering, first of all, that each laboratory, as soon as they received the samples, had divided them into several parts, each of which was then subjected to a dating which, as a result, had given its age value, accompanied by its tolerance. Data that were not expressed, as we all usually do, in calendar years, but, in BP years, Before Present. As, by convention, carbonists must always do."

"Many have raised reservations on the reliability of the radiocarbon dating system, considering it too influenced by all those pollutions to which an ancient cloth such as the Shroud may have inevitably been subjected. At least at first, however, I wanted to consider that system fully reliable. Therefore also taking into account the verification performed by dating, in addition to the Shroud, three old cloths, the so-called control samples, of which the age was perfectly known. Obtaining certainly positive results. So trust in the system. Provided, however, to be sure of the identity of the samples analyzed and the correctness of the various processes, to which the results obtained should have been subjected. Important in this regard, certainly was the division of samples to which I mentioned. Important, in that, if you wanted to know the age of the individual sample, you had to average all the partial results concerning him, subjecting them, therefore, to a series of arithmetic operations that, if they had not been carried out correctly, they could have altered it. Averages that were prescribed were weighted, with the weight function performed by the inverse of the square of tolerance. Not only. The averages of the age values were also to be made independently of those of the tolerance values. And, while the partial results were all collected in table 1 of the report, those of the averages were in table 2. Obviously, I tried to focus *Radiocarbon's* attention on the fact that one of these was not correct. To demonstrate this, I started with the calculation made by the British Museum to define the average of the partial results that the Oxford laboratory had obtained by dating the Shroud sample. Which had been divided into three parts, the dating of which, according to table 1, had led to the following results:"

"795 ± 65, 730 ± 45, and 745 ± 55"

"The 'weighted' average of which, according to table 2, was equal at 750 ± 30 years. Value which, when checked, was correct. That 750, in fact, actually corresponded to the weighted average of the three ages, 795, 730, and 745, and resulted from the use of a formula (which I will save you) which, in the end, was reduced to the following fraction rounded then to 750 years, 0794940 / 0.00106109 = 749.13 years. Also that ±30 which, according to table 2, was the average of the tolerances, the average, that is, between 65, 45, and 55, was correct, as it corresponded to the root square of the inverse of the result of the expression $1/65^2 + 1/45^2 + 1/55^2 = 0.00106109$. That 0.00106109, therefore, which previously had assumed the function of denominator in the fraction that had led to the average of the age, that is to 750, then it had also determined the ± 30 constituting the average tolerances. As a result, if it was claimed that the average age was 750 years, the average tolerance had to be equal to ±30."

"The Zurich laboratory, on the other hand, had divided its Shroud sample into five parts and had obtained the five following results:"

"733 ± 61, 722 ± 56, 635 ± 57, 639 ± 45, and 679 ± 51"

"whose average, always according to table 2, was equal to 676 ± 24 years."

"The formula that led to the weighted average of the ages, i.e., that 676, on average between 733, 722, 635, 639, and 679, was such as to become, simplifying, the following rounded fraction,"
"1.199273397/0.001773704 = 676.14 years. Rounded then to 676 years."

"The ±24, on the other hand, the average between the tolerances, the average, that is, between 61, 56, 57, 45, and 51, was obtained by making the square root of the inverse of the result of the expression:"

"$1/61^2 + 1/56^2 + 1/57^2 + 1/45^2 + 1/51^2 = 0.001773704$"

"Figure that, also in this case, is the same as that which led to that 676 years old, corresponding to the average age. So the same rule also applied to Zurich. If the average age was 676 years, the average tolerance had to be ±24 years. Finally, the Tucson laboratory had divided its sample into four parts, obtaining the following four partial results:"

"591 ± 30, 690 ± 35, 606 ± 41, and 701 ± 33"

"the average of which, always according to table 2, turned out to be equal to 646 ± 31 years. Again, the average of the four ages, i.e., the average between 591, 690, 606, and 701, that is, that 646 years was correct and was obtained with a formula that, in the end, assumed the aspect of the following fraction, 2.224957/0.00344595 = 646.67. By analogy with the previous cases, if the average of the ages had to be equal to the square root of the inverse of 0.00344595 be, that is, equal to ± 17. While, instead, in table 2 it turned out to be ±31. A result in full contradiction with that 646, which was indicated as the average age. A remarkable diversity that, at the moment, I just couldn't explain. For greater safety, I also checked the 18 results (ages and tolerances) obtained by the three laboratories when they had dated the three control samples. 15 of these averages, the 12 concerning age and tolerances detected by the laboratories of Zurich and Oxford and the three concerning the Tucson age measurements were correct: which confirmed, among other things, the validity of the verification method adopted. The three averages of tolerances regarding the measurements made at Tucson on the control samples, however, were all incorrect."

"Based on my checks, they should all have been equal to ± 20, while, according to table 2, they were equal to ± 32, ± 46, and ± 43. Why? Especially since, even in these cases, the value of the average age had been reached by passing from that number which would have allowed us to obtain the correct average of the tolerances. For example, the average age, 927 years, made by Tucson on the cloth of Nubia, the control sample number 2, had been obtained with a fraction whose divisor was the number 0.00344595, the square root of the inverse of which gave rise to that ± 20, the correct average of the tolerances. How come, then, that ± 32 of table 2? And why, then, for control sample 3 the ±20 becomes ±46, and ± 43 for the sample 4? It seems evident that the results were purposely changed to try to confuse the ideas in case of possible verifications, to raise the doubt that, who knows why, the measurements made at Tucson required special systems for calculating the means. An attempt to justify changing that ± 17 to ± 31? There could be no more doubts, therefore, that this was an indisputable alteration of the result."

"And the Editor Manager of *Radiocarbon*, replying to me with an email dated December 7 last *[2005]*, had to admit the error, writing I agree it is not exactly 31 years (I recognize that they are not exactly 31 years old). It was the first time that a carbonist, a carbonist moreover important, has recognized that the report in *Nature*, when it concerned the dating of the Shroud was incorrect. Even if, immediately afterwards, he closes the email, pointing out that this error, in any case, should be considered minimal. According to McClure, in fact, the scientists of the Tucson laboratory, being free to adopt the calculation process they preferred, had taken into consideration the standard deviation which brought the average of the tolerances to 29 and not to the 17 years indicated on *Nature*. So the error instead of being made up of the difference between 17 and 31 years was reduced to that, more modest, between 29 and 31. I could not agree and I immediately wrote to *Radiocarbon* in my second email of December 6. I could not agree, as the counting had not been carried out, as they claimed, by the laboratory, but by the British Museum in London, to which Tucson, like Zurich and Oxford, had sent, as soon as they were detected, its partial results. And in London they had adopted a single calculation criterion for everyone, doing the averages, as it appeared in *Nature*, without ever taking into consideration that standard deviation to which *Radiocarbon* was referred. Valid, therefore was the ± 17 and not the ± 29."

"And I added, also, that it had not been difficult to realize the reason that the British Museum could have induced them to make 31 that 17. It is always a good practice, when operating, as in our case, in the field of statistics, to make sure that the values to be averaged are not very different from each other. And, in fact, the dating protocol of the Shroud entailed the obligation to carry out a specific verification, the test called by the chi square. Which is considered exceeded if it gave, as a result, a significance level equal, at least, to 5%. Interestingly, in this regard, note that table 2, *Nature* reports all the significance levels achieved by doing that dating. They show that, while the control samples had reached values of 90, 50, and 30%, the dating of the Shroud was at a minimum; that is, a significance level of just 5% was reached. Thanks to which, however, the British Museum could consider admissible to average and conclude the dating. When, however, I redid the calculation, considering, instead of the wrong 646 ± 31, of which the British Museum had taken into account, the correct 646 ± 17, I realized that the significance level was it went from 5 to about 1%. In other words, making the right accounts, it appeared that the test with whoever could not be passed, so, since it was not lawful to make the final average, the conditions for concluding the dating were not met."

"The reason why they had changed that 17 to 31 was therefore evident. They had to make the test with the chi square appear, which would have allowed the execution of the final average and, therefore, to complete the dating. It is a further reason that replacement is considered not an accidental error but a measure taken on purpose and to understand the importance it had on the outcome of the dating. About significance levels -- when I found out that ±17 had become ± 31, I also asked myself: how come ± 31? Then, I realized that this was the minimum tolerance that had to be reached in order to have a significance level at least equal to 5%, a significance level that would allow us to say that the chi square test had been passed. Even if, taking a closer look at those counts, if we took into account another result that *Nature* provided, that is the value of chi square equal to 6.4, we would have noticed that not even that 5% was correct. In fact, it was only 4%. But, given the rest, this, we can consider it an irrelevant detail. Consequently, *Radiocarbon* Editor Manager Mike McClure, when he replied to me for the second time on December 11, had to admit: we agree that the chi square test gives a result greater than one [...]. With that, after acknowledging that the average of the tolerances was not correct, he returned to admit that the result of the chi square test was such as not to allow the execution of the final average and consequently to regularly conclude the dating. So, given that *Radiocarbon* magazine can be considered the spokesperson for scientists specializing in radiocarbon dating, these two admissions become the official recognition of the invalidity of that dating. That official acknowledgment that everyone expected and that we hope for definitively closes the dating chapter. If *Radiocarbon* agreed that dating wasn't serious, he was also determined not to publish it. Adding the excuse that, however, that was an old practice about which much had already been written and discussed. It seems to me that such an attitude is more than ever questionable. It is true that much has been written about it, but it has always been done on the basis of rather vague hypotheses."

"Now, however, we were faced with very clear conclusions which had been reached on the basis of certain motivations, of which they themselves had recognized the validity. It seemed to me that this should make them feel obliged to make public the new situation that had arisen. As, in their time, they had triumphantly announced that this sheet, being medieval, could not have wrapped the body of Jesus, today, all the more reason, they had to feel the duty to confess that they were wrong. In his first letter, by the way, McClure accused me of defaming the authors of the dating and asked me to formally withdraw from such a position. Having never, however, impressed single people, I did not seem to have to please him. I limited myself to denouncing, proofs in hand, of the sure misconduct and the very fact that *Radiocarbon* admitted them, demonstrates the validity of my accusations. Interestingly, however, that request was not renewed in the second letter. Although there is no doubt that the alteration has occurred, even if the reasons that may have induced those gentlemen to make use of it are evident, it remains to be explained why the conditions that forced laboratories and the British Museum to behave that way. Considering that this occurred during an important work, which was arousing interest all over the world and which also exposed them to all kinds of criticism. To answer this last question, however, I can only resort to hypotheses. The first of which is that, at a certain moment, they were forced to replace the authentic samples given to them by the Cardinal with pieces of cloth from who knows where, but medieval. If we consider what the *Nature* report told us about the shape and size of the samples, we must admit that it may have materialized. Why? They could have resorted to such an expedient to make sure they were able to achieve the goal they can actually think of: to prove that the Shroud was a fake."

"I do not think they feared that the result of a regularly conducted dating would do them wrong, as that those priests, resorting to some cheating, would make them date the samples of the cloth of the era of Jesus that did not come, however, from the Shroud. In theory, their task should have been to date correctly what would have been delivered to them. Others, first of all the British Museum, the guarantor of the test, [...] should have guaranteed the authenticity of the samples [...]. Nothing to do: the fear of replacement must have been such as not to even hesitate to publicly question even the correctness of the Cardinal. In short, that dating was a proof that had to prove, at all costs, that, not being the cloth of the time of Jesus, the Shroud had to be a fake. And it may have been the fear that those priests might replace the samples, seriously inducing them to think that it was more convenient to resort to that trick themselves. To arrange for them to replace the samples. They would certainly have behaved in an ethically incorrect way, but they would have been sure of obtaining the desired result, the same one they would have had by dating the authentic sample. This is how we ended up being grossly cheated. Without our being able to notice. We were sure that the samples taken had ended up in the hands of the laboratories. What we never managed to verify was whether the laboratories later dated them. We have never been able to do this, as Vatican representatives have never been allowed access to the laboratories. The troubles that then put Laboratories and the British Museum in difficulty could be due to the fact that the medieval fabric used as a substitute for the Shroud cloth, once dated, turned out to be a little too young to be the Shroud. If in the history of the Shroud of Turin there is a certain fact, this is that, in the years around 1350, it was publicly exhibited in Lirey, France. And those carbonists must have known it well, if they remembered it in the fourth line of their report in *Nature*. And that 691 ± 31 years obtained with dating, if it seems to be compatible with 1350, when taken into consideration a little carefully, is not always convincing. When translated from BP years into normal calendar years, in fact, that 691 ± 31 (see table 3 in *Nature*), doubles, becoming a first period ranging from 1262 to 1342, and a second from 1353 to 1384. But how could they have exhibited, in 1350, a cloth made in the period from 1353 to 1384? Furthermore, is a fabric that is 13th century dating back to the time of Jesus? So only the first interval, that between 1262 and 1384, would remain valid, even if not really convincing."

"Provided, however, that the correct result of dating can actually include this period. It should not be forgotten, in this regard, that the 691 ± 31 was obtained by averaging the results of Zurich and Tucson with that 750 ± 30 of Oxford, so dissimilar from the others that it can be considered responsible for the failure to pass the chi square test. What would 691 ± 31 have become if Oxford's response had also been such as to allow the test to be passed with whom? Had it been, that is, similar to those of Zurich and Tucson? The Shroud would have appeared less old, would have been less than the 691 BP years that resulted in that one manufacturing period compatible with the Lirey years. There is no suspicion, then, that that 750 ± 30 of Oxford, besides being a result defended with the teeth, was also a result obtained by, perhaps, resorting to a somewhat special dating, performed by the last laboratory entered in function, from that English laboratory which, being also physically very close to the British Museum, could have been kept as the last means of arranging things? Suspicions that got stronger when we knew what actually happened. When we learned, to begin with, about the delay with which Oxford presented its result. The three labs had received the samples at the same time: end of April in Turin. Tucson started immediately and finished his analysis, sending his results to London on June 8th. Zurich started on May 6th and ended in mid-July. That was the period in which the carbonists met in Dubrovnik for their congress, the period in which Prof. Gove, who was also present, came to know that in Oxford they had not yet dealt with that dating. The same professor tells us about it in his book[1]. That, he explains how, after making several calls, only on September 9th he learned that the three laboratories had agreed to accept that result which the English laboratory had finally reached."

"How to explain such a delay? And, also, why, while Zurich and Tucson had sent their results to London without asking anyone's opinion, did Oxford have to have the approval of the colleagues? Perhaps because, since it could not be compatible with the chi-square test, that result would have required recourse to those messes that could have damaged the good name of everyone who was doing that test? And how to explain the sensational reaction of relief expressed by Professor Gove, in his book, to that announcement:? 'Donahue said all three labs were in agreement [...] *I was considerably relieved.*' Everything seems to confirm that the hypothesis that, in fact, they have not dated the Shroud, but pieces of medieval cloth, which, with dating, would have proved a little too young to have already existed in 1350, in the years when the Shroud had been displayed at Lirey. Hence the need to make older a few years the result obtained, so as to make that cloth appear not only as medieval, but also manufactured at least a few years before the time of Lirey. But if that result solved one problem, it created another, as it would not have made it possible to pass the chi square test. Unless resorting to the alteration of the average of Tucson's tolerances, an artifice that, if it had been discovered, would have been really a trouble for the good name of those Institutes. But that, perhaps, one could risk adopting. Also because, not directly interested in the final result, if noticed, it would not have been considered so influential."

"All maneuvers, however, that, making everything appear regular, in the end always had the task of hiding the replacement of the samples. I believe that my hypotheses, supported by many real data, can be considered plausible. If, however, those who read me were not of my opinion, the alteration of the result that still completely invalidates the dating remains indisputable. It is clear, in any case, that the purpose of my communication is not to denounce this scandal, nor to open processes. I am only interested in one thing: no longer being forced to say that, scientifically speaking, the Shroud must be considered medieval and, therefore, a forgery. And I hope that the Curia of Turin will take note of it, not only by continuing to venerate it, as it rightly does, but also by denying what has been said, in that official statement issued in October 1988 which provided what was then believed to be the result of dating [...].

[1]Harry Gove, professor emeritus of the University of Rochester (N: Y) - <u>Relic, Icon or Fake? Carbon dating the Shroud of Turin-</u>University of Philadelphia - 1996 - Prof. Gove, in Rochester, had developed the dating process that made it possible to date very small samples, the process that, therefore, had allowed the dating of a precious cloth like the Shroud. He felt offended when, by reducing the laboratories that would make the dating to just three, the Vatican excluded his, that of Rochester, from that test.

Source: Brunati, Ernesto. "La corrispondenza con *Radiocarbon* sulla datazione della Sindone." *Collegamento Pro Sindone*, February 2006. http://www.sindone.info/BRUNATI2.PDF.

Comments: As with several articles by Van Haelst, there is an overlap in material in several of Brunati's articles but also additional material. Once again, to be thorough, I have reproduced most of this article despite the duplication. Also, there is frankly so much complex information to digest, it's probably a good thing to read some data more than once.

Note that Brunati pointed out that *Radiocarbon* made two admissions to him that indicated to him "these two admissions become the official recognition of the invalidity of that dating" but that they didn't publish anything about it. Given the fact that the journal is published by the University of Tucson, one of the labs involved in the dating, that is not a surprise.

2008. Christopher Ramsey, who was involved in the 1988 dating and is currently the head of the lab, wrote a general article on C-14 dating and stated, "When radiocarbon date outliers (i.e., dates that do not make sense archaeologically compared with other data) are encountered, these are sometimes due to some measurement problem, but much more often they are due to misinterpretation of the sample context."

Source: Ramsey, C. Bronk. "Radiocarbon Dating: Revolutions in Understanding." *Archaeometry* **50**:2 (2008): 249–275, on pg. 263.

Comments: There have been several indications that there were outliers in the dates that the three labs had. Ramsey also stated, "There is a lot of other evidence that suggests to many that the Shroud is older than the radiocarbon dates allow and so further research is certainly needed. It is important that we continue to test the accuracy of the original radiocarbon tests as we are already doing. It is equally important that experts assess and reinterpret some of the other evidence. Only by doing this will people be able to arrive at a coherent history of the Shroud which takes into account and explains all of the available scientific and historical information" (http://greatshroudofturinfaq.com/Science/Dating/ramsey-on-shroud.html). It's unfortunate that the Vatican has been, apart from one major statement in 1990, unwilling to play its part in testing the 1988 dating accuracy as Ramsey has suggested.

Interestingly, a similar statement to the one above had been posted on the Oxford web site per a posting on www.shroudstory.com of December 31, 2008 but when one now clicks on that link, it is dead. However, it is still available via a site that stores pages no longer in its original place. The link for this statement is: https://web.archive.org/web/20090322182548/ http://c14.arch.ox.ac.uk/embed.php?File=news.html.

Dr. Jull from Arizona, who had been involved in the 1988 Shroud dating (and is now the head of the lab) confirmed in an email to Dr. Ramsey of Oxford that there's likely a "sample context" problem with the Shroud. Sue and I exchanged various emails with scientists from the three labs in 2008 and 2009. In an email of August 21, 2008 to Ramsey and copied to Georges Bonani of Zurich and to Sue, Jull said, "I think Sue Benford's paper in *Chemistry Today* (http://www.shroud.com/pdfs/benfordmarino2008.pdf) has summarized of *[sic]* a lot of interesting information (although I don't agree with many of the conclusions), however their Fig. 7 shows a picture of the original 'Raes threads' which are clearly flax (and not cotton). Hence, if someone finds cotton in it, there's a problem there." (Rogers found cotton per his 2005 article in *Thermochimica Acta*.)

Just a few days before the aforementioned Jull email of August 21st, he said in another email, referring to the presence of cotton in the sampling area, "OK, lets *[sic]* suppose we accept it might have some cotton. Why does that invalidate the date?" That's a mystery to him, given that the Shroud is linen? As Rogers pointed out in his *Thermochimica Acta* paper, the C-14 sample was not representative of the main cloth and was thus invalid for determining an accurate date for the Shroud. Look at what transpired, though, in several days: on the 18th Jull was questioning why cotton in the sample would affect the date, and on the 21st he emailed Ramsey that [...]. [I]f someone finds cotton in it, there's a problem there." Jull also added in his email of the 18th, "It seems to me that redating a new piece of the Shroud is the most effective solution to this question" *[my emphasis]*.

2008. STURP chemist Ray Rogers wrote in a posthumously-published book, "In many cases where questions arise, an appeal is made to 'authority.' There can be no question about the authority of the radiocarbon investigators; however, true scientists like to see all loose ends questioned and tested. With the Shroud, neither the radiocarbon investigators nor the authorities in Turin have cooperated in attempts to resolve the 'dating problem.' The Church officials appear to be content to have society view the Shroud as a medieval hoax, and the radiocarbon laboratories have refused to consider the possibility that they were given a spurious sample. In a manner uncharacteristic of rigorous scientists, they refuse to allow observations on retained samples. They also refuse to do their own simple chemical observations. They refuse to discuss or show any photomicrographs of samples they might have. This kind of action is all too characteristic of Shroud studies. Emotions tend to overwhelm science [...]."

"[...] [T]he sample was approved at the time of sampling by two textile experts, Franco Testore, professor of Textile Technology at the Turin Polytechnic, and Gabriel Vial, curator of the Ancient Textile Museum, Lyon, France. No chemical or microscopic investigations were made to characterize the sample. I believe that was a major disaster in the history of Shroud studies. Control samples should always be retained to enable confirmation of results at a later date. Retained samples, if any, have not been made available for study. This leads one to question the ethics or rigor of any 'scientists' involved in the process. Is something being hidden?"

Source: Rogers, Raymond N. A Chemist's Perspective on the Shroud of Turin. Edited by Barrie M. Schwortz, (Lulu.com), 2008, pg. 63.

Comments: Rogers was temperate in his criticism of the Turin authorities, despite the fact that he had tried to contact them several times in regards to his findings and he never even received the courtesy of a single reply. One would have thought that after the time, effort (and no doubt some of his own money) that Rogers put in on the Shroud, the authorities would have given him the consideration of a reply, but they did not, which is a sad commentary. Rogers concluded in his book (pg. 76), "A rigorous application of scientific method would demand a confirmation of the date with a better selection of samples."

Regarding the labs' samples, Barrie Schwortz was allowed in 2012 to photograph one of Arizona's leftover samples, although he had been promised he would be able to photograph two (http://www.shroud.com/pdfs/arizona.pdf). Photos of Oxford's samples were released in

2014 after Pam Moon of England put in a "Freedom of Information" request (https://archdams.arch.ox.ac.uk/?c=1203&k=1bcdc90a8b).

2008. The late Robert Villareal, a physicist from Los Alamos Laboratories, who had been given Raes samples from Ray Rogers, presented new evidence at an international Shroud conference based on his work with eight other researchers. Villarreal had studied a spliced fiber from the Raes sample (Thread #1) at Ray Rogers' request. The two ends of the fiber appeared to be different in color and amounts of coating. Rogers had asked if Villarreal could use his highly sensitive lab instrumentation to analyze the thread. In addition, Villarreal was also asked (by Rogers' colleague, Barrie Schwortz of STURP, and Benford) to analyze two other threads (Threads #7 & 14) from John Brown's lab in Marietta, Georgia. Sadly, Rogers died before the work was completed. Villareal primarily used a Time-of-Flight Secondary Ion Mass Spectrometry (ToF-SIMS) and a Fourier Transform Infrared Spectroscopy (FTIR) with Reflectance Mode Capability. The ToF-SIMS results showed that the spectra from the two ends were similar to cotton rather than linen (flax). After several scans of individual fibers from Thread #1, the FTIR data demonstrated that the two ends were definitely cotton and not linen (flax). The crust appeared to be an organic-based resin, perhaps a terpene species, with cotton as a main sub-component. The final results of the FTIR analysis on all three fibers taken from the Raes sampling area (adjacent to the C-14 sampling corner) led to identification of the fibers as cotton and definitely not linen (flax).

Source: Villarreal, Robert with Barrie Schwortz and M. Sue Benford. "Analytical Results On Thread Samples Taken From The Raes Sampling Area (Corner) Of The Shroud Cloth". Presented at "The Shroud of Turin: Perspectives on a Multi-Faceted Enigma" conference in Columbus, Ohio on August 14-17, 2008. Video at: https://www.youtube.com/watch?v=86wWOMGqsWQ&feature=youtu.be.

Comments: Villarreal pointed out that one of the first rules of radiocarbon dating is that any sample analyzed to characterize an area or population must necessarily be representative of the whole. Villarreal's analyses of the three thread samples taken from the Raes and C-14 sampling corner led him to conclude that this was not the case.

"Terpenes" are defined as a "large and diverse class of organic compounds, produced by a variety of plants, particularly conifers, and by some insects."

2008. In a story released by the University of Arizona, magnified fibers from a Shroud sample was shown. The caption read, "Polarized Light Microscopy was used to confirm that the *major* fiber content of the sample is linen" *[my italics]*.

Source: "Art and Science Converge in State Museum Exhibit," by University Communications, November 3, 2008. Accessible at http://uanews.org/node/22384.

Comments: The fact that the word "major" is used in conjunction with the fiber content seemingly could be a reference to cotton being a minor content. Arizona was contacted for additional pictures but no response was received.

2008. In the introduction to an article in about dating textile relics of the medieval period, using the same C-14 technique used on the Shroud, the authors wrote, "Dating of materials connected to faith is always a delicate matter; however the goal of this paper is not the one of emphasizing the results themselves (i.e., whether the dates of the relics are compatible or not with their believed attribution), but the methodology used, and especially the importance of a correct strategy of sampling."

In their "Results and Conclusions" section, they stated, "[T]he most important aspect of this work, from the point of view of physicists working in the field of AMS dating, is the one concerning sampling strategy. First, sampling should always be done in agreement with and under the guidance of scholars and people involved in the historical or archaeological problem. In addition, whenever possible, collecting several samples from the object to be dated (as we did in the case of the two frocks) is definitely the right approach in order to reduce the possibility of ambiguities."

Source: Fedi, M.E., et al. "AMS radiocarbon dating of medieval textile relics: The frocks and the pillow of St. Francis of Assisi. *Nuclear Instruments and Methods in Physics Research* ***B* 266** (2008) 2251-2254, on pages 2251 and 2254.

Comments: Sadly, the Shroud C-14 scientists did not work collaboratively with scholars who were familiar with the unique problems associated with the Shroud, as they pressed to work independently, nor did the Turin authorities choose to take more than one sample to test.

2009 *(January)*. I emailed Jull in late January 2009 about many of the issues being detailed in this book. He emailed me back on the 26th, "Surely, it's 21 years ago and we need to move beyond the process and **politics** *[my bolding]* of the original event. The only question is the validity of the sample, which in my opinion is still valid. The idea of science is that measurement & hypotheses can be tested." But, as recounted previously, Jull had told Barrie Schwortz it "was a shoddy affair and should be redone." Also, he was quoted in an Italian book in 2011: "This is a bad level. Normally, with such a result, I make the measures again." I emailed back to Jull on the 27th, "I don't agree that the politics are irrelevant. They definitely affected how the test was conducted and because the politics were so rampant, with many of the sordid details laid out in our two articles, there are valid reasons on the politics alone to be able to question the results. But more importantly, the sample is scientifically suspect. The one sample used for all 3 labs was an area that several people warned was problematic. The authorities also were advised to take samples from several different areas of the cloth; this was not done. Quite frankly, I don't understand how one can put one's confidence in the scenario that unfolded."

Source: Personal emails among Benford and/or author and members of the three labs involved in the 1988 dating -- various members of the labs also having been copied.

2009 (December). Christopher Ramsey of Oxford authored a paper titled "Dealing with Outliers and Offsets in Radiocarbon Dating," in which he stated four reasons why C-14 dates could be incorrect:

* The radiocarbon measurement of a particular sample might not be correct.

* The radiocarbon ratio of a sample might be different from that of the associated reservoir.

* A whole set of radiocarbon measurements might be biased in some way relative to the calibration curve -- either because the measurements themselves are biased or because the reservoir from which the sample draws its carbon might not have the expected radiocarbon isotope ratio.

* The sample measured might not relate to the timing of the event being dated.

Source: *Radiocarbon* **51** (3) (December 2009): 1023-1045.

Comments: One got the impression from all of the labs right after the C-14 dating that none of these possibilities could have applied to the Shroud, much less even existed.

2010 (December). Timothy Jull was co-author of an article for which the abstract began, "We present a photomicrographic investigation of a sample of the Shroud of Turin, split from one used in the radiocarbon dating study of 1988 at Arizona." See the comment below.

Source: Freer-Waters, Rachel A. and A.J. Timothy Jull "Investigating a Dated Piece of the Shroud of Turin." *Radiocarbon*, **52**:4 (2010): 1521-1527.

Comments: The 1989 *Nature* report gave the impression that Arizona had used up all of the samples they had been given. As previously shown, Damon had admitted they had kept a piece. We've also seen that the recollections of the various scientists in Arizona conflicted on various points. On October 17, 2001, my late wife, Sue, emailed Dr. Larry Schwalbe of STURP regarding the possibility of Schwalbe communicating with some of the C-14 scientists. Sue wrote to Schwalbe, "My previous interactions were with Doug Donahue. I had dealt with him on some non-Shroud testing questions so was able to get to him a second time when the question came up about the Shroud. Needless to say, this was a sore subject with him. He got very defensive when I suggested that Arizona had reserved the smaller piece as suggested by Walsh. He repeated several times (quite loudly I might add) that they did no such thing and that they used everything they were given. I know Bryan *[Walsh]* disputes this but I'm just telling you what Donahue said in no uncertain terms." Barrie Schwortz published an article in 2012 (https://www.shroud.com/pdfs/arizona.pdf), detailing photographing that year an extra piece that Arizona had. Conflicted recollections, indeed.

Radiocarbon is a peer-reviewed journal. Jull is the current editor and was co-author on this paper. The question can be asked if it was given the same treatment as any other paper.

For an excellent response to the Freer/Jull paper by New Zealand's Mark Oxley, see: http://www.shroud.com/pdfs/oxley.pdf. Oxley's article also has some detailed information regarding the discrepancy regarding sizes and weights of the samples.

Even Italian Shroud skeptic Gian Marco Rinaldi questioned the data put forth in the Freer/Jull article. He wrote at http://sindone.weebly.com/articoli.html, (google translation): "The authors do not publish the article by *Radiocarbon* a photograph of the entire fragment (just a small detail), but you can see the entire picture of the obverse and reverse in a video that is on site of the Arizona Museum. At the minute 02.31 there is a picture of the right, where they are in greater evidence threads warp (horizontal). [At] 01.23 there is the reverse with the weft threads (vertical). Gather photos from 11.30 minute for a couple of minutes."

"Although with a video screen that is very small, the number of threads, for the warp as in the weft, it can *[be]* count*[ed]* with ease. You see very well that the thread density is greater for the warp and the weft, contrary to what reported by Freer and Jull. To calculate the number of threads per centimeter, it is necessary to know the size of the fragment, which is in the shape of rectangle. In the article dimensions are such as about 5 to 10 mm, but it is approximated numbers and taken to round figures, because *[in]* the photo it is seen that the ratio between the size is not of 1:2. Leveraging a scale shown in the photographs of video, where a segment indicated as equivalent to a millimeter, can be traced back to an approximate estimate of the fragment size is inserted, which are approximately 6 to 8/9 mm. Using these measures, after counting the threads of the fragment we arrive at a valuation well compatible with the values of about 38/cm and 25/cm found by various experts but totally incompatible with those of Freer and Jull."

"So we must fear that the radio-date sample was different from the rest of the shroud. We can fairly ask how it is that Freer and Jull have fallen into such serious mistake. It should be noted that it is not difficult to count the threads. You need not be textile experts. You do not need provide a sample of the cloth just a close-up photograph, it is very small as well, such as the one shown in the video. Anyone can do it, just who can count, in this case, by one until recently more than twenty. Perhaps in Tucson *[they]* do not know how to count to twenty?"

Perhaps it's not so much the inability to count as perhaps another example of how Gove had said (see last "1988 *(May)*" entry referenced above) that results can be steered by scientists.

The Italian agency ENEA (New Technologies, Energy and Sustainable Economic Development), made news in 2010 with experiments that duplicated on linen via excimer lasers the color and penetration depth of the Shroud image. (See http://www.acheiropoietos.info/proceedings/proceedings.php).

2011. Cardinal Tarcisio Bertone stated, "The analysis of carbon-14 seems to have been a mistake, particularly because of prejudices, of which it is useless to speak, because the verdict was decided even before performing the analyses."

Source: Marinelli, Emanuela. "The Setting for the Radiocarbon Dating of the Shroud." Presented at 1st International Congress on the Holy Shroud in Spain -- Valencia -- Centro Español de Sindonologia (CES), April 28-30, 2012, pg. 16. www.shroud.com/pdfs/marinelliv.pdf.

Comments: Prominent Italian historian and Shroud skeptic Andrea Nicolotti admitted in one of his books, "With the benefit of hindsight, we can say that the undertaking of radiocarbon dating was handled abnormally from an organizational point of view [The Shroud of Turin: The History and Legends of the World's Most Famous Relic (Waco: Baylor University Press, 2019, pg. 417)]. Nicolotti also mentioned (pg. 203) some information that lends support to the idea that there could have been an invisible reweave: "A second restoration of the Shroud is worthy of mention, for which on December 18, 1595, papal authorization was requested: it was necessary in fact to adjust some parts of the Shroud that were damaged in the fire of 1532, 'which, with lapse of time and with folding and unfolding, have become so extensive that they have great need of mending as soon as possible'." This would have been sixty-one to sixty-two years after the repairs done in 1534, two years after the fire. At my conference presentation in Ancaster, Canada in 2019, I showed a slide by the late archaeologist Maria-Grazia Siliato, who detailed various repairs that had been done on the Shroud through the centuries; this one initiated in 1595 was not listed.

2011. Emanuela Marinelli said that when Tite was asked if there was an official report on the sample taking, he said to "Ask Gonella -- that wasn't my job."

Source: *The Night of the Shroud (La Notte de la Sindone)*, documentary directed by Francesca Saracino, 2011. In 2016, it was revised and retitled "Cold Case: The Shroud of Turin," which is available at amazon.com. I have a review copy of the original version, which has an English voiceover. The revised version has English subtitles. Material pertaining to the post sample-taking period starts at about the thirty-six-minute mark.

Comments: Tite was the overseer of the project. He should have been ultimately responsible for an official report.

2012. Cambridge art historian Thomas De Wesselow, who specializes in art of the 14th century, the period when Shroud debunkers say the cloth was made, believes the Shroud does not fit into the context of 14th century art. It is also worth noting that De Wesselow is not a Christian. He had some blunt words about the Shroud C-14 dating process in his chapter "The Carbon-Dating Fiasco": "Doubting *Nature*, the voice of Science, is quite a proposition. It is tempting, therefore, to bow to the authority of this scientific pronouncement and to give up the complex and difficult struggle to understand the Shroud. But on reflection, we know that all scientists can err, and even the most polished scientific article can mask errors and false assumptions. So anyone who is serious about comprehending the Shroud will want to subject the carbon-dating result to rigorous scrutiny -- the sort of scrutiny used to evaluate all scientific evidence [...]."

"[...] Other factors can introduce a significant degree of uncertainty into the interpretations of the data [...]. Contamination is a major problem. Although various potential sources of contamination are known, including volcanic activity and carbon exchange with the surrounding environment (air, smoke, groundwater, etc.), it is not always possible to explain the cause of an erroneous reading. Due to the ever-present possibility of contamination, no radiocarbon date is absolutely certain. A recent review of the history of carbon dating concludes, with direct reference to the Shroud, that 'the issue of organic reactions and non-contemporaneous contamination of ancient materials can be a very serious and complex matter, deserving quantitative investigation of the possible impacts on measurement accuracy[1].' In other words, the problem of contamination is severe and difficult to quantify [...]."

"[...] Unfortunately, the interpretation of the 1988 carbon-dating results was left to the physicists who performed the tests, men who knew little about the Shroud and had no experience in interpreting such a complex artifact."

"The problems with carbon dating are most starkly revealed when the results produced by different labs differ among themselves. In 1989, for instance, a year after the Shroud test, the Greek archaeologist Spyros Iakovidis was confronted by a totally incoherent result: 'I sent to two different laboratories in two different parts of the world a certain amount of the same burnt grain. I got two readings differing by 2,000 years, the archaeological dates being right in the middle. I feel that this method is not exactly to be trusted[2]'."

"That there are general problems with the technique is acknowledged by carbon-dating scientists themselves. Consider, for instance, the following caution in a 1985 conference paper, one of whose joint authors was Willy Wolfli, one of the professors responsible, three years later, for the carbon dating of the Shroud: 'The existence of significant indeterminate errors can never be excluded from any age determination. No method is immune from giving grossly incorrect datings when there are non-apparent problems with the samples originating in the field. The results illustrated [in this paper] show that this situation occurs frequently'."

"This is a startling admission. According to Wolfli and his colleagues, in the field of carbon dating gross errors occurs frequently. But, while the scientists discuss these problems among themselves, they are less ready to dent the prestige of their discipline in public[3] [...]."

In Part 1, details of the 1983 "laboratory intercomparison test" were recounted. De Wesselow commented, "The inter-laboratory comparison exercise shows how unreliable the carbon dating of cloth was prior to the 1988 Shroud test. It was still unreliable immediately afterwards. In 1989 Britain's Science and Engineering Research Council (S.E.R.C.) decided to conduct a trial in which the carbon dating technique itself would be tested. Thirty-eight laboratories were involved in the trial, each being asked to date artefacts whose age was already known. (Oxford did not participate because they claimed during the time the intercomparison was done, engaged in a research and development program of our system that prevented them from using their typical dating methods.) The findings, reported in *New Scientist* under the headline 'Unexpected errors affect dating techniques', were salutary. It was found that 'The margin of error with radiocarbon dating [...] may be two to three times as great as practitioners of the technique have claimed [...].' 'Of the thirty-eight *[laboratories]*, only seven produced results that the organizers of the trial considered to be satisfactory[4].' In other words, about 80 percent of the labs failed the test. The three laboratories that dated the Shroud the previous year employed a technique known as Accelerator Mass Spectrometry (AMS), which 'came out of the survey badly'. According to one of the organizers of the trial, 'some of the accelerator laboratories were way out when dating samples as little as 200 years old[5]'. So, just a year after the Shroud was damned by AMS, the authority of this carbon-dating technique itself took a severe blow."

"There is a vast discrepancy, then, between the popular perception of carbon dating as infallible and its true scientific status. The fact is that carbon-dating results are often wrong, that the claims made on behalf of carbon dating are often inflated, and that the AMS technique used in 1988 to date the Shroud is (or was) particularly error prone. The purveyors of any technology, carbon dating included, are inclined to exaggerate its power and usefulness. Also, being physicists, so not embroiled in the business of making historical sense of their findings, they probably have a tendency to underestimate the method's rate of failure. Those responsible for the historical interpretation of ancient artefacts, usually archaeologists, are the ones who decide whether or not to reject carbon-dating results. But, because archaeologists were excluded from the 1988 testing of the Shroud, scientific caution was thrown to the winds when the results of this high-profile test were announced[6]."

"How was this situation allowed to occur? The answer lies in the sorry history of the project. People tend to envisage the carbon-dating result as a nice, neat number churned out by a machine, an impersonal, objective answer to a human query. If only the contents of science journals were so straightforward. All scientific work is conditioned by human concerns, and the Shroud carbon dating, in particular, was the product of a lengthy, messy process of **politicking** [my bolding] that resulted in a deeply flawed procedure, dictated by the Vatican. The public is for the most part ignorant of the quagmire of self-interest and scientific compromise on which the 'fact' of the Shroud's carbon dating tests [...]."

"[...] Keen to be involved in the carbon dating of the Shroud, should it be permitted, was Harry Gove, one of the inventors of AMS. A rather egotistical character, Gove was interested in the project not because he wanted to find out about the Shroud, but because he reckoned it would provide 'a highly public demonstration of the power of carbon dating by AMS[7]'. In the 1980s he assumed leadership of a group of carbon-dating scientists which began lobbying the Catholic Church for the opportunity to date the Shroud. All were conscious of the potential publicity value of such a test. Prominent among them was Teddy Hall, a professor at Oxford, who was trying to raise funds to endow a chair at the university, a cause he knew would be well served by the high-profile Shroud project.

Gove, meanwhile, saw STURP as a biased, Christian organization and a rival to his own group, and agitated to have them excluded from the carbon-dating exercise, despite their detailed knowledge of the cloth [...]."

Regarding the 1986 Turin Protocol, De Wesselow commented: "Remarkably, however, the Turin Protocol contained another clause that compromised the stated need for the lab representatives to have 'complete knowledge' of the sampling process. To enhance the credibility of the test with the general public, the lab representatives undertook to receive the Shroud samples and control samples blind: 'These shroud samples will be distributed to the seven laboratories in such a way as to ensure that the seven laboratories are not aware of the identification of their individual sample[8].' Obviously, this meant that they could not witness the entire process of which they were supposed to have 'complete knowledge'."

"There was no scientific justification for this decision. In fact, it was a sham. Everyone at the Turin workshop understood that it was impossible for the tests to be conducted blind, for the simple reason that no control cloth could be found matching the distinctive weave of the Shroud. They knew that the Shroud samples would be recognized the moment they were unpacked." However, the majority of the delegates at the workshop -- all except Gove and Meacham -- were concerned that the carbon-dating test should be *seen* to be done blind. And so they settle on a *faux*-blind sampling procedure, designed to reassure people that the test was 'objective', even though it meant that they themselves would not be able to keep track of the samples, jettisoning a crucial bulwark against any imputation of fraud. Frankly, it beggars belief that a group of eminent scientists should agree to compromise and misrepresent a scientific test for the purposes of propaganda [...]."

"[...] How much faith should we have in the 1988 carbon-dating result? Not as much as is generally assumed. Given the patchy record of the scientific technique and the shenanigans of the Shroud carbon-dating project itself, it would hardly be surprising if an error was made. On what grounds can this badly organized test be considered immune to the many problems that afflicted the science of carbon dating in the 1980s? Recognizing the potential for error is one thing, though; deciding that something actually did go wrong is another. What reasons are there in this particular case for disbelieving the carbon-dating result?"

"First of all, dating the Shroud to the Middle Ages makes it literally incomprehensible. For over a century mainstream scholars have viewed the Shroud, *a priori*, as a medieval artefact, and for over a century they have completely failed to make sense of it. This is unsurprising, for, as we have seen, the Shroud is inconceivable as a medieval work of art and can be understood neither as a deliberate 'recreation' of Christ's burial cloth nor as a bizarre accident. The onus is on those who uphold the carbon dating result to integrate it into a full and adequate description of the Shroud's origin -- just as archaeologists would do with any other carbon dating result. This they have been conspicuously unable to do. The poverty of the carbon-daters' own understanding of the problem is illustrated by Teddy Hall's comment at the London press conference that someone in the fourteenth century 'just got a bit of linen, faked it up and flogged it[9]'."

"[...] The carbon dating of the Shroud will probably go down in history as one of the greatest fiascos in the history of science. It would make an excellent case study for any sociologist interested in exploring the ways in which science is affected by professional biases, prejudices and ambitions, not to mention religious (and irreligious) beliefs. And it should certainly serve as a warning to practitioners of any discipline tempted to see their work as more important and 'fundamental' than any other. Research on the Shroud is like a microcosm of all human knowledge, a great multidisciplinary effort to describe a perplexing phenomenon as elegantly and comprehensively as possible. It so happens that in the case of the Shroud, carbon dating has so far turned out to be less useful than a study of needlework. (Stitches are easier to observe and interpret than atom ratios, which makes them a relatively reliable source of information about old textiles.) Carbon dating may still make a valuable contribution to sindonology, if the Catholic Church ever allows further tests, and if those tests are integrated into a full, interdisciplinary research programme, as Professor Ramsey recommends[10]. In the meantime, we can safely ignore it and concentrate on more productive avenues of research."

Footnotes *[Note: I have adjusted and/or supplemented De Wesselow's original endnotes with additional information to make it easier for the reader to find the resource, including online.]*

[1] Currie, L. A., "The remarkable metrological history of radiocarbon dating (II)," *Journal of Research of the National Institute of Standards and Technology*, **109**/2, pp. 185-217. Accessible at http://hbar.phys.msu.su/gorm/dating/histc14.pdf.

[2] Quoted in Wilson, Ian. The Blood and the Shroud. (London: Free Press), 1998, pg. 193.

[3] Meacham, William, editor. "Radiocarbon Measurement and the Age of the Turin Shroud: Possibilities and Uncertainties" in *Turin Shroud: Image of Christ? Symposium and Exhibition of Photographs*, March 3-9, 1986, Proceedings. Hong Kong: Turin Shroud Photographic Exhibition Organizing Committee pp. 41, 42, 43 and 53. An abbreviated version of this paper was published in *Shroud Spectrum International*, No. 19, 1986, which is accessible at http://www.shroud.com/pdfs/ssi19part4.pdf.

[4] Coghlan, A., "Unexpected errors affect dating techniques." *New Scientist*, September 30, 1989, pg. 26.

[5] Ibid.

[6] Meacham and two Italian archaeologists offered to supervise the Shroud C-14 dating but their offer was rejected. See Meacham's book, pg. 83.

[7] See Gove's book, pg. 14.

[8] Quoted in Antonacci, Mark. The Resurrection of the Shroud: new scientific, medical and archaeological evidence. (New York: M. Evans), 2000, pg. 177.

[9] Quoted in Wilson, Ian. The Blood and the Shroud. (London: Free Press), 1998, pg. 7.

[10]See the Ramsey statement above under "2008."

Source: De Wesselow, Thomas. The Sign: The Shroud of Turin and the Secret of the Resurrection. (London: Penguin Books), 2012, pp. 160-172.

2014. Australian Shroud blogger Stephen Jones started a series in which he put forth the hypothesis that the labs results were the result of a computer hacking. He summarized his findings to me in an email of September 4, 2016 as shown below.

"My hacker theory began in 2007 when I read in David Sox's book, "The Shroud Unmasked" (1988), the account provided by an eyewitness Prof. Harry Gove, of very first radiocarbon dating of the Shroud at Arizona laboratory. That the 'calculations were *produced on the* [AMS] *computer, and displayed on the screen*.' Sox was not at that time told by Gove the date on the screen (except that the Shroud was closer to 1000 than 2000 years old) but Gove in his 1996 book "Relic, Icon or Hoax?: Carbon Dating the Turin Shroud," revealed that it was '1350 AD'."

"In the late 1980s/early 1990s I was the Systems Administrator of a wide area network of 7 Western Australian hospitals' UNIX computer systems. As part of my job interest in computer security, I read Clifford Stoll's book, 'The Cuckoo's Egg' in which he recounted his part in discovering in 1986 the hacking of university and military computers by German hacker Markus Hess. Coincidentally Stoll had worked at Arizona University and Hess was in the same small German hacker ring as Karl Koch, whom I allege had installed Timothy W. Linick's program on Zurich and Oxford's AMS computers."

"So I realised in 2007 that it was not the actual radiocarbon dating of the Shroud that those in Arizona's laboratory were seeing, but what the AMS *computer* was displaying. That between the actual carbon dating by the AMS system and those watching the computer screen, was a *computer program*! So one explanation of why the authentic first-century Shroud had a 1260-1390 radiocarbon date, is that a hacker had installed a program in the three laboratories' AMS computers which substituted* the Shroud's actual radiocarbon date with bogus dates, which when combined and averaged made it appear the Shroud dated shortly before its first undisputed appearance at Lirey, France in *[c.]* 1355."

"However, it was not until 2014, when I read again page 264 of Gove's book, which stated of that first Arizona dating of the Shroud that: '*All this was under computer control* and the *calculations produced by the computer* were displayed on a cathode ray screen,' that I posted my first blog post which asked, 'Were the radiocarbon dating laboratories duped by a computer hacker'?"

"I then in 2014 did a Google search on '1989' and 'hacker' and discovered that a German hacker Karl Koch had been inexplicably murdered in May/June 1989, and his murder made to look like suicide [...]."

"According to my first post of 22 February 2014, 'Were the radiocarbon dating laboratories duped by a computer hacker?' It was in 2007, after reading Sox's account of Arizona's first C14 dating run: At 9.50 a.m. what matters to the layman was available -- the results of the measurements, the first carbon dating test on the Turin Shroud [...]. The night before the test Damon told Gove he would not be surprised to see the analysis yield a date around the fifth-century, because after that time the crucifixion was banned and a forger would not have known of the details depicted so accurately on the Shroud. Timothy Linick, a University of Arizona research scientist, said: `If we show the material to be medieval that would definitely mean that it is not authentic. If we date it back 2000 years, of course, that still leaves room for argument. It would be the right age -- but is it the real thing?' [...]. Shirley Brignall [...] and Gove had a bet. Gove said 1000 years although he hoped for twice that age. Whoever lost was to buy the other a pair of cowboy boots. The calculations were produced on the computer, and displayed on the screen. Even the dendrochronological correction was immediately available. All eyes were on the screen. The date would be when the flax used for the linen relic was harvested. Gove would be taking cowboy boots back to Rochester." (Sox, H.D., 1988, "The Shroud Unmasked: Uncovering the Greatest Forgery of All Time," Lamp Press: Basingstoke UK, pp. 146-147) that I first realised that it was not the actual carbon dating results that those in Arizona's laboratory were seeing, but *what the computer* was displaying" and "I put two and two together back then in 2007 and realised that [...] one explanation of its 1260-1390 radiocarbon date is that a hacker had [...] substituted the Shroud's actual dates coming from the AMS machine for bogus dates [...]."

Source: Email from Stephen Jones to author on September 4, 2016.

Comments: If you'll recall the curious phone call that "Harry" had received from a seemingly troubled caller back in the spring of 1989, the person said he had thrown the sample in the trash, but Jones indicated to me in an email of September 1, 2016 that the "German sounding distraught phone caller who said he had 'trashed' the Shroud is consistent with him being Koch."
*In emails of July 3, 2020, Jones indicated that instead of "substituted" he would now say *"built on"* or *"enhanced"* them, and that it makes his hacker theory compatible with carbon contamination and medieval repair theories.

2015 *(March-June)*. "Arnaud-Aaron Upinsky, mathematician, epistemologist and linguist, had exposed the compelling reasons which had prompted him, on March 29, 2015, to send to Pope Francis, in the name *'Et vous qui de la Science,'* an Open Letter asking him to proceed to the official recognition of the authenticity of the Shroud of Turin concealing scientific proof of the founding fact of our era -- to return to the status of true Relic of the Passion and the Resurrection which had not ceased to be that recognized by the Church, from 1473 to 1988, and to put an end to the intolerable ambiguity attributed to the Shroud, authentic for some, false for some, criminal object for others. 'The doubt on the Shroud of Turin must stop, the Church must decide, will you worship or worship it?' Despite censorship, this Open Letter to the Pope, distributed worldwide, translated into Italian, Spanish, German, English, Russian ... and hotly debated in the Vatican, created a veritable low-noise electroshock. But during his face-to-face meeting with the Shroud, in the darkness of the cathedral of Turin, on June 21, 2015. for his exceptional visit ending the ostension of 2015, what response did Pope Francis give? [...]. 'I do not recognize this Holy Shroud as a relic or the real linen that wrapped Jesus during the descent from the Cross; but as an icon'."

Upinsky further stated, "The most important question today that we must never stop repeating; it is that no scientist engaged in the world in research affirms that the Shroud of Turin is a fake and more nobody claims that it is a fake because no one can claim that it is a fake. This is one of the results of one of the two symposia in Paris on September 7 and 8, 1989 which ridiculed the fake thesis so much that no scientist at the risk of being ridiculed can claim that it is a fake. [...] A few days after this symposium, Dr Tite, who was director of research in the laboratory of the British Museum, which coordinated all research relating to C14 in 1988, was forced to write to Luigi Gonella a scientific advisor to the Cardinal of Turin, and therefore custodian of the Shroud that he never said it was a forgery [...]. Consequently if an archaeological piece is not a forgery and it has no signature of forgery, it is therefore authentic. The Turin Shroud is not reproducible. It is unique, a piece is paradoxical, it is an enigma and a provocation'."

Source: Ojeda-Mari, Victor, Linceul De Turin -- L'Imposture Du C14, (Les Edition le Gant et la Plume, 2018), Kindle locations 3175, 3226 and 3637.

Comments: No doubt Pope Francis was trying to be "politically correct" in his remarks, but it seems a shame that the Church, if the Shroud is authentic, seems to be unable to use it to full advantage and bring people closer to the divine through it. Considering that the Church is swimming in various problems, the Shroud does not seem to be an object to be downplayed. (There is also no doubt he was trying to be "religiously correct" -- the Church has almost two-thousand years of complex theology and dogma with which to contend.)

2016. English film-maker David Rolfe produced a made-for-the-Internet documentary titled "A Grave Injustice: an Investigation into the 'First Selfie'." It is an excellent twenty-seven-minute production, and covers many aspects of the dubious aspects of the Shroud C-14 dating.

Source: www.shroudenigma.com.

2019 *(March)*. After a Freedom of Information request in 2017 to get the raw data from the British museum, a paper authored by French and Italian researchers was published in a peer-reviewed journal that stated, "A statistical analysis of the *Nature* article and the raw data strongly suggests that homogeneity is lacking in the data and that the procedure should be reconsidered." It also mentioned the Oxford lab and the Arizona lab having found cotton in their samples. It was yet another indication that repairs had been made.

Source: Casabianca, Tristan *et al.* "Radiocarbon Dating of the Turin Shroud: New Evidence from Raw Data." *Archaeometry*, 2019, doi: 10.1111/arcm.12467.

Comments: Casabianca, the lead author and the person who had made the Freedom of Information Act request, attended a Shroud conference in Ancaster, Ontario, Canada in August 2019; I was also in attendance. I asked him if he had made a Freedom of Information request to find out who the "rich businessmen" were who had made the one-million-pound donation to Oxford after the Shroud dating. He said he made the request at the same time he made the request about the raw data. He indicated that not only did he not receive the information, he hadn't (and apparently still hasn't as of August 2020) even yet received a reply to his request! No one should be shocked at that.

Bryan Walsh (see entry immediately below) emailed me on January 17, 2020 and wrote, "Here are some of my observations about the raw data:"

* Oxford performed 5 measurements which somehow resulted in 3 radiocarbon dates;

* Two of the five Zurich date measurements reported in *Nature* were much higher than the raw data;

* In total, six of the error terms associated with the data reported in *Nature* were increased above the raw data while one was decreased;

* Two of the Tucson raw measurements (A5 and A6) were combined when they statistically were beyond what the reported uncertainties would allow;

* All of the Oxford uncertainties reported in *Nature* were increased over the raw data.

"I'm hoping that at least some of this may be explained in the 700 pages that Casabianca *et al*. received from the British Museum as a result of their FOIA and follow-up."

Ironically, *Archaeometry* is published by Oxford University!

2019 *(December).* A peer-reviewed article was published, for which one of the co-authors was a member of STURP; it had similar findings to the *Archaeometry* paper that had been published in March. The authors were critical of some elements of the *Nature* paper. For example they wrote (pg. 1), "In conducting the experiment, the authors stated, 'The underlying principle of the statistical analysis has been to assume that, unless there is strong evidence otherwise, the quoted errors fully reflect all sources of error and that weighted means are therefore appropriate.' Unfortunately, the report was weak in details about how the data were reduced and analyzed. However, it seems the authors failed to conduct standard statistical tests for homogeneity among the inter- and intralaboratory data, a necessary step to justify combining the data as they did. Had they conducted these tests, the statistical heterogeneity of the data should have served as a qualifier to their conclusions and led to a further exploration of the underlying cause of the discrepancies."

And on pages 5-6 Walsh and Schwalbe wrote, "The overall conclusion is that Damon *et al.** did not follow the W-W** recommendation to reconsider the data. Rather, they chose to weight equally each of the three means – the scatter-weighted Tucson data and the quoted error-weighted Zurich and Oxford data -- to find their arithmetic mean. They then estimated the standard error of that mean by combining the standard errors of those means as if all the data were drawn from the same population. This procedure is inappropriate since it deliberately ignores the heterogeneous nature of the data uncovered by the analysis and introduces error into the statistical analysis. We reject the *Damon et al.** approach as outlined above. Instead, we acknowledge the data heterogeneity and explore potential underlying causes for the effect. We do this, first by quantifying the systematic offsets in terms of how they relate to the original sample locations, and second by how they possibly relate to differing cleaning procedures among the laboratories."

Source: Walsh, Bryan and Schwalbe, Larry. "An instructive inter-laboratory comparison: The 1988 radiocarbon dating of the Shroud of Turin." *Journal of Archaeological Science: Reports,* v.**29** (2020) 102015. https://www.sciencedirect.com/science/article/pii/S2352409X19301865.

Comments: The multiple criticisms of the methodology of the three labs in 1988 reflect the lack of rigor in the Shroud dating.
 *Damon *et al.* refers to the 1989 *Nature* paper. **The "W-W" refers to a test in radiocarbon circles acknowledged as a test of the homogeneity of samples.
 Walsh added this in a comment in an email to Barrie Schwortz published in the January 21, 2020 update at www.shroud.com about the paper: "I think it's important to physically and chemically characterize the sample using non-destructive means before performing a radiocarbon dating on it since you can't know exactly what you're dating unless you have first characterized it. Simple microscopic evaluation will only detect very gross contaminants. So, the desire to radiocarbon date the char and debris removed in the 2002 'restoration' may be premature since there's no way to know just what's being radiocarbon dated along with the linen fabric. Given what we noted in the paper we wrote, both Larry and I believe it's important to first assess the sample for its physical and chemical characteristics before attempting to radiocarbon date it."

2020 *(July)*. An excellent new video, titled "The Shroud of Turin 1988 Carbon Dating: Triumph or Travesty?" was released. Containing incredible graphics, it was produced by Michael Lewis Kowalski, who studied physics at the University of Manchester in England. It covered some of the major problems of the Shroud C-14 dating process and can be found at https://www.youtube.com/watch?v=JBDuKZSgDSI&t=19s.

CONCLUSION

The late Prof. E.T. Hall wrote in an article in *Textile Horizons* in 1990, "It would seem a pity if ill-informed skeptics contrived to denigrate the Shroud experiment without some valid evidence." One thing I am not, is ill-informed. And anyone who has read this book is not ill-informed. I believe I have presented irrefutable proof that politics, and not the pursuit of truth, was the main theme of the C-14 dating of the Shroud. I further believe the results must thus be seriously questioned, if not outrightly discarded, due to a lack of rigor. There is little doubt now that the AD 1260-1390 date does not accurately reflect the true calendar age of the Shroud of Turin. Frankly, the totality of the information here should turn heads. Consider this: much of the material presented here has been compiled from public sources. The rare material I have also utilized paints a consistent picture of dubious behavior by the major players. There is no doubt that there is additional secret data that have never seen the light of day. In one sense, it almost doesn't matter, since there is already enough data to cast doubt upon the 1988 results. But wouldn't it be interesting to know the rest of the details??

However, I harbor no illusions that at a certain point, there would be enough evidence to convince everyone. Humans have a remarkable ability to be able to avoid the truth. Pilate's question to Jesus, "What is truth?" is just as relevant today as it was back then. But in the 21st century, mainly due an offshoot of political persuasions, many people's answer to the question of "what is truth?" is: "whatever I want it to be." In this stance, no amount of factual evidence is enough to convince a person of the veracity of something they simply don't want to believe in. Such will be the case with all the evidence presented here that the 1988 Shroud C-14 dating is, at <u>best</u>, unreliable. I'm still hoping, though, that people within the Catholic Church, in their roles of caretaker of the Shroud, will care enough about the truth to do more than just to leave in its container most of the time and bring out periodically for public exhibitions.

Dr. Jull had emailed me on January 26, 2009 saying, […] "[W]e need to move beyond the process and **politics** *[my emphasis]* of the original event. They are not relevant now." I couldn't disagree more. If the politics affected the rigor of the experiment, and there is absolutely no doubt that they did, it is most relevant even though we are beyond three decades after the original testing.

The late Dr. Robert Dinegar made the following wise suggestions in a conference paper in 1991 (see https://www.youtube.com/watch?v=NGRrt3d9Rao&feature=youtu.be&list=UUSv7BD9sKjIcA24ct1Hz_Aw): "I end this paper by suggesting (1) an unbiased scientific appraisal of all data that has been collected on the Shroud; (2) a continuing effort to clear up through experiments of every kind, apparently contradictory facts; and (3) charity and respect towards all investigators as well as an appreciation of the efforts of everyone, including those with whom we now disagree."

We can continue to hope that the Catholic Church will in the near future allow some new testing, so that we can learn all we can about this enigmatic cloth and the person who was wrapped in it. (See for example, www.testtheshroud.org.) The New Testament says that Jesus "gave himself as a ransom for all people" (1 Tim. 2:6). The price of that ransom is shown on The Shroud of Turin. The living Pope may be the legal owner of the Shroud, BUT THE CLOTH BELONGS TO THE WHOLE WORLD! It's time to test it further, so that all of its secrets can be probed with the aim of the betterment of all humankind!

KEY FIGURES

Adler, Alan (Dr.): Chemistry professor at Western Connecticut State University with expertise in blood chemistry. Was a member of the Shroud of Turin Research Project (STURP). Attended the 1986 workshop on dating the Shroud. Died 2000.

Ballestrero, Anastasio (Cardinal): Archbishop of Turin during the 1988 dating. Was custodian of the Shroud. Hosted the 1986 workshop on dating the Shroud. Died 1998.

Bollone, Baima (Dr.): Was affiliated with the Institute of Forensic Medicine of the University of Turin and also a member of the Turin Centro.

Bonnet-Eymard (Bro.): Member of the French monastic group "Catholic Counter-Reformation in the XXth (now XXIst) Century." Believed samples of another cloth was switched with the Shroud samples.

Bracaglia, Giorgio: Overseer of the archives of the Holy Shroud Guild, which contains many important documents and letters among participants in the 1988 C-14 dating of the Shroud.

Bray, Anthos (Prof.): Engineer at G. Colonetti Institute in Turin, who was involved in the statistical analysis of the C-14 dating data. Died 2016.

Brignall, Shirley: Assistant to Dr. Harry Gove of the C-14 lab at University of Rochester. Attended the 1986 workshop on dating the Shroud. Died 2019.

Brunati, Ernesto: Italian engineer who wrote various articles questioning the results of the 1988 Shroud C-14 test. Died 2011.

Bulst, Werner (Fr. and Dr.): German Jesuit priest who wrote various books and articles on the Shroud. He believed the Shroud samples that had been dated had been switched with one of the control samples. Died 1995.

Canuto, Vittorio (Victor) (Dr.): A theoretical astrophysicist from NASA, who was a science advisor to the president of the Pontifical Academy of Sciences, the Pope's scientific advisory group. Assisted Chagas and Gove in organizing the 1986 workshop on dating the Shroud, which he attended.

Casaroli, Agostino (Cardinal): Vatican secretary of state at the time of the Shroud dating. Died 1998.

Cavaglia, Giuseppe (Fr.): Secretary to Cardinal Ballestrero. Died 2017.

Celli, Celestino (Monsignor): Vatican ambassador to the United Nations at the time of the Shroud dating.

Chagas, Carlos (Dr.): Medical doctor from Brazil, who was president of the Pontifical Academy of Sciences. He was chairman of the 1986 workshop on dating the Shroud. Pope John Paul II removed him from the position shortly after the dating results were announced. Died 2000.

Coero-Borga, Piero (Fr.): Secretary of the Turin Centro. Died 1986.

Cottino, Jose (Monsignor): Press spokesman and caretaker of Shroud for Cardinal Ballestrero. Died 1983.

Damon, Paul (Dr.): Professor and co-director of the C-14 lab at the University of Arizona in Tucson, one of the three labs that dated the Shroud. Attended the 1986 workshop on dating the Shroud. Died 2005.

Dardozzi, Renato (Monsignor): An electrical engineer who was ordained a priest at age 52. He was a deputy director at the Pontifical Academy of Sciences. Died 2003.

Dinegar, Robert (Fr. and Dr.): Research chemist from Los Alamos National Laboratory. He was an Episcopalian priest, member of STURP and chair of their radiocarbon dating committee. Died 2005.

D'Muhala, Thomas: President of Nuclear Technology Corporation (Connecticut, USA) and a member of STURP.

Di Rovasenda, Enrico (Fr.): Turinese priest who was past director of the Pontifical Academy of Sciences. Attended the 1986 workshop on dating the Shroud. Died 2007.

Druzik, James: Art conservator at the J.P. Getty Museum in Los Angeles, California and member of STURP.

Duplessy, Jean-Claude: Senior scientist at the C-14 facility in Gif-sur-Yvette, France. Attended the 1986 workshop on dating the Shroud.

Dutton, Denis: School of Fine Arts, University of Canterbury, Christ Church, New Zealand. Wrote several letters to *Nature* questioning the procedures of the Shroud dating.

Evin, Jacques (Dr.): Was head of the C-14 lab at University of Claude Bernard, Lyon. Attended the 1986 workshop on dating the Shroud. Involved in providing an extra control sample that was not planned in the protocols.

Fleming, Stuart (Dr.): Worked at Hall's lab in Oxford and later in the museum at the University of Pennsylvania. Asserted that a medieval restorer could have rewoven the sample area such that it was invisible to the naked eye.

Flury-Lemberg, Mechthild: Head of the Textile Workshop of the Abegg-Siftung Institute in Switzerland. Originally was supposed to excise the sample for the Shroud dating. Ultimately, it was taken by Giovanni Riggi di Numana. Flury-Lemberg was present at the sample-taking. Attended the 1986 workshop on dating the Shroud.

Ghiberti, Giuseppe (Msgr.): President of the Diocesan Commission for the Shroud.

Gonella, Luigi (Dr.): Professor of metrology (Physics) at Turin Polytechnic Institute. Scientific advisor to Cardinal Ballestrero of Turin. Attended the 1986 workshop on dating the Shroud. Died 2007.

Gove, Harry (Dr.): Co-inventor of the C-14 dating method used on the Shroud: accelerator mass spectrometry. Was Professor of Physics at the University of Rochester and director of the Nuclear Structure Research Laboratory from 1963 to 1988. He assisted Canuto and Chagas in organizing the 1986 Turin workshop on dating the Shroud. By his own admission, he worked to get STURP from being involved in the dating and was successful, although his own lab was not chosen to participate in the dating. Died 2009.

Hall, Edward (Dr.): Was Professor and head of the Research Laboratory for Archaeology and the History of Art, which included the Radiocarbon Accelerator Unit, at Oxford University, which was one of the labs to date the Shroud. Attended the 1986 workshop on dating the Shroud. Was on the board of Trustees for the British Museum, from which was chosen the overseer of the Shroud C-14 process, Dr. Michael Tite. Shortly after the dating, a one-million-pound donation was given to Oxford to establish a chair, Hall resigned, and Tite replaced him. Hall died 2001.

Harbottle, Garman (Dr.): Was a senior scientist at Brookhaven National Laboratory Chemistry Department and was the head of the C-14 facility there. He worked with STURP but also worked closely with Gove. Attended the 1986 workshop on dating the Shroud. Died 2016.

Hedges, Robert (Dr.): Senior scientist at C-14 lab in Oxford. Attended the 1986 workshop on dating the Shroud. Appeared with Dr. Tite and Dr. Hall at the press conference held at the British Museum on Oct 14, 1988 to announce the results of the Shroud dating.

Heller, John (Dr.): Medical doctor, biophysicist, and blood expert. Member of STURP. Died 1995.

Ibarrola, Faustino Cervantes: Ecclesiastical assistant of the Mexican Center of Sindonology. Died 1995.

Jackson, John (Dr.): Physicist and co-founder of STURP along with Eric Jumper, while both were stationed at Air Force Academy in Colorado Springs, Colorado.

Jull, Timothy (Dr.): Scientist at the University of Arizona C-14 lab. Took part in the measurements that lab took after dating its Shroud sample. Jull is currently the head of the lab.

Jumper, Eric (Dr.): Physicist and co-founder of STURP along with John Jackson, while both were stationed at Air Force Academy in Colorado Springs, Colorado.

Libby, William (Dr.): Inventor of the radiocarbon method of dating. Died 1980.

Lukasik, Stephen (Dr.): Was vice-president of Technology division at Northrup Corporation in Los Angeles California. He was in charge of experiments that STURP had hoped to carry out in conjunction with the 1988 Shroud dating. Attended the 1986 workshop on dating the Shroud. Died 2019.

Maloney, Paul: Archaeologist and General Project Director of the "Association of Scientists and Scholars International for the Shroud of Turin." Died 2018.

Marinelli, Emanuela. Italian Shroud researcher and author for over forty years. Co-author of two of the main book sources used in this book, and author of a conference paper also used as a major source.

McCrone, Walter (Dr.): Microscopist and head of microscopy/chemistry lab in Chicago, Illinois. Wanted to be involved in the Shroud C-14 dating. Based on his studies, he believed both the Vinland Map and the Shroud were fakes. Died 2002.

Meacham, William: American archaeologist based in Hong Kong. Attended the 1986 workshop on dating the Shroud. Warned before the Shroud dating of potential pitfalls and of putting too much stock in the dating. Accurately predicted many of the outcomes that ensued after the testing.

Minor, Mike: Lawyer for STURP. Died 2008.

Nitowski, Eugenia (Dr.) (a.k.a. Sr. Damian of the Cross): American archaeologist and Carmelite Nun. She performed image-formation experiments in Jerusalem in 1986 and also was concerned with Shroud conservation. Died 2007.

Oeschger, Hans (Dr.): the founder of the Division of Climate and Environmental Physics at the Physics Institute of the University of Bern in 1963 and director until his retirement in 1992. Died 1998.

Otlet, Robert (Dr.): Was head of the C-14 dating facility in Harwell, England. Attended the 1986 workshop on dating the Shroud.

Otterbein, Adam (Fr.): President of the Holy Shroud Guild in Esopus, New York in the period when the Shroud was dated. Died 1998.

Parlato, Charles: American businessman who headed the Shroud organization "American Shroud of Turin Association" (ASTA).

Petrosillo, Orazio: Vatican correspondent for the Italian newspaper *Il Messaggero*. Co-author of two of main book sources used in this book. Died 2007.

Polach, Henry: Considered the father of radiocarbon dating in Australia, and a significant player on the worldwide stage of radiocarbon. Attended the 1985 C-14 meeting in Trondheim that had a huge impact on the Shroud C-14 dating in 1988. Died 1996.

Pope John Paul II: Became Pope only a few days after STURP ended their 1978 investigation. Was the Pope during the 1988 C-14 dating and also during the controversial 2002 restoration of the Shroud, both of which he authorized. Died 2005.

Raes, Gilbert (Dr.): Was professor and textile expert at Ghent University in Belgium. Was given a small sample of the Shroud in 1973 for examination. The 1988 C-14 sample was literally right next to the area from which the Raes sample was taken. Died 2001.

Raffard de Briene, Daniel: French journalist. Died 2007.

Riggi di Numana (a.k.a. Riggi), Giovanni: Italian scientist who cut the sample from the Shroud in 1988 for the C-14 dating. Along with Franco Testore, gave multiple versions of the sizes and weights of the various sub-samples given to the three labs. Died 2008.

Rinaldi, Peter (Fr.): Catholic priest born in Turin. Vice-President of the Holy Shroud Guild. Was pastor of Church in Port Chester, New York but had dual citizenship. Mediator between STURP and the Turinese authorities. Died 1993.

Rogers, Raymond: Was a chemist at Los Alamos National Laboratory and a member of STURP. Shortly before his death, he had an important paper published in *Thermochimica Acta* that stated that the Shroud sample tested for the C-14 dating was not representative of the main cloth. Died 2005.

Schwortz, Barrie: STURP documenting photographer. President of Shroud of Turin Education and Research Association (STERA), which publishes the premier Shroud website: www.shroud.com.

Souverain, Raymond. Was vice-president of the French Shroud organization, Centre International d'Études sur le Linceul de Turin (CIELT).

Schwalbe, Larry (Dr.): Physicist and member of STURP.

Somolo, Martinus (Cardinal): Vatican undersecretary of state at the time of Shroud dating.

Sox, David (Fr.): Was an American Episcopalian priest stationed in England and a former secretary of the British Society for the Turin Shroud. Wrote several books on the Shroud, including one called Unmasking the Shroud, which was printed two weeks before the official announcement of the Shroud C-14 dating results. Died 2016.

Stevenson, Kenneth (now both Rev. and Dr.): Handled public relations for STURP during the 1978 examination of the Shroud. Has co-authored or authored several books on the Shroud.

Tessiore, Giorgio: Teacher of Natural Science in a lyceum of Turin. Member of the Turin Polytechnic University. Died 2000.

Testore, Franco (Prof.): Italian textile expert from the Department of Material Science at the Turin Polytechnic. He assisted Riggi with the sample-taking and along with Riggi, gave multiple versions of the sizes and weights of the various sub-samples given to the three labs. Died 2018.

Tite, Michael (Dr.): Had been head of the Research Laboratory of the British Museum and was appointed as the overseer of the Shroud dating process. Attended the 1986 workshop on dating the Shroud. After Edward Hall from Oxford retired and the Oxford lab had received a one-million-pound donation, Tite left the British Museum and replaced Hall at Oxford.

Umberto II (King): the last king of Italy, who was exiled to Portugal. He belonged to the House of Savoy, which owned the Shroud. It was he who had given permission for STURP to conduct the 1978 examination. Before he died, he willed the Shroud to the living Pope before the former's death in 1983. After legal proceedings, the cloth officially became the property of the living Pope starting in 1985.

Upinksy, Arnaud-Aaron (Dr.): French mathematician. Called into question the labs' statistics regarding the C-14 dating results.

Van Haelst, Remi: Belgian chemist who wrote many articles questioning the data produced by the three labs after the 1988 Shroud C-14 dating

Vial, Gabriel: Textile expert from the Historical Museum of Fabrics, Lyon, France. Was involved in various aspects of the 1988 Shroud dating. Died 2005.

Walsh, Bryan: American mathematician and statistician. Wrote or co-wrote several articles critiquing the 1988 Shroud dating results.

Weisskopf, Victor (Dr.): Austrian-born American theoretical physicist. Member of the Pontifical Academy of Sciences. Died 2002.

Whanger, Alan (Dr.): Professor of Psychiatry and Shroud researcher. Was an amateur photographer. Believed that photographic and video evidence of the C-14 sample area indicated that the area had been rewoven. Died 2017.

Wilson, David (Sir): Director of the British Museum. It was he to whom Gove wrote to say that the Shroud dating had become "a shoddy enterprise."

Wilson, Ian: Historian originally from England who studied at Oxford. Authored several books on the Shroud, including the seminal The Shroud of Turin in 1978, which helped to make the Shroud better known, especially in the Western world.

Woefli (a.k.a. Wolfli, Wölfli), Willy: Was director of the C-14 lab in Zurich, Switzerland, one of the labs to have dated the Shroud. Attended the 1986 workshop on dating the Shroud. Died 2014.

REFERENCES

Adler, Alan D. and Alan and Mary Whanger, "Concerning the Side Strip on the Shroud of Turin," http://www.shroud.com/adler2.htm. Accessed May 29, 2020.

Adler, Alan D. Updating Recent Studies on the Shroud of Turin. In M.V. Orna (Ed.), *Archaeological Chemistry: Organic, inorganic and biochemical analysis* (pg. 225) *ACS Symposium Series*, vol. **625**, 1996. Washington, DC: American Chemical Society.

"Art and Science Converge in State Museum Exhibit," by University Communications, November 3, 2008. Accessible at http://uanews.org/node/22384. Accessed May 29, 2020.

Ballestrero, Anastasio. Letter of July 26, 1988 to Paul Damon, Douglas Donahue and Michael Tite, from British Museum archives, released in 2019 via Freedom of Information Act request.

Benford, M.S. and Joseph G. Marino. "Discrepancies in the radiocarbon dating area in the Turin shroud." *Chemistry Today*, v.**26**,no.4 (July/August 2008): 4-12. Accessible at: https://www.shroud.com/pdfs/benfordmarino2008.pdf. Accessed May 29, 2020.

Bollone, Pier Luigi Baima. Sindone O No, Torino: Societa Editrice Internazionale, 1990.

Bonnet-Eymard, Bruno. "The Crime Committed Against the Holy Shroud." *Catholic Counter-Reformation in the XXth Century*, February-March 1996, No. 283.

Bonnet-Eymard Bruno (Brother). "The Holy Shroud Is As Old As The Risen Christ." *The Catholic Counter-Reformation in the XXth Century*, May 2000, No. 330.

Bonnet-Eymard, Bruno. "The Holy Shroud Is Authentic." *The Catholic Counter-Reformation in the XXth Century,*" (Christmas 1988/Easter 1989), No. 219.

Bonnet-Eymard, Bruno. "The Holy Shroud, Silent Witness." *The Catholic Counter-Reformation in the XXth Century*, April 1997, No. 295.

Bonnet-Eymard, Bruno. "The Shroud Daters." *Catholic Counter-Reformation in the XXth Century*, June 1989, No. 220.

Bonnet-Eymard, Bruno. "The Victory of the Holy Shroud Won By Science." *Catholic Counter-Reformation in the XXth Century*, September-October 1989, No. 223.

Bonnet-Eymard, Bruno and Georges de Nantes. "Appeal to the Twenty-One Co-authors of the Report on the Carbon 14 Dating of the Holy Shroud." *Catholic Counter-Reformation in the XXth Century*, November-December 1989, No. 224.

Bonnet-Eymard, Bruno and Georges de Nantes. "The Carbon 14 Dating: In Pursuit of the Forgers." *Catholic Counter-Reformation in the XXth Century*, April 1991, No. 238.

Bonnet-Eymard, Bruno and Georges de Nantes. "Holy Shroud -- The Turin Tricksters in Disarray." *Catholic Counter-Reformation in the XXth Century*, November-December 1989, No. 224.

Boyce, David. "Dr. Tite in Surrey." *Catholic Counter-Reformation in the XXth Century*, December 1990, No. 234.

Bracaglia, Giorgio. Uncovering the Paradox within the Archives of the Holy Shroud Guild, (Honeoye, N.Y.: Holy Shroud Guild), 2019.

Brunati, Ernesto. "Altro che rammendi! La datazione della Sindone e tutta un falso." *Collegamento Pro Sindone*, May 2005. http://www.sindone.info/BRUNATI1.PDF. Accessed May 29, 2020.

Brunati, Ernesto. "La corrispondenza con *Radiocarbon* sulla datazione della Sindone." *Collegamento Pro Sindone*, February 2006. http://www.sindone.info/BRUNATI2.PDF. Accessed May 29, 2020.

Brunati, Ernesto. "Lettera Aperta Al Prof. Franco Testore." *Collegamento Pro Sindone*, November/December 1989, pp. 41-45.

Brunati, Ernesto. "Testimoni, Non Accusati." *Collegamento Pro Sindone*, January/February 1990, pp. 45-51.

Bulst, Werner, S.J. "L'Enigma Del Test radiocarbonico Della Sindone, Proposta Della Soluzione." *Collegamento Pro Sindone*, September/October 1990, pp. 37-44.

Burleigh, R., M. Leese and M. Tite. "An Intercomparison of Some AMS and Small Gas Counter Laboratories." *Radiocarbon* **28**, No. 2A, 571-577 (1986).

Busson, Pierre. "Sampling error?" [Letter], *Nature*, **352**, 187 (18 July 1991).

"Cardinal Ballestrero challenges the carbon 14 dating test of 1988." *Revue Internationale Du Linceul De Turin*, (Autumn 1997), No. 6, pg. 28.

Casabianca, Tristan *et al*. "Radiocarbon Dating of the Turin Shroud: New Evidence from Raw Data." *Archaeometry*, 2019, doi: 10.1111/arcm.12467. https://www.academia.edu/38607635/Radiocardon_Dating_of_the_Turin_Shroud_New_Evidence_From_Raw_Data. Accessed May 26, 2020.

Case, T.W. The Shroud of Turin And The C-14 Dating Fiasco: A Scientific Detective Story, (Cincinnati: White Horse Press), 1996.

Clark, Kenneth R. "Shroud of Turin Controversy Resumes." *Chicago Tribune*, October 14, 1988, pp. 1, 4.

Coghlan, Andy. "Unexpected errors affect dating techniques." *New Scientist*, September 30, 1989, pg. 26.

[Crispino, Dorothy]. "Radio Courtoise." *Shroud Spectrum International*, Issue #32/33, September/December 1989, pp. 36-37. https://www.shroud.com/pdfs/ssi3233part6.pdf. Accessed May 29, 2020.

Damon, Paul. Letter of August 26, 1988 to Luigi Gonella, from Luigi Gonella archives.

De Nantes, Georges (Abbe) and Brother Bruno Bonnet-Eymard. "The Scientific Rehabilitation of the Holy Shroud of Turin: Part Four -- the Carbon Affair," *Catholic Counter-Reformation in the XXth Century*, May 1989, no. 219.

De Wesselow, Thomas. The Sign: The Shroud of Turin and the Secret of the Resurrection, (London: Penguin Books), 2012.

"DECLARATION OF THE SCIENTIFIC COMMITTEE OF THE PARIS INTERNATIONAL SCIENTIFIC SYMPOSIUM" *Shroud Spectrum International*, No. 32/33 (September/December) 1989, pp. 33-35. https://www.shroud.com/pdfs/ssi3233part5.pdf. Accessed May 29, 2020.

Di Lazzaro Paolo *et al.* "Revisione propositiva dei risultati di radio-datazione della Sindone di Torino" ("Proposal revision of the radio-dating results of the Turin Shroud"). ENEA Technical Report, 2020. https://www.academia.edu/43936359/Revisione_propositiva_dei_risultati_di_radio_datazione_della_Sindone_di_Torino.

Dinegar, Robert. Personal Notes taken at Trondheim C-14 conference, June 1985. Copy sent by Dinegar to author.

Dutton, Denis. "Protocols for Turin Shroud." [Letter], *Nature* **331**, 108, (1988).

Dutton, Denis. "The Shroud of Turin." [Letter], *Nature*, **332**, 300, (1988).

Evin, J. "Materials of Terrestrial Origin Used for Radiocarbon Dating." *PACT* **8**: 235-276 (1983).

Farkas, Ilona. "Notizie Varie" in *Collegamento Pro Sindone*, July-August 1990, pp. 53-59. Translated by Dr. Daniel Scavone.

Fedi, M.E., *et al.* "AMS radiocarbon dating of medieval textile relics: The frocks and the pillow of St. Francis of Assisi. *Nuclear Instruments and Methods in Physics Research B,* **266** (2008): 2251-2254.

Feuerherd, Peter. "Shroud expert from Turin hits scientists' methods." *The [Albany] Evangelist,* October 20, 1988, pg. 8A.

Freer-Waters, Rachel A. and A.J. Timothy Jull "Investigating a Dated Piece of the Shroud of Turin." *Radiocarbon*, **52**:4 (2010): 1521-1527.

"Formal Proposal For Performing Scientific Research On The Shroud of Turin." Submitted By The Shroud of Turin Research Project, Inc. http://www.mondosindone.com/Site/documenti/DSS001_02%20-%20Sturp%201984.pdf.

"From Emeritus Professor Gilbert Raes." *Shroud News*, (December 1988), No. 50, pg. 20.

Glass, Robert. "Modern Technology May Finally Fix Age of the Shroud of Turin." *Chicago Sun Times*, April 8, 1988, pg. 4.

Gonella, Luigi. Archives -- Letters and documents related to the Shroud collected between 1978 and 2001.

Gonella, Luigi and Riggi di Numana, Giovanni. Sindone: il mistero continua, (Milan: Fondazione 3M), 2005.

Gove, Harry. "Radiocarbon-dating the shroud," [Letter], *Nature*, **333**, 110 (12 May 1988).

Gove, Harry. Relic, Icon or Hoax?: Carbon Dating the Turin Shroud, (Bristol and Philadelphia: Institute of Physics Publishing), 1996.

Gove, Harry. "Turin Shroud," [Letter], *Nature*, **327**, 652 (25 June 1987).

Haas, Herbert. Letter of November 11, 1986 to William Meacham, from Luigi Gonella archives.

Halisey, Robert. "More on the Shroud." [Letter], *Nature* **346**, 100, (12 July 1990). Reply from Dr. Michael Tite immediately followed.

Harbottle, Garman. "Carbon Dating the Shroud of Turin." Unpublished article originally written for "Perspectives" Section of *SCIENCE* Magazine" by the Senior Chemist, Brookhaven National Laboratory, Upton, NY.

Holy Shroud Guild archives CD. (Honeoye, N.Y.: Holy Shroud Guild).

International Study Group. "An Inter-Laboratory Comparison of Radiocarbon Measurements in Tree Rings." *Nature*, **298**, 619-623.

"Intervista Al Prof. Tite Del British Museum." Interview of Michael Tite on September 8, 1989 by Orazio Petrosillo and Emanuela Marinelli. Transcribed and translated with comments by Ernesto Brunati. *Collegamento Pro Sindone*," January/February 1990, pp. 38-44.

Jasper, William. "Science and Faith," *The New American*, **6** (26), December 17, 1990.

Jennings, Peter. "Art historian not convinced the Shroud is a fake." *Our Sunday Visitor*, October 23, 1988, pg. 24.

Jennings, Peter. "Shroud of Turin to Undergo Radiocarbon Testing." *Our Sunday Visitor*, February 14, 1988, pg. 3.

"John Cornwell Interviews Edward Hall." *The Tablet,* 14 January 1989, pp. 36 and 38.

Kava, Brad. "Scientist Protests Vatican Changes in Shroud Testing." *Corpus Christi Caller-Times*, April 23, 1988, pp. 14A-15A.

Kersten, Holger and Elmar R. Gruber. Jesus Conspiracy: The Turin Shroud & The Truth About The Resurrection, (Rockport, MA: Element), 1994.

Laverdiere, H. "The Socio-Politic of a Relic: Carbon Dating of the Turin Shroud," (Doctoral Thesis) 1989. Doctoral thesis. Accessible via free download at http://ethos.bl.uk/OrderDetails.do?did=1&uin=uk.bl.ethos.531916. Accessed May 29, 2020.

Lukasik, S.J. "Draft Protocol For The Next Examination Of The Shroud of Turin." http://freepages.rootsweb.com/~wmeacham/religions/sturp87b.pdf.

[Maloney, Paul.] Press Release sent by ASSIST to UPI: "The Carbon Date for the Shroud of Turin," October 14, 1988.

Maloney, Paul. ASSIST document "A Brief Evaluation of Current Plans to Carbon Date the Shroud of Turin," 3rd revised version, April 23, 1988.

Maloney, Paul. "What Went Wrong With the Shroud's Radiocarbon Date? Setting it all in Context," presented August 16, 2008 at "The Shroud of Turin: Perspectives on a Multifaceted Enigma" Shroud conference in Columbus, Ohio. Accessible at https://shroudofturin.files.wordpress.com/2014/12/maloneywhatwentwrongwiththeshroudversionfive2014.pdf. Accessed May 29, 2020.

Marinelli, Emanuela. "The Setting for the Radiocarbon Dating of the Shroud." Presented at 1st International Congress on the Holy Shroud in Spain -- Valencia -- Centro Español de Sindonologia (CES), April 28-30, 2012. Accessible at www.shroud.com/pdfs/marinelliv.pdf. Accessed May 29, 2020.

Marino, Joseph G. "The Invisible Reweave and Other Challenges to the Turin Shroud's C-14 Medieval Dating: A Review," originally presented at "Science, Theology and the Turin Shroud International conference in Ancaster, Ontario CANADA on August 15, 2019. Accessible at https://www.academia.edu/40272184/The_Invisible_Reweave_and_Other_Challenges_to_the_Turin_Shrouds_C-14_Medieval_Dating_A_Review. The corresponding PowerPoint is at: https://www.academia.edu/40271611/_Powerpoint_for_The_Invisible_Reweave_and_Other_Challenges_to_the_Turin_Shrouds_C-14_Medieval_Dating_a_Review. Each slide in the PowerPoint is numbered at the lower right and can be matched with the slide number shown in the text. Accessed June 19, 2020.

Marino, Joseph G. Wrapped Up in the Shroud: Chronicle of a Passion (Joseph G. Marino), 2020.

McClellan, Bill. "Secrets of the Shroud." *St. Louis Post-Dispatch*, May 15, 1988, pp. 1, 13-14.

Meacham, William. "Thoughts on the Shroud ^{14}C Debate." In The Turin Shroud: past, present and future. International Scientific Symposium. Torino, 2-5 March 2000. Silvano Scannerini and Piero Savarino, editors. Published jointly by *Sindon*, journal of the Centro Internazionale di Sindonologia and Effata Editrice, pp. 441-453.

Meacham, William. "A Proposal to Cardinal Ballestrero, Archbishop of Turin, for the Planning and Implementation of Radiocarbon Dating of the Holy Shroud."

Meacham, William. "Radiocarbon Measurement and the Age of the Turin Shroud: Possibilities and Uncertainties." Accessible at www.shroud.com/meacham.htm. Accessed May 20, 2020.

Meacham, William. The Rape of the Turin Shroud: How Christianity's most precious relic was wrongly condemned and violated, (Lulu.com), 2005.

Meacham, William. "Turin Shroud Carbon Dating." Unpublished manuscript, October 1988.

Meacham, William. Letter of November 29, 1986 to Dr. Herbert Haas, from Luigi Gonella archives.

Meacham, William. Letter of December 15, 1987 to Steven Lukasik. http://freepages.rootsweb.com/~wmeacham/religions/wmslcorr.pdf, pg. 6.

Meacham, William. Letter of October 5, 1988 to Rev. Kim Dreisbach, from Luigi Gonella archives.

Meacham, William. Letter of December 18, 1988 to Cardinal Anastasio Ballestrero, from Luigi Gonella archives.

Meacham, William. Letter of December 18, 1988 to Luigi Gonella, from Luigi Gonella archives.

Night of the Shroud (La Notte de la Sindone), documentary directed by Francesca Saracino, 2011. In 2016, it was revised and retitled "Cold Case: The Shroud of Turin," which is available at www.amazon.com.

[Nitowski, Eugenia.] "The Shroud of Turin and Carbon Dating." *The Wanderer,* November 24, 1988, in "the Forum" section.

Nydal, Reidar. "Optimal Number of Samples and Accuracy in Dating Problems." *PACT* **8**: 107-121 (1983).

Official Press Release. "AGREEMENT ON DATING THE SHROUD OF TURIN (From the British Museum)." (Issued January 22, 1988.)

Ojeda-Mari, Victor, Linceul De Turin – L'Imposture Du C14, (Les Edition le Gant et la Plume), 2018.

Paci, Stefano M. "All Those Carbon Errors." *30 Days in the Church and in the World*, No. 9, 1993, pp. 60-63.

Petrosillo, Orazio. "La Sindone? Un Falso Da Nascondere." *Collegamento Pro Sindone*, January/February 1991, pp. 4-7.

Petrosillo, Orazio and Emanuela Marinelli. The Enigma of the Shroud: A Challenge to Science, (San Gwann, Malta: Publishers Enterprises Group), 1996.

Petrosillo Orazio and Emanuela Marinelli. "Le Esternazioni Contro I Sindonologi." *Collegamento Pro Sindone*, January/February 1992, pp. 17-22.

Petrosillo, Orazio and Emanuela Marinelli. La Sindone: Storia Di Un Enigma. (Milan: Rizzoli, 3rd edition), 1998.

Pourrat, Olivier. Shroud dating still questioned." [Letter], *Nature*, **349**, 558 (14 February 1991).

Raloff, J. "Controversy Builds as Shroud Tests Near." *Science News*, April 16, pg. 345.

Ramsey, C. Bronk. "Radiocarbon Dating: Revolutions in Understanding." *Archaeometry* **50**:2 (2008): 249–275.

Ramsey, Christopher. "Dealing with Outliers and Offsets in Radiocarbon Dating." *Radiocarbon* **51** (3) (December 2009): 1023-1045.

Riani, Marco, *et al.* "Carbon Dating of the Shroud of Turin: Partially Labelled Regressors and the Design of Experiments." *Stat Comput* DOI 10.1007/s11222-012-9329-5, 2012. Accessible at http://www.riani.it/pub/RAFC2012.pdf. Accessed August 22, 2020.

Rogers, Raymond N. A Chemist's Perspective on the Shroud of Turin. Edited by Barrie M. Schwortz, (Lulu.com), 2008.

Rogers, R.N. Studies on the Radiocarbon Sample of the Shroud of Turin. *Thermochimica Acta*, Vol. **425**, No. 1/2, 20 January 2005. Accessible at http://www.shroud.it/ROGERS-3.PDF. Accessed May 29, 2020.

Rogers, Raymond N. and Anna Arnoldi. "Scientific Method Applied To The Shroud of Turin: A Review." Accessible at www.shroud.com/pdfs/rogers2.pdf. Accessed May 29, 2020.

Savarino, Piero. "La radiodatazione dell Sindone." In Sindone E Scienza -- Atti III Congresso Internazionale di studi sulla Sindone -- a cura di Pier Luigi Baima Bollone -- Maruizio Lazzero -- Carolina Marino (CD-ROM).

Savarino, Piero and Bruno Barberis. Shroud, Carbon Dating and Calculus of Probabilities, (London: St. Paul's, 1998).

Schwortz, Barrie: (2003-2011). "Some Details About The STURP Quad Mosaic Images." Accessible at http://www.shroud.com/pdfs/quad.pdf. Accessed May 29, 2020.

Siliato, Maria-Grazia. Contre Enquete Sur Le Saint Suaire, (Paris: France Loisirs), 1998.

Souverain, Raymond. "Dating of the Turin Shroud: situation in 2001," *Revue Internationale Du Linceul De Turin*, (February 2002), no. 22, pp. 15-31.

Sox, David. The Shroud Unmasked, (Basingstoke, Hampshire: The Lamp Press), 1988.

Tite, M.S. "Turin shroud." [Letter], *Nature* **327**, 456; (11 June 1987).

Tite, M.S "Turin Shroud." [Letter], *Nature* **332,** 482 (7 April 1988).

Tite, Michael. Letter of September 23, 1988 to Cardinal Anastasio Ballestrero, from British Museum archives, released in 2019 via Freedom of Information Act request.

Tite, Michael. Letter of September 14, 1989 to Luigi Gonella. British Society for the Turin Shroud Newsletter, No. 24, January 1990, pg. 7. Accessible at https://www.shroud.com/pdfs/n24part5.pdf. Accessed May 29, 2020.

Tite, Michael. Letter of April 18, 1990 to Luigi Gonella. Luigi Gonella archives.

Tite, Michael. "TURIN SHROUD -- Thursday 21 April 1988." Official notes from release of British Museum documents in 2017 due to Freedom of Information Act request.

Upinksy, Arnaud-Aaron. "Lettera Al Direttore Del British Museum." *Collegamento Pro Sindone*, November/December 1990, pp. 39-42.

Van Haelst, Remi. "Il British Museum All Fine Aprira' I Suoi Archivi Sindonici?" *Collegamento Pro Sindone*, July/August 1993, pp. 29-33.

Van Haelst, Remi. "Un <u>Caveat</u> Riguardante La Datazione Radiocarbonica Della Sindone," *Collegamento Pro Sindone*, November/December 1994, pp. 41-43.

Van Haelst, Remi. "La Datazione: Al Radiocarbonio Rivista Dagli Esperti." *Collegamento Pro Sindone*, July/August 1991 pp. 44-48.

Van Haelst, Remi. "Onesta' Scientifica." *Collegemento Pro Sindone*, September/October 1991, pp. 52-53.

Van Haelst, Remi. "Quando Gli Esperti Del Radiocarbonio Diventano Esperti Tessli." *Collegamento Pro Sindone*, January/February 1992, pp. 39-41.

Van Haelst, Remi. "Una Domando Per La Verita." *Collegamento Pro Sindone*, July/August 1990, pp. 45-47.

Van Haelst, Remi. "UNA RECENSIONE DELLE RELAZIONI DI PARIGI Sul. Prelievo Di Campione E LA Datazione Al Radiocarbono Della Sindone Di Torino." *Collegamento Pro Sindone*, January/February 1992, pp. 31-38.

Villarreal, Robert with Barrie Schwortz and M. Sue Benford. "Analytical Results On Thread Samples Taken From The Raes Sampling Area

(Corner) Of The Shroud Cloth". Presented at "The Shroud of Turin: Perspectives on a Multi-Faceted Enigma" conference in Columbus, Ohio on August 14-17, 2008. Video at: https://www.youtube.com/watch?v=86wWOMGqsWQ&feature=youtu.be. Accessed May 29, 2020.

Walsh, Bryan and Schwalbe, Larry. "An instructive inter-laboratory comparison: The 1988 radiocarbon dating of the Shroud of Turin." *Journal of Archaeological Science: Reports,* v.**29** (2020) 102015. Accessible at https://www.sciencedirect.com/science/article/pii/S2352409X19301865. Accessed May 29, 2020.

[Wilson, Ian.] "Who's Who in the Shroud World Ray Rogers." *British Society for the Turin Shroud Newsletter*, No. 60 – December, 2004, pp. 16-23. Accessible at https://www.shroud.com/pdfs/n60part3.pdf. Accessed May 29, 2020.

Wilson, Ian. The Shroud: The 2000-Year-Old Mystery Solved, (Sydney: Bantam Press), 2010.

Wilson, Ian. Letter of April 30, 1988 to Luigi Gonella. Accessible at https://www.shroud.com/pdfs/gonellaltr.pdf. Accessed May 29, 2020.

Wilson, Ian. Letter of May 30, 1988 to Michael Tite. Luigi Gonella archives.

Wilson, Ian. "On the recent 'Leaks'…" Memo to BSTS members on September 23, 1988. Accessible at https://www.shroud.com/pdfs/bstsleaks.pdf. Accessed May 29, 2020.

Wright, Pearce. "New Dispute on Dating Tests." *London Times*, January 16, 1988, pg. 3.

INDEX

Numerical

3M Company (Milan) 28, 64, 65, 71, 79, 84, 98, 103, 105, 111, 116, 121, 125, 156, 206, 215, 219, 222, 228, 230, 242, 246, 255, 268, 283, 285, 295, 304, 310, 319, 345, 363, 417, 434, 441, 471, 496, 510, 517, 577, 589, 727

5th National Congress of Shroud, Cagliari, Italy, (1990) 581

12th International Radiocarbon Conference 49, 54

30 Days in the Church and in the World 520

'78 (1978) Examinations xiii, xv, xxiii, xxvii, 6, 8, 9, 12, 16, 24, 44, 50, 61-63, 66, 79, 84, 104, 113, 123, 124, 128, 138, 144, 150, 157, 159, 161, 183, 191, 233, 234, 267, 277, 279, 288, 290-292, 310, 314, 355, 356, 369, 375, 383, 447, 448, 450, 459, 492, 577, 582, 584, 588, 619, 636, 642, 656, 661

1532 Fire xv

A

Accelerator Mass Spectrometry [see "Carbon dating, accelerator method (AMS)"]

Accetta, Dr. August v

Adler, Alan (A.D.) 42, 77, 114, 121, 127, 135-137, 140, 144, 145, 167, 199, 213, 244, 287-292, 365, 368, 371, 372, 400, 412, 477, 494, 497, 558, 570, 571, 639, 640, 644, 647-649, 653, 713, 723

Air Force Academy 9, 41, 199, 717

Alonso, Marcel 368

Aluminum oxide mordant 652, 653

AMS (a.k.a. "TAMS") [see "Carbon dating, accelerator method (AMS)"]

American Shroud of Turin Association (ASTA) 556, 718

Ancaster, Ontario, CANADA Shroud Conference (2019) 468, 697, 708, 730

Anderson, Ian 281

Anderson, P.J. 2

ANSA Agency 498, 525, 542, 544

Antipope Clement VII 606

Antonacci, Mark v, 497, 703

Archaeology Chemistry 395

Archaeometry 25, 50

Archaeometry (journal) 688, 708, 709, 725, 732

Archdiocese of Turin xxiii, 105, 163, 164, 174, 194, 195, 656

Archiepiscopal Ordinariate of Turin 100

Arnoldi, Dr. Anna 652, 653, 733

513, 522, 523, 525, 527, 537, 539-541, 543-546, 568, 572-574, 580, 581, 599-601, 604, 605, 609, 615, 619, 663, 665, 676, 713, 723, 724

Bottema, Dr. Sietse 531, 532

Bourcier de Carbon, Philippe 460, 528, 535, 553, 627

Bowman, Dr. Sheridan 48, 49, 628, 629, 674, 675

Boyce, David 578, 579, 724

Boyd, Dr. Janice 48, 49, 60

Bracaglia, Giorgio xvii, xxv, 12, 15-19, 24, 25, 30, 37, 41, 44, 45, 66, 97, 110, 149, 202, 204, 215, 220, 224, 237, 242-244, 257-259, 274, 275, 303, 306, 331, 413, 418, 426, 468, 524, 542, 558, 569-571, 713, 724

Bray, Prof. Anthos 175, 431-433, 445, 456, 458, 471, 501, 526, 531, 536, 556, 621, 627, 630, 713

Brew, Prof. 3

Brignall, Shirley 48, 49, 64, 67, 80, 114, 116, 117, 121, 127, 194, 388, 402-404, 705, 713

British Museum xiii, 20, 21, 22, 28, 37-42, 44, 48-56, 58-61, 64-67, 69, 72, 73, 80, 82, 87, 88, 93, 100, 103, 105, 107, 114, 117, 120-134, 136-139, 141, 142, 146, 147, 153, 164, 165, 171, 173-175, 181, 182, 222, 224-228, 230-232, 237, 239, 240, 247, 251, 254, 255, 257, 262, 263, 271, 275, 293, 296, 304-308, 312, 320, 321, 333, 334, 336, 340, 351, 352, 357, 358, 361, 364, 370, 372, 391-393, 396-398, 405, 408, 413, 414, 420, 425, 431, 432, 434-438, 443, 445, 453, 456-458, 464, 465, 481, 488, 490, 495, 496, 499, 500, 504, 508, 511, 512, 519, 521, 524, 526, 529, 531, 536, 543, 546, 549, 551, 552, 555, 578, 585, 587, 591-594, 612, 618, 619, 621, 623, 625, 628, 630-634, 639, 640, 645, 665-668, 670, 672, 674-676, 678, 679, 682, 684-686, 707, 709, 716, 720, 721, 723, 728, 731, 733, 734

British Museum Shroud protocol (January 1988) 225, 232, 286, 292, 303, 305-315, 325, 330, 387, 395, 410, 414, 441, 601

British Society for the Turin Shroud (BSTS) 7, 24, 29, 96, 115, 248, 319, 328, 401, 426, 434, 436, 473, 484, 489, 490, 558, 564, 635, 659, 720, 733, 735

Bromley, D. Allan 1, 8, 10, 106, 144

Brookhaven National Laboratory 8, 11, 14, 15, 21, 24, 28, 29, 33, 49, 50, 54, 59, 60, 77, 80, 87, 91, 105, 107, 121, 122, 128, 224, 255, 268, 269, 282, 295, 297, 332, 395, 402, 454, 716, 728

Brown, Dr. John 691

Brunati, Ernesto 376, 377, 535, 536, 546, 552, 565, 567, 568, 575, 576, 578, 640, 641, 662, 664, 676, 677, 687, 688, 713, 725, 728

BSTS (see "British Society for the Turin Shroud")

Bucklin, Dr. Robert 14

Bulst, Fr. Werner 523, 565, 593, 600, 714, 725

Burleigh, Dr. Richard 21, 48, 67, 103, 634, 725

Busson, Dr. Pierre 610, 725

Byzantine art 223, 595, 607

C

C-14 dating [see "Carbon dating (general)"]

Cameron, Neil 397, 407

Canuto, Dr. Vittorio (Victor) 26, 28, 31, 33, 37, 43-47, 56, 59, 60, 63, 64, 66-68, 72-74, 84, 96, 99, 101, 102, 104, 107-117, 119, 121, 125, 127, 135, 149, 155-157, 161-166, 168, 172-174, 176, 178-181, 186, 193-196, 201-203, 205-207, 210-212, 214, 215, 220, 223, 224, 240, 241, 245, 246, 258-260, 264-266, 268, 278-280, 294, 295, 315, 316, 318, 322, 323, 325, 714, 716

Carbon 12 (C12) 270, 384

Carbon 13 (C13) 153, 154, 270, 384, 587, 621, 629

Carbon 14 (C14) [see "Carbon dating (general)"]

Carbon dating, accelerator method (AMS) xxv, 3, 4, 6, 8, 10, 21, 23, 32, 49, 50, 74, 84, 100, 101, 129, 132-134, 136, 138, 160, 177, 178, 220, 222, 224, 227, 229, 231, 245, 250, 255, 258, 259, 269, 282, 306, 310, 313, 324, 370-372, 383, 384, 392, 393, 398, 410, 415, 420, 425, 454, 462, 469, 470, 480, 485, 493, 528-532, 534, 541, 553-555, 591-593, 610-613, 620, 622, 623, 628, 632, 640, 644, 654, 656, 658, 659, 673, 692, 693, 699, 700, 703-705, 707, 725, 716, 727, 731

Carbon dating, chi-square test 530, 627, 630, 670, 671, 672, 675, 682, 683, 685-687

Carbon dating, control samples (see also "Fourth Sample") 82, 100, 124, 130, 132, 137-139, 142, 146, 174, 182, 209, 232, 251, 255, 306, 309, 311, 327, 328, 336, 337, 339, 340, 344, 345, 352, 353, 357, 359, 360, 363, 364, 370, 379, 396, 405, 408, 410, 415, 417, 419, 492, 498, 511, 524, 529, 542, 543, 545-547, 550, 595-597, 603, 609, 627, 638, 679, 681, 682, 690, 701, 714

Carbon dating, environmental effects 168

Carbon dating (general) vi, xiv, xix, xxii, 2, 3, 10, 11, 16, 18-32, 34-46, 49, 50, 54, 57, 59, 60, 62, 63, 71, 72-76, 78, 80, 85-87, 89-91, 93-101, 106, 108, 112-115, 120, 123, 126, 129-131, 133, 134, 136, 137, 139-146, 149-154, 157, 158, 160, 162, 169, 175, 176, 184, 187, 189-191,

193, 195, 197, 200, 201, 203-208, 212-214, 216, 218, 220-224, 228, 230, 233-235, 238-241, 242-245, 248, 249, 254, 257-259, 261, 263, 270, 275, 277, 278, 284, 285, 287-289, 292-295, 299-301, 303, 311, 314, 315, 317, 318, 321, 322, 324-327, 329-331, 333, 336, 338, 341, 349, 363, 367-370, 375, 384, 386, 390, 395, 396, 398, 402, 406, 408, 413, 414, 418, 423, 425, 426, 429, 438-440, 442, 446-448, 452, 453, 457, 459-485, 492, 493, 496, 500-506, 508-511, 513-515, 517, 527, 530-532, 534, 537, 538-540, 544, 552-554, 556-558, 561-563, 565, 569-571, 573-575, 584, 587, 588, 590, 591, 604, 605, 617, 620, 624, 627, 632, 633, 637, 639, 640, 642, 643-650, 696, 707, 710, 711, 713-721, 724-726, 730

Carbon dating, proportional counter (a.k.a. "Small Counter") 21, 49, 84, 88, 39, 101, 129, 132, 134, 138, 146, 177, 178, 224, 249, 256-258, 262, 263, 269, 270, 274, 282, 286, 335, 388, 399, 413, 425, 439, 443, 481

Carbon dating, rogue dates or samples 75, 76, 95, 135, 269, 272, 426, 427, 476, 641, 645

Carbon dating, small counter (see "Carbon dating, proportional counter")

Carpocratians 272

Casabianca, Tristan 78, 708, 709, 725

Casaroli, Cardinal Agostino 103, 105, 107, 108, 110, 117, 121, 154, 156, 165, 185, 202, 205, 215, 228, 229, 240-242, 245, 248, 250, 254, 263, 268, 278, 283, 295, 317, 319, 331, 429, 433, 440, 452, 510, 512, 516, 521, 522, 588, 714

Case, T.W. 78, 146, 639, 640, 725

Cathedral of St. John the Baptist 8,

Catholic Counter-Reformation in the XXth Century (CRC), xxvi, 22, 23, 130, 327, 328, 365, 418, 469, 498, 499, 513, 523, 525, 527, 540, 541, 543-546, 564, 565, 571, 568, 572-574, 578-581, 604, 609, 723, 724,

Catholic Herald 578

Caviglia, Fr. Giuseppe 429, 507, 642

CBC Man Alive (TV program) 10

Celli, Msgr. Celestino 47, 72, 93, 107, 116, 117, 119

Centre International d'Etude des Textiles Anciens of Lyon 310

Centro Internazionale di Sindonologia in Turin, ("Centro") xxiii, xxv, 3, 15-18, 24, 74, 100, 110, 131, 257, 372, 512, 514, 593, 651, 655, 656, 713, 714, 730

Chevalier, Canon Ulysse 607

Chi-square test (see "Carbon dating, chi-square test")

Chicago Tribune 294, 302, 325, 327, 725

Ciarla, Roberto 239, 647

Clark, Kenneth R. 301, 302, 726

Cloutiér, Annette vii

Coero-Borga, Fr. Piero ("Coero") xxiii, 15, 106, 220, 221, 257, 259, 714

Coghlan, Andy 527, 726

Cold Case: The Shroud of Turin (Documentary) (see "*La Notte de la Sindone*", the original in Italian)

Collegamento Pro Sindone xxvi, 392, 523, 552, 567, 568, 578, 585, 588, 593, 600, 606-608, 613, 616, 618, 623, 625, 631, 635, 676, 687, 724, 725, 728, 732, 734

Committee for the Scientific Investigation of Claims of the Paranormal (CSICOP) 226, 304

Colonetti Institute [Istituto di Metrologia G. Colonetti (IMGC)] 120, 121, 133, 134, 228, 254, 255, 309, 311, 406, 430, 431, 432, 456, 478, 499, 510, 519, 520, 526, 556, 609, 627, 713

Conard, Nicholas 48

Congregation for the Doctrine of the Faith 26, 28

Conti, Pierluigi 641

Cooper, Elisabeth 549

Cope of St. Louis d'Anjou (see also "Control samples" and "Fourth sample") 328, 336, 343, 357, 371, 378, 405, 419, 437, 488, 512, 524, 525, 539, 543, 551, 596, 597, 598

Cornwell, John 492, 494, 527, 728

Corrispondenza Romana 641

Corti, M. Consolata 582

Cottino, Msgr. Jose 16, 714

CRC (see "Catholic Counter-Reformation in the XXth Century")

Crispino, Dorothy 509, 527, 530, 726

CSICOP (see "Committee for the Scientific Investigation of Claims of the Paranormal")

D

Daily Mail 435

Daily Telegraph 41, 293, 305

Dallas Shroud Conference (2005) 590

Damon, Dr. Paul 121, 123, 127, 173, 174, 213, 240, 241, 247, 248, 250, 260, 265, 266, 280-282, 286, 297, 299, 301, 307, 308, 312, 313, 321, 323, 346, 367, 369, 394, 397, 403, 404, 408, 416, 420, 421, 426-428, 489, 490, 497, 574, 601, 603, 608, 610, 654, 655, 694, 705, 709, 710, 714, 723, 726

Danin, Prof. Avinoam 594

D'Amato, Senator Al 305

d'Arcis, Bishop Pierre 386, 466, 607

Dardozzi, Msgr. Renato 119, 121, 127, 183, 203, 204, 206, 211, 212, 215, 216, 218, 219, 228, 316, 343, 346, 355, 714

de Briene, Daniel Raffard 419, 420, 719

De Giovanni, Prof. Livia 641

De Nantes, (Abbe) Georges 23, 537, 540, 568, 572-574, 600, 601, 604, 724, 726

De Wesselow, Thomas 697-699, 701, 704, 726

di Lalaing, Antoine 459

Di Lazzaro, Paolo 79, 230, 631, 726

Di Rovasenda, Fr. Enrico 98, 108, 121, 124, 128, 715

Dichtl, Dr. Rudy 287

Die Welt 641

Dinegar, Dr. Robert xvi, 13, 15, 18, 19, 24, 29, 33, 36-41, 43, 44, 47-50, 56, 60, 62-65, 68-74, 87, 88, 96, 97, 99, 101, 108-110, 121, 127, 129, 130, 140, 141, 185, 201, 244, 266, 267, 275, 277-279, 281, 287, 288, 290, 291, 292, 403, 404, 428, 475, 542, 546, 712, 715, 726

Diocesan Commission for the Holy Shroud 661

D'Muhala, Thomas 275, 276, 279, 281, 287

Donahue, Dr. Doug 48, 49, 59, 105, 107, 121, 125, 127, 149, 152, 156, 165, 185, 213, 235, 236, 240, 250, 253, 254, 257, 258-260, 264-267, 279, 280, 286, 307, 308, 312, 317, 318, 321, 323, 325, 334, 346, 358, 367, 394, 396, 402-404, 408, 416, 420, 469, 493, 496, 567, 574, 600, 601, 603, 609, 619, 628, 686, 695, 723

Donovan, Fr. Vincent 12,

Double-blind test (for Shroud C-14 dating) 545, 627

Dreisbach, Fr. Albert (Kim) xx, 273, 438, 440, 569, 570, 731

Drews, Dr. Robert 272

Druzik, James 14, 199, 715

Dubarle, Fr. A-M. 528

Dubrovnik Radiocarbon Conference (1988) 262, 406, 414, 416, 417, 427, 686

Duplessy, Dr. J.C. (Claude) 112, 113 121, 127, 131, 136, 138, 149, 241, 260, 280, 319, 323, 484, 715

Dupont, Chantal 528

Dutton, Denis 224, 225-227, 230-232, 237, 259, 294, 304, 332, 334, 715, 726, 727

E

Eastwood, Gillian 412

Egyptian linen control sample (for Shroud dating) (see also "Control samples" and "Fourth sample") 21, 296, 334, 437, 458

Elmore, Dr. David 48

ENEA (Institute, "New Technologies, Energy and Sustainable Economic Development") 79, 319, 696, 726

Ercoline, Bill 287

ETH Zurich (C-14 lab) (see "University of Zurich")

Evin, Dr. Jacques 22, 23, 112, 114, 121, 128, 327, 328, 336, 343, 347, 352-354, 359, 362-364, 370, 405, 418, 421, 427, 453, 484, 513, 524-526, 528, 529, 534, 534, 538, 542-545, 547, 549, 550, 572, 573, 587, 596, 612, 624, 640, 654, 715, 727

F

Faia, Franco 33, 126, 153

Farkas, Ilona 585, 727

Fedi, M.E. 693, 727

Feuerherd, Peter 467, 727

Firpo, Prof. 431

Flaherty, Tom 556

Fleming, Dr. Stuart 258, 259, 269-273, 476, 483

Flury-Lemberg, Mechthild 117, 121, 123, 124, 126, 128-132, 137-141, 143, 147 171, 176, 181, 182, 197, 227, 233, 255, 257, 262, 263, 305, 311, 317, 354, 364, 395, 414, 715

Fossati, Rev. Luigi 2, 95, 150, 459, 607

Fourier Transform Infrared Spectroscopy (FTIR) 367, 644, 691

Fourth Sample (for Shroud dating) (see also "Cope of St. Louis d'Anjou" and "Control samples") 352-354, 357, 361, 364, 378, 405, 524, 525, 529, 539, 542, 543, 549-551

Foye, Dr. James R. 272

Fractionation 270, 384, 385

Freedom of Information Act (FOIA) 19, 398, 413, 417, 420, 423, 438, 511, 556, 631, 708, 723, 733, 734

Freemasonry 511, 512

Freer, Rachel 694-696, 727

FTIR (see "Fourier Transform Infrared Spectroscopy")

G

Garwin, Dr. Richard 612

Garza-Valdes, Dr. Leoncio 612, 625, 626, 649

General Ionex Corporation 4

Gervasio, Riccardo 3, 16

Getty Museum (see "J.P. Getty Museum")

Ghiberti, Msgr. Giuseppe 345, 590, 626, 637, 655, 656, 661, 662, 716

Gif-sur-Yvette (French C-14 lab) 112, 120, 122, 131, 136, 715

Glasgow University 274, 282, 330, 391, 509, 611

Glass, Robert 338, 727

Gonella, Dr. Luigi xvi, xxiv, xxvi, 7, 14-18, 23-26, 28, 31, 32, 34, 37, 39, 41, 44, 46, 47, 56, 59, 60, 62-65, 68, 70-74, 79-81, 83, 84, 91-93, 97-107, 110-140, 142, 143, 145, 147, 150, 152, 154-157, 174, 184-186, 198, 201-203, 206-208, 210-215, 218, 219, 222-225, 227, 228-230, 232, 235, 237-246, 248, 249, 255-258, 260, 261, 264-266, 268, 269, 273-276, 278-283, 285-287, 292, 294-297, 299, 300, 303-306, 308-325, 327, 329, 333-339, 341, 343-347, 349, 350, 352-355, 357, 359, 360-366, 368, 371, 372, 379, 380, 382, 387-389, 391-394, 399, 401-403, 405, 407, 408, 415-417, 421-423, 426-436, 440-444, 446, 448, 454, 458, 461, 463-467, 469, 471-473, 478-480, 485, 488, 490, 495, 496, 499, 501-507, 509-511, 513-515, 517-520, 522-526, 532, 533, 541, 542, 546-549, 552, 558, 559, 561, 563, 564, 569, 570, 573, 578-580, 582-589, 593, 595, 598, 599, 603, 614-618, 625, 626, 633, 636, 645, 656, 659, 662, 697, 707, 716, 726-728, 731, 733-735

Gove, Dr. Harry xxv, xxv, 1, 4-15, 20, 21, 26-29, 31-34, 37-51, 56, 57, 59, 60, 62-65, 67-74, 77, 80, 83, 84-93, 96, 98-109, 112-126, 128-133, 135-145, 148-150, 152, 153, 155, 156, 157, 159-163, 165-167, 169, 170, 172-176, 179, 183, 185-187, 191, 193-198, 200-203, 205, 206, 209-216, 220-225, 227, 228, 231-237, 240-250, 254-268, 275, 276, 279-282, 285, 286, 293-295, 299-301, 303-305, 312, 315-325, 328, 332-336, 341, 358, 362, 369-371, 379, 382-385, 387-389, 396-398, 402-404, 407-411, 413-416, 420-423, 425-430, 444, 472, 473, 475, 480-482, 489, 490, 493, 496, 498, 506, 509, 520, 574, 599, 619, 620,

623, 625, 646, 647, 649, 650, 651, 654, 673, 674, 686, 687, 696, 700, 701, 704, 705, 713, 714, 716, 721, 727, 728

Gove/Harbottle C-14 proposal (1979) 11, 14, 19, 136, 299, 304, 646, 647

Gowlett, J.A.J. 48, 49

Gruber, Elmar (see "Kersten and Gruber")

H

Haas, Dr. Herbert 469-471, 476, 478, 481, 484, 728, 731

Habermas, Dr. Gary v, 6, 85

Halisey, Robert 585, 586

Hall, E.T. [Dr. Edward (Teddy)] 25, 29, 32, 33, 39, 41, 48, 59, 67, 105, 107, 115, 121-123, 128, 129, 132, 137, 138, 139, 142, 144, 152, 156, 159, 165, 173, 185, 224, 227, 228, 236, 240-242, 245-250, 253, 254, 261, 262, 264-268, 279-283, 286, 299, 304, 305, 307, 310, 313, 318, 320, 321, 323, 333, 334, 345, 346, 358, 360, 361, 367, 382, 383, 392, 394, 396-398, 408, 421-424, 426, 429, 430, 434, 435, 440, 441, 445, 453, 455, 457, 468, 481, 482, 489, 491-496, 499, 501, 504, 505, 511, 512, 528, 533, 542, 546, 567, 569, 572, 581, 595, 602, 604, 605, 609, 624, 677, 700, 711, 716, 720

Harbottle, Dr. Garman 19, 24, 25, 28, 29, 33, 48, 49, 51, 56, 59, 60, 63, 83, 87, 89, 91, 102, 105, 107, 121, 123-125, 128, 138, 139, 149, 152, 156, 165, 176, 182, 203, 209, 224, 235, 236, 242, 255, 256, 259, 260, 263, 267, 268, 275, 276, 278-281, 294, 295, 299-301, 303-305, 313, 318, 323, 332, 341, 348, 392, 393, 395, 396, 475, 476, 509, 646, 647, 649, 716, 728

Harpers Magazine 410

Harwell Laboratory (AERE Harwell) 2, 20, 21, 28, 38, 49-51, 54, 59, 80, 121, 122, 128, 224, 249, 250, 254-256, 269, 282, 285, 297, 301, 303, 332, 389, 393, 399, 400, 412, 413, 454, 481, 484, 490, 504, 718

Haynes, Vance 655

Hedges, Dr. Robert 48, 49, 121, 128, 213, 246, 307, 308, 323, 338, 346, 367, 394, 418, 430, 434, 453, 457, 493, 496, 532, 602, 622, 624, 625, 628, 629, 632, 635, 648, 716

Heller, Dr. John 14, 17, 18, 23, 24, 43, 400, 412, 494, 717

Herzberg, Dr. G. 48

Holy See 43, 66, 67, 87, 104, 106, 113, 120, 127, 244, 245, 250, 251, 255, 262, 295, 298, 310, 312, 343, 400, 433, 444-446, 450, 452, 456, 461, 506, 521, 606-608, 636

Holy Shroud Guild vii, xxiii, xxv, 12, 15,

Horowitz, Prof. Aharon 594

Hungarian Pray Manuscript 386

Hutchins, J.K. (Jane) 199, 287-290

I

Iakovidis, Spyros 698

Ibarrola, Faustino Cervantes 625, 626, 717

Il Giornale 546

Il Messaggero 512, 718

Il Telo 671

Independent (*The*) 396

Inside the Vatican 368, 661

Institut Euclide (Paris) 538

Institute for Near and Far East (IsMeo) 207, 647, 484

Intercomparison test (British Museum) (1983) 20, 21, 25, 28, 37, 38, 40, 50, 53, 296, 392, 393, 634, 699

Intercomparison test (*Nature* International Study Group) (1982) 19, 20, 28, 56, 300, 302, 728

Intercomparison test (Scotland) (1989) 527

International Radiocarbon Calibration Programme 282, 509 (see also "Glasgow University")

International Sindonology Center in Turin (see "Centro Internazionale di Sindonologia")

Invisible Reweave (see also "Mending" and "Repairs") vi, xiii, xvii, xix, 6, 16, 31, 75, 76, 349, 375, 468, 643, 660, 661, 697, 708, 730

IsMeo (see "Institute for Near and Far East")

IsoTrace (C-14 lab in Toronto, Canada) 329, 370, 384

Istituto di Metrologia G. Colonetti (IMGC) (see "Colonetti Institute")

J

J P Getty Museum 114, 128, 199, 715

Jackson, Dr. John 9, 14, 24, 84-86, 96, 161, 201, 279, 287-292, 528, 558, 717

Jaggi, Diego 405

Jasper, William 548, 605, 728

Jennings, Peter 99-101, 102, 106, 107, 115, 167, 300, 302, 468, 728

Jet Propulsion Laboratory 9, 41

Jones, Stephen 488, 516, 704, 706

Judica-Cordiglia, G. 319

Jull, Dr. A.J.T. (Timothy) 48, 49, 321, 580, 601, 603, 641, 653, 689, 693, 694-696, 712, 717,

Jumper, Dr. Eric xi, xv, 9, 10, 287, 717

K

Kava, Brad 301, 302

Kersten, Holger (see also "Kersten and Gruber") xxv, 26, 27, 44, 46, 99, 101, 145, 215, 255, 256, 455, 473, 474, 485, 486, 488-492, 495, 522, 523, 539, 542, 564, 565, 604, 729

Kersten and Gruber (see also "Kersten, Holger") xxv, 26, 27, 44, 46, 99, 101, 145, 215, 255, 256, 455, 474, 488, 490, 492, 495, 522, 523, 539, 565, 729

King Umberto II (see "Umberto II, King")

Kowalski, Michael Lewis vii, 710

L

La Notte de la Sindone (The Night of the Shroud) (Documentary) (a.k.a. "*Cold Case*", renamed and updated), 25, 34, 46, 47, 153, 202, 372, 575, 697, 731

La Stampa 149, 150, 223, 224, 238, 265, 293, 295, 297, 323, 372, 373, 393, 431

Laghi, Cardinal Pio 274

Lamont Geological Laboratory (U.S.A.) 612, 620

Laverdiere, H. xxvi, 29, 77, 113, 237, 285, 315, 329, 370, 371, 381, 385, 425, 729

Lavoie, Dr. Gilbert 58

Lawrence Berkeley National Laboratory (see "Berkeley National Laboratory")

Leclercq, Michel 418, 544

Leese, Dr. Morven 21, 587, 612, 621, 628-630, 634, 725

Lejeune, Prof. Jerome 547, 548, 575, 632, 633

Levi, Prof. R. 431

Libby, Dr. Willard 2, 598, 610, 717

Lignin 36, 660

Lindow Man 224, 480, 481, 504

Linick, Dr. Timothy 469, 705

Lirey, France 225, 386, 458, 583, 606, 669, 685, 686, 704

Litherland, Dr. Ted 333, 370, 371

London Evening Standard 423, 426, 434

London, Dr. R. J. 15

London Times (a.k.a. "*The Times*" and "*Sunday Times*") 302, 304, 357, 418, 422, 430, 434, 435, 735

Long, W.H. 29, 292

Lorre, Dr. Jean 656

Los Alamos National Laboratory 9, 41, 288, 691, 715, 719

Luckett, Dr. Richard 423, 424, 426, 434, 435

Lucotte, Prof. Gérard 319

Lukasik, Dr. Steven 106, 112, 121, 122, 138, 140, 143, 144, 149, 183, 202, 213, 219, 237, 238, 247, 248, 257, 275, 284, 324, 444, 717

M

Madder root dye 455, 492, 652, 653, 660

Maloney, Paul C. 258, 268, 273-275, 282, 283, 301, 314, 329, 331, 371, 372, 391, 392, 395, 413, 462, 463, 559, 612, 619, 631, 656, 662, 717, 729

Marcilacy, Prevto 526

Marinelli, Emanuela (see also "Petrosillo and Marinelli") xxv, xxvi, 2, 11, 26, 98, 99, 104, 105, 131, 134, 154, 155, 224, 225, 229, 248, 249, 262, 263, 264, 281, 282, 285, 299, 300, 311, 312, 315, 317, 322, 335, 337,

338, 346, 363, 369, 371, 403, 408, 414, 415, 417, 421-424, 426, 428, 430, 444-446, 449, 454, 455, 472, 478, 492, 497, 499, 508, 511, 515, 518, 520, 531, 532, 538, 548-552, 581, 584, 589, 594, 608, 613, 614, 618, 625-628, 633, 636, 638, 639, 641, 696, 697, 717, 728, 729, 732

Marino, Carolina 643, 733

Marino, Joseph G. i, ii, vi, xiii, 78, 657, 660, 665, 723, 729, 730

Marino and Benford (see "Benford and Marino")

Masonic Plot (in Shroud dating) xix, 638, 641

Mass Spectrometry 131, 649, 660, 691

Materials Issues in Art and Archaeology 292

Mattingly, Dr. Steven 636

McCarrick, Cardinal Theodore 656, 657

McClellan, Bill 402, 403, 408, 409, 730

McClure, Mike 682-684

McCrone, Dr. Walter 4-7, 9, 12, 13, 70, 158, 159, 161, 162, 185, 196, 197, 336, 400, 412, 473-475, 540, 541, 647, 717

Meacham, William xxv, 29, 31, 33, 34, 36, 74-76, 93-96, 106, 121, 128, 130, 131, 135-137, 140, 144, 145, 152, 153, 154, 203-207, 211-216, 218, 219, 237, 239, 240, 248, 270, 284, 301, 302, 325, 327, 328, 338, 345, 365, 368, 398, 403, 438, 440, 443, 444, 455, 461, 463-466, 469, 471, 476-480, 485, 510, 590, 618, 644-647, 650, 651, 655-658, 662, 676, 677, 701, 718, 728, 730, 731

Mending (see also "Invisible Reweave" and "Repairs") vi, xiii, xvii, xix, 6, 16, 30-31, 75, 76, 337, 349, 375, 378, 468, 643, 660, 661, 665, 697, 708, 730

Messori, Vittorio 453, 584

Micro-Raman 131, 649

Ministry of Cultural Heritage and Activities and Tourism xxiv, 343

Minor, Mike 276, 558, 718

Moon, Pam 531, 691

Morris, Dr. R.A. 15

Mottern, Dr. R.W. 15

Moynihan, Senator Daniel 323, 334, 402

Muldoon, Shena 661

Mummy of Cleopatra of Thebes control sample (for Shroud dating) (see also "Control samples") 351, 378

Murphy, Cullen 410

Museum of Applied Science Center for Archaeology (MASCA) 258, 269

N

National Science Foundation (NSF) 65, 67-69, 91, 98, 99, 101, 109, 137, 167, 314, 394

National Textile Authority (France) 508

Nature (see also "International Study Group") xix, 19, 20, 224-227, 230-232, 236, 237, 255, 263, 293, 294, 302, 304, 307, 312, 332, 334-336, 338-341, 348, 350, 353, 360, 361, 384, 386, 404, 410, 411, 414, 419, 421, 427, 437, 458, 471, 478, 486, 493, 494, 496, 497, 500, 501, 510, 513, 518, 521-523, 526, 527, 535, 536, 540, 542-544, 546, 548, 551, 553, 556, 572-574, 577, 579, 580, 585-587, 597, 598, 601, 602, 608-610, 612, 613, 618, 620-622, 624, 627-631, 634, 638, 664, 666, 668, 669, 671, 672, 674-678, 682-685, 694, 698, 708-710, 715, 725, 727, 728, 732, 733

Natuur en Techiek 531

Navarro-Valls (a.k.a. "Navarro"), Dr. Joaquín 433, 444

Nelson, Dr. Erle 270

New Scientist 281, 527, 622, 623, 699, 703

New York Times 4, 468

Nicolotti, Dr. Andrea 697

Night of the Shroud (see "*La Notte de la Sindone*") (Documentary) (see also "*Cold Case*")

Nitowski, Eugenia (a.k.a. "Sister Damian") 45, 475, 718, 731

Non-statistical errors (C-14) 253, 254, 296, 297

NSF (see "National Science Foundation")

Nuclear Instruments and Methods in Physics Research 60, 148, 231, 247, 693, 727

Nydal, Dr. Reider 20, 49, 300, 302

O

October 13, 1988 (Announcement of Shroud C-14 dating results) [See "Press conference at British Museum (October 13, 1988)"]

Oeschger, Dr. Hans 38, 41, 48, 59, 718

Ojeda-Mari, Victor 420, 462, 707, 731

Olsson (author last name) 3

Osservatore Romano 356, 370, 408, 440, 446, 450, 455, 472

Otlet, Dr. Robert 20, 21, 38, 39, 48, 49, 59, 121, 124, 128, 130, 149, 213, 240, 249, 250, 256, 259, 261, 280, 301, 304, 305, 312, 318, 323, 333, 396, 442, 484, 509, 611, 648, 718

Otterbein, Fr. Adam xxiii, 17-19, 23, 24, 41, 96, 97, 110, 242, 243, 245, 273, 438, 528, 558, 559, 561, 563, 718

Our Sunday Visitor 101, 103, 107, 167, 302, 468, 728

Oxford Star 635

P

Paci, Stefano 632, 633, 731

Paglia, Guido 295

Paolicchi, M. 596

Paris International Scientific Symposium (1989) 347, 348, 376, 377, 380, 460, 525-528, 530, 532, 534, 535, 538-547, 552, 556, 557, 559, 563, 566, 569, 572, 573, 575, 576, 581, 587, 591, 592, 593, 599, 608, 612, 619, 620, 627, 628, 640, 663, 707

Paris Match 418, 525, 526, 544, 545

Parlato, Charles 556, 558, 718

Pearson curve (C-14 statistic) 154, 437, 460, 535

Pellegrino, Cardinal Michele 191, 458

Perry Chemical Engineers' Handbook 629

Petrosillo, Orazio (see also "Petrosillo and Marinelli") xxv, 2, 98, 99, 104, 105, 131, 134, 154, 155, 224, 225, 229, 248, 249, 262-264, 281, 282, 285, 299, 300, 312, 315, 317, 322, 335, 337, 338, 346, 363, 369, 371, 403, 408, 414, 415, 417, 421-424, 426, 428, 430, 444-446, 449, 450, 454, 472, 478, 499, 508, 512, 513, 515, 520, 532, 538, 548, 549-552, 581, 583, 584, 594, 605, 606, 608, 613, 614, 616, 618, 625, 626, 633, 636, 639, 718, 728, 731, 732

Petrosillo and Marinelli (see also "Petrosillo, Orazio" and "Marinelli, Emanuela") xxv, 2, 98, 99, 104, 105, 131, 134, 154, 155, 224, 225 229, 248, 249, 262-264, 281, 282, 285, 299, 300, 312, 315, 317, 322, 335, 337, 338, 346, 363, 369, 371, 403, 408, 414, 415, 417, 421-424, 426, 428, 430, 444-446, 449, 450, 454, 472, 478, 499, 508, 515, 520, 532, 538, 548-552, 581, 584, 594, 613, 614, 618, 625, 626, 633, 636, 638, 639, 728, 732

Phillips, Dr. Thomas 385, 386, 496

Piczek, Isabel 661

Piltdown Man 22, 457, 632

Plummer, Mark 220, 221

Polach, Henry 509, 718

Polarized Light Microscopy 692

Poletto, Cardinal Severino 655-658, 662

Polytechnic of Turin 32, 80, 84, 90, 93, 102, 104, 126, 189, 210, 295, 310, 315, 357, 402, 431, 445, 456, 505, 523, 552, 617, 626, 663, 690, 716, 720

Pontifical Academy of Sciences xxiii, 18, 25, 26, 29, 31, 34, 37, 39, 43, 45, 48, 53, 54, 56, 59-61, 64-67, 79, 80, 93, 98-101, 105, 107, 113, 114, 118, 120, 122, 124, 126-128, 133, 139, 147, 148, 161, 164, 166-169, 171, 174-176, 179-182, 186, 187, 189, 193-195, 197, 198, 205, 209-212, 214, 219, 223, 226. 227, 230, 231, 235, 236, 240, 244-246, 251, 255, 259, 261, 263, 294, 295, 298, 307, 316, 320, 326, 343, 355, 364, 410, 441, 472, 503, 506, 507, 514, 517, 536, 547, 555, 559, 561, 575, 632, 633, 714, 715, 721

Poor Clare Nuns 129

Pope Benedict XVI (see also "Ratzinger, Cardinal Joseph") 47

Pope Francis 706, 707

Pope John Paul II (see also "Wojtyla, Cardinal Karol") 107, 117, 126, 273, 331, 441, 445, 450, 451, 455, 513, 606, 616, 651, 655, 662, 714, 719

Press conference at British Museum (October 13, 1988) xix, 5, 397, 433, 440, 441, 444, 450, 453, 456, 465, 471, 492, 495, 616, 702, 716

Prior, Ed vi, vii, xix, 111, 368

Proportional Counter (see "Carbon dating, proportional counter") 21, 49, 101, 129, 132, 134, 138, 224, 249, 258, 262, 263, 269, 270, 274, 282, 335, 399, 413, 425

Pyrolysis-Mass-Spectrometry 660

R

Radiation 149, 152, 188, 191, 194, 199, 212, 292, 385, 505, 554, 650

Radio Courtoisie (France) 455, 527, 528, 547

Radiocarbon [see "Carbon dating (general)"]

Radiocarbon (journal) 21, 54, 76, 154, 229, 398, 437, 500, 551, 612, 620, 634, 653, 654, 678, 679, 681-684, 687, 688, 694, 695, 725, 727, 732

Raes, Prof. Gilbert xvi, 3, 6, 14, 232, 234, 365, 373-375, 461, 473-475, 478, 479, 486, 489, 540, 541, 569, 619, 625, 630, 631, 719, 727

Raes sample (located next to C-14 sample) xvi, 4, 5, 14, 15, 74, 88, 136, 270, 273-275, 313, 344, 347, 348, 350, 365, 367, 369, 373-375, 398, 461, 473-475, 482, 489, 510, 540, 541, 569, 625, 631, 647, 652, 653, 660, 662, 689, 691, 692, 719, 734, 735

RAI TV (Italy) 664

Raloff, J., 301, 302, 732

Ramsey, Dr. Christopher Bronk (a.k.a. "C. Bronk Ramsey") 497, 688, 689, 694, 702, 704, 732

Rapporto Sindone (Book) 134, 353, 542, 544, 577

Ratzinger, Cardinal Joseph (Pope Benedict XVI) 47, 72, 123, 157, 215

Repairs (see also "Invisible Reweave" and "Mendings") vi, xiii, xvii, xix, 3, 6, 16, 31, 75, 76, 349, 375, 468, 643, 660, 661, 697, 708, 730

Resurrection of Jesus (possible effect on Shroud dating results) v, 27, 44, 101, 126, 252, 256, 386, 452, 474, 488, 490, 492, 495, 496, 522, 529, 530, 539, 555, 564, 565, 589, 637, 676, 703, 704, 706, 726, 729

INDEX

Riani, Marco 78, 732

Riggi di Numana, Giovanni, (a.k.a. "Riggi") xxi, xxiv, 28, 44, 64, 65, 71, 72, 79, 84, 98, 102, 103, 105, 106, 111, 116, 124, 125, 134, 143, 156, 176, 184, 199, 201, 206, 215, 219, 222, 228, 230, 237, 240, 242, 246, 255, 268, 283, 285, 287-291, 295, 304, 310, 311, 314, 319, 337, 338, 343-357, 359-369, 371-373, 375-378, 380, 395, 398, 404, 417, 422, 425, 434, 441, 455, 458, 471, 496, 500, 502, 505, 510, 514, 517, 522-526, 539, 540, 542-544, 546, 550, 557, 567, 568, 570, 571, 573, 575-579, 582, 586, 589, 593, 595, 600, 603, 604, 618, 619, 620, 624-626, 631, 633, 634, 639, 659, 664, 715, 719, 720, 727

Rinaldi, Fr. Peter xxiii, 7, 15-17, 96, 97, 215, 220, 221, 226, 242-244, 257-259, 265-268, 273, 278, 279, 286, 312, 315, 316, 324, 333, 338, 341, 399, 416, 418, 426, 463, 465, 472, 473, 558, 559, 561, 719

Rinaldi, Gian Marco 695

Robinson, Bishop John 10

Rogers, Ray iii, xiv, xvii, xxvii, 78, 367-369, 492, 558, 640, 644, 652-661, 663, 689, 690, 691, 719, 732, 733

Rolfe, David 215, 223, 707

Rolling Stone article (on Shroud) 93

Rosetum (hall in Milan, Italy) 518, 568, 584, 617

Rubio, Meyer 43

S

Saclay Laboratory (France) 332

Saldarini, Cardinal Giovanni 350, 452, 518, 520, 588, 590 625, 626, 630, 633, 636

Sample switch hypothesis 224, 225, 230, 337, 340, 565

Sandia Laboratory 9, 41

Saracino, Francesca xxvi, 25, 34, 46, 47, 153, 202, 372, 575, 697

Savarino, Piero 95, 327, 352, 363, 364, 642, 643, 650, 656, 657, 730, 733

Save-Soderbergh (author last name) 3

Saxer, Msgr. Victor 606, 607

Scannerini, Silvano 95, 650, 730

Scanning Electron Microprobe Spectroscopy 644

Scanning Electron Microscope (see also "SEM") 405, 458

Schwalbe, Dr. Larry 79, 237, 286-288, 290-292, 558, 695, 709, 710, 719, 735

Schwortz, Barrie M. vi, 14, 292, 321, 409, 444, 655, 656, 690, 691, 693, 695, 710, 719, 732-734

Science and Engineering Research Council (SERC) 699

Science News 267, 275-279, 302, 341, 732

Scott, Dr. Marian (E.M.) 274, 282, 330, 391, 509, 611, 612, 620

SEARCH (Australian Shroud group) 593

Second International Congress on the Shroud (Turin, 1978) 8

Sega, R.M. 292

SEM (see also "Scanning Electron Microscope") 405, 649

SERC (see "Science and Engineering Research Council")

Seven-Lab Protocol (of the Turin Workshop in 1986) 156, 253, 257, 258, 260, 265, 266, 269, 270, 274, 275, 279-281, 300, 301, 313, 320, 321, 326, 334, 395, 410, 411, 441

Shroud, secret 1982 C-14 testing 406, 461, 482

Shroud, TV exhibition (1973) 3

Shroud blog (www.shroudstory.com) 497, 653, 688

Shroud Conference Columbus, Ohio (2008) 274, 656, 691, 729, 735

Shroud Custodian (see also "Ballestrero, Cardinal Anastasio") xxiii, 105, 111, 113, 125, 126, 155, 159, 161, 162, 164, 168, 170, 171, 174, 176, 179, 180-182, 187, 189-194, 196, 197, 200, 215, 251-253, 278, 307, 339, 356, 445, 456, 515, 517, 521, 561, 570, 588, 636, 657

Shroud of Turin Research Project (see "STURP")

Side strip on the Shroud 644, 723

Siliato, Maria-Grazia 4, 16, 697, 733

Simon Frazer University AMS laboratory 270

Sindon 3, 651, 730

Sister Damian (see "Nitowski, Eugenia")

Small Counter (see "Carbon dating, proportional counter")

Smith, P.R. 293

Somolo, Cardinal Martinus 106, 108, 117, 720

South, Peter 426, 501, 620

Souverain, Raymond 652, 719, 733

Sox, Rev. David xxv, 4-7, 12-14, 27, 43, 46, 64, 92, 93, 99, 102, 104, 106, 113, 130, 220, 223, 224, 304, 305, 312, 320, 335, 336, 347, 357-359, 362, 370, 379, 382, 403-408, 416, 417, 422, 423, 425, 426, 429, 434, 435, 454, 473-475, 489, 490, 540, 541, 550, 551, 599, 647, 704, 705, 720, 733

Speck, Richard 287-292

St. John the Baptist Cathedral (Turin) xxiv, 8, 343

St. Maximin (Basilica in Provence, France) 328, 337, 343, 344, 378, 526, 539, 596

Stampa Sera 373

Stevenson, Rev. Kenneth xiii, 6, 9, 10, 85, 86, 92, 604, 720

Strontium 308, 359

Stuiver, Dr. Minze 59, 61, 437, 509

STURP ("Shroud of Turin Research Project") xiii, xiv, xx, xxiii, xxv, xxvii, 1, 6, 8-10, 12-19, 23-31, 33, 36, 37-47, 49-52, 55, 56, 58-63, 65-74, 77, 79, 84-88, 90-93, 96-101, 103, 104, 106, 109-115, 121-125, 128-131, 135-142, 144, 145, 149, 150, 152, 154-164, 167, 169, 171, 176, 181-183, 185-191, 193-204, 206, 213-217, 220-224, 227, 228, 233-236, 239, 240, 243, 244, 246-248, 257, 259, 263, 264, 266, 267, 268, 274, 275-279, 281, 284, 286, 287, 289, 290, 292, 295, 310, 314, 316, 319, 321, 322, 333, 337, 341, 342, 348, 350, 362, 363, 367-369, 383, 397, 399, 408, 409, 414, 415, 428, 438, 442-444, 455, 463, 465, 472, 473, 475, 477, 478, 484, 499, 504, 536, 555, 558, 559, 569, 593, 595, 604,

618, 619, 627, 639, 644-647, 649, 656, 657, 661, 689, 691, 695, 701, 709, 713, 715-720, 733

STURP meeting, Albuquerque, New Mexico (1989) 569, 570

STURP meeting, Rye, NY (1987) 219, 258, 266, 275, 279, 287

STURP meeting, Santa Barbara, CA (1979) 12, 39

STURP Protocol (1987) xvi, xvii, 87, 159, 224, 226, 235, 236, 337

STURP Tests Proposal (1984) xvi, xvii, 87, 159, 224, 226, 235, 236

Superintendency for the Artistic and Historical Heritage of Piedmont 310, 374

T

Tablet (The) 492, 494, 542, 546, 729

TAMS [see also "Carbon dating, accelerator method (AMS)"] 4, 50, 393

Tassot, Dominique 528, 534, 535

Terpene 691, 692

Tessiore, Giorgio 393, 523, 524, 720

Testore, Prof. Franco 300, 310, 311, 315, 337, 345, 347-351, 353, 354, 357, 358, 360, 362, 364, 365, 368, 369, 371, 372, 376, 377, 524, 538-540, 542, 543, 550, 552, 563, 565, 567-569, 571, 572, 575-578, 584, 593, 600, 604, 605, 618-620, 624, 631, 634, 654, 663, 664, 666, 667, 690, 719, 720, 725

Three-Lab Protocol [linked to "British Museum Shroud protocol (January 1988)"] 303, 305, 306, 313, 314, 325, 330, 387, 410, 414, 441

Time Magazine 4, 612, 620

Time-of-Flight Secondary Ion Mass Spectrometry (ToF-SIMS) 691

Tite, Dr. Michael 20, 21, 37, 41, 48, 59, 77, 93, 121, 123, 137, 142, 147, 174, 206, 213, 219, 222, 225, 227, 230, 231, 236, 237, 239-243, 245, 246, 250, 254, 255, 257, 258, 262, 265, 268, 275, 280, 293, 305, 306-313, 318, 320, 321, 323, 327, 328, 333-337, 340, 341, 343-347, 351-354, 357-361, 363-365, 367, 371, 372, 374, 380, 391, 394, 398, 405, 406-408, 410, 411, 413, 415-417, 420, 422, 423, 429, 430-435, 437, 438, 440, 441, 445, 453-458, 460, 471, 478, 482, 485, 487, 488, 490-492, 498-502, 504, 505, 510-513, 521, 523, 525-530, 532, 533, 536, 537, 539, 542-544, 546-553, 563, 564, 572, 573, 577-581, 585-588, 592-597, 599-605, 608, 609, 612, 614, 618, 622-625, 628, 634, 635, 666, 667, 674, 697, 707, 716, 720, 723-725, 728, 733-735

Toolin, Dr. L.J. 600, 601

Tosatti, Marco 372

Tosi, Maurizio 239, 647

Trondheim (Norway) C-14 Conference (1985) 21, 27, 31, 32, 37, 39, 42-44, 46-57, 60-63, 65-67, 72-74, 86-89, 91, 94, 96, 98, 101, 136, 137, 159, 186, 197, 392, 611, 718, 726

Trumbore, Susan 406

Turin Centro (see "Centro Internazionale di Sindonologia")

Turin Congress (1978) 7-11

Turin Workshop Protocol (1986) 119-121, 125, 128-149, 152, 156, 228, 239, 253, 257, 258, 260, 265, 266, 269, 270, 272, 274, 275, 279-281, 300, 301, 303, 305, 306, 308, 313, 320, 321, 326, 334, 336, 395, 410, 411

U

Umberto (II), King 6, 25, 29, 43, 126

United Nations 24, 47, 66, 67, 93, 161, 295, 318, 714

University of Arizona (Tucson) 20, 29, 49, 50, 54, 59, 77, 80, 81, 87, 91, 105, 107, 121, 127, 223, 240-244, 247, 250, 251, 254, 257, 259, 264-267, 281, 286, 295, 297, 301, 306-308, 318, 320, 321, 332, 334, 335, 339, 366, 370, 372, 374, 378, 384, 387, 388, 402, 404, 408, 413-417, 420, 421, 426, 431, 432, 435, 437, 445, 452, 456, 464, 469, 489, 490, 493, 494, 497, 524, 526, 531, 536, 553, 554, 574, 587, 600, 601, 603, 604, 612, 619, 621, 623, 624, 628-631, 639, 641, 653-655, 664, 669, 670, 672, 675, 678, 689, 692, 694, 695, 704, 705, 708, 714, 717

University of Oxford 20, 25, 29, 32, 39, 41, 49-51, 54, 59, 67, 80, 105, 107, 121, 122, 128, 129, 137, 173, 223, 224, 227, 229, 241-243, 247, 249-251, 254, 255, 258, 259, 262, 264, 265, 269, 279, 297, 304-308, 314, 317, 320, 329, 332, 333, 334, 338, 339, 358, 360, 361, 370, 374, 382, 384, 387, 392-396, 404, 405, 408, 409, 415, 417, 418, 421-423, 426, 430, 431, 434, 435, 437, 445, 455, 456, 460, 461, 464, 468, 493, 495, 497, 499, 501, 504, 505, 511, 517, 524, 526, 527, 531, 532, 536, 553, 554, 569, 580, 586, 595, 597, 602, 605, 607, 612, 619, 620-624, 627-631, 634, 635, 664, 669-674, 677, 679, 681, 682, 685, 686, 688, 689, 694, 699, 700, 708, 709, 715, 716, 720, 721

University of Rochester 4, 8, 11, 14, 15, 20, 21, 28, 29, 49, 50, 54, 59, 64, 72, 77, 80, 91, 92, 98, 99, 101, 102, 121, 124, 127, 128, 139, 141, 143, 163, 169, 170, 224, 232, 259, 267-269, 280, 282, 295, 297, 332, 389, 402, 411, 620, 674, 687, 705, 713, 716

University of Toronto 4

University of Zurich (ETH Zurich) [see also "Bern/Zurich (C-14 lab)"] 20, 21, 29, 39, 49, 50, 54, 56, 59, 80, 81, 87, 121, 122, 128, 129, 138, 223, 224, 227- 229, 242, 243, 249, 251, 254-256, 258-260, 262, 269, 271, 280, 297-299, 306-308, 312, 320, 332, 334, 339, 358, 361, 362, 364,

370, 374, 379, 384, 386, 388, 389, 393-397, 403-407, 415-417, 421, 424, 425, 432, 435, 437, 445, 456, 458, 460, 461, 463, 464, 469, 470, 485, 488, 493, 526, 536, 551, 553, 554, 564-566, 595, 598, 599, 601, 603, 604, 607, 613, 619, 621, 623, 628, 631, 664, 665, 669, 671-673, 680-682, 685, 686, 689, 704, 708, 709, 721

Upinsky, Arnaud-Aaron 419, 538, 591, 593, 706

Ultraviolet (UV) irradiation 149

UV (see "Ultraviolet (UV) irradiation")

V

Valfre, Blessed Sebastian

Van Haelst, Remi 361, 459, 460, 494, 500, 501, 525, 526, 531, 537, 547, 586, 587, 588, 610, 613, 614, 618, 619, 623-625, 631-633, 635, 639, 641, 687, 721, 734

Van Oosterwyck-Gastuche, Dr. Marie Claire 461, 462

Vanillin 660

Vatican Bank 316

Vatican Press Office 356, 375, 433, 589

Vial, Gabriel 300, 310, 311, 315, 328, 336, 337, 343-345, 347, 352-354, 357, 359, 362, 364, 365, 368-370, 378, 525, 526, 529, 539, 543, 544, 550, 569, 571, 572, 594, 596, 619, 654, 690, 721

Villareal, Dr. Robert 691

Vinland Map 5, 185, 399, 718

W

Walsh, Bryan 79, 695, 708-710, 721

Weisskopf, Dr. Victor 45, 48, 66, 73, 74, 96, 316, 721

Whanger, Dr. Alan 538, 556, 557, 644, 721, 723

Whanger, Mary 644, 723

White, Dr. Cheryl vi,

Wilcox, Robert 409

Wild, Dr. John P. 412

Wild, Fr. Robert 95

Wilson, Ian 3, 10, 96, 115, 248, 285, 286, 312, 315, 316, 319, 324, 328, 341, 361, 362, 399, 401, 411, 413, 425, 434, 436, 454, 458, 491, 492, 529, 533, 548, 659, 703, 721

Wilson, Sir David 105, 107, 165, 185, 304, 320, 333, 423, 591

Wilson-Ward (C-14 statistical measurement) 460, 531, 630, 670, 671

Windish, D. 292

Winter, Michael 460, 462

Wojtyla, Cardinal Karol (see also "Pope John Paul II") 606

Wolfli, Willi (a.k.a "Woelfli" or "Wölfli") 48, 49, 121, 128, 129, 138, 144, 145, 149, 153, 154, 177, 182, 213, 224, 227, 228, 235, 242, 249, 253, 254, 259, 260, 264, 266, 280, 281, 304, 307, 308, 321, 323, 346, 362, 367, 382, 386, 394, 397, 405, 406, 407, 416, 417, 424, 425, 463, 469,

470, 476, 477, 478, 481, 484-488, 493, 497, 539, 542, 564, 567, 573, 580, 595, 596, 598-602, 604, 610, 620, 631, 645, 699, 721

Wright, Pearce 301, 302, 735

X

X-ray fluorescence 128, 129, 191, 237, 275, 291, 308, 584

Z

Z-twist of the Shroud 306